VOLUME ONE HUNDRED AND ONE

Annual Reports on NMR Spectroscopy

VOLUME ONE HUNDRED AND ONE

ANNUAL REPORTS ON NMR SPECTROSCOPY

Guest Editor

PROF. ATTA-UR-RAHMAN, FRS

UNESCO Science Laureate
Professor Emeritus
International Centre for Chemical and Biological Sciences
University of Karachi
Karachi 75270
Pakistan

ELSEVIER

ACADEMIC PRESS
An imprint of Elsevier

Academic Press is an imprint of Elsevier
125 London Wall, London, EC2Y 5AS, United Kingdom
The Boulevard, Langford Lane, Kidlington, Oxford OX5 1GB, United Kingdom
50 Hampshire Street, 5th Floor, Cambridge, MA 02139, United States
525 B Street, Suite 1650, San Diego, CA 92101, United States

First edition 2020

Notices
Knowledge and best practice in this field are constantly changing. As new research and experience broaden our understanding, changes in research methods, professional practices, or medical treatment may become necessary.

Practitioners and researchers must always rely on their own experience and knowledge in evaluating and using any information, methods, compounds, or experiments described herein. In using such information or methods they should be mindful of their own safety and the safety of others, including parties for whom they have a professional responsibility.

To the fullest extent of the law, neither the Publisher nor the authors, contributors, or editors, assume any liability for any injury and/or damage to persons or property as a matter of products liability, negligence or otherwise, or from any use or operation of any methods, products, instructions, or ideas contained in the material herein.

ISBN: 978-0-12-820974-5
ISSN: 0066-4103

For information on all Academic Press publications
visit our website at https://www.elsevier.com/books-and-journals

Publisher: Zoe Kruze
Acquisitions Editor: Jason Mitchell
Editorial Project Manager: Shellie Bryant
Production Project Manager: Abdulla Sait
Cover Designer: Victoria Pearson-Esser

Typeset by SPi Global, India

Contents

Contributors

Marko Bertmer
Leipzig University, Felix Bloch Institute for Solid State Physics, Leipzig, Germany

Mattias Edén
Physical Chemistry Division, Department of Materials and Environmental Chemistry, Arrhenius Laboratory, Stockholm University, Stockholm, Sweden

Shang-Bin Liu
Institute of Atomic and Molecular Sciences, Academia Sinica, Taipei, Taiwan

Leszek Pazderski
Faculty of Chemistry, Nicolaus Copernicus University, PL, Toruń, Poland

Xianfeng Yi
State Key Laboratory of Magnetic Resonance and Atomic and Molecular Physics, National Center for Magnetic Resonance in Wuhan, Key Laboratory of Magnetic Resonance in Biological System, Wuhan Institute of Physics and Mathematics, Innovation Academy for Precision Measurement Science and Technology, The Chinese Academy of Sciences, Wuhan, PR China

Anmin Zheng
State Key Laboratory of Magnetic Resonance and Atomic and Molecular Physics, National Center for Magnetic Resonance in Wuhan, Key Laboratory of Magnetic Resonance in Biological System, Wuhan Institute of Physics and Mathematics, Innovation Academy for Precision Measurement Science and Technology, The Chinese Academy of Sciences, Wuhan, PR China

Preface

Exciting developments in the field of NMR spectroscopy continue to find wide ranging applications in organic chemistry, organometallics, material sciences, and in a number of other fields. Volume 101 of this long standing series contains reviews contributed by eminent experts in some of these areas.

Metal–organic frameworks (MOFs) have attracted much attention during recent years because of the structural variety encountered as well as the numerous potential applications. These porous materials with ordered structures lend themselves to characterization by multinuclear solid-state NMR (SSNMR). The porosity of these structures allows small molecules to be adsorbed on them. The orientation, adsorption sites, and dynamics of these small molecules in the porous MOF structure can be readily studied by SSNMR. The first chapter by Marko Bertmer describes studies on certain adsorbed molecules in MOFs.

In Chapter 2 by Liu et al, some of the recent developments on acidity characterization of solid acid catalysts, particularly those using phosphorus-containing molecules as probes, using ^{31}P solid-state nuclear magnetic resonance (SSNMR) spectroscopy, are discussed. The use of trimethylphosphine (TMP), phosphines, diphosphines, and trialkylphosphine oxides (R_3PO) for characterization of distribution, accessibility, concentration, and strength of acid sites in various solid acids is described. The third chapter by Leszek Pazderski presents ^{15}N NMR coordination shifts in transition metal complexes and in organometallics with heterocycles containing nitrogen. The ^{15}N NMR data for 407 complexes or organometallics of such transition metal ions are described. Particular attention is paid to ^{15}N NMR coordination shifts in relation to some structural features of the reviewed coordination compounds.

Aluminosilicate glasses and melts are of great scientific and commercial importance. They exhibit a high resistance to thermal shock, and are therefore used as sealing materials for hot systems. If coated with an electrically conductive film, they also find applications in manufacture of resistors for electronic circuitry. In the last chapter contributed by Mattias Edén, the use of magic-angle spinning (MAS) ^{27}Al NMR to study the structure of such compounds is described. The readers will find the description of practical aspects of performing MAS and triple-quantum MAS NMR experiments of particular interest.

I would like to thank all the authors for their excellent contributions which should be of considerable interest to the readers. I am also most grateful to my associates Dr. Taqdees Malik and Mr. Mahmood Alam for the secretarial support extended to me in the editing of the manuscripts.

Prof. Atta-ur-Rahman, FRS
UNESCO Science Laureate
Professor Emeritus
International Centre for Chemical and Biological Sciences
University of Karachi
Karachi 75270
Pakistan

CHAPTER ONE

Solid-state NMR of small molecule adsorption in metal–organic frameworks (MOFs)

Marko Bertmer*
Leipzig University, Felix Bloch Institute for Solid State Physics, Leipzig, Germany
*Corresponding authors: e-mail address: bertmer@physik.uni-leipzig.de

Contents

Abstract

Metal–organic frameworks (MOFs) have gained significant attraction in the scientific community in the last two decades for their structural variety and because of that their multiple potential applications. These ordered, typically porous materials are well-suited for characterization with methods of solid-state NMR (SSNMR). The different elements involved in the MOF structure can be studied by multinuclear solid-state NMR. Furthermore, the porous structure allows for the adsorption of various small molecules. The characterization of orientation, adsorption sites, and dynamics of these small molecules in the porous MOF structure is perfectly suited for SSNMR methods ranging from relaxation time analysis, spectroscopy to diffusion studies.

Annual Reports on NMR Spectroscopy, Volume 101
ISSN 0066-4103
https://doi.org/10.1016/bs.arnmr.2020.07.003

In this report, the focus is on adsorption studies in MOFs with solid-state NMR methods. After a general introduction to the field, typical aspects of certain MOFs and certain adsorbed molecules will be discussed. The combination with calculations and the different experimental approaches are also discussed.

Keywords: Metal–organic frameworks, MOFs, Adsorption, Adsorption sites, Molecule dynamics, Diffusion, NMR, Solid-state NMR, PFG, Xenon, H_2O, CO_2, CH_4, H_2, NO

1. Introduction

Metal–organic frameworks (MOFs) represent a rather new class of materials made by linking metal ions or metal clusters with organic linker molecules to construct a 3D network. Because of the various possible metal ions and bi- or trifunctional organic ligands, a huge variety of materials has been prepared in the last 20 years. One important property of these materials is that they are most often porous with setting record-values for the surface area of more than 7000 m^2g^{-1} [1]. This means that small molecules can be incorporated. Applications such as gas mixture separation, gas storage, catalysis, and others can be envisioned.

Typical characterization methods of MOFs are X-ray diffraction and determination of surface area via BET measurements. Typical structural motifs have been determined and classified. To get more information on adsorbed species often other characterization techniques are required, such as Fourier transform infrared spectroscopy (FTIR) or UV/VIS spectroscopy, among others.

The organic part of the MOF as well as some of the metal ions also qualify for solid-state NMR investigations. On the order of 1000 papers have been published so far dealing with solid-state NMR characterization of MOFs. Further information on the host structure can be obtained regarding the electronic properties in MOFs with paramagnetic metals, ligand ordering in unsymmetrical linker molecules or in mixed-linker MOFs. Especially in case of low crystallinity XRD provides no detailed structural information, solid-state NMR can give details on the local structure.

The local information that NMR provides makes this method also especially suitable for characterization of adsorbed molecules in the pores. Information on adsorption sites, adsorption strength, orientation in the pore, and dynamics of adsorbed molecules can principally be obtained. Additionally, diffusion studies are possible using pulsed field gradients. About 500 papers have been published so far with reference to adsorbed molecules in MOFs.

There exist several reviews on the use of solid-state NMR for the characterization of MOFs and also of adsorbed molecules in MOFs [2–6]. This report focusses on the characterization of adsorbed molecules in MOFs by both studying the molecules itself, host–guest interaction as well as indirect effects of adsorbed molecules on the host structure. At first, some general considerations on the NMR relevant parameters in describing adsorbed molecules are discussed in Section 2. Section 3 mentions briefly some general facts on MOF materials with focus on NMR related properties. Few selected MOFs, that have been studied in several papers by NMR, are discussed as well. The major part deals with the NMR results obtained for individual molecules (Section 4). The importance of linking solid-state NMR measurements with calculations is illustrated in Section 5. Finally, selected techniques are discussed in Section 6.

2. Solid-state NMR of adsorbed molecules in MOFs

In this section, no revision of the different NMR interactions will be done. Nevertheless, it will be briefly summarized in which way a certain NMR interaction can be used to obtain information on the properties of adsorbed molecules. The different examples will then be given in Section 4.

An NMR spectrum can be categorized based on signal intensity, signal frequency and linewidth. Additionally, the spin-lattice relaxation time T_1 and spin-spin relaxation time T_2 (additionally spin-lattice relaxation time in the rotating frame $T_{1\rho}$) are characteristic for time and frequency ranges of a sample. These characteristics are the result of internal NMR interactions, the chemical shift, quadrupolar (for spins $I > \frac{1}{2}$) and dipolar coupling. The latter can be subdivided into homo- and heteronuclear dipolar coupling and direct and indirect (scalar) coupling. Furthermore, hyperfine coupling between nuclear and electron spins can occur in materials with unpaired electrons like some transition metals. Using magnetic field gradients, also diffusion of molecules can be determined. All these interactions can in principle be used to get information on adsorbed molecules in MOFs.

- Binding sites, host–guest interactions

  The identification of binding sites for the guest molecules is of special importance with consequences on the adsorption and diffusion properties of loaded molecules. For separation purposes between different adsorbed molecules, the binding strength is a characteristic value to explain adsorption selectivity. Solid-state NMR can provide different methodologies to trace binding sites. One is the effect of guest molecules on the NMR spectra of host nuclei. As often the metal ions are most probable adsorption sites, changes to the isotropic chemical shift and/or quadrupolar coupling (see below) can give detailed information on adsorption sites.

Another aspect is the presence of paramagnetic species as often transition metals with unpaired electrons are present inside MOFs. Here the hyperfine coupling can significantly affect the NMR spectra of the loaded guest molecules and by that indicate host–guest interactions. This approach is also important if the loaded molecules cannot be detected directly, as for instance in the case of nitric oxide (NO) (see Section 4.6).

This is not the place for a full discussion of the hyperfine interaction, the reader is referred to more general literature, e.g., Refs. [7, 8, 9] and references cited therein. Some general information will nevertheless be given here. The hyperfine coupling stems from the interaction of a nuclear with an electron spin. It can be divided into two parts,

$$\delta_{\text{hyperfine}} = \delta_{\text{contact}} + \delta_{\text{PC}}. \tag{1}$$

The Fermi-contact interaction refers to the electron spin density at the site of the nucleus and is an isotropic interaction. The pseudocontact (PC) part is a dipolar electron–nucleus interaction which is anisotropic. Not only spin-spin interaction, but also the nuclear spin-electron orbit interaction contribute to this [10]. It depends on the electron-nucleus distance and the orientation with respect to the magnetic field, additionally to the electron spin and g-tensor. In many cases, for the influence on the chemical shift, the contact term is dominating. It depends on the size of the isotropic part of the hyperfine interaction tensor, the electron spin, and the electron g-tensor. Huge effects of hundreds or thousands of ppm are known. Since both contact and pseudocontact parts have a $\frac{1}{T}$ dependence, the two contributions cannot be separated by variable temperature measurements and prediction of shifts is not straightforward. Often, NMR spectra are recorded at different temperature and extrapolation to infinite temperature in a $\frac{1}{T}$ plot yields the isotropic chemical shift and aids in assignment. In recent years, progress in the calculation of paramagnetic shifts has been done, clearly helping in the interpretation of NMR spectra.

The Curie spin has a significant effect on the linewidth of a signal. In MAS spectra (see below), this gives rise to a number of spinning sidebands from which the hyperfine coupling can be extracted. Furthermore, under MAS conditions, the residual linewidth comes from the anisotropic bulk magnetic susceptibility. This can have severe effects for spectrum resolution. Furthermore, paramagnetic substances also give rise to a contribution to shielding that is positionally dependent and due to the bulk magnetic anisotropy of the object (see, e.g., [11, 12]). However, a significant linebroadening in the interaction between the MOF and the adsorbed molecule in a paramagnetic system can therefore be interpreted in terms of a significant host–guest interaction. Its absence then let's one conclude on a weak host–guest interaction. Examples are given in Section 4.

In a more direct way, one can use the dipolar coupling (homo- or heteronuclear) to assign interactions between host and guest atoms. In some cases, isotopic labelling might be required. In general, though, interatomic distances can be determined. Several experimental strategies exist, ranging from standard cross-polarization measurements to HETCOR- (heteronuclear correlation) or REDOR-type (rotational echo double resonance) experiments.

- Chemisorption or physisorption

  One important question in adsorption of small molecules in MOFs and other porous materials is, if the molecules are free to move or if there are special places they stay at the binding site. In this case, chemisorption means a chemical reaction of the guest molecule with the host structure, physisorption an interaction without reaction. One characteristic to decide on chemisorption or physisorption is the isotropic chemical shift of the adsorbed molecule.

  In case of chemisorption, a different isotropic chemical shift is expected to that of the free molecule. The shift itself might already give an idea of the type of reaction. Examples are the reaction of carbon dioxide with amino functions to form carbamic acid [13] or of nitric oxide (NO) with amines to form NONOates [14]. Further details are given in Section 4.

  In the physisorbed state often no changes to the isotropic chemical shift are observed. However, influences from the host structure can be still visible, e.g., by linebroadening due to hyperfine interaction with transition metals in the host structure or restricted mobility (see below). Linebroadening can in principle also be induced by dipolar couplings between guest and host molecules (see below). In other cases, coordination of adsorbed molecules to open metal sites could be detected by a change in coordination number of the metal ion. Changes to the coordination number give rise to different chemical shifts, e.g., for ^{27}Al in NO adsorption in MIL-100(Al) [15]. As the metal ions are often quadrupolar nuclei, changes to the quadrupolar coupling can also be interpreted in terms of physisorption.

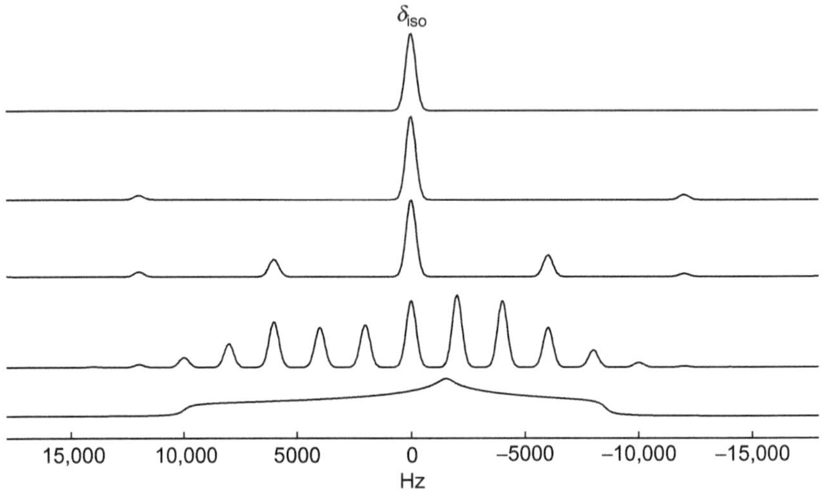

Fig. 1 Effects of MAS on a typical lineshape of a powdered sample; from *bottom* to *top*: static spectrum, 2 kHz MAS, 6 kHz MAS, 12 kHz MAS, 20 kHz MAS.

As in most cases, high resolution is advantageous, the most important narrowing technique is magic-angle spinning (MAS). Here, the sample is rotated at an angle of 54.7 degrees (deriving from the angular dependence $3\cos^2\theta - 1$, see below) with respect to the magnetic field that removes anisotropic broadening (to first order). If the rotation frequency is not large enough, MAS spectra will also show spinning sidebands (Fig. 1).

- Small molecule dynamics

  The pore characteristics of the different MOFs (pore size, connectivity) may confine adsorbed molecules and by that restrict their mobility. This can have strong effects on the NMR lineshape and linewidth of the adsorbed molecules. In case of free mobility with isotropic translation and rotation, narrow lines similar to that of the free gas are observed. In other situations, characteristic lineshapes result that can be simulated assuming certain motions (see Section 5). Well known is the quadrupolar coupling in case of ^{2}H (for example H_2, Section 4.4).

The quadrupolar interaction $\overleftrightarrow{Q}$ between the nuclear quadrupole moment and the surrounding electric field generated by the surrounding atoms can be described by a tensor for the electric field gradient (EFG) $\overleftrightarrow{V}$.

$$\overleftrightarrow{Q} = \frac{eQ\overleftrightarrow{V}}{2I(2I-1)h} \tag{2}$$

Here, Q is the quadrupolar moment, e the electric charge, I the nuclear spin and h Planck's constant. As a result, a line splitting will result according to

$$\Delta\nu_Q = \frac{1}{2}\nu_Q[3\cos^2\theta - 1 - \eta\sin^2\theta\cos 2\phi], \tag{3}$$

with

$$\nu_Q = \frac{3e^2Qq_{zz}}{2I(2I-1)h} \tag{4}$$

being the quadrupolar coupling frequency. eq_{zz} corresponds to the largest principal value of the EFG, V_{zz}. η is the asymmetry parameter

$$\eta = \frac{V_{yy} - V_{xx}}{V_{zz}}, \tag{5}$$

describing the deviation of the EFG from axial symmetry. θ and ϕ are the horizontal and azimuthal angle relating the laboratory axis system with respect to the principal axis system (PAS) of the EFG tensor.

In powder samples, a typical Pake-type spectrum results. Motion of the molecule under study leads to a characteristic variation of the resulting lineshape, as can be seen for example in Fig. 2. As also different types of motion can be present at the same time, simulation programmes can be used to identify different types of motion (Section 5).

In case of spins with $I = \frac{1}{2}$, the chemical shift anisotropy (CSA) will be affected by the presence of molecular mobility (for example, CO_2, Section 4.3). The chemical shift results from the interaction between the nuclear spin and the surrounding electron orbitals that can shield or deshield the external magnetic field from the nuclear spin. The shielding is also described by a tensor, $\overleftrightarrow{\sigma}$. As this is defined relative to the bare nucleus, it is accessible in calculations but not in experiments. Therefore the chemical shift $\overleftrightarrow{\delta}$ is given with respect to the shielding of a reference compound.

Similar to the quadrupolar coupling, the tensor can be described by the three diagonal elements δ_{11}, δ_{22}, and δ_{33} in the PAS. From that the isotropic chemical shift is defined from the trace of the tensor:

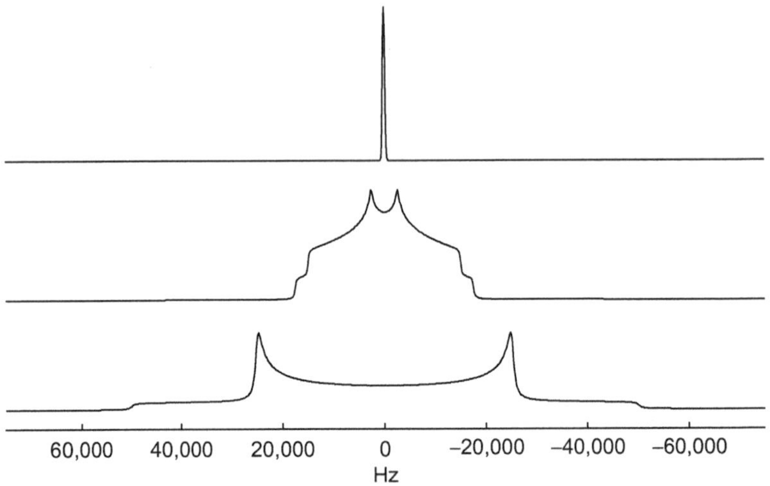

Fig. 2 Arbitrary powder lineshape for a ^{2}H nucleus under static conditions (*bottom*), with specific type of motion (*middle*), isotropic line obtained from free mobility of the ^{2}H nucleus (*top*).

$$\delta_{iso} = \frac{1}{3}(\delta_{11} + \delta_{22} + \delta_{33}) \tag{6}$$

Other conventions exist for describing the chemical shift anisotropy. In the Herzfeld-Berger [16] notation the terms span Ω and skew κ are defined:

$$\Omega = \delta_{11} - \delta_{33} \tag{7}$$

$$\kappa = \frac{3(\delta_{22} - \delta_{iso})}{\Omega} \tag{8}$$

The span is a good measure to indicate the width of the line. κ ranges between -1 and $+1$.

Also, the Haeberlen convention [17] is frequently in use. Here, the terms anisotropy $\Delta\delta$ and asymmetry η are used:

$$\delta_{CS} = \delta_{zz} - \delta_{iso} \tag{9}$$

$$\Delta\delta = \delta_{zz} - \frac{\delta_{xx} + \delta_{yy}}{2} = \frac{3}{2}\delta_{CS} \tag{10}$$

$$\eta_{CS} = \frac{\delta_{yy} - \delta_{xx}}{\delta_{CS}} \tag{11}$$

δ_{CS} is also called the reduced anisotropy, η_{CS} ranges between 0 and 1, for 0 stating axial symmetry. The principal tensor components are reassigned based on the separation from the centre of gravity, according to

$$|\delta_{zz} - \delta_{iso}| \leq |\delta_{xx} - \delta_{iso}| \leq |\delta_{yy} - \delta_{iso}|. \tag{12}$$

For a powder sample, the following expression is obtained:

$$\delta = \delta_{11}(\cos\phi\sin\theta)^2 + \delta_{22}(\sin\phi\sin\theta)^2 + \delta_{33}\cos^2\theta. \tag{13}$$

In the Haeberlen convention, this reads as:

$$\delta = \delta_{iso} + \frac{1}{2}\delta_{CS}(3\cos^2\theta - 1 - \eta_{CS}\sin^2\theta\cos 2\phi). \tag{14}$$

Motion will render the CSA and lead to a reduction in span Ω and—depending on the type of motion—to a different lineshape. Explanation of the different parameters and the effects of motion are illustrated in Fig. 3.

Finally, the dipolar coupling between nuclei in the adsorbed molecule could also be a significant source of information. It originates from the interaction of nuclear spins and can be described in a similar way as the quadrupolar coupling without the asymmetric part.

$$\nu_D = \pm aD(3\cos^2\theta - 1) \tag{15}$$

with D being the dipolar coupling constant defined as

$$D = -\frac{\mu_0\gamma_I\gamma_S\hbar}{8\pi^2 r_{IS}^3}, \tag{16}$$

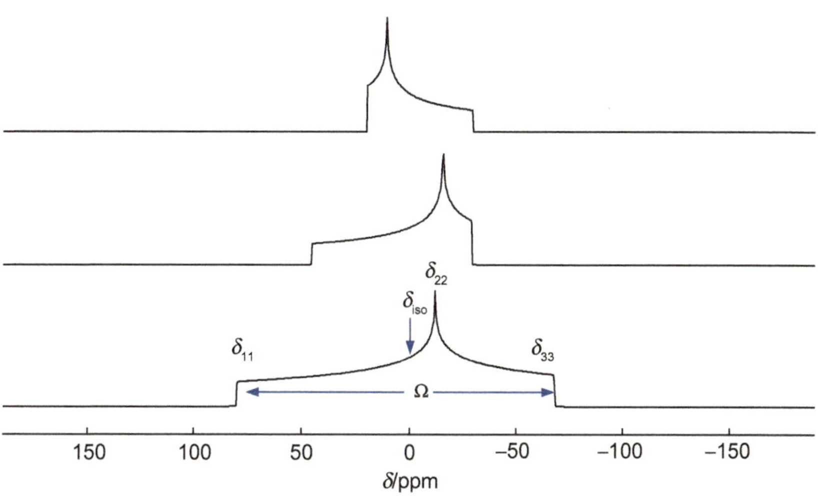

Fig. 3 Arbitrary powder lineshape for a site with chemical shift anisotropy, indicating the different vales for characterizing the tensor (*bottom*). The two other lineshapes are characteristic for the effect of certain motions on the lineshape.

with r_{IS} being the distance of spins. a is 1 for heteronuclear and $\frac{1}{2}$ for homonuclear spins. In a powder sample this will also lead to a Pake-like spectrum (see above, Fig. 2).

Motion will also render the frequency splitting time-dependent. Often, however, multiple couplings are present leading to a Gaussian-like shape. Then, predominantly, the linewidth alone can give some information on the dynamics of a molecule adsorbed in a MOF.

- Diffusion [18, 19]

  In order to track spatial displacements of adsorbed molecules in MOFs, pulsed field gradients are needed. With that, a spatial position is identified by a specific frequency:

$$B(x) = B_0 + x^* G \tag{17}$$

$$\omega = \gamma B(x) \tag{18}$$

Here, for example, a gradient $G = \frac{\partial B}{\partial x}$ is applied along the x-axis; γ is the gyromagnetic ratio. Diffusion measurements can be performed, e.g., using a stimulated echo sequence, see Fig. 4. Other variants also exist, but omitted here for clarity.

Molecular diffusion during the time interval of measurement leads to a reduction in echo signal intensity which is therefore sensitive to the projection of the mean square displacement $\langle r^2(\Delta) \rangle$ onto the x-axis. The reduction in signal intensity can be described by

$$\frac{S(q)}{S_0} = e^{-q^2 D\Delta} e^{-\frac{2\tau}{T_2}} e^{-\frac{\Delta'}{T_1}}, \tag{19}$$

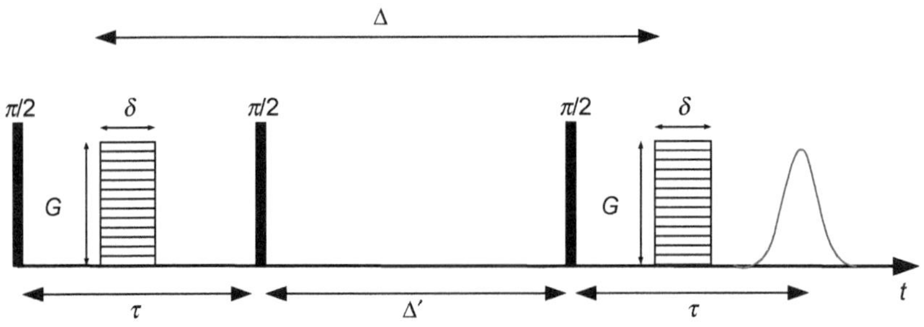

Fig. 4 Stimulated spin-echo experiment, Δ is the diffusion time, G the strength, and δ the width of the gradient pulses.

where $q = \gamma\delta G$. T_1 and T_2 are the spin-lattice and the spin–spin nuclear magnetic relaxation times. D is the self-diffusion coefficient. For τ and Δ' see Fig. 4.

The temperature dependence of D can be described by an Arrhenius equation:

$$D(T) = D_0 e^{-\frac{E_A}{RT}}, \tag{20}$$

where D_0 is the diffusivity at infinite temperature, E_A the activation energy and R the ideal gas constant.

This description is for the case of homogenous, one-phase diffusion. In real situations, as typically occurring in MOFs, several effects could be present such as intra- and intercrystalline diffusion in powder samples, interaction with the pore walls (physisorption), or even multicomponent systems with more than one diffusion coefficient. These then typically require molecular dynamics (MD) simulations (Section 5). These can provide the mean square displacement $\langle r^2(t) \rangle$ which is related to the self-diffusion coefficient via the Einstein-relation

$$D = \frac{\langle r^2(t) \rangle}{6t}. \tag{21}$$

It has to be noted, though, that MD simulations yield results in the range of $\langle r^2(t) \rangle$ of up to several tens of nm, while PFG experiments work in timescales of more than 1 μm. Nevertheless, comparison of experimental and theoretical values can provide information on, e.g., host–guest interactions.

At last, a note should be given on the sample preparation as this varies with regard to the information to be obtained. If static NMR spectra are measured (including diffusion measurements), a certain amount of gas is adsorbed in a glass ampule that fits in the NMR coil. Afterwards, the ampule is flame sealed so that no modifications happen. In other cases, if various gas pressures are wanted, a special pressure apparatus can be designed (see, for example, Section 4.7.1) so that the pressure of the gas can be varied during the measurements. Section 6 also discusses the continuous-flow mode.

Otherwise, if magic-angle spinning (MAS) is required for high-resolution spectra (can also be used in diffusion measurements), the requirements are different. Special glass ampules can be prepared as described before that tight-fit into the MAS rotors (see, for example, Refs. [20, 21]). For this, a well-defined amount of gas can be adsorbed though the sample size is much smaller compared to a full MAS rotor. In other cases, loading the full MAS rotor with the to be studied small molecule and closing the rotor by the cap might also work, especially when the adsorption is strong, as used for example in Ref. [22]. In that case, however, the exact loading is not easy to determine.

To describe the amount of adsorbed molecules, typically the molar ratio with respect to the amount of metal ions is given. In the case of open metal sites as primary adsorption site (see below), this is often most straightforward. Otherwise, the amount of adsorbed molecules can also be expressed with respect to the amount of ligand [14]. Alternatively, especially for pressure-dependent studies, the actual gas pressure is used to indicate the amount of gas inside the MOF [23].

3. Remarks on MOF materials

Clearly, there is a huge variety of different MOFs existing. Some have probably not been investigated in many details if the PXRD pattern indicates low crystallinity or amorphous structures. Nevertheless, these materials could hold high potential for small molecule adsorption and could be studied by SSNMR. MOFs can be subdivided in different categories which is not pursued here. The focus in this section is on the effects on NMR properties for small molecule adsorption.

Several MOFs contain a potential adsorption site at the metal atom, also called coordinatively unsaturated metal site (CUS). These positions are typically created after synthesis by removing solvent molecules by evacuation and are normally the primary sites for adsorption. For MOFs without CUS, the assignment of adsorption sites for small molecules is often not straightforward. Here interaction with the organic ligand is possible or small molecules reside somewhere in the pore volume. Calculations often predict binding sites, see Section 5.

A classification of MOF structures with regard to NMR properties is the electron configuration of the metal atom. As often transition metals are used in MOF synthesis, many of those contain unpaired d-electrons. Examples are Fe, Ni, Cu, or V containing MOFs. The hyperfine interaction of the unpaired electrons with the NMR active nuclei of both the MOF ligand as well as the adsorbed molecules can have significant effects on the resulting spectra, as described in Section 2. Often the presence of unpaired electrons is disadvantageous because of linebroadening effects. However, it can also be used in favour of the characterization of host–guest interaction. If the interaction of the adsorbed molecule with the framework is weak, little effects on linewidth and relaxation times are expected for the small molecule. Otherwise for strong interaction, significant effects result for the adsorbed molecule.

For diamagnetic MOFs the metal atom can be accessible by solid-state NMR. Examples of accessible nuclei are ^{27}Al, $^{69/71}$Ga, ^{113}Cd or ^{207}Pb. Even more complicated to measure nuclear spins contained in MOFs like ^{25}Mg or ^{67}Zn have been successfully studied by solid-state NMR. In those cases, typically highly symmetric surrounding of the metal sites and high magnetic fields are necessary. Examples are given in Section 4.

Finally, the preparation of mixed-metal MOFs has been successfully done for varying the magnetic properties or the variation of the adsorption strengths of loaded molecules. Also the investigation by magnetic resonance methods is feasible to characterize, e.g., the distribution of the different metal atoms in the MOF. On the one hand, small amounts of paramagnetic ions are included in diamagnetic MOFs for investigation with electron paramagnetic resonance (EPR) and SSNMR is used to compare spectra with the undoped MOF. On the other hand the effects of higher amounts of paramagnetic ions (or of diamagnetic ions in normally magnetic MOFs) can also be studied by SSNMR. Examples are, e.g., Al/Fe in MIL-100 [24], Cu/Zn in $Cu_3(btc)_2$ [25, 26] or Al/V in MIL-53/MIL-47 [27].

It becomes clear that SSNMR investigations of MOFs and molecules adsorbed in them allow for different strategies to obtain the relevant information from adsorbed molecules in MOFs. As few MOFs have been studied more intensively with NMR methods (see Section 4), those should be briefly described here. This selection is arbitrary but should give some insight into the variability of the class of MOFs. For further details on MOF structures the reader is referred to the individual literature cited.

- $Cu_3(btc)_2$ (also know as HKUST-1) [28]

  This MOF, $Cu_3C_{18}H_6O_{12}$, contains 1,3,5-benzenetricarboxylate (btc) as organic linker to form a 3D network. The metal cluster is made up of a so-called paddle-wheel of two copper atoms bound to the carboxylate groups of the linker (see Fig. 5). BET surface area is around 1500 m^2g^{-1}, specific pore volume about 0.7 cm^3g^{-1}. It is not water stable (see below) and because of large pores accessible for a number of adsorbed molecules with high capacity.

  The antiferromagnetic coupling of the two Cu^{2+} ions leads to an $S=0$ electronic ground state. However, at temperatures above 70 K, also the $S=1$ state is populated. In NMR spectra this shows up in a way similar to paramagnetic behaviour at elevated temperatures, characterized by hyperfine interactions.

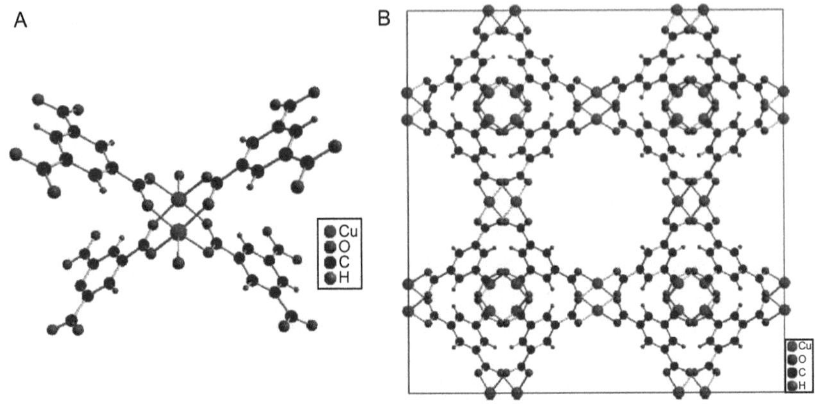

Fig. 5 Crystal structure of $Cu_3(btc)_2$; (A) the paddle-wheel unit, (B) the pore network. *Reprinted (adapted) with permission from B. Jee, K. Eisinger, F. Gul-E-Noor, M. Bertmer, M. Hartmann, D. Himsl, A. Pöppl, Continuous wave and pulsed electron spin resonance spectroscopy of paramagnetics framework cupric ions in the Zn(II) doped porous coordination polymer $Cu_{3-x}Zn_x(btc)_2$, J. Phys. Chem. C 114 (2010) 16630–16639. Copyright 2010 American Chemical Society.*

- MIL-53

  MIL-53 (=Material Institute Lavoisier), $AlC_8H_5O_5$, consists of 1,4-benzene dicarboxylate (bdc)/terephthalate linkers connecting aluminium metal centres [29]. It can also be prepared with chromium [30, 31], iron [32], gallium [33], scandium [34], and indium [35]. With vanadium as metal ion it is called MIL-47 [36]. Special in this MOF is the fact that it is a so-called flexible MOF existing in a narrow pore (np) and a large pore (lp) form (see Fig. 6). Transition from the lp to the np form can be induced by adsorbed gases or by reducing the temperature. It is air and water stable.
- ZIF-8

  ZIF-8 (Zeolitic Imidazolate Framework), $ZnC_8H_{10}N_4$, consists of zinc ions coordinated by 2-methylimidazole (Fig. 7). BET surface area is about 1200 m^2g^{-1} and the specific pore volume about 0.65 cm^3g^{-1}. Because of the binding angles between the metals and the imidazole ring, zeolite-like topologies are present. Several similar structures can be obtained with various ligands (and different metals). They show a robust porosity, good thermal resistance and have high chemical stability against various substances, including water.
- MOF-74

  MOF-74, $M_2C_8H_4O_6$, uses 2,5-dihydroxyterephthalate as organic linker (Fig. 8). It is also known as M_2(dobdc) or CPO-27. It can be

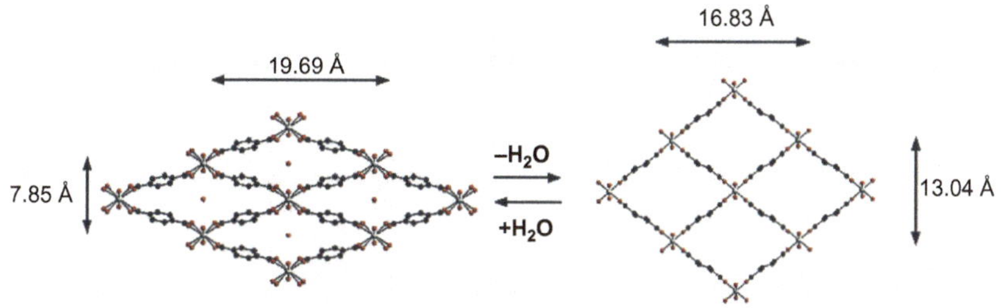

Fig. 6 Crystal structure of (A) MIL-53 (np) and (B) MIL-53 (lp). *Reprinted (adapted) with permission from C. Serre, F. Millange, C. Thouvenot, M. Noguès, G. Marsolier, D. Louer, G. Férey, Very large breathing effect in the first nanoporous chromium(III)-based solids: MIL-53 or Cr(OH)·OC-CH-CO·HOC-CH-COH· HO, J. Am. Chem. Soc. 124 (2002) 13519–13526. Copyright 2002 American Chemical Society.*

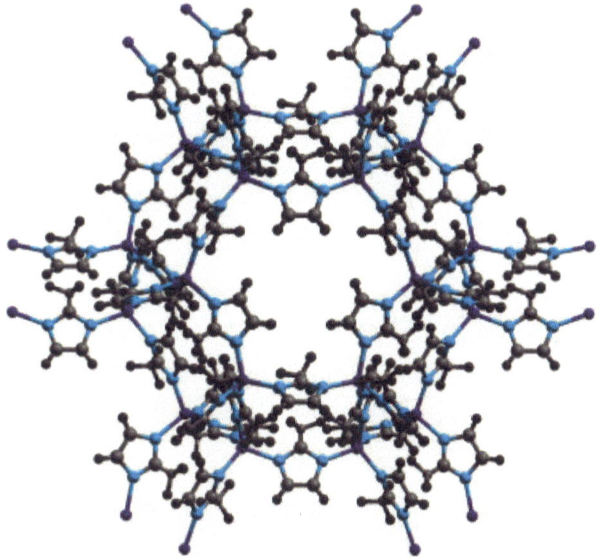

Fig. 7 Crystal structure of ZIF-8. *Reprinted (adapted) with permission from O. Karagiaridi, M.B. Lalonde, W. Bury, A.A. Sarjeant, O.K. Farha, J.T. Hupp, Opening ZIF-8: a catalytically active zeolitic imidazolate framework of sodalite topology with unsubstituted linkers, J. Am. Chem. Soc. 134 (2012) 18790–18796. Copyright 2002 American Chemical Society.*

prepared with several metal ions including Mg^{2+}, Ca^{2+}, Sr^{2+}, Ba^{2+}, Zn^{2+}, Co^{2+}, Mn^{2+}, Ni^{2+}, Fe^{2+}, Cd^{2+} [37]. For MOF-74(Ni) the BET surface area is about 670 m^2g^{-1} and the specific pore volume is 0.38 cm^3g^{-1}. It is a hydrophobic network and also air and water stable.

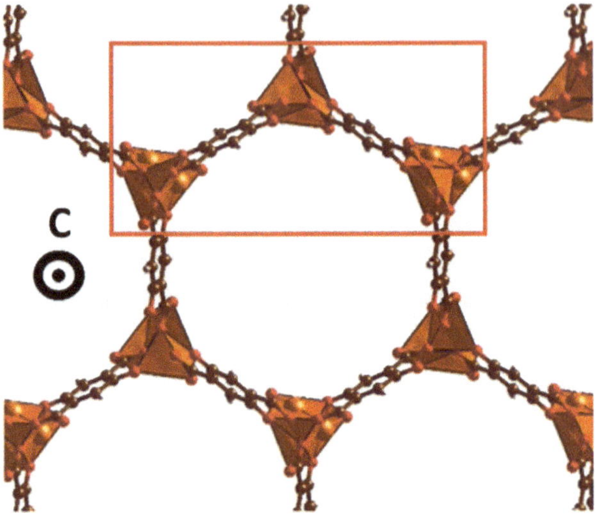

Fig. 8 Crystal structure of MOF-74, projected along the *c*-axis. *Reprinted (adapted) with permission from R.M. Marti, J.D. Howe, C.R. Morelock, M.S. Conradi, K.S. Walton, D.S. Sholl, S.E. Hayes, CO_2 dynamics in pure and mixed-metal MOFs with open metal sites, J. Phys. Chem. C 121 (2017) 25778–25787. Copyright 2017 American Chemical Society.*

4. Adsorbed molecules

In this section, different adsorbed molecules and the SSNMR results obtained with their incorporation into MOFs is discussed, separated by the type of adsorbed molecule.

4.1 Xenon, Xe

The inertness of xenon gas and the large chemical shift range make ^{129}Xe NMR a very useful tool to study porous materials. The chemical shift depends on several factors such as the interactions among xenon atoms or of xenon atoms with the pore walls. Furthermore, adsorption sites as well as magnetic and electric fields affect the chemical shift. ^{129}Xe NMR has therefore been used often to study porous materials, including MOFs. The first NMR studies of adsorbed molecules in MOFs dealt with ^{129}Xe NMR [38]. Depending on synthesis, two isotropic signals were found and assigned to two different pore sizes in the MOF $Cu_3(btc)_2$. In a mixture of ethylene and xenon the authors could also show that xenon tends to occupy the large pores. Ueda et al. investigated the MOF $[M_2(O_2CPh)_4(pyz)]_n$ (M = Rh, Cu) with both ^{129}Xe NMR and xenon adsorption isotherm measurements [39]. They determined the maximum loading of xenon per molecular unit. The narrow one-dimensional channels result in anisotropic lineshapes of ^{129}Xe. Together with temperature-dependent measurements, the presence of xenon dimers in the channels was concluded.

Pawsey et al. were the first to use hyperpolarized (HP) xenon for the investigation of MOFs [40]. The large polarization allows for fast measurements and high sensitivity. However, one has to be aware that it could happen that not all possible xenon sites in the MOF are accessed in a given timespan from hyperpolarized xenon atoms. The average pore diameter was estimated from the chemical shift. Variable temperature data allowed the extraction of adsorption enthalpies being lower than published data for just physisorption of xenon. The high sensitivity of hyperpolarized xenon allowed for 2D EXSY measurements that revealed slow exchange of xenon in the pore with the free gas (Fig. 9). From $^1H \rightarrow {}^{129}Xe$ cross-

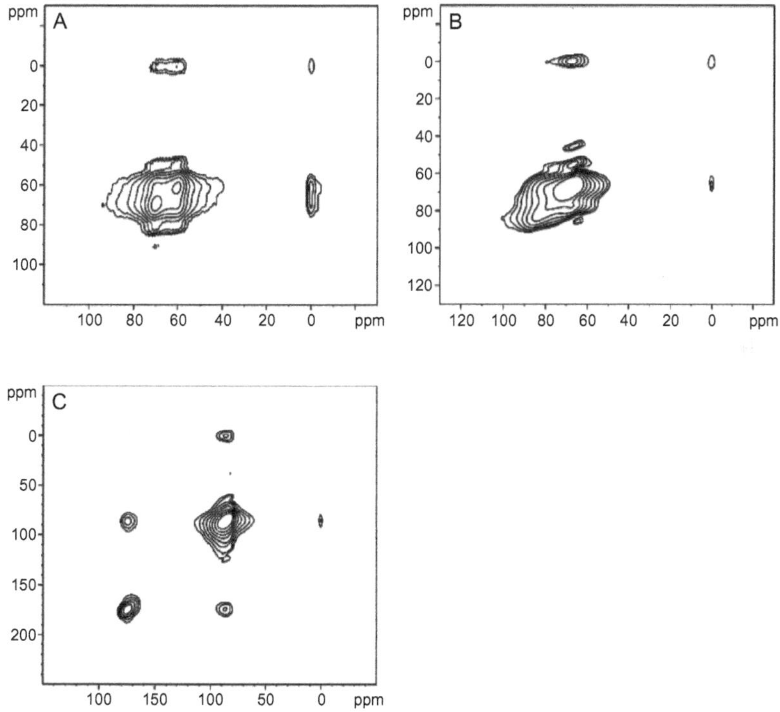

Fig. 9 ^{129}Xe 2D EXSY NMR spectra of IRMOF-2 at temperatures of (A) 35, (B) − 30, and (C) − 100°C, recorded with a mixing time of 50 ms. At room temperature exchange between adsorption sites and free Xe gas is less than 50 ms. This exchange becomes weaker at −30°C. At the lowest temperature, still exchange between the adsorption sites can be seen. *Reprinted (adapted) with permission from S. Pawsey, I. Moudrakovski, J. Ripmeester, L.-Q. Wang, G.J. Exarhos, J.L.C. Rowsell, O.M. Yaghi, Hyperpolarized ^{129}Xe nuclear magnetic resonance studies of isoreticular metal-organic frameworks, J. Phys. Chem. C 111 (2007) 6060–6067. Copyright 2007 American Chemical Society.*

polarization experiments it was concluded that the preferred adsorption site was near the phenylene ring of IRMOF-6. The experiment though had low signal intensity and was run at low temperature (– 50°C). By that, lower mobility and with that also larger dipolar coupling allowed for successful cross polarization. In a related study, four IRMOF samples with cubic cages of different sizes were also investigated by HP xenon [41]. It was found that the ^{129}Xe chemical shift is not only dependent on the cage size. Exchange of xenon atoms between different adsorption sites also contributes to the shift.

In one of the much-investigated flexible MOFs, MIL-53(Al), ^{129}Xe NMR was used to reveal the transition from the large pore to the narrow pore phase that is induced by temperature or loaded gases. It was shown that two lines were observed in the ^{129}Xe NMR spectra, indicating the coexistence of the two phases at pressures above $5 \cdot 10^4$ Pa. Complete transformation was possible only by also reducing temperature. The hysteresis of xenon adsorption was also studied [42]. In a later study of some of the same authors [43] a more detailed analysis of this system was done including cross-polarization and 2D EXSY experiments.

In line with those results, also for the flexible MOF $Ni_2(2,6\text{-ndc})_2(dabco)$ (DUT-8, Dresden University of Technology) ^{129}Xe NMR experiments were performed to follow the so-called gate-pressure effect [44, 45]. This means that pore opening happens only when a certain gas pressure is applied. Chemical shift as well as linewidth were demonstrated to be sensitive for the transformation process. In a follow-up study [23], in situ ^{129}Xe high-pressure NMR was used together with DFT calculations including van der Waals interactions. In the open form a high xenon density compared to the liquid phase was observed together with fast exchange of xenon atoms inside the pores with the surrounding gas phase. Fig. 10 illustrates both the high-pressure setup as well as the pressure-dependent spectra.

Another flexible system, ZIF-8, was investigated by HP ^{129}Xe NMR. In this MOF, the flexibility stems from a reorientation of the organic linker at lower temperature. By that, a large increase in chemical shift is observed which is accompanied by an increase in xenon loading [46].

In two papers dealing with another DUT-MOF, DUT-49, also a flexible framework material, the rare case of a negative gas adsorption (NGA) was observed [47]. Especially in this case, direct observation of the guest molecule was possible. In the open-pore (op) phase, xenon is adsorbed until a critical relative pressure of 0.15 when contraction of the framework to a closed-pore (cp) phase occurs. This results in release of xenon. Further pressure increase leads to a recovery. ^{129}Xe NMR allowed to characterize this process as a collective, stepwise one with the op phase containing much more xenon, see Fig. 11.

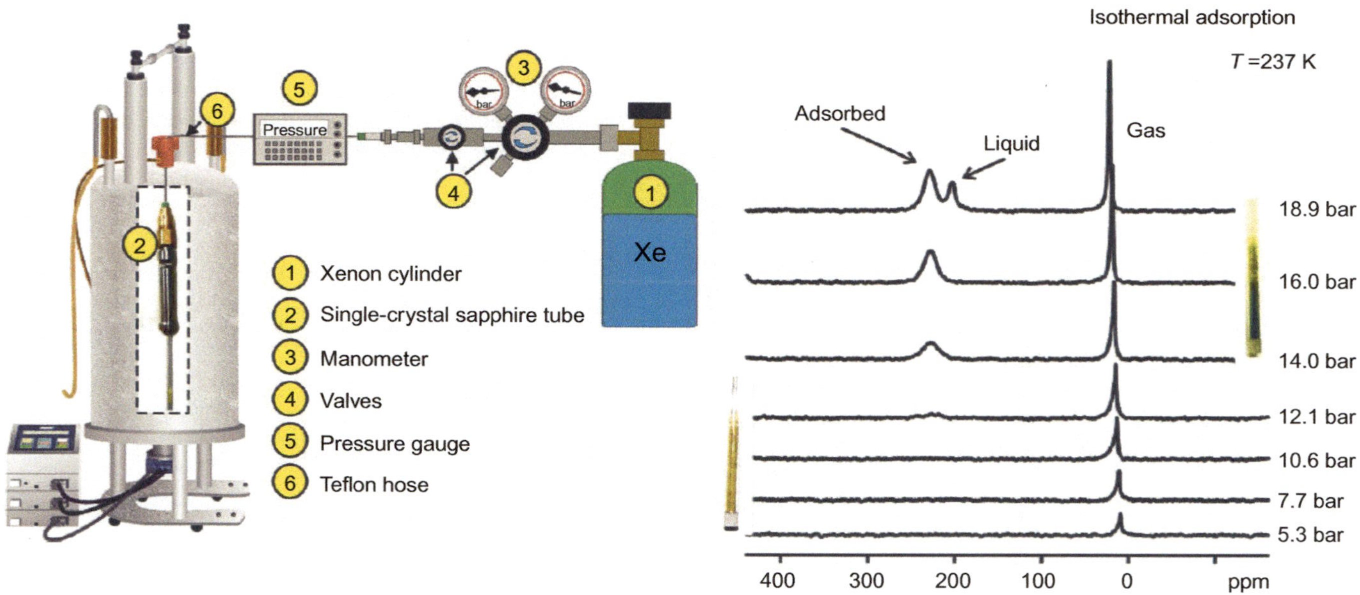

Fig. 10 *Left*: setup for high-pressure in situ experiments; *right*: ^{129}Xe NMR spectra recorded at 237 K with different pressures. *Reprinted (adapted) with permission from H.C. Hoffmann, B. Assfour, F. Epperlein, N. Klein, S. Paasch, I. Senkovska, S. Kaskel, G. Seifert, E. Brunner, High-pressure in situ ^{129}Xe NMR spectroscopy and computer simulations of breathing transitions in the metal-organic framework $Ni_2(2,6\text{-}ndc)_2(dabco)$ (DUT-8(Ni)), J. Am. Chem. Soc. 133 (2011) 8681–8690. Copyright 2011 American Chemical Society.*

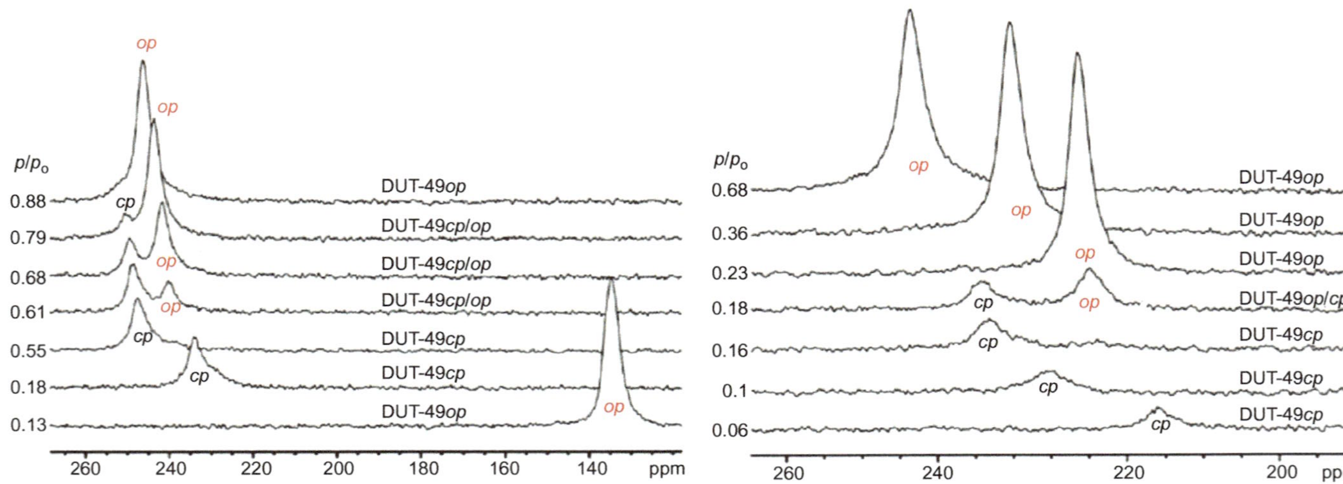

Fig. 11 ^{129}Xe NMR spectra recorded at 200 K and different relative pressures; *left*: adsorption, *right*: desorption experiment. *Reprinted (adapted) with permission from J. Schaber, S. Krause, S. Paasch, I. Senkovska, V. Bon, D.M. Többens, D. Wallacher, S. Kaskel, E. Brunner, In situ monitoring of unique switching transitions in the pressure-amplifying flexible framework material DUT-49 by high-pressure* ^{129}*Xe NMR spectroscopy, J. Phys. Chem. C 121 (2017) 5195–5200. Copyright 2017 American Chemical Society.*

These examples show the importance of ^{129}Xe NMR for the characterization of the pore space in MOFs. Especially for flexible MOFs, important information has been obtained. The experiments can also be quite variable using constant loading with gas and sealing, variable pressures in in situ experiments as well as using hyperpolarization techniques for fast measurements.

4.2 Water, H_2O

Many MOFs are unstable in the presence of water. In other MOFs, water has a strong binding affinity preferably to metal ions. Therefore, numerous studies deal with the characterization of adsorption of water in MOFs and MOF stability.

Lieder et al. studied water in the narrow pore form of MIL-53 [22]. They found two signals for water in the ^{1}H NMR spectra at 2.6 and 5.0 ppm. These originate from two types of water molecules with different hydrogen bonding with oxygen atoms from the linker (preferably at the carboxylic oxygen). Additionally, an increase of the quadrupolar coupling of ^{27}Al is observed upon hydration ($C_Q = 10.7$ MHz compared to 8.7 MHz in the dehydrated form). This is mainly due to the strain from the narrowing of the pores.

$Cu_3(btc)_2$ as one of the prominent water-unstable MOFs was in the centre of several studies. In a study with different amounts of H_2O loading, a varying ^{1}H shift was observed [21]. This is indicative of a dynamic process of water adsorbed at the copper site and free mobility in the pore at higher loadings. Additionally, the variation of spectra was monitored over the period of several days. Together with ^{13}C NMR spectra, decomposition was observed. Especially the ^{13}C spectra showed additional signals typical for diamagnetic samples that hint at a decomposition of the (paramagnetic) MOF at the carboxylate linker. In a related study, the same authors studied the time-dependent water uptake by ^{1}H MAS spectra in a deuterated $Cu_3(btc)_2$ to omit background ^{1}H signals [48]. Here, a packed MAS rotor was just revealed to atmospheric water vapour and spectra taken at certain times. The rather slow ingress of water and the movement of water within the MAS rotor from top to bottom revealed three different water signals developing over time (see Fig. 12). They were assigned to one and two water molecules per paddle-wheel appearing early in the hydration stage and the third signal when an excess of water is present. At this situation there is a permanent exchange of water adsorbed at the copper sites and free in the pores.

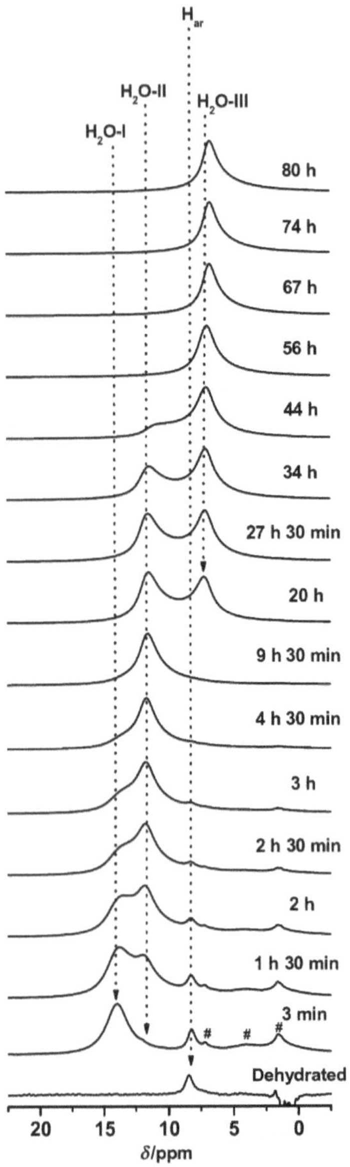

Fig. 12 ^{1}H MAS NMR spectra of d-$Cu_3(btc)_2$ after different exposure times to water vapour in air. # refers to residual solvent signal (ethanol). *Reprinted from F. Gul-E-Noor, D. Michel, H. Krautscheid, J. Haase, M. Bertmer, Time dependent water uptake in $Cu_3(btc)_2$ MOF: identification of different water adsorption states by ^{1}H MAS NMR, Microporous Mesoporous Mater. 180 (2013) 8–13. Copyright 2013, with permission from Elsevier.*

An interesting finding on the stability of $Cu_3(btc)_2$ was observed by Giovine et al. [49]. They observed that at high temperatures (above 150°C) the MOF is stable against water vapour. Both ^1H and ^{13}C MAS spectra revealed decomposition of the MOF after steam-flow treatment at 100°C and below, but not at higher temperatures. This observation was explained by the time available for interaction of water with the framework. At lower temperatures condensation of water in the framework can take place allowing enough time for interaction and hydrolysis. At higher temperature, though, only a transient interaction is happening that does not last long enough to start decomposition.

Stabilization against decomposition by water in a paddle-wheel structure was achieved by using variants of the btc linker [50]. The linker in STAM-1 (St. Andrews MOF) is the monomethyl ester of btc and in STAM-17 OEt the linker is 5-ethoxy isophthalate. In these materials, water adsorption leads to changes in the lattice parameters. This has significant effects on the ^{13}C spectra, see Fig. 13. The flexibility of the MOF allows for a regeneration of the original structure after regeneration.

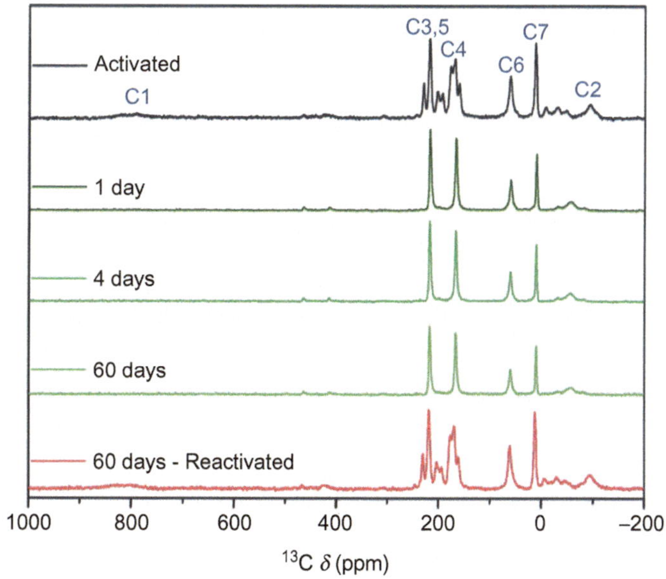

Fig. 13 ^{13}C MAS NMR spectra of STAM-1 dehydrated and after certain times of hydration. *Reprinted (adapted) with permission from A. Terracina, L.N. McHugh, M. Todaro, S. Agnello, P.S. Wheatley, F.M. Gelardi, R.E. Morris, G. Buscarino, Multitechnique analysis of the hydration in three different copper paddle-wheel metal-organic frameworks, J. Phys. Chem. C 123 (2019) 28219–28232. Copyright 2019 American Chemical Society.*

Another approach used pulsed field gradients (PFG) NMR in combination with molecular dynamics (MD) simulations to characterize the type of diffusion of water in the MOF MIL-100(Al) [51, 52]. The experimental data for different water loadings indicated the presence of intracrystalline diffusion and exchange between water in the crystals and the gas phase surrounding them [51]. In a follow-up paper the same authors revisited the experiments and used a two-site exchange model to fit the complete spin-echo attenuation. With that, the intracrystalline diffusion coefficient and the mean residence time was obtained [52].

4.3 Carbon dioxide, CO_2

Climate change because of increasing levels of carbon dioxide probably represents the most challenging problem in our times. One option to minimize the effects of climate change—removal of carbon dioxide from air or from power plant exhausts—is widely discussed. MOFs could contribute in this direction. For this and other reasons, most NMR studies with adsorbed molecules in MOFs deal with the adsorption properties of CO_2. Besides the general importance of CO_2 research, for NMR the easy isotopic ^{13}C labelling is a clear benefit for studying carbon dioxide adsorption. Recently, Witherspoon et al. reviewed SSNMR studies of CO_2 in MOFs [3]. A paper from Bermin et al. discusses NMR studies on CO_2 adsorption in MOFs and other materials like silica [18].

Therefore, here not all studies will be mentioned. The different objectives in SSNMR characterization of CO_2 adsorption are treated separately. In several publications ^{13}C solid-state NMR spectroscopy is used as one of several methods for characterization of MOFs. There, differences in the spectra with and without adsorbed CO_2, sometimes at different loadings and pressures, are discussed. These publications will not be discussed here, the focus is on those publications with a more detailed analysis of CO_2 adsorption in MOFs.

4.3.1 Effect of CO_2 adsorption on host structure

In MIL-53(Ga), the metal nuclear spins ^{69}Ga and ^{71}Ga in their NMR spectra are very sensitive towards changes to the GaO_6 octahedron [53]. From WURST-CPMG spectra at different CO_2 loadings, it was observed that above 0.1 CO_2/Ga two crystallographic gallium sites are present (Fig. 14). At low loadings, only some of the empty pores are occupied by CO_2 and with increasing loading all empty channels are occupied giving rise to two ^{71}Ga

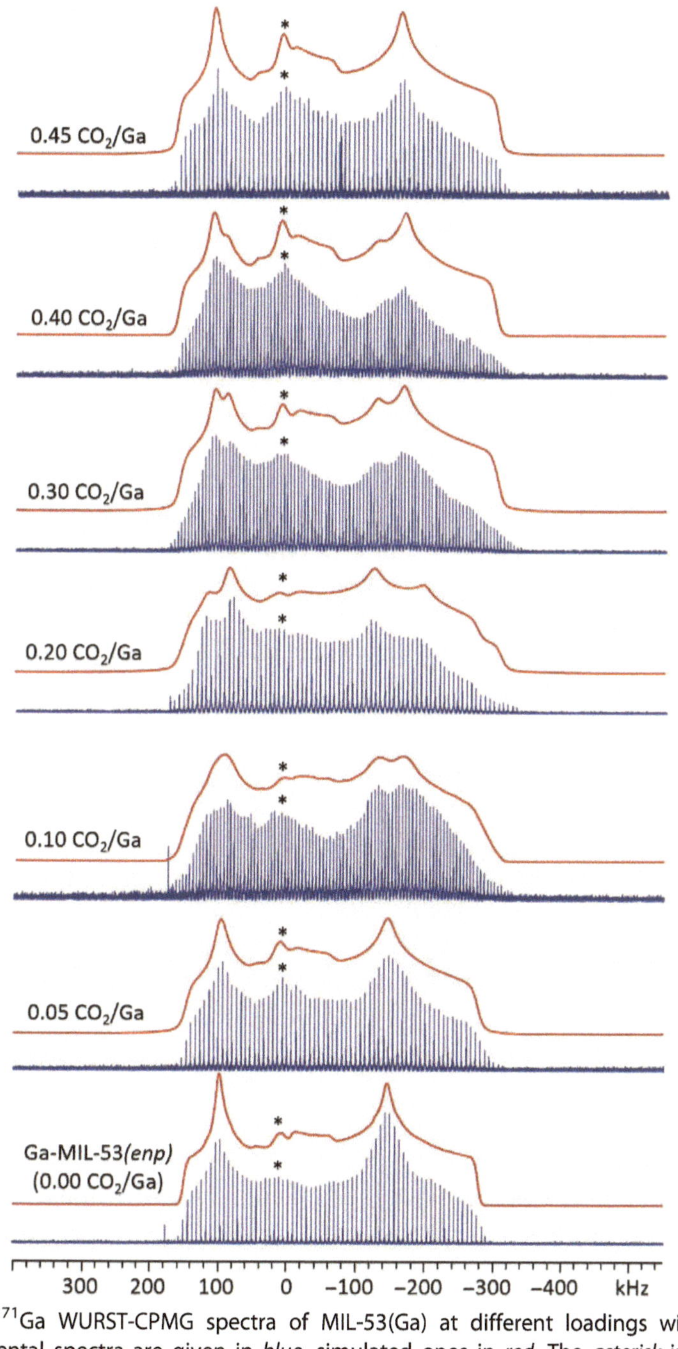

Fig. 14 ^{71}Ga WURST-CPMG spectra of MIL-53(Ga) at different loadings with CO_2. Experimental spectra are given in *blue*, simulated ones in *red*. The *asterisk* indicates an impurity. *Reprinted from Y. Zhang, B.E.G. Lucier, V.V. Terskikh, R. Zheng, Y. Huang, Tracking the evolution and differences between guest-induced phases of Ga-MIL-53 via ultra-wideline* $^{69/71}$*Ga solid-state NMR spectroscopy, Solid State Nucl. Magn. Reson. 84 (2017) 118–131. Copyright 2017, with permission from Elsevier.*

sites. In this study also the effect of adsorption of different organic compounds was monitored by $^{69/71}Ga$ SSNMR.

In SIFSIX-3-Zn, the same group used ^{67}Zn NMR spectra to track effects of CO_2 loading on the MOF [54]. In this case, no changes due to carbon dioxide adsorption compared to the activated material are observed. Further information is obtained from $^{13}C\{^{19}F\}$ CPMAS spectra that show the signal of CO_2 only at long contact times. $^{13}C\{^{1}H\}$ REDOR experiments indicate a rather strong interaction between framework protons and CO_2 which increases further at lower temperatures due to decreased mobility of CO_2.

Another approach was followed by Li et al. [55] In MFM-305-CD_3, ^{2}H spin-lattice relaxation times were determined as a function of temperature comparing the pure MOF with the MOF with adsorbed CO_2 to check influences on the rotation of the -CD_3 group with gas loading. In both cases two types of motion were observed, with the motion becoming slower with CO_2 loading assumed due to random collisions.

4.3.2 CO_2 dynamics

To study the motion of CO_2 inside the MOF pores, variable temperature ^{13}C lineshape analysis has been done from several groups in different MOFs. Especially the group of Yining Huang, University of Western Ontario, Canada, studied several MOFs with variable temperature NMR to deduce interaction and mobility in MOFs with CO_2 both with and without open metal sites [54, 56–61].

Kong et al. were probably the first to use variable temperature ^{13}C NMR together with relaxation time measurements to analyze MOFs, in this case Mg_2(dobdc) (MOF-74(Mg)) [62]. They used a large temperature range from 12 to 390 K and different CO_2 loadings together with lineshape simulations of uniaxial rotation (see Fig. 15). From relaxation time analysis, activation energies and correlation times for the rotational motion were obtained. These results laid the basis for the combination with DFT and molecular dynamics simulations that have been used further on (see also Section 5).

An interesting study involved the use of ^{17}O NMR of CO_2 adsorbed in the MOF CPO-27 (MOF-74(Mg)) [56]. In this case, the isotopic labelling is clearly more complicated/expensive, though the quadrupolar coupling of the ^{17}O nucleus provides an additional source of information for CO_2 dynamics in MOFs. In the temperature range 150–403 K the combined presence of a wobbling and hopping motion was determined with simulations using the EXPRESS software [63] (Fig. 16).

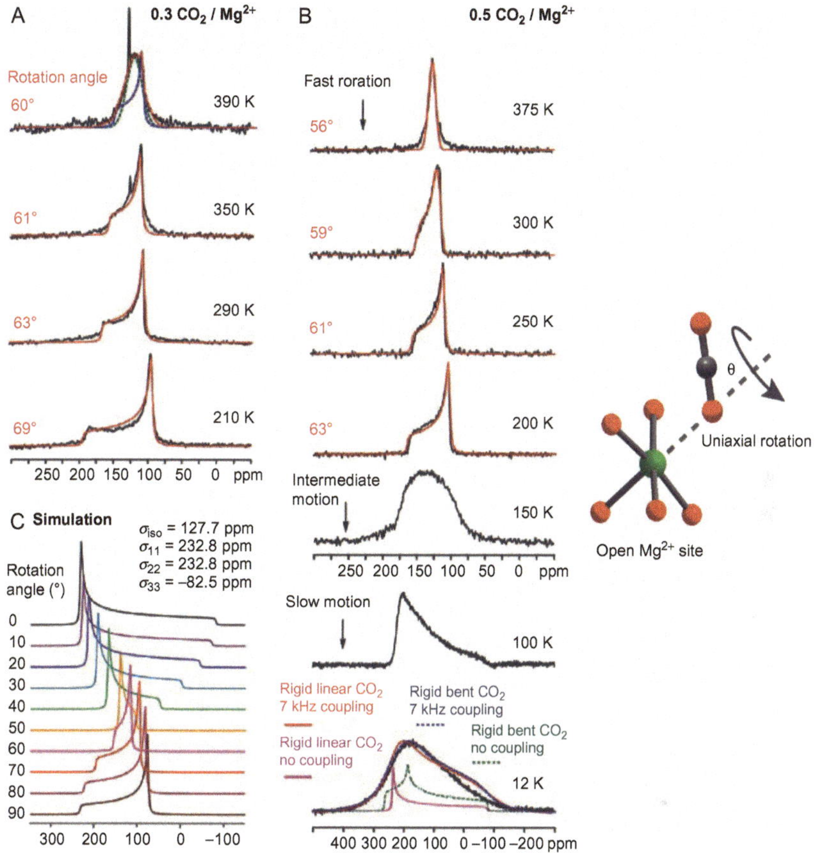

Fig. 15 Variable temperature ^{13}C NMR lineshapes of CO_2 in $Mg_2(dobdc)$. (A) 0.3 CO_2 per Mg^{2+}, (B) 0.4 CO_2 per Mg^{2+}, (C) lineshape simulation for uniaxial rotation of CO_2 at a fixed rotation angle. The rotation angle is defined on the *right. Reprinted (adapted) with permission from X. Kong, E. Scott, W. Ding, J.A. Mason, J.R. Long, J.A. Reimer, CO_2 dynamics in a metal-organic framework with open metal sites, J. Am. Chem. Soc. 134 (2012) 14341–14344. Copyright 2012 American Chemical Society.*

4.3.3 PFG NMR

^{13}C PFG NMR is by far less common than ^{1}H PFG NMR. However, the ease to use ^{13}C-enriched CO_2 allows for these measurements with good signal-to-noise ratio. In the MOF $Cu_3(btc)_2$, self-diffusion coefficients of CO_2 were determined as a single component and in a mixture with methane, CH_4 [64]. The results were compared with the values for methane from ^{1}H PFG measurements. Results are shown in Fig. 17. The smaller diffusion coefficient is explained by the stronger adsorption of CO_2.

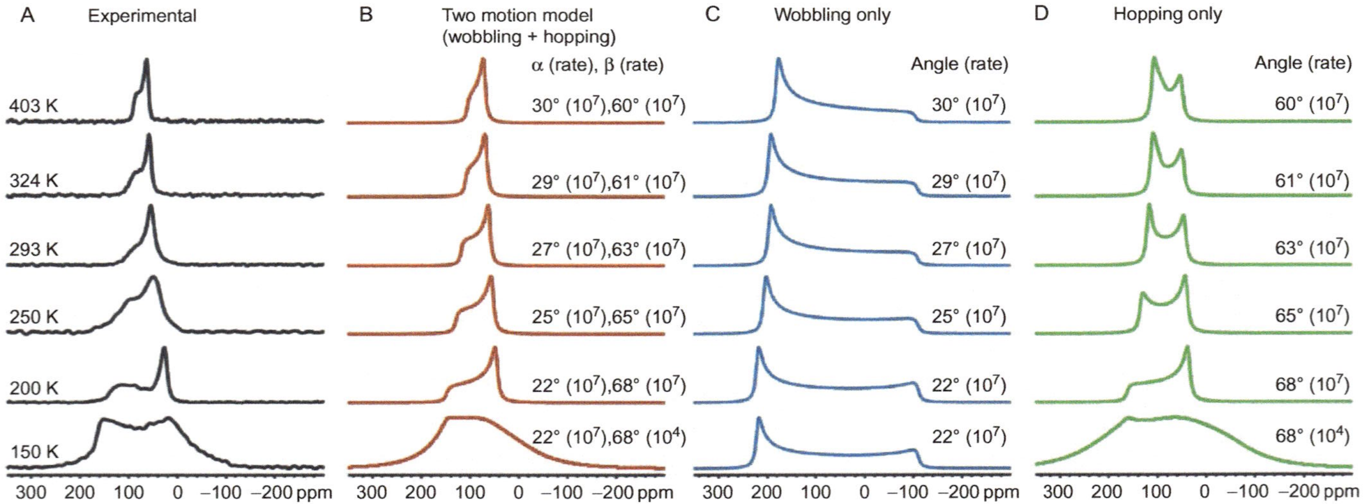

Fig. 16 Variable temperature ^{17}O NMR spectra of $C^{17}O_2$ adsorbed in CPO-27 with 0.1 CO_2 per Mg site. (A) Experimental lineshapes, (B) combined simulated lineshapes using EXPRESS [63] by a combination of a wobbling motion (C) and a hopping motion (D). *Reprinted (adapted) with permission from W.D. Wang, B.E.G. Lucier, V.V. Terskikh, W. Wang, Y. Huang, Wobbling and hopping: studying dynamics of* CO_2 *adsorbed in metal-organic frameworks via* ^{17}O *solid-state NMR, J. Phys. Chem. Lett. 5 (2014) 3360–3365. Copyright 2014 American Chemical Society.*

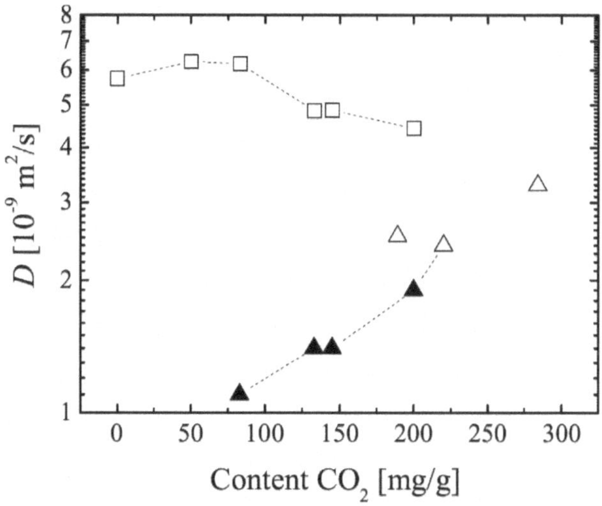

Fig. 17 Self-diffusion coefficients of a mixture of CH_4 (*open squares*) and CO_2 (*filled triangles*) as a function of CO_2 content. For comparison, *open triangles* show the values for only CO_2. *Reprinted from S. Schlayer, A.-K. Pusch, F. Pielenz, S. Beckert, M. Peksa, C. Horch, L. Moschkowitz, W.-D. Einicke, F. Stallmach, X-nuclei NMR self-diffusion studies in mesoporous silica foam and microporous MOF CuBTC, Materials 5 (2012) 617–633.*

With higher loading, adsorption sites are filled and the remaining CO_2 molecules show a faster diffusion.

In two papers, Forse et al. investigated the anisotropic diffusion of CO_2 in the MOF Zn_2(dobpdc) by PFG NMR [65, 66]. This MOF contains 1D-channels with a diameter of 2 nm. This gives rise to an anisotropic lineshape. By measuring the lineshape at different gradient strengths, self-diffusion coefficients for different orientations of the CO_2 tensor could be obtained.

4.4 Hydrogen, H_2

Reduction of the use of fossil fuels and with that the search for alternative transportation devices also calls for the use of hydrogen in fuel cells. Several MOFs allow for significant uptake of hydrogen—although only at fairly low temperatures. Understanding the mechanism of hydrogen uptake can lead to improved materials with higher uptake and/or uptake at higher temperatures. Solid-state NMR is especially suited for studying hydrogen adsorption as ^1H is the most receptive nonradioactive isotope. Furthermore, investigation of D_2 can be used selectively where the quadrupolar coupling of ^2H (spin $I = 1$) can yield important information. A mixed HD molecule has also been investigated.

Albeit these advantages for the use of solid-state NMR, only few studies have employed this possibility for hydrogen adsorption studies. In the metal–organic framework Z-1200, a relaxation time study using HD was done [67]. In the temperature range between 2 and 50 K a strong temperature dependence of the spin-lattice relaxation time close to those of quantized translational energy levels of a constrained HD molecule in the supercage of the MOF was found. The spin–spin relaxation time indicates a high mobility at these temperatures. Unfortunately, neither the magnetic field strength nor the measured nucleus, ^1H or ^2H (^2D), is indicated.

In α-Mg-formate, Lucier et al. used ^2H NMR to characterize the adsorption of D_2 in the temperature range from 293 to 123 K [68]. In that paper, a full solid-state NMR analysis of the MOF is given with the D_2 adsorption being only one part of it. They observed two equally populated adsorption sites together with a signal for free D_2. There is some indication for a third site at low temperatures which would be in accordance with theoretical calculations [69]. From dynamic simulations, two motions of D_2 in the MOF are deduced, a localized rotational wobbling and a two-site hopping between adsorption sites (Fig. 18). Also, the quadrupolar coupling increases with decreasing temperature.

In another paper by the same group, six MOFs are compared with respect to their hydrogen adsorption strength (i.e., D_2) with the help of ^2H NMR [70]. UiO-66 and Zn-MOF-74 yield narrow resonances for weak adsorption from temperatures of 133 K and higher. In opposition to this, Mg-MOF-74, Ni-MOF-74, α-$Mg_3(COOH)_6$, and α-$Zn_3(COOH)_6$ show broader powder patterns at lower temperatures. Among those, also different lineshapes and different onsets of broadening are apparent. For the nickel MOF, additional influence of the paramagnetic nickel centre shifts the resonance frequency. Here, broadening and shifting of the line indicates interaction with the open metal site.

There are few other studies involving hydrogen adsorption measurements and the use of solid-state NMR. Volkringer et al. reported on gallium and indium frameworks of the MIL-68 type. They used various ^1H NMR data including 2D correlations to study the framework structure. Hydrogen sorption data were supported, but—unfortunately—no NMR data on hydrogen adsorption were given [71]. Yang et al. used postsynthetic exchange by Li^+ in the pore and characterized that by ^7Li NMR, but did

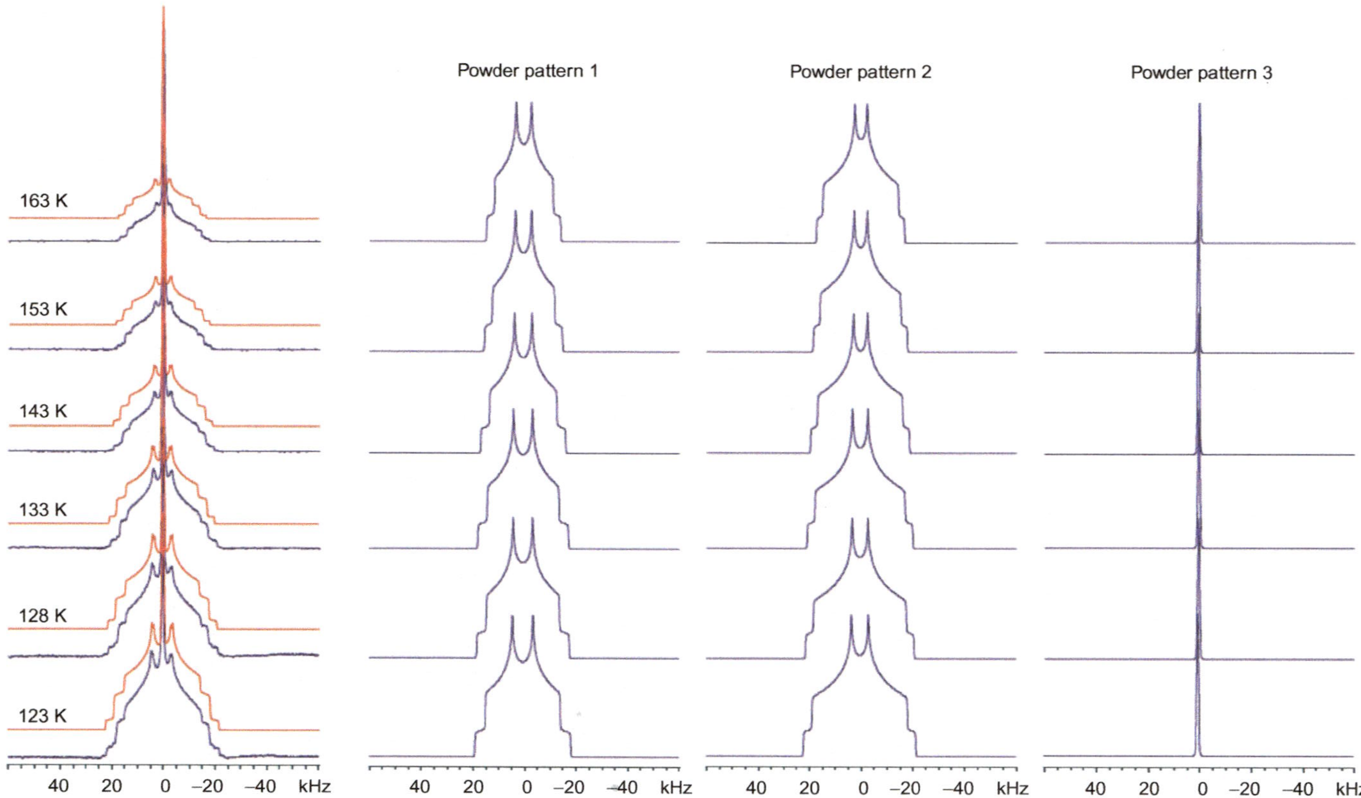

Fig. 18 (A) Experimental (*blue*) and simulated (*red*) ^{2}H NMR patterns for D_2 in α-$Mg_3(COOH)_6$ as a function of temperature. (B) The individual powder patterns for the three types of D_2. *Reprinted with permission from B.E.G. Lucier, Y. Zhang, Y. Huang, Complete multinuclear solid-state NMR of metal-organic frameworks: the case of α-Mg-formate, Concepts Magn. Reson. A 45 (2016) 21410. Copyright 2019 John Wiley & Sons, Inc.*

not follow H_2 or D_2 adsorption via NMR [72]. The same is true for an investigation of porous aromatic frameworks (PAFs) [73].

It seems that the potential of solid-state NMR to study hydrogen adsorption in MOFs has not been used to full extent. Clearly, typically low temperatures are needed to identify adsorption states but information can be obtained that is not easy to obtain otherwise. Often theoretical calculations are used to provide information on binding site and strength.

4.5 Carbon monoxide, CO

Few studies deal with the adsorption of carbon monoxide in MOFs. Especially the similarities and differences to the adsorption of carbon dioxide (Section 4.3) in the same material is of interest. As an exhaust from burning fossil fuels extraction from air would also be of interest. Furthermore, CO is also a precursor in chemical synthesis but is typically produced in a gas mixture and separation of CO is required. For NMR measurements, similar to CO_2, easy and fairly cheap isotopic enrichment with ^{13}C allows for easy detection and discrimination from carbon signals from the ligands. Variable temperature measurements allow for the determination of mobilities of CO in the pores.

In the system $Cu_3(btc)_2$ (also with partial replacement of Cu by Zn), different mobilities of CO were detected at different temperatures [74]. Already at a temperature of 30 K there exists a mobile, isotropic fraction of CO together with immobilized CO characterized by a CSA pattern. At 100–110 K the anisotropic lineshape has completely vanished. Furthermore, above 200 K, signals for gaseous CO and adsorbed, but mobile CO in the MOF can be distinguished, see Fig. 19.

The interaction with the unpaired electrons at the copper site reflecting the adsorption of CO is visible from the isotropic shift values of CO that undergo a variation by 12 ppm in the measured temperature range due to the increasing population of the $S=1$ electronic state with increasing temperature. The increased mobility at higher temperatures then leads to the effect of reversing the shift change above 200 K, as shown in Fig. 20.

Further investigation of the ^{13}C spin-lattice relaxation time of CO as a function of temperature allowed for a determination of the activation energy for motion [75]. A jump like motion is concluded with the jump length being longer than the distance between adsorption sites.

In Zn- and Mg-MOF-74 [76], variable temperature measurements revealed that isotropic motion of CO sets in at much higher temperatures as for $Cu_3(btc)_2$, above 173 K for the Mg-MOF and above 193 K for the

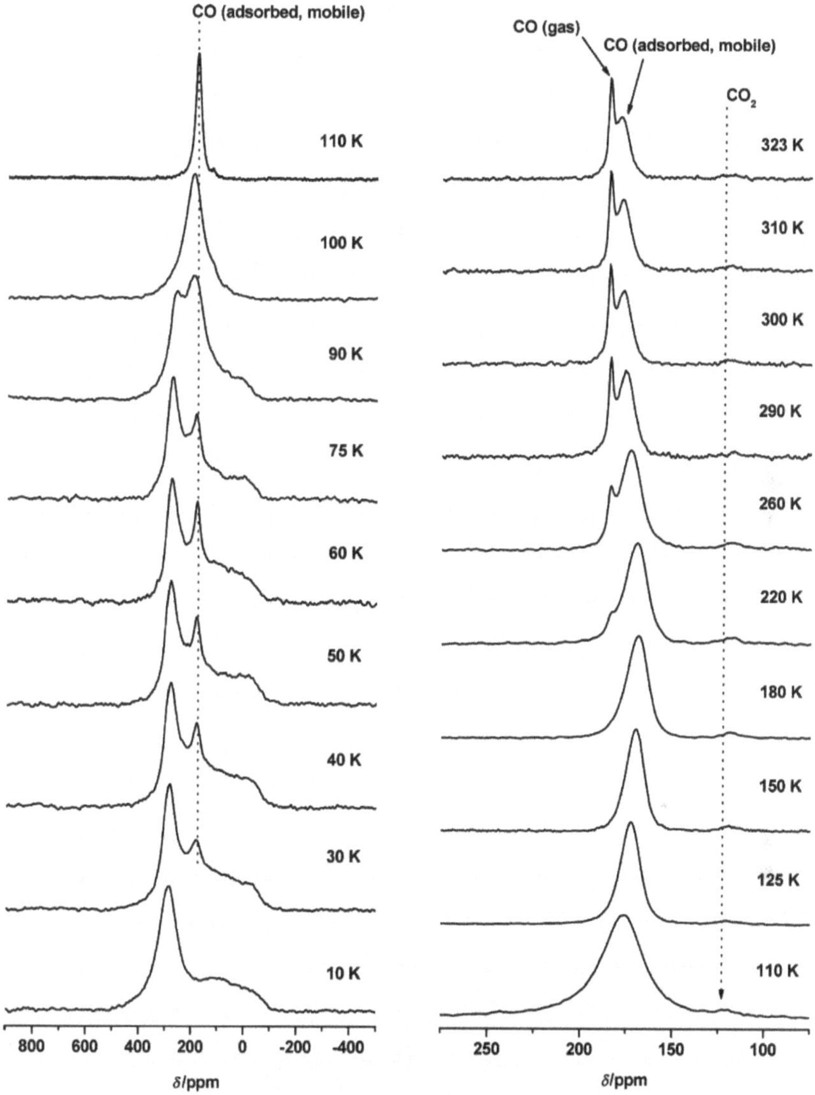

Fig. 19 ^{13}C static spectra of 2 CO/Cu in $Cu_3(btc)_2$ recorded in the temperature range from 10 to 323 K. *Reprinted (adapted) with permission from F. Gul-E-Noor, M. Mendt, D. Michel, A. Pöppl, H. Krautscheid, J. Haase, M. Bertmer, Adsorption of small molecules on $Cu_3(btc)_2$ and $Cu_{3-x}Zn_x(btc)_2$ metal-organic frameworks (MOF) as studied by solid-state NMR, J. Phys. Chem. C 117 (2013) 7703–7712. Copyright 2013 American Chemical Society.*

Zn-MOF. Still at 373 K an anisotropic lineshape is observed though with a smaller span than at lower temperatures. This difference to the results in $Cu_3(btc)_2$ are a result of the stronger adsorption of CO in MOF-74. With more detailed analysis of the CO dynamics using the EXPRESS

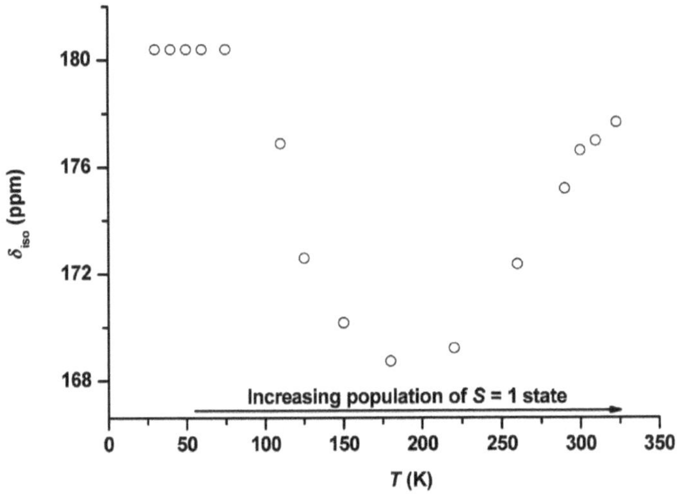

Fig. 20 ^{13}C isotropic chemical shift of the mobile CO in $Cu_3(btc)_2$ as a function of temperature. *Reprinted (adapted) with permission from F. Gul-E-Noor, M. Mendt, D. Michel, A. Pöppl, H. Krautscheid, J. Haase, M. Bertmer, Adsorption of small molecules on* $Cu_3(btc)_2$ *and* $Cu_{3-x}Zn_x(btc)_2$ *metal-organic frameworks (MOF) as studied by solid-state NMR, J. Phys. Chem. C 117 (2013) 7703–7712. Copyright 2013 American Chemical Society.*

software [63], a combination of two types of motions happen simultaneously: a wobbling motion at the metal site and a hopping motion along the channel interior containing six metal clusters (Fig. 21). For the two different systems, in the Mg-MOF the wobbling angles are smaller but the hopping angles are larger than in the Zn-MOF. This indicates weaker binding of CO in the Zn-MOF.

In another study by some of the same authors, CO dynamics in CdSDB and PbSDB MOFs was analyzed (SDB = 4,4′-sulfonyldibenzoate) [77]. Main difference is that these MOFs have no open metal sites. Although there are similarities to the previous example, the dynamics is more complex in these MOFs. The anisotropic lineshape for instance changes as a function of temperature as illustrated by a large change in skew value from 0.3 at low temperature to 0.9 at high temperatures for the Pb-MOF. For the Cd-MOF two CO sites are present with significantly different span values resulting from different adsorption strengths, see Fig. 22. Due to the lack of open metal sites, interaction with the π-bonds of the aromatic rings is expected to be the primary interaction. Further analysis revealed a rotation around a threefold axis together with a twofold jump. These results are similar to those observed for CO_2 adsorption [57].

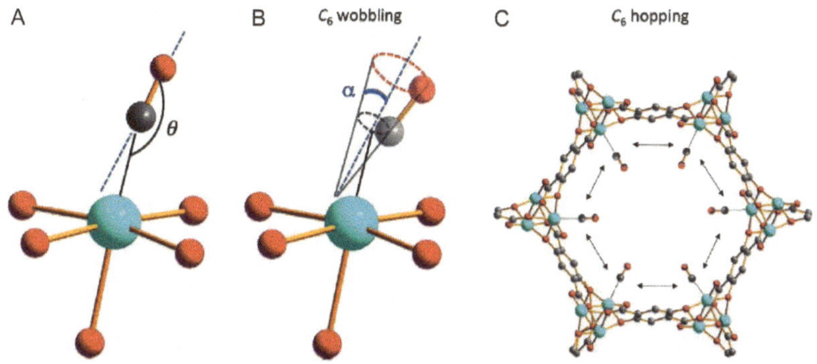

Fig. 21 Illustration of the present motions of CO in MOF-74. (A) Illustration of the angle θ, (B) wobbling angle α, (C) hopping motion on the circumference of the MOF channel. *Reprinted with permission from B.E.G. Lucier, H. Chan, Y. Zhang, Y. Huang, Multiple modes of motion: realizing the dynamics of CO adsorbed in M-MOF-74 (M = Mg, Zn) by using solid-state NMR spectroscopy, Eur. J. Inorg. Chem. 2016 (2016) 2017–2024. Copyright 2019 John Wiley & Sons, Inc.*

4.6 Nitric oxide, NO

There have been only few studies dealing with the adsorption of nitric oxide. However, the interest in this molecule stems from the function of nitric oxide in the human body. It is endogenously produced and functions in platelet activation, contributing to wound healing and is a signalling agent in the immune system. There are also anticancer therapies discussed including NO. For introduction of NO in the human body, suitable drug-delivery systems are needed. MOFs could also play an important role in this. Especially, the chemisorption of NO in MOFs as NONOates (see below) is of interest as this is a reversible reaction under pH conditions existing in the human body.

From the NMR point of view, NO is quite complicated to study due to its paramagnetic nature at normal conditions. This assigned NO as being "undetectable" by NMR [78]. However, recently Khan et al. were able to detect signal from a pure NO sample at temperatures where NO is liquid [79]. This is most probably due to dimerization at these conditions which represents a diamagnetic system.

Because of this, effects of NO adsorption can normally be detected only indirectly via NMR active nuclei of the MOF due to the paramagnetism of NO. In MIL-100(Al), NO adsorption expected on the aluminium site in the Al_3 trimer showed a change of coordination number of aluminium from

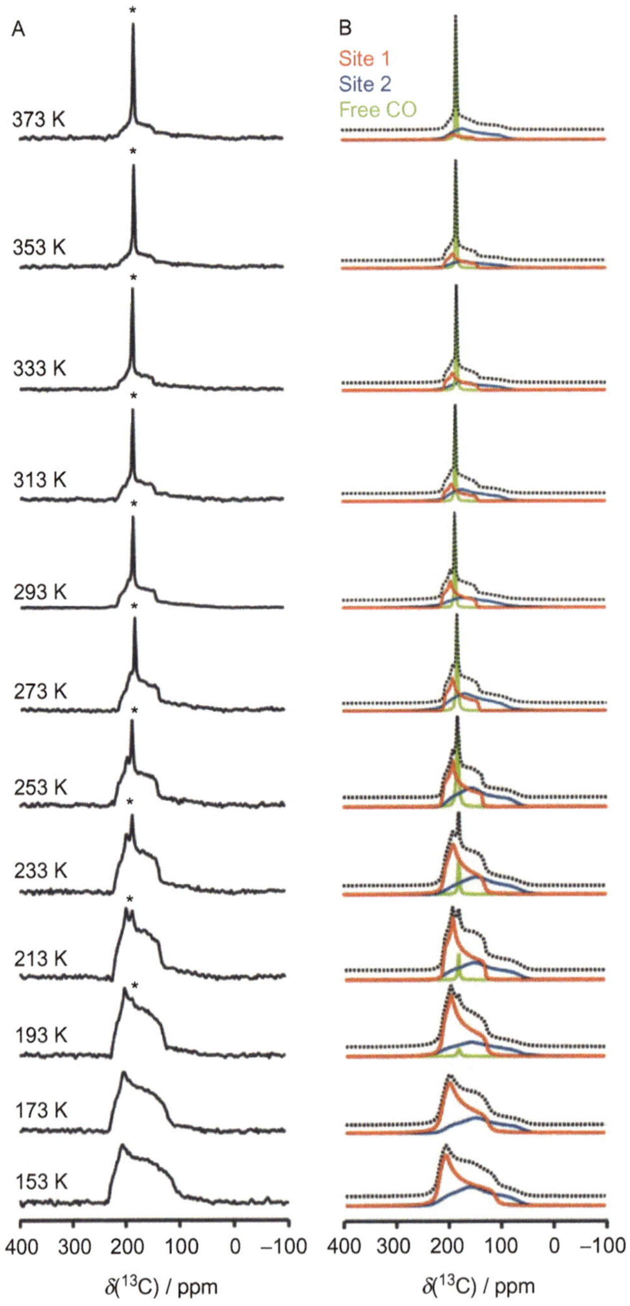

Fig. 22 Variable temperature ^{13}C spectra of CO adsorbed in CdSDB; *left*: experimental, *right*: simulation with two adsorption sites together with free CO (marked by an *asterisk* in the experimental spectra). *Reprinted (adapted) with permission from Y.T.A. Wong, T.K. Babcock, S. Chen, B.E.G. Lucier, Y. Huang, CO guest interactions in SDB-based metal-organic frameworks: a solid-state nuclear magnetic resonance investigation, Langmuir 34 (2018) 15640–16549. Copyright 2018 American Chemical Society.*

5- to 6-coordinated as a function of NO loading [15]. In ^{13}C CPMAS spectra with increasing amount of NO a line-narrowing especially for the carboxylate carbon was found similar to a hydrated MIL. This was explained with some stress release on the framework due to coordination change of aluminium. As was also shown by FT-IR measurements, the presence of NO^+ is assumed in accordance with NMR data. Otherwise, ^{27}Al NMR spectra should be affected by the paramagnetic NO adsorption leading to shift effects and linebroadening.

An opposing effect of NO loading could be demonstrated on $Cu_3(btc)_2$ and its derivatives [80]. As these materials behave paramagnetic themselves at elevated temperatures (copper paddle-wheel structure has $S = 0$ electronic state at low temperature, but $S = 1$ is populated at temperatures above 70 K), the adsorption of paramagnetic NO at the copper sites leads to some quenching of unpaired electron density. The consequences can be followed by ^1H MAS NMR. The aromatic ^1H shift decreases with increasing NO loading. Additionally, the hyperfine coupling strength decreases with NO loading due to the quenching of unpaired electron density. Because of this, also the ^1H spin-lattice relaxation time increases (Fig. 23).

Besides these indirect effects, chemical reactions of NO with modified linker molecules allow for a direct detection of NO chemisorption. This is enabled by the presence of amino functionalities that react with two molecules of NO to form N-diazeniumdiolates (so-called NONOates) [81]. Since two molecules react, the reaction product is diamagnetic and can be detected. Whereas in amino-functionalized $Cu_3(btc)_2$ hints for NONOate formation had been proposed from indirect measurements (no changes in hyperfine coupling and relaxation time with NO loading, data included in Fig. 23) [80], a detailed investigation using isotopically labelled ^{15}NO on an ethylamino functionalized $Cu_3(btc)_2$ (UHM-37, University of Hamburg Materials) showed clear evidence [14]. Besides effects on ^1H and ^{13}C shifts, ^{15}N NMR spectra showed the existence of NONOates at shifts of about -75 ppm (see Fig. 24).

Miller et al. reported on a Ca-MOF suitable for the controlled release of NO [82]. They showed the ^{13}C and even rare ^{43}Ca NMR. However, only the pure material without adsorbed NO was analyzed with NMR. Given the toxicity of copper-containing MOFs, the Ca-MOF is probably more suitable for medicinal application. In the future, the investigation of nitric oxide adsorption in MOFs might offer several possibilities for studies with solid-state NMR.

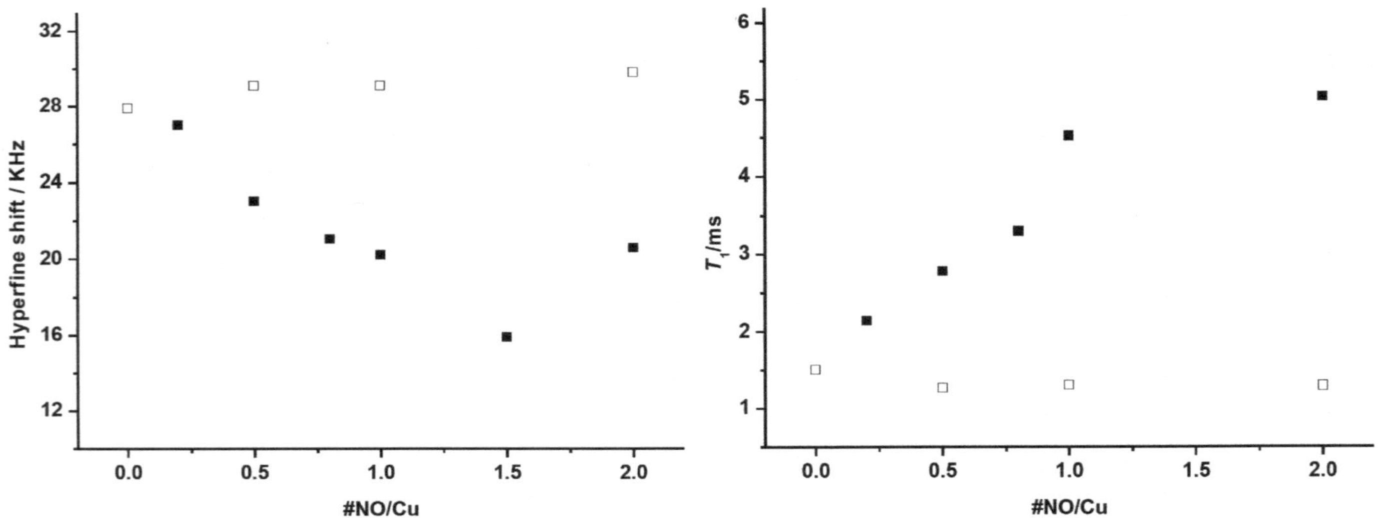

Fig. 23 *left*: ^1H hyperfine coupling, and *right*: ^1H T_1 relaxation time for $Cu_3(btc)_2$ (*filled squares*) and $Cu_3(NH_2btc)_2$ (*open squares*) as a function of NO loading. *Reprinted (adapted) with permission from A.H. Khan, K. Peikert, F. Hoffmann, M. Fröba, M. Bertmer, Nitric oxide adsorption in Cu_3btc_2-type MOFs–physisorption and chemisorption as NONOates, J. Phys. Chem. C 123 (2019) 4299–4307. Copyright 2019 American Chemical Society.*

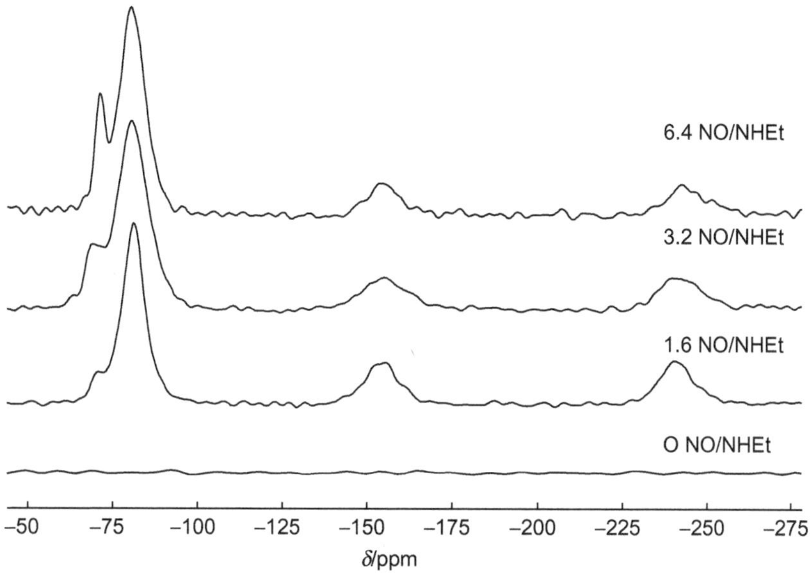

Fig. 24 ^{15}N NMR spectra of UHM-37 with different NO loading. Signals at −155 and −240 ppm are assigned as N_2O impurities in the original gas cylinder. *Reprinted (adapted) with permission from A.H. Khan, K. Peikert, F. Hoffmann, M. Fröba, M. Bertmer, Nitric oxide adsorption in Cu_3btc_2-type MOFs—physisorption and chemisorption as NONOates, J. Phys. Chem. C 123 (2019) 4299–4307. Copyright 2019 American Chemical Society.*

4.7 Hydrocarbons

4.7.1 Methane, CH_4

Several papers discuss the interaction of methane gas with the MOF framework. It is of special interest for separation processes between methane and carbon dioxide or other small alkanes. Different strategies have been employed to get information about methane in MOFs. Diffusion studies using pulsed field gradients have been done to get information about the self-diffusivities and to relate different adsorbents to each other [83, 84]. ^2H NMR measurements aimed at the temperature-dependent dynamics of methane inside the pores [59, 85], and various spectroscopic measurement techniques—including relaxation time analysis—have been used to get information about the adsorption sites of methane inside the MOF [84, 86–88]. Different sample preparations for NMR measurements have been used, including high-pressure and in situ NMR measurements.

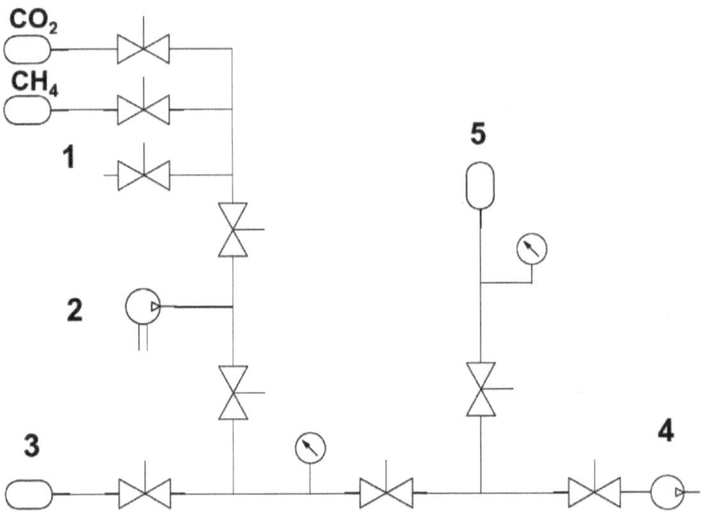

Fig. 25 Set-up of the high-pressure loading station: gas inlets (1), high-pressure generator (2), gas reservoir (3), vacuum pump (4), and NMR pressure cell (5). *Reprinted from C. Horch, S. Schlayer, F. Stallmach, High-pressure low-field ^1H NMR relaxometry in nanoporous materials, J. Magn. Reson. 240 (2014) 24–33. Copyright 2014, with permission from Elsevier.*

With low-field ^1H relaxometry, relaxation time distributions of methane alone and in a mixture with carbon dioxide were determined for $Cu_3(btc)_2$ and ZIF-8 at different pressures. For this purpose, a combined system for sample activation, loading with gas or gas mixtures and subsequent NMR measurement was designed (Fig. 25). Significant changes on the relaxation time distribution were observed (Fig. 26).

A comparison of the diffusivities of methane, ethane, and ethene in ZIF-8 as single component and in mixtures was done by ^1H PFG MAS NMR. [83] Fig. 27 shows the results as a function of temperature. Both as single component and in a mixture methane showed the highest diffusivities among the three components.

In a detailed investigation of the MOF Cu(INAIP) by a combination of single-crystal XRD, SSNMR, and DFT calculations, methane adsorption was studied together with carbon dioxide and ethyne [59]. Particular focus was on the question which part of the amide functionalization was responsible for the adsorption, the carbonyl, or the amine group. It was found that the carbonyl group acts a primary binding site for the gases. For methane and ethyne, temperature-dependent ^2H NMR spectra were recorded. In the case of methane, guest disorder was observed at all investigated temperatures. The linewidth is the result of multiple binding interactions and also the influence from paramagnetic Cu^{2+}.

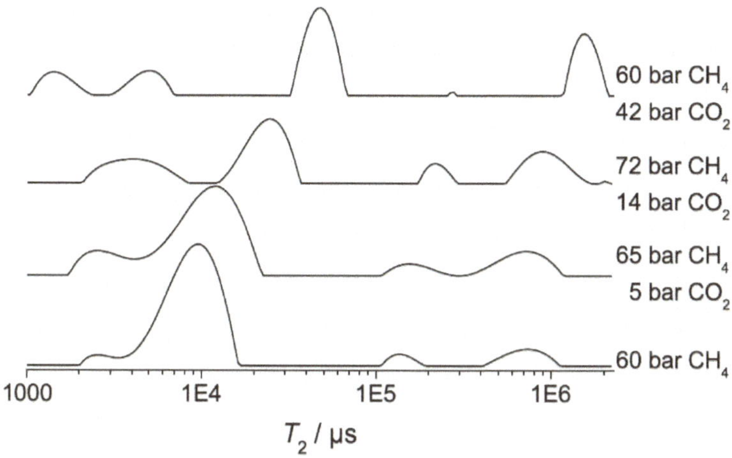

Fig. 26 Pressure-dependent T_2 relaxation time distributions of mixtures of CH_4 and CO_2. *Reprinted from C. Horch, S. Schlayer, F. Stallmach, High-pressure low-field ^{1}H NMR relaxometry in nanoporous materials, J. Magn. Reson. 240 (2014) 24–33. Copyright 2014, with permission from Elsevier.*

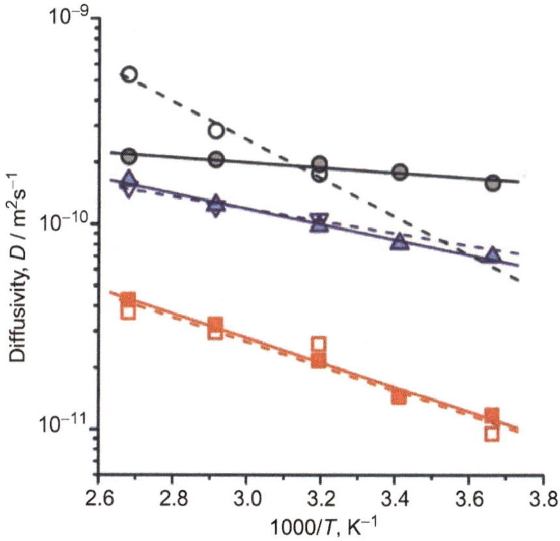

Fig. 27 Arrhenius plots for diffusion coefficients of methane (*circles*), ethane (*squares*), and ethene (*triangles*) adsorbed in ZIF-8. Single-component curves are given by open, in mixture by filled symbols. *Reprinted (adapted) with permission from N. Dvoyashkina, D. Freude, S.S. Arzumov, A.G. Stepanov, Monitoring the diffusivity of light hydrocarbons in a mixture by magic angle spinning pulsed field gradient NMR: methane/ethane/ethene in ZIF8, J. Phys. Chem. C 121 (2017) 25372–25376. Copyright 2017 American Chemical Society.*

In a similar manner, the same group investigated the methane (CH_3D) adsorption of several MOFs, α-$Mg_3(HCO_2)_6$, α-$Zn_3(HCO_2)_6$, SIFSIX-3-Zn, and M-MOF-74 (M=Mg, Zn, Ni, Co), by a combination of single-crystal XRD, DFT calculations and ^2H NMR. In each MOF only one adsorption site was found. A dynamic exchange between adsorbed methane and free methane in the pores occurs. Interestingly, the ^2H CSA has a significant contribution to the lineshape. Adsorption is stronger in Mg-MOF-74 compared to Zn-MOF-74. For the cobalt and nickel form of MOF-74, paramagnetic broadening from the metal centres complicated the analysis.

In MTV-UiO-66, besides standard ^1H and ^{13}C NMR, the $^1H\{^{13}C\}$ *J*-HMQC experiment was used to select the signal from the guest molecule, in this case methane, ethane, and propane [88]. Combined with ^1H-^1H spin diffusion, buildup-curves could be recorded that indicate the adsorption site of the small gases at the host structure (Fig. 28). Together with numerical simulations, distance information could be extracted. The preferential adsorption site is near the BDC-2OH site.

4.7.2 Ethene/ethane, C_2H_4/C_2H_6

In chemical industry, there is high demand for the separation of ethene and ethane as well as of ethene and carbon dioxide. Metal–organic frameworks are thought to be suitable candidates for this process. This topic has also been addressed by NMR [59, 83, 88–93].

In ZIF-8, ethene showed a higher diffusivity compared to ethane, as determined by ^1H MAS PFG NMR [83, 89]. This is true for a single component as well as in a mixture. The reason for this is the smaller size of ethene. Differences in host–guest interaction have no impact on the mobility of the molecules.

PFG NMR was also used to investigate a mixed-linker MOF (ZIF-7-8) composed of 2-methylimidazolate and benzimidazolate linkers. Ethane self-diffusion [91] as well as ethene/ethane mixtures (besides other gases) were analyzed [93]. A distribution of ethane diffusivities was found originating from different ZIF-7-8 crystals. This result was also confirmed by IR microscopy data. In the mixed-linker MOF the diffusivities are always smaller than in ZIF-8 containing only the 2-methylimidazolate. In ZIF-7-8, similar diffusivities for ethene and ethane were determined. In the system CO_2/C_2H_4, however, an improved selectivity compared to ZIF-8 was found. This was explained by the smaller pore aperture for ZIF-7-8 compared to ZIF-8.

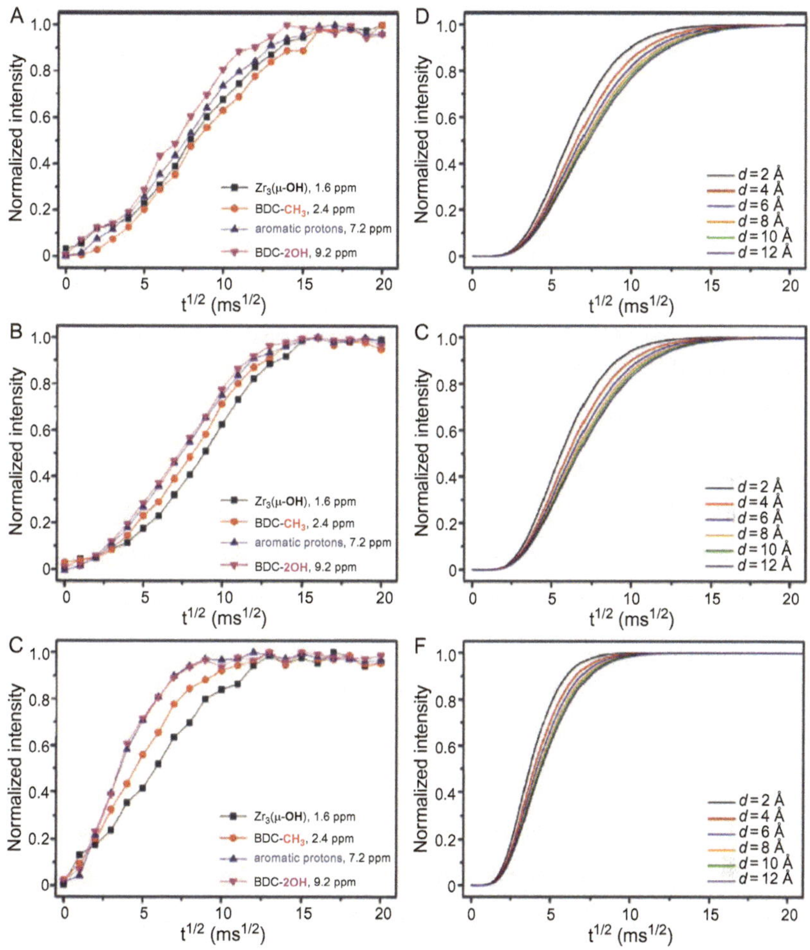

Fig. 28 1D ^{1}H-{^{13}C}*J*-HMQC-SD build curves of MTV-UiO-66 after adsorption of methane (a), ethane (b), and propane (c). On the right are the results of numerical simulations using different distances between the corresponding alkanes. *Reprinted (adapted) with permission from Y. Xiao, Y. Chu, S. Li, Y. Su, J. Tang, J. Xu, F. Deng, Primary adsorption sites of light alkanes in multivariate UiO-66 at room temperature as revealed by solid-state NMR, J. Phys. Chem. C 124 (2020) 3738–3746. Copyright 2020 American Chemical Society.*

An interesting system has been presented by Fan and coworkers [90]. Here, the selectivity between ethene and ethane can be varied by an external stimulus like temperature or light resulting in a gate-opening effect of a diarylethene MOF. Another system using a cyclodextrin-based MOF showed an unusual preference of carbon dioxide compared to ethene useful for selective adsorption [92]. In terms of NMR, however, both systems have (only) been characterized by ^{13}C MAS NMR to demonstrate the regeneration of the MOF after UV irradiation in the former, and the formation of a carbonic acid function after CO_2 adsorption in the latter case.

4.7.3 Aromatics

The field of aromatic compounds is clearly huge, and several papers deal with the adsorption of aromatic compounds in MOFs. Mainly benzene adsorption is studied in terms of diffusion measurements and ^2H lineshape analysis. Another application of solid-state NMR is the determination of the acid strength of MOF sites using pyridine.

PFG NMR was used to determine the self-diffusion of benzene in MOF-5 at different loadings [19]. They observed that the diffusion drops significantly (to $\frac{1}{3}$) when increasing the loading from 10 to 20 molecules per unit cell. This was in opposition to molecular dynamic simulations that predicted an increase in the diffusion. Explained is this fact by a clustering of adsorbates at certain temperatures. This observation should help in improving MD simulations.

Kolokolov and coworkers used ^2H NMR together with results from quasi-elastic neutron scattering as well as molecular dynamics simulations to study benzene-d_6 (besides other aromatic compounds) in the different MOFs MIL-53(Cr) [94], UiO-66(Zr) [95], and ZIF-8 [96]. In the flexible MIL-53, benzene adsorption leads to the narrow pore form which limits diffusion consisting of 1D jump translation together with uniaxial rotation around the C_6 axis [94]. In UiO-66 at low temperatures, parts of benzene is confined in the tetrahedral cages leading to an anisotropic C_6 rotation [95]. In the octahedral cages, rotation around the C_2 symmetry axis together with local isotropic rotation is found. Limiting diffusion is the passage through the windows between the cages, the jump being anisotropic.

Adsorption of the three xylene isomers together with toluene and isobutane was investigated in the MOF ZIF-8 [96]. ^2H lineshape and relaxation times revealed three different states for the guest molecules, i.e., highly mobile species in the cage, molecules bound to the cell walls and those passing through the window.

Another objective was followed by Lieder et al. [22]. Here, the focus was on the effect of *o*-xylene (next to water and nitrogen-containing bases) on the AlOH groups of MIL-53. With different loading of *o*-xylene the ^{27}Al quadrupolar coupling changed in two steps. This could be correlated with the double-sigmoidal adsorption isotherm. A further increase of C_Q was found at high loading which indicates the presence of two local structures around the $AlO_4(OH)_2$ centres.

The interaction of styrene and ethylbenzene with the MIL-53 host was studied by multinuclear SSNMR [97]. As styrene is an important reagent for polymer production, good separation from ethylbenzene is needed, but generally complicated. MIL-53 shows good selectivity for this process. The host–guest interactions were responsible for this selectivity. Various sophisticated SSNMR techniques were used to shed light on the host–guest interactions. Among them are—besides ^{1}H, ^{13}C, and ^{27}Al MAS NMR spectra, ^{1}H-^{13}C HETCOR, ^{13}C-^{27}Al S-RESPDOR [98] (Fig. 29), ^{13}C DIPSHIFT

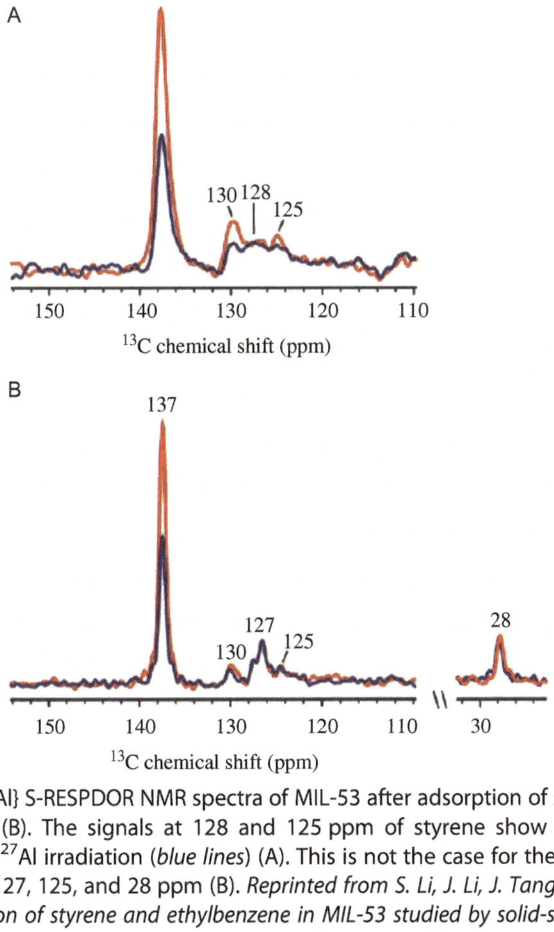

Fig. 29 ^{13}C{^{27}Al} S-RESPDOR NMR spectra of MIL-53 after adsorption of styrene (A) and ethylbenzene (B). The signals at 128 and 125 ppm of styrene show some reduced intensity after ^{27}Al irradiation (*blue lines*) (A). This is not the case for the signals of ethylbenzene at 127, 125, and 28 ppm (B). *Reprinted from S. Li, J. Li, J. Tang, F. Deng, Host-guest interaction of styrene and ethylbenzene in MIL-53 studied by solid-state NMR, Solid State Nucl. Magn. Reson. 90 (2018) 1–6. Copyright 2018, with permission from Elsevier.*

[99], and ^1H-^1H correlation spectra with spin diffusion. Combining the results of these techniques, it is concluded that styrene has less mobility than ethylbenzene because of π-π interactions with the aromatic ring of the MIL-53 linker.

Motion of benzene (together with *n*-decane) was also studied by ^1H NMR in IRMOF-1 [100]. This direct approach required no labelling and from temperature-dependent spin-lattice relaxation time measurements and ^1H wideline NMR spectra two different phases were differentiated. In the low-temperature phase, benzene molecules show some restricted mobility located in one cavity. Above 150 K, isotropic motion is observed indicating intracavity motion. Above 216 K, self-diffusion of benzene accompanies the phase transition.

Information on the acidity strength of binding sites in MOFs have been obtained by adsorption of pyridine-d_5 in MIL-100(Al) [101] and (functionalized) UiO-66 [102]. However, two different approaches were taken. Tang et al. investigated different forms of UiO-66-X (X=H, 2COOH, SO_3H) together with adsorption of 2-^{13}C acetone and pyridine-d_5. For acetone as adsorbent, the resulting ^{13}C shift as a result of an aldol condensation can be correlated with the acid strength of the MOF. For pyridine, adsorption on the Brønsted site results in a ^1H shift in the range of 12–20 ppm, with a smaller shift indicating a stronger acidity. Fig. 30 shows the results for the differently functionalized UiO-66. There are differences in the acid strength, though not very significant.

Khudozhitkov et al. used temperature-dependent ^2H NMR wideline spectra and relaxation time analysis to study both mobility of pyridine-d_5 as well as interaction with binding sites [101]. At lower temperatures, bound and mobile fractions of pyridine are present. Above 333 K, only the mobile state remains. Binding is via the open metal or Al-OH site. At the aluminium site fast π-flips around the C_2 symmetry axis are present. At the Al-OH group a free uniaxial rotation around the C_2 axis is present. Relaxation data revealed activation energies for the motional modes. Exchange between the two adsorption sites is limited by the chemical exchange between bound and mobile states.

4.7.4 Methanol, CH_3OH

Methanol and other small alcohols are also interesting adsorbates for MOFs. They are in some way comparable to water and typically occupy open metal sites.

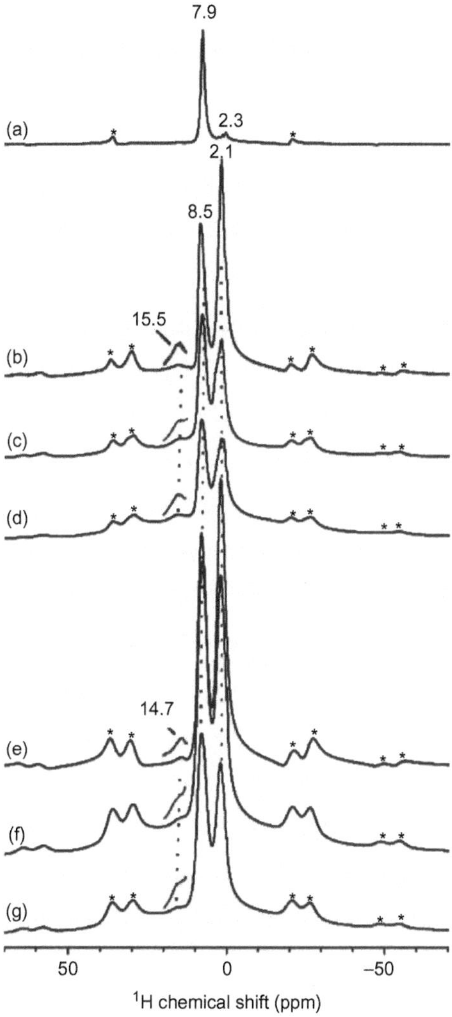

Fig. 30 ^{1}H MAS NMR spectra of UiO-66 based MOFs after pyridine-d_5 adsorption. (a) UiO-66, (b) UiO-66-2COOH, (c) UiO-66-(BDC-2COOH)$_{0.8}$-(BDC)$_{0.2}$, (d) UiO-66-(BDC-2COOH)$_{0.7}$-(BDC)$_{0.3}$, (e) UiO-66-SO_3H, (f) UiO-66-(BDC-SO_3H)$_{0.8}$-(BDC)$_{0.2}$, (g) UiO-66-(BDC-SO_3H)$_{0.7}$-(BDC)$_{0.3}$. The peaks amplified in *blue* indicate the acid site shifts. *Reprinted with permission from J. Tang, S. Li, Y. Chu, Y. Xiao, J. Xu, F. Deng, Solid-state NMR studies of the acidity of functionalized metal-organic framework UiO-66 materials, Magn. Reson. Chem. (2019) 1–8, https://doi.org/10.1002/mrc.4923. Copyright 2019 John Wiley & Sons, Inc.*

In ZIF-8 and ZIF-90, adsorption and selectivity of ethanol versus methanol was studied by Grand Canonical Monte Carlo (GCMC) simulations with two different force fields [103]. Afterwards, self-diffusivities were measured using PFG NMR. A diffusion selectivity of methanol over ethanol was found. This could be of interest for biofuel purification.

A rare lead-containing MOF was synthesized and tested for selective adsorption of methanol [104]. ^{13}C NMR showed the reversible adsorption of methanol that is desorbed after dissolution in water (Fig. 31), thereby regenerating the original spectrum.

A combination of solid-state NMR techniques and DFT calculations identified binding sites for methanol in UIO-66, also in comparison with acetone and cyclohexane as adsorbates [105]. In- and ex-pore resonances for the adsorbates were found (Fig. 32) as well as different effects on chemical shift for the different adsorbates. DFT calculations indicated three reasons for a shift value, ring-current effects due to interaction with the aromatic linker, bonds of the adsorbates with the Zr-OH groups and hydrogen bonding. For methanol, interaction with the Zr-OH groups seems to dominate the shift changes.

In the well-known MOF $Cu_3(btc)_2$ with partial replacement of Cu by Zn, a combination of various solid-state NMR experiments (2D ^{1}H-^{1}H and ^{1}H-^{13}C correlation spectra, ^{1}H-^{1}H spin diffusion) were used to identify

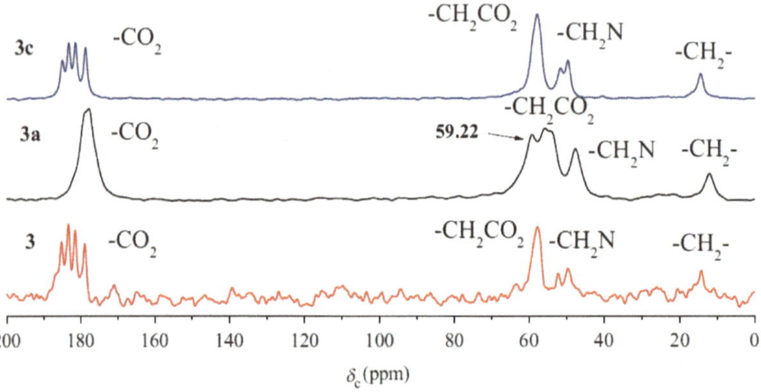

Fig. 31 Solid ^{13}C NMR spectra of $\{[Pb_2(1,3\text{-pdta})(H_2O)_4]\cdot 4H_2O\}_n$ (3), after adsorption with methanol (3a) and re-dissolution in water (3c). *Reprinted from J.-W. Dai, X. Li, J.-M. Zheng, X. Dong, Z.-H. Zhou, Transformations of lead 1,3-propylenediaminetetraacetate to its MOF products for the selective adsorption of methanol, J. Solid State Chem. 237 (2016) 364–370. Copyright 2016, with permission from Elsevier.*

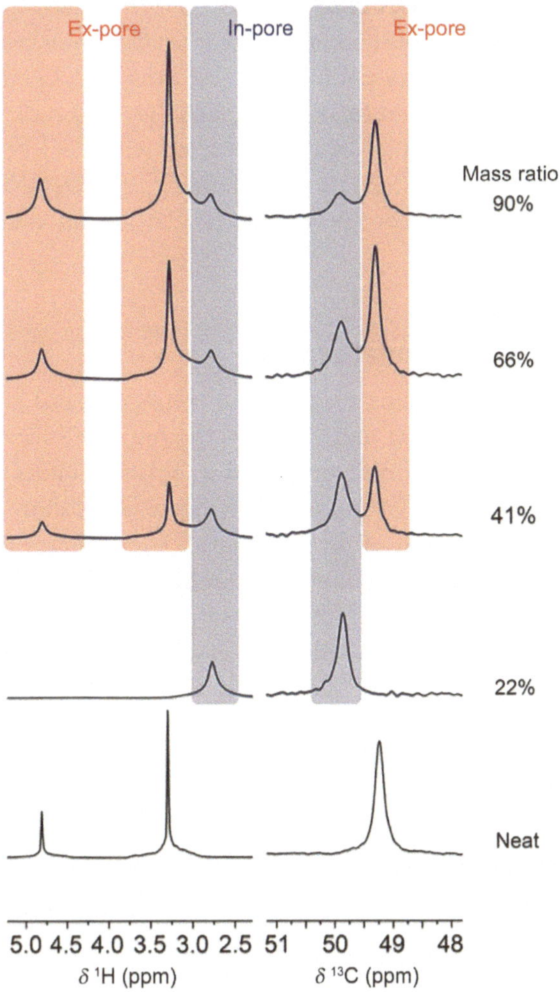

Fig. 32 ^{1}H (*left*) and ^{13}C NMR spectra (*right*) of methanol in UiO-66(Zr) at different loadings. *Reprinted (adapted) with permission from A. Nandy, A.C. Forse, V.J.Witherspoon, J.A. Reimer, NMR spectroscopy reveals adsorbate binding sites in the metal-organic framework UiO-66(Zr), J. Phys. Chem. C 122 (2018) 8295–8305. Copyright 2018 American Chemical Society.*

adsorption sites for methanol [106]. Experiments were also conducted with pure copper and pure zinc containing MOFs. It was concluded that Zn^{2+} is the primary adsorption site instead of Cu^{2+}. Spin-lattice relaxation time measurements together with homonuclear correlation experiments hint at the formation of domains of Zn^{2+} and Cu^{2+} ions.

4.7.5 Other organic compounds

There are few other organic compounds whose adsorption has been studied by SSNMR. These are only summarized here briefly. PFG NMR was used to study DMSO diffusion in MIL-53-NH_2 (Al) [107]. Adsorption of C_3 to C_6 hydrocarbons in $Cu_3(btc)_2$ was also studied with PFG NMR together with MD simulations [108]. Motkuri et al. used ^{19}F MAS NMR to identify the presence of different fluorocarbons in MIL-101 by shift differences to the free gases [109].

4.8 Molecules with metal atoms

Several papers discuss the incorporation of metals or metal precursors into MOFs. Here, the catalytic activity is of interest. The metal-containing catalysts or precursors were incorporated in small amounts and the high porosity and high surface area of MOFs is expected to allow for fast input of reactants and fast output of products.

A platinum precursor was introduced in MOF-177 [110]. After reduction, Pt nanoparticles were present in the MOF. Hydrogen adsorption up to 2.5 wt% was achieved but decreased largely in subsequent cycles. Furthermore, successful oxidation of alcohols in air was realized. ^{13}C NMR data were obtained to show changes due to incorporation of metal (precursor) and to identify stability of the MOF. The sensitivity to water was mentioned, requiring other MOFs for these applications.

Also in MOF-177, different organometallics, mainly based on cyclopentadiene, and iron, copper, zinc and palladium were adsorbed via the gas phase [111]. ^{1}H and ^{13}C MAS NMR showed the intactness of the framework. In case of the palladium incorporation, narrow ^{1}H resonances for the guest molecules hinted at some mobility inside the pore. The same group used MOF-5 to include Cu and ZnO nanoparticles to be used as catalyst for methanol synthesis [112]. Besides ^{1}H and ^{13}C NMR, they used ^{17}O enrichment of either the linker terephthalic acid or of water to label the linker and/or the Zn_4O-cluster. From that it was concluded that no oxygen exchange with the ZnO nanoparticles takes place. In a similar way, solid-state NMR has been used on ZnO inside ZIF-8 [113] and metallocenes in MIL-53/MIL-47 [114].

Another class interesting for catalysis are polyoxometalates (POM) containing tungsten, derived from the heteropolyacid $H_3PW_{12}O_{40}$ with Keggin structure. These have been incorporated into different MOFs and tested with regard to oxidation reactions [115–118]. In MIL-101 it was used for alkene epoxidation [115], in the Zr-MOF NU-1000 the oxidation of chloroethyl ethyl sulphide [118]. In MIL-101 it was used for desulfurization of model oil [116]. In all cases the oxidative reagent was H_2O_2. MIL-100 together with the polytungstate was used for esterification and acetalization reactions [117]. ^{31}P NMR was the method of choice—among other characterization techniques—to document the presence of the polyoxotungstate in the MOF. No or only a minor shift effect was visible after incorporation, indicating no or only minor changes to the heteropolyacid. However, typically an increased linewidth was found for the incorporated molecule. This was interpreted as an indication of immobilization or a chemical shift distribution by different amounts of protons in the POM or distortion due to interaction with the MOF. Fig. 33 shows an example.

In the polyoxotungstate, the ^{31}P nucleus is clearly the most easiest to measure. As there is only the one phosphorous atom in the centre of the quasi spherical molecule, not too much information can be obtained in terms of, e.g., host–guest interaction. This typically limits the solid-state NMR applications of this incorporated molecule.

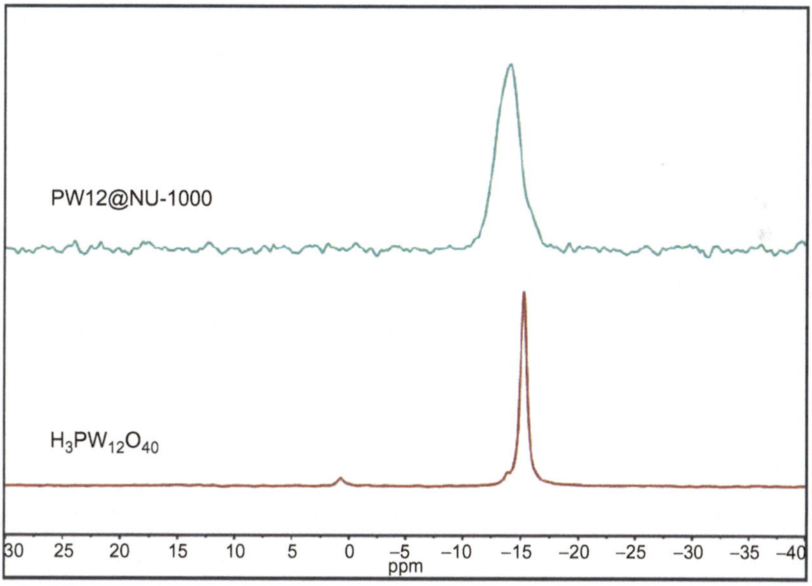

Fig. 33 ^{31}P MAS NMR spectra of the pure polyoxotungstate (*bottom*) and incorporated in NU-1000 (*top*). *Reprinted (adapted) with permission from C.T. Buru, P. Li, B.L. Mehdi, A. Dohnalkova, A.E. Platero-Prats, N.D. Browning, K.W. Chapman, J.T. Hupp, O.K. Farha, Adsorption of a catalytically accessible polyoxometalate in a mesoporous channel-type metal-organic framework, Chem. Mater. 29 (2017) 5174–5181. Copyright 2017 American Chemical Society.*

4.9 Other systems

The many possibilities to use adsorption in MOFs is also documented by a multitude of different molecules that have been studied. Here, some additional molecules not discussed before are summarized.

Different water stable MOFs were tested for the removal of boron from aqueous solutions [119]. ZIF-8 showed the highest boron adsorption capacity. With ^{11}B MAS spectroscopy interaction with the zinc site of the MOF is proposed. In two cadmium based MOFs showing selectivity in O_2/N_2 and CO_2/N_2 separation, ^{113}Cd NMR was used as a sensitive probe for structural changes [120]. While no NMR data were recorded on gas-adsorbed samples, changes due to solvation by water from synthesis or methanol are clearly evident.

Ammonia (NH_3) reaction with $Cu_3(btc)_2$ was studied by ^{1}H and ^{13}C MAS NMR. Under dry conditions it is supposed that a diammine copper complex is formed. On the other hand, in humid conditions a decomposition of the MOF occurs in forming $Cu(OH)_2$ and $(NH_4)_3$btc.

Polysiloxane is an impurity in biogas, complicating purification of methane. Adsorption of siloxane D4 (octamethylcyclotetrasiloxane) as a major component of impurities in biogas was tested on different porous coordination polymers [121]. ^{29}Si NMR was used to get information on the adsorption of siloxane D4 in the porous materials. The isotropic shift in the adsorbed state with respect to that of pure siloxane D4 is an indicator of the strength of the host–guest interaction. Among the studied materials, DUT-4 showed the strongest interaction.

5. The importance of calculations

In order to get detailed information about the behaviour of adsorbed molecules in MOFs and to verify experimental results, theoretical calculations are definitely highly important as has also been demonstrated by the several examples given in Section 4. Experimental and theoretical results clearly influence each other and contribute to an improvement of the methodology. As this report focuses on solid-state NMR spectroscopy, no detailed discussion of theoretical methods will be done. Nevertheless, three theoretical methods are briefly mentioned that have different goals for experimental verification.

- Lineshape simulation

  Types of motion of adsorbed species have specific influence on the NMR lineshape as demonstrated in Section 4. It can affect both the width (span) and the symmetry. By studying the temperature dependence of the lineshape, the dynamics of adsorbed molecules (types of motion) in MOFs can be studied in detail.

Especially for ^{2}H NMR (via the quadrupolar coupling) but also for CO_2 dynamics (via the chemical shift anisotropy), different simulation programmes exist. It is not the purpose to give a list of programmes or to classify them. One often used programme for lineshape simulations is EXPRESS [63]. It is a MATLAB based programme that can simulate various types of experiments and includes quadrupolar coupling, chemical shift anisotropy as well as relaxation time anisotropy. Jump dynamics with various orientation angles can be calculated and compared with the experiment. It also includes an optimization procedure to a best-fit of the experiment.

- Density functional theory (DFT)

  Quantum chemical calculations provide tensor parameters for chemical shift and quadrupolar coupling. Mainly the isotropic chemical shift of an adsorbed species as a result of the position at the host structure is used. Agreement with experiment will then be a strong indicator to prove the presence of the adsorption site.

  An ab initio DFT calculation including van-der-Waals interactions was used to predict shielding of adsorbed molecules (H_2, CO_2, H_2O) in MOF-74-Mg [122]. By placing the adsorbents at different positions inside the MOF, including the calculated adsorption site, the shielding is calculated for different loadings. Rotation of adsorbed molecules and its influence on energy and shielding are also discussed.

  In case of the paramagnetic MIL-101 (Cr), DFT calculations are on the one hand more complicated due to the hyperfine interaction, on the other hand also really important to understand the observed chemical shifts. Wittmann et al. used different adsorbents, H_2O, diethylamine and 2- and 3-aminopyridine to get information on the binding strengths of these molecules at the open metal site [123]. Assignments were aided by $^{13}C\{^1H\}$ REDOR [124] experiments which also showed influences from linker dynamics and the adsorbed molecules.

  In variants of Mg_2(dobdc) functionalized by alcoholamine and alkoxyalkylamine, van der Waals corrected DFT calculations were used to investigate the mechanism of CO_2 adsorption [13]. From five possible variants, the formation of hydrogen bond stabilized carbamic acid showed the best agreement with ^{13}C MAS shifts and $^{13}C\{^1H\}$ HETCOR spectra.

- Molecular dynamics (MD) simulations

  Several papers used MD simulations together with PFG NMR experiments [84, 103, 125–127]. MD simulations are used in this regard to calculate the diffusion of small molecules inside the framework. Typically the self-diffusivities are calculated and compared to the experimental diffusion measurements using PFG NMR. Appropriate force fields are necessary to best describe both the framework and the adsorbed molecules. From that the probability distribution of the guest molecule and/or their velocities is calculated in certain time steps. From those the self-diffusivity is calculated. Clearly, the time steps for the calculation are much smaller than the observation time for the NMR experiment. Examples include methane and ethane diffusion in ZIF-8 and ZIF-90 [103], C3 to C6 hydrocarbons in $Cu_3(btc_2)$ [108], methane in M_2(dobdc) [84], xylenes in MOF-5 [127], and ferrocene in MOF-5 [125].

6. Selected techniques—highlights

The different strategies and techniques to characterize small molecule adsorption in MOFs with solid state NMR have been described in Section 4. Here, two of them are mentioned that are thought of having highest potential. This is clearly a personal selection but might induce further investigations in these directions.

- Variable pressure measurements—in situ NMR

  Most studies are performed with adsorption of a certain amount of the small molecule or at a certain gas pressure. For pressure-dependent studies then, several samples are needed. A variable pressure setup would be superior in that case, also as measurements can be done with increasing and decreasing pressure to find out if some kind of hysteresis is present. This can then directly be compared to standardly performed sorption isotherm studies. Performing these type of experiments in combination with MAS for high resolution is typically not possible. Under certain conditions static measurements might be sufficient. This is the case when isolated spectra of only the adsorbed molecules can be measured like for xenon or isotopically labelled $^{13}CO_2$. Few studies have been done at variable pressures. In this respect also the term in situ NMR is used. Although this term can in general be used more widely, for gas adsorption in MOFs it typically refers to variable pressure studies or continuous-flow mode of gases. To mention are here variable pressure studies of xenon gas by the group of Eike Brunner, Dresden University of Technology, Germany, on the flexible MOFs DUT-8 [23] and DUT-49 [47]. They also did variable pressure ^{13}C NMR studies for CO_2 adsorption in the MOF $[Zn_2(BME\text{-}bdc)_x(DB\text{-}bdc)_{2-x}dabco]_n$ [128]. The low-field setup for ^{1}H relaxation measurements were used to characterize the adsorption of methane and carbon dioxide also in mixtures of the two molecules in $Cu_3(btc)_2$ and ZIF-8 [86] by the group of Frank Stallmach, Leipzig University, Germany.

- Dipolar coupling related techniques

  Of the different possibilities to study the binding sites and binding strength, using the dipolar coupling is the most direct one as it is directly related to the distance of involved spins. Combined with variable temperature studies information on the mobility of the adsorbed molecule can be obtained. Certainly, a suitable system of host and guest is required to allow for a suitable selection of spins. Whereas in the first years, NMR experiments dealt with studies of chemical shift and quadrupolar coupling, in recent years experiments employing the dipolar coupling have been published. To mention are here the REDOR experiment [54], the *J*-HMQC experiment together with ^{1}H-^{1}H spin diffusion [88], and the S-RESPDOR experiment. As there are many possible pulse sequences for the determination of dipolar couplings, for each pair of MOF-adsorbed molecule a suitable variant can be selected. It can be expected that this approach will be used more often in the future.

7. Conclusions and outlook

The class of metal–organic frameworks contains a vast number of different structures, containing different elements and geometries. Solid-state NMR offers a high potential in studying these materials to obtain different types of information. Especially the characterization of adsorbed molecules represents a strength of SSNMR compared to other techniques. This report focussed on the adsorption of small molecules in different MOFs, demonstrating the different options to obtain information on the small molecules. These include studies of molecular mobility and diffusion in the pores as well as characterization of adsorption sites and adsorption strength.

In the future, it can be expected that SSNMR will be one of the central methods for characterization of adsorbed molecules in MOFs. With technical and experimental advancements, the applicability of this technique can even be widened. Here, higher sensitivity by higher magnetic fields and polarization techniques are only few aspects. First examples have already been shown using dynamic nuclear polarization (DNP) for the characterization of the MOF host structure. The signal enhancement allowed for faster acquisition of, e.g., ^{15}N CPMAS spectra [129, 130], 2D correlation spectra [129, 131], ^{27}Al-^{13}C RESPDOR curves [131] and ^{195}Pt wideline spectra [132]. It will be of course of interest how this approach can be expanded also to characterize adsorbed molecules without interference with the involved radicals in the pores. As many MOFs contain transition metals with unpaired electrons, comparison of NMR and ESR investigations of adsorbed molecules in MOFs can also provide some interesting synergies, for the NMR representing bulk properties whereas ESR is focussed on the sites with unpaired electrons.

Furthermore, especially in situ measurements hold promise for real-time investigation of the adsorption process that can be directly compared to bulk adsorption measurements. Then, also characterization of XRD-amorphous materials could be envisioned that could hold higher potential for certain applications like separation than the current crystalline MOFs. Solid-state NMR then would be the method of choice.

References

[1] O.K. Farha, I. Eryazici, N.C. Jeong, B.G. Hauser, C.E. Wilmer, A.A. Sarjeant, R.Q. Snurr, S.T. Nguyen, A.Ö. Yazaydin, J.T. Hupp, Metal-organic framework materials with ultrahigh surface areas: is the sky the limit? J. Am. Chem. Soc. 134 (2012) 15016–15021.

[2] H.C. Hoffmann, M. Debowski, P. Müller, S. Paasch, I. Senkovska, S. Kaskel, E. Brunner, Solid-state NMR spectroscopy of metal-organic framework compounds (MOFs), Materials 5 (2012) 2537–2572.

[3] V.J. Witherspoon, J. Xu, J.A. Reimer, Solid-state NMR investigations of carbon dioxide gas in metal-organic frameworks: insights into molecular motion and adsorptive behavior, Chem. Rev. 118 (2018) 10033–10048.

[4] B.E.G. Lucier, S. Chen, Y. Huang, Characterization of metal-organic frameworks: unlocking the potential of solid-state NMR, Acc. Chem. Res. 51 (2018) 319–330.

[5] M. Bertmer, Solid-state NMR studies of small molecule adsorption in metal-organic frameworks (MOFs), in: G. Webb (Ed.), Modern Magnetic Resonance, Springer, Cham, 2018, pp. 635–649.

[6] Y.T.A. Wong, V. Martins, B.E.G. Lucier, Y. Huang, Solid-state NMR spectroscopy: a powerful technique to directly study small gas molecules adsorbed in metal-organic frameworks, Chem. Eur. J. 25 (2019) 1848–1853.

[7] J. Mao, Y. Zhang, E. Oldfield, Nuclear magnetic resonance shifts in paramagnetic metalloporphyrins and metalloproteins, J. Am. Chem. Soc. 124 (2002) 13911–13920.

[8] M. Bertmer, Paramagnetic solid-state NMR of materials, Solid State Nucl. Magn. Reson. 81 (2017) 1–7.

[9] I. Bertini, C. Luchinat, G. Parigi, E. Rivera, NMR of Paramagnetic Molecules, second ed., Elsevier, Amsterdam, 2015.
[10] G. Parigi, L. Benda, E. Ravera, M. Romanelli, C. Luchinat, Pseudocontact shifts and paramagnetic susceptibility in semiempirical and quantum chemistry theories, J. Chem. Phys. 150 (2019) 144101.
[11] R. Dickinson, A.T. Royappa, F. Tone, L. Ujj, G. Wu, Distribution of non-uniform demagnetization fields in paramagnetic bulk solids, J. Appl. Phys. 110 (2011) 013902.
[12] R. Pigliapochi, L. O'Brien, A.J. Pell, M.W. Gaultois, Y. Janssen, P.G. Khalifah, C.P. Grey, When do anisotropic magnetic susceptibilities lead to large NMR shifts? Exploring particle shape effects in the battery electrode material $LiFePO_4$, J. Am. Chem. Soc. 141 (2019) 13089–13100.
[13] V.Y. Mao, P.J. Milner, J.-H. Lee, A.C. Forse, E.J. Kim, R.L. Siegelman, C.M. McGuirk, L.B. Porter-Zasada, J.B. Neaton, J.A. Reimer, J.R. Long, Cooperative carbon dioxide adsorption in alcoholamine- and alkoxyalkylamine-functionalized metal-organic frameworks, Angew. Chem. Int. Ed. 59 (2020) 2–12.
[14] A.H. Khan, K. Peikert, F. Hoffmann, M. Fröba, M. Bertmer, Nitric oxide adsorption in Cu_3btc_2-type MOFs–physisorption and chemisorption as NONOates, J. Phys. Chem. C 123 (2019) 4299–4307.
[15] A.H. Khan, B. Barth, M. Hartmann, J. Haase, M. Bertmer, Nitric oxide adsorption in MIL-100(Al) MOF studied by solid-state NMR, J. Phys. Chem. C 122 (2018) 12723–12730.
[16] J. Herzfeld, A.E. Berger, Sideband intensities in NMR spectra of samples spinning at the magic angle, J. Chem. Phys. 73 (1980) 6021–6030.
[17] U. Haeberlen, High resolution NMR in solids selective averaging, Suppl. 1, in: J.S. Waugh (Ed.), Advances in Magnetic Resonance, Academic Press, New York, 1976.
[18] D. Bernin, N. Hedin, Perspectives on NMR studies of CO_2 adsorption, Curr. Opin. Coll. Interface Sci. 33 (2018) 53–62.
[19] S. Hertel, M. Wehring, S. Amirjalayer, M. Gratz, J. Lincke, H. Krautscheid, R. Schmid, F. Stallmach, NMR studies of benzene mobility in metal-organic framework MOF-5, Eur. Phys. J. Appl. Phys. 55 (2011) 20702.
[20] D. Freude, M. Hunger, H. Pfeifer, Study of brønsted acidity of zeolites using high-resolution proton magnetic resonance with magic-angle spinning, Chem. Phys. Lett. 91 (1982) 307–310.
[21] F. Gul-E-Noor, B. Jee, A. Pöppl, M. Hartmann, D. Himsl, M. Bertmer, Effects of varying water adsorption on a $Cu_3(BTC)_2$ metal-organic framework (MOF) as studied by ^1H and ^{13}C solid-state NMR spectroscopy, Phys. Chem. Chem. Phys. 13 (2011) 7783–7788.
[22] C. Lieder, S. Opelt, M. Dyballa, H. Henning, E. Klemm, M. Hunger, Adsorbate effect on $AlO_4(OH)_2$ centers in the metal-organic framework MIL-53 investigated by solid-state NMR spectroscopy, J. Phys. Chem. C 114 (2010) 16596–16602.
[23] H.C. Hoffmann, B. Assfour, F. Epperlein, N. Klein, S. Paasch, I. Senkovska, S. Kaskel, G. Seifert, E. Brunner, High-pressure in situ ^{129}Xe NMR spectroscopy and computer simulations of breathing transitions in the metal-organic framework $Ni_2(2,6\text{-ndc})_2(\text{dabco})$ (DUT-8(Ni)), J. Am. Chem. Soc. 133 (2011) 8681–8690.
[24] G. Mali, M. Mazaj, I. Arčon, D. Hanžel, D. Arčon, Z. Jagličić, Unraveling the arrangement of Al and Fe within the framework explains the magnetism of mixed-metal MIL-100(Al,Fe), J. Chem. Phys. Lett. 10 (2019) 1464–1470.
[25] B. Jee, K. Eisinger, F. Gul-E-Noor, M. Bertmer, M. Hartmann, D. Himsl, A. Pöppl, Continuous wave and pulsed electron spin resonance spectroscopy of paramagnetics framework cupric ions in the Zn(II) doped porous coordination polymer $Cu_{3-x}Zn_x(btc)_2$, J. Phys. Chem. C 114 (2010) 16630–16639.

[26] F. Gul-E-Noor, B. Jee, M. Mendt, D. Himsl, A. Pöppl, M. Hartmann, J. Haase, H. Krautscheid, M. Bertmer, Formation of mixed metal $Cu_{3-x}Zn_x(btc)_2$ frameworks with different zinc contents: incorporation of Zn^{2+} into the metal-organic framework structure as studied by solid-state NMR, J. Phys. Chem. C 116 (2012) 20866–20873.
[27] O. Kozachuk, M. Meilikhov, K. Yusenko, A. Schneemann, B. Jee, A.V. Kuttatheyil, M. Bertmer, C. Sternemann, A. Pöppl, R.A. Fischer, A solid-solution approach to mixed-metal metal-organic frameworks-detailed characterization of local structures, defects and breathing behaviour of Al/V frameworks, Eur. J. Inorg. Chem. 2013 (2013) 4546–4557.
[28] S.S.-Y. Chui, S.M.-F. Lo, J.P.H. Charmant, A.G. Orpen, I.D. Williams, A chemically functionalizable nanoporous material $[Cu_3(TMA)_2(H_2O)_3]_n$, Science 283 (1999) 1148–1150.
[29] T. Loiseau, C. Serre, C. Huguenard, G. Fink, F. Taulelle, M. Henry, T. Bataille, G. Férey, A rationale for the large breathing of the porous aluminum terephthalate (MIL-53) upon hydration, Chem. Eur. J. 10 (2004) 1373–1382.
[30] C. Serre, F. Millange, C. Thouvenot, M. Noguès, G. Marsolier, D. Louer, G. Férey, Very large breathing effect in the first nanoporous chromium(III)-based solids: MIL-53 or Cr(OH)·OC-CH-CO·HOC-CH-COH· HO, J. Am. Chem. Soc. 124 (2002) 13519–13526.
[31] F. Millange, C. Serre, G. Férey, Synthesis, structure determination and properties of MIL-53as and MIL-53ht: the first Cr^{III} hybrid inorganic-organic microporous solids: $Cr^{III}(OH)\cdot O_2C\text{-}C_6H_4\text{-}CO_2\cdot HO_2C\text{-}C_6H_4\text{-}CO_2H_x$, Chem. Commun. (2002) 822–823.
[32] P. Horcajada, C. Serre, G. Maruin, N.A. Ramsahye, F. Balas, M. Vallet-Regí, M. Sebban, F. Taulelle, G. Férey, Flexible porous metal-organic frameworks for a controlled drug delivery, J. Am. Chem. Soc. 130 (2008) 6774–6780.
[33] C. Volkringer, T. Loiseau, N. Guillou, G. Férey, E. Elkaïm, A. Vimont, XRD and IR structural investigations of a particular breathing effect in the MOF-typ gallium terephthalate MIL-53(Ga), Dalton Trans. (2009) 2241–2249.
[34] J.P.S. Mowat, S.R. Miller, A.M.Z. Slawin, V.R. Seymour, S.E. Ashbrook, P.A. Wright, Synthesis, characterisation and adsorption properties of microporous scandium carboxylates with rigid and flexible frameworks, Microporous Mesoporous Mater. 142 (2011) 322–333.
[35] E.V. Anokhina, M. Vougo-Zanda, X. Wang, A. Jacobson, $In(OH)BDC\cdot 0.75BDCH_2$ (BDC = Benzenedicarboxylate), a hybrid inorganic-organic vernier structure, J. Am. Chem. Soc. 127 (2005) 15000–15001.
[36] K. Barthelet, J. Marrot, D. Riou, G. Férey, A breathing hybrid organic-inorganic solid with very large pores and high magnetic characteristics, Angew. Chem. Int. Ed. 41 (2002) 281–284.
[37] L.J. Wang, H. Deng, H. Furukawa, F. Gándara, K.E. Cordova, D. Peris, O.M. Yaghi, Synthesis and characterization of metal-organic framework-74 containing 2, 4, 6, 8, and 10 different metals, Inorg. Chem. 53 (2014) 5881–5883.
[38] W. Böhlmann, A. Pöppl, M. Sabo, S. Kaskel, Characterization of the metal-organic framework compound Cu_3(benzene 1,3,5-tricarboxylate)$_2$ by means of ^{129}Xe nuclear magnetic and electron paramagnetic resonance spectroscopy, J. Phys. Chem. B 110 (2006) 20177–20181.
[39] T. Ueda, K. Kurokawa, T. Eguchi, C. Kachi-Terajima, S. Takamizawa, Local structure and xenon adsorption behavior of metal-organic framework system $[M_2(O_2CPh)_4(pyz)]_n$ (M = Rh and Cu) as studied with use of single-crystal X-ray diffraction, adsorption isotherm, and Xenon-129 NMR, J. Phys. Chem. C 111 (2007) 1524–1534.

[40] S. Pawsey, I. Moudrakovski, J. Ripmeester, L.-Q. Wang, G.J. Exarhos, J.L.C. Rowsell, O.M. Yaghi, Hyperpolarized ^{129}Xe nuclear magnetic resonance studies of isoreticular metal-organic frameworks, J. Phys. Chem. C 111 (2007) 6060–6067.
[41] K.J. Ooms, R.E. Wasylishen, ^{129}Xe NMR study of xenon in iso-reticular metalorganic frameworks, Microp. Mesop. Mater. 103 (2007) 341–351.
[42] M.-A. Springuel-Huet, A. Nossov, Z. Adem, F. Guenneau, C. Volkringer, T. Loiseau, G. Férey, A. Gédéon, ^{129}Xe NMR study of the framework flexibility of the porous hybrid MIL-53(Al), J. Am. Chem. Soc. 132 (2010) 11599–11607.
[43] R. Giovine, C. Volkringer, M.-A. Springuel-Huet, A. Nossov, F. Blanc, J. Trébosc, T. Loiseau, J.-P. Amoureux, O. Lafon, F. Pourpoint, Study of xenon mobility in the two forms of MIL-53(Al) using solid-state NMR spectroscopy, J. Phys. Chem. C 121 (2017) 19262–19268.
[44] N. Klein, C. Herzog, M. Sabo, I. Senkovska, J. Getzschmann, S. Paasch, M.R. Lohe, E. Brunner, S. Kaskel, Monitoring adsorption-induced switching by ^{129}Xe NMR spectroscopy in a new metalorganic framework $Ni_2(2,6\text{-ndc})_2(dabco)$, Phys. Chem. Chem. Phys. 12 (2010) 11778–11784.
[45] N. Klein, H.C. Hoffmann, A. Cadiau, J. Getzschmann, M.R. Lohe, S. Paasch, T. Heydenreich, K. Adil, I. Senkovska, E. Brunner, S. Kaskel, Structural flexibility and intrinsic dynamics in the $M_2(2,6\text{-ndc})_2(dabco)$ (M = Ni, Cu, Co, Zn) metal-organic frameworks, J. Mater. Chem. 22 (2012) 10303–10312.
[46] M.-A. Springuel-Huet, A. Nossov, F. Guenneau, A. Gédéon, Flexibility of ZIF-8 materials studied using ^{129}Xe NMR, Chem. Commun. 49 (2013) 7403–7405.
[47] J. Schaber, S. Krause, S. Paasch, I. Senkovska, V. Bon, D.M. Többens, D. Wallacher, S. Kaskel, E. Brunner, In situ monitoring of unique switching transitions in the pressure-amplifying flexible framework material DUT-49 by High-Pressure ^{129}Xe NMR Spectroscopy, J. Phys. Chem. C 121 (2017) 5195–5200.
[48] F. Gul-E-Noor, D. Michel, H. Krautscheid, J. Haase, M. Bertmer, Time dependent water uptake in $Cu_3(btc)_2$ MOF: identification of different water adsorption states by ^{1}H MAS NMR, Microporous Mesoporous Mater. 180 (2013) 8–13.
[49] R. Giovine, F. Pourpoint, S. Duval, O. Lafon, J.-P. Amoureux, T. Loiseau, C. Volkringer, The surprising stability of $Cu_3(btc)_2$ metal-organic framework under steam flow at high temperature, Cryst. Growth Des. 18 (2018) 6681–6693.
[50] A. Terracina, L.N. McHugh, M. Todaro, S. Agnello, P.S. Wheatley, F.M. Gelardi, R.E. Morris, G. Buscarino, Multitechnique analysis of the hydration in three different copper paddle-wheel metal-organic frameworks, J. Phys. Chem. C 123 (2019) 28219–28232.
[51] T. Splith, E. Pantatosaki, P.D. Kolokathis, D. Fröhlich, K. Zhang, G. Füldner, C. Chmelik, J. Jiang, S.K. Henninger, F. Stallmach, G.K. Papadopoulos, Molecular dynamics phenomena of water in the metalorganic framework MIL-100(Al), as revealed by pulsed field gradient NMR and atomistic simulation, J. Phys. Chem. C 121 (2017) 18065–18074.
[52] T. Splith, F. Stallmach, Restricted self-diffusion of adsorbed water in MIL100(Al), Magn. Reson. Imaging 56 (2019) 52–56.
[53] Y. Zhang, B.E.G. Lucier, V.V. Terskikh, R. Zheng, Y. Huang, Tracking the evolution and differences between guest-induced phases of Ga-MIL-53 *via* ultra-wideline $^{69/71}Ga$ solid-state NMR spectroscopy, Solid State Nucl. Magn. Reson. 84 (2017) 118–131.
[54] B.E. Desveaux, Y.T.A. Wong, B.E.G. Lucier, V.V. Terskikh, P.D. Boyle, S. Jiang, Y. Huang, CO_2 behavior in a highly selective ultramicroporous framework: insights from single-crystal Xray diffraction and solid-state nuclear magnetic resonance spectroscopy, J. Phys. Chem. C 123 (2019) 17798–17807.

[55] L. Li, I. da Silva, D.I. Kolokolov, X. Hang, J. Li, G. Smith, Y. Cheng, L.L. Daemen, C.G. Morris, H.G.W. Godfrey, N.M. Jacques, X. Zhang, P. Manuel, M.D. Frogley, C.A. Murray, A.J. Ramirez-Cuesta, G. Cinque, C.C. Tang, A.G. Stepanov, S. Yang, M. Schroder, Post-synthetic modulation of the charge distribution in a metal-organic framework for optimal binding of carbon dioxide and sulfur dioxide, Chem. Sci. 10 (2019) 1472–1482.
[56] W.D. Wang, B.E.G. Lucier, V.V. Terskikh, W. Wang, Y. Huang, Wobbling and hopping: studying dynamics of CO_2 adsorbed in metal-organic frameworks via ^{17}O solid-state NMR, J. Phys. Chem. Lett. 5 (2014) 3360–3365.
[57] S. Chen, B.E.G. Lucier, P.D. Boyle, Y. Huang, Understanding the fascinating origins of CO_2 adsorption and dynamics in MOFs, Chem. Mater. 28 (2016) 5829–5846.
[58] Y. Zhang, B.E.G. Lucier, Y. Huang, Deducing CO_2 motion, adsorption locations and binding strengths in a flexible metalorganic framework without open metal sites, Phys. Chem. Chem. Phys. 18 (2016) 8327–8341.
[59] M. Chen, S. Chen, W. Chen, B.E.G. Lucier, Y. Zhang, A. Zheng, Y. Huang, Analyzing gas adsorption in an amide-functionalized metal organic framework: are the carbonyl or amine groups responsible? Chem. Mater. 30 (2018) 3613–3617.
[60] Y. Zhang, B.E.G. Lucier, S.M. McKenzie, M. Arhangelskis, A.J. Morris, T. Friščić, J.W. Reid, V.V. Terskikh, M. Chen, Y. Huang, Welcoming Gallium- and Indium-fumarate MOFs to the family: synthesis, comprehensive characterization, observation of porous hydrophobicity, and CO_2 dynamics, ACS Appl. Mater. Interfaces 10 (2018) 28582–28596.
[61] B. Wu, Y.T.A. Wong, B.E.G. Lucier, P.D. Boyle, Y. Huang, Exploring host-guest interactions in the α-$Zn_3(HCOO)_6$ metal-organic framework, ACS Omega 4 (2019) 4000–4011.
[62] X. Kong, E. Scott, W. Ding, J.A. Mason, J.R. Long, J.A. Reimer, CO_2 dynamics in a metal-organic framework with open metal sites, J. Am. Chem. Soc. 134 (2012) 14341–14344.
[63] R.L. Vold, G.L. Hoatson, Effects of jump dynamics on solid state nuclear magnetic resonance line shapes and spin relaxation times, J. Magn. Reson. 198 (2009) 57–72.
[64] S. Schlayer, A.-K. Pusch, F. Pielenz, S. Beckert, M. Peksa, C. Horch, L. Moschkowitz, W.-D. Einicke, F. Stallmach, X-nuclei NMR self-diffusion studies in mesoporous silica foam and microporous MOF CuBTC, Materials 5 (2012) 617–633.
[65] A.C. Forse, S.A. Altobelli, S. Benders, M.S. Conradi, J.A. Reimer, Revisiting anisotropic diffusion of carbon dioxide in the metal-organic framework Zn_2(dobpdc), J. Phys. Chem. C 122 (2018) 15344–15351.
[66] A.C. Forse, M.I. Gonzalez, R.L. Siegelman, V.J. Witherspoon, S. Jawahery, R. Mercado, P.J. Milner, J.D. Martell, B. Smit, B. Blümich, J.R. Long, J.A. Reimer, Unexpected diffusion anisotropy of carbon dioxide in the metal-organic framework Zn_2(dobpdc), J. Am. Chem. Soc. 140 (2018) 1663–1673.
[67] Y. Ji, Y. Tang, J.A. Hamida, N.S. Sullivan, NMR study of HD adsorbed in a Z-type metal-organic framework, J. Phys. Conf. Ser. 400 (2012) 012024.
[68] B.E.G. Lucier, Y. Zhang, Y. Huang, Complete multinuclear solid-state NMR of metal-organic frameworks: the case of α-Mg-formate., Concepts Magn. Reson. A 45 (2016) 21410.
[69] T. Pham, K.A. Forrest, E.H.L. Falcao, J. Eckert, B. Space, Exceptional H_2 sorption characteristics in a Mg^{2+}-based metal-organic framework with small pores: insights from experimental and theoretical studies, Phys. Chem. Chem. Phys. 18 (2016) 1786–1796.
[70] B.E.G. Lucier, Y. Zhang, K.J. Lee, Y. Lu, Y. Huang, Grasping hydrogen adsorption and dynamics in metal-organic frameworks using ^2H solid-state NMR., Chem. Commun. 53 (2016) 7541–7544.

[71] C. Volkringer, M. Meddouri, T. Loiseau, N. Guillou, J. Marrot, G. Férey, M. Haouas, F. Taulelle, N. Audebrand, M. Latroche, The Kagomé topology of the gallium and indium metal-organic framework types with a MIL-68 structure: synthesis, XRD, solid-state NMR characterizations, and hydrogen adsorption, Inorg. Chem. 47 (2008) 11892–11901.
[72] S. Yang, G.S.B. Martin, J.J. Titman, A.J. Blake, D.R. Allan, N.R. Champness, M. Schröder, Pore with gate: enhancement of the isosteric heat of adsorption of dihydrogen via postsynthetic cation exchange in metal-organic frameworks, Inorg. Chem. 50 (2011) 9374–9384.
[73] T. Ben, C. Pei, D. Zhang, J. Xu, F. Deng, X. Jing, S. Qui, Gas storage in porous aromatic frameworks (PAFs), Energy Environ. Sci. 4 (2011) 3991–3999.
[74] F. Gul-E-Noor, M. Mendt, D. Michel, A. Pöppl, H. Krautscheid, J. Haase, M. Bertmer, Adsorption of small molecules on $Cu_3(btc)_2$ and $Cu_{3-x}Zn_x(btc)_2$ metal-organic frameworks (MOF) as studied by solid-state NMR, J. Phys. Chem. C 117 (2013) 7703–7712.
[75] F. Gul-E-Noor, D. Michel, H. Krautscheid, J. Haase, M. Bertmer, Investigation of the spin-lattice relaxation of ^{13}CO and $^{13}CO_2$ adsorbed in the metal-organic frameworks $Cu_3(btc)_2$ and $Cu_{3-x}Zn_x(btc)_2$, J. Chem. Phys. 139 (2013) 034202.
[76] B.E.G. Lucier, H. Chan, Y. Zhang, Y. Huang, Multiple modes of motion: realizing the dynamics of CO adsorbed in M-MOF-74 (M = Mg, Zn) by using solid-state NMR spectroscopy, Eur. J. Inorg. Chem. 2016 (2016) 2017–2024.
[77] Y.T.A. Wong, T.K. Babcock, S. Chen, B.E.G. Lucier, Y. Huang, CO guest interactions in SDB-based metal-organic frameworks: a solid-state nuclear magnetic resonance investigation, Langmuir 34 (2018) 15640–16549.
[78] J. Wu, S.C. Larsen, ^{15}N solid state NMR study of the reactions of propane or propene, ^{15}NO and oxygen on Na-, H- and CuZSM-5, Catal. Lett. 70 (2000) 43–50.
[79] A.H. Khan, D. Michel, M. Bertmer, Temperature dependent ^{15}N NMR study of nitric oxide, Solid State Nucl. Magn. Reson. 101 (2019) 38–43.
[80] A.H. Khan, K. Peikert, M. Fröba, M. Bertmer, NO adsorption in amino-modified $Cu_3(btc)_2$-type MOFs studied by solid-state NMR, Microporous Mesopoous Mater. 216 (2015) 111–117.
[81] M.J. Ingleson, R. Heck, J.A. Gould, M.J. Rosseinsky, Nitric oxide chemisorption in a postsynthetically modified metal-organic framework, Inorg. Chem. 48 (2009) 9986–9988.
[82] S.R. Miller, E. Alvarez, L. Fradcourt, T. Devic, S. Wuttke, P.S. Wheatley, N. Steunou, C. Bonhomme, C. Gervais, D. Laurencin, R.E. Morris, A. Vimont, M. Daturi, P. Horcajada, C. Serre, A rare example of a porous Ca-MOF for the controlled release of biologically active NO, Chem. Commun. 49 (2013) 7773–7775.
[83] N. Dvoyashkina, D. Freude, S.S. Arzumov, A.G. Stepanov, Monitoring the diffusivity of light hydrocarbons in a mixture by magic angle spinning pulsed field gradient NMR: methane/ethane/ethene in ZIF8, J. Phys. Chem. C 121 (2017) 25372–25376.
[84] V.J. Witherspoon, R. Mercado, E. Braun, A. Mace, J. Bachmann, J.R. Long, B. Blümich, B. Smit, J.A. Reimer, Combined nuclear magnetic resonance and molecular dynamics study of methane adsorption in M_2(dobdc) metal-organic frameworks, J. Phys. Chem C 123 (2019) 12286–12295.
[85] Y. Zhang, B.E.G. Lucier, M. Fischer, Z. Gan, P.D. Boyle, B. Desveaux, Y. Huang, A multifaceted study of methane adsorption in metal-organic frameworks by using three complementary techniques, Chem. Eur. J. 24 (2018) 7866–7881.
[86] C. Horch, S. Schlayer, F. Stallmach, High-pressure low-field ^1H NMR relaxometry in nanoporous materials, J. Magn. Reson. 240 (2014) 24–33.
[87] M. Sin, N. Kavoosi, M. Rauche, J. Pallmann, S. Paasch, I. Senkovska, S. Kaskel, E. Brunner, In situ ^{13}C NMR spectroscopy study of CO_2/CH_4 mixture adsorption by metal-organic frameworks: does flexibility influence selectivity? Langmuir 35 (2019) 3162–3170.

[88] Y. Xiao, Y. Chu, S. Li, Y. Su, J. Tang, J. Xu, F. Deng, Primary adsorption sites of light alkanes in multivariate UiO-66 at room temperature as revealed by solid-state NMR, J. Phys. Chem. C 124 (2020) 3738–3746.
[89] C. Chmelik, D. Freude, H. Bux, J. Haase, Ethene/ethane mixture diffusion in the MOF sieve ZIF-8 studied by MAS PFG NMR diffusometry, Microporous Mesoporous Mater. 147 (2012) 135–141.
[90] C.B. Fan, L.L. Gong, L. Huang, F. Luo, R. Krishna, X.F. Yi, A.M. Zheng, L. Zhang, S.Z. Pu, X.F. Feng, M.B. Luo, G.C. Guo, Significant enhancement of C_2H_2/C_2H_4 separation by a photochromic diarylethene unit: a temperature- and light-responsive separation switch, Angew. Chem. Int. Ed. 56 (2017) 7900–7906.
[91] S. Berens, C. Chmelik, F. Hillman, K. Kärger, H.-K. Jeong, S. Vasenkov, Ethane diffusion in mixed linker zeolitic imidazolate framework-7-8 by pulsed field gradient NMR in combination with single crystal IR microscopy, Phys. Chem. Chem. Phys. 20 (2018) 23967–23975.
[92] L. Li, J. Wang, Z. Zhang, Q. Yang, Y. Yang, B. Su, Z. Bao, Q. Ren, Inverse adsorption separation of CO_2/C_2H_2 mixture in cyclodextrin-based metal-organic frameworks, ACS Appl. Mater. Interfaces 11 (2019) 2543–2550.
[93] S. Berens, F. Hillman, H.K. Jeong, S. Vasenkov, Self-diffusion of pure and mixed gases in mixed linker zeolitic imidazolate framework-7-8 by high field diffusion NMR, Microporous Mesoporous Mater. 288 (2019) 109603.
[94] D.I. Kolokolov, H. Jobic, S. Rives, P.G. Yot, J. Ollivier, P. Trens, A.G. Stepanov, G. Maurin, Diffusion of benzene in the breathing metal–organic framework MIL-53(Cr): a joint experimental-computational investigation, J. Phys. Chem. C 119 (2015) 8217–8225.
[95] D.I. Kolokolov, A.G. Maryasov, J. Ollivier, D. Freude, J. Haase, A.G. Stepanov, H. Jobic, Uncovering the rotation and translational mobility of benzene confined in UiO-66 (Zr) metal-organic framework by the ^{2}H NMR-QENS experimental toolbox, J. Phys. Chem. C 121 (2017) 2844–2857.
[96] A.E. Khudozhitkov, S.S. Arzumanov, D.I. Kolokolov, A.G. Stepanov, Mobility of aromatic guests and isobutane in ZIF8 metal-organic framework studied by ^{2}H solid state NMR spectroscopy, J. Phys. Chem. C 123 (2019) 13765–13774.
[97] S. Li, J. Li, J. Tang, F. Deng, Host-guest interaction of styrene and ethylbenzene in MIL-53 studied by solid-state NMR, Solid State Nucl. Magn. Reson. 90 (2018) 1–6.
[98] F. Pourpoint, J. Tréebosc, R.M. Gauvin, Q. Wang, O. Lafon, F. Deng, J.-P. Amoureux, Measurement of aluminumcarbon distances using S-RESPDOR NMR experiments, ChemPhysChem 13 (2012) 3605–3615.
[99] M. Hong, J.D. Gross, C.M. Rienstra, R.G. Griffin, K.K. Kumashiro, K. Schmidt-Rohr, Coupling amplification in 2D MAS NMR and its application to torsion angle determination in peptides, J. Magn. Reson. 129 (1997) 85–92.
[100] T. Ueda, K. Kurokawa, Y. Kawamura, K. Miyakuba, T. Eguchi, ^{1}H NMR study of molecular motion of benzene and *n*-decane confined in the nanocavities of metal-organic frameworks, J. Phys. Chem. C 116 (2012) 1012–1019.
[101] A.E. Khudozhitkov, A.V. Toktarev, S.S. Arzumanov, A.A. Gabrienko, D.I. Kolokolov, A.G. Stepanov, ^{2}H solid-state NMR spectroscopy reveals the dynamics of a pyridine probe interacting with coordinatively unsaturated metal sites of MIL-100(Al) metal-organic frameworks, Chem. Eur. J. 25 (2019) 10808–10812.
[102] J. Tang, S. Li, Y. Chu, Y. Xiao, J. Xu, F. Deng, Solid-state NMR studies of the acidity of functionalized metal-organic framework UiO-66 materials. Magn. Reson. Chem. (2019) 1–8. https://doi.org/10.1002/mrc.4923.
[103] J.A. Gee, J. Chung, S. Nair, D.S. Sholl, Adsorption and diffusion of small alcohols in zeolitic imidazolate frameworks ZIF-8 and ZIF-90, J. Phys. Chem. C 117 (2013) 3169–3176.

[104] J.-W. Dai, X. Li, J.-M. Zheng, X. Dong, Z.-H. Zhou, Transformations of lead 1,3-propylenediaminetetraacetate to its MOF products for the selective adsorption of methanol, J. Solid State Chem. 237 (2016) 364–370.
[105] A. Nandy, A.C. Forse, V.J. Witherspoon, J.A. Reimer, NMR spectroscopy reveals adsorbate binding sites in the metal-organic framework UiO-66(Zr), J. Phys. Chem. C 122 (2018) 8295–8305.
[106] J. Tang, S. Li, Y. Chu, Y. Xiao, J. Xu, F. Deng, Host-guest interaction between methanol and metal-organic framework $Cu_{3-x}Zn_x(btc)_2$ as revealed by solid-state NMR, J. Phys. Chem. C 123 (2019) 24062–24070.
[107] R. Thoma, J. Kärger, N. de Sousa Amadeu, S. Nießing, C. Janiak, Assessing guest-molecule diffusion in heterogeneous powder samples of metal-organic frameworks through pulsed-field-gradient (PFG) NMR spectroscopy., Chem. Eur. J. 23 (2017) 13000–13005.
[108] M. Wehring, J. Gascon, D. Dubbeldam, F. Kapteijn, R.Q. Snurr, F. Stallmach, Self-diffusion studies in CuBTC by PFG NMR and MD simulations, J. Phys. Chem. C 114 (2010) 10527–10534.
[109] R.K. Motkuri, H.V.R. Annapureddy, M. Vijaykumar, H.T. Schaef, P.F. Marting, B.P. McGrail, L.X. Dang, R. Krishna, P.K. Thallapally, Fluorocarbon adsorption in hierarchical porous frameworks, Nature Commun. 5 (2014) 4368.
[110] S. Proch, J. Herrmannsdörfer, R. Kempe, C. Kern, A. Jess, L. Seyfarth, J. Senker, Pt@MOF-177: synthesis, room-temperature hydrogen storage and oxidation catalysis, Chem. Eur. J. 14 (2008) 8204–8212.
[111] M. Müller, O.I. Lebedev, R.A. Fischer, Gas-phase loading of $[Zn_4O(btb)_2]$ (MOF-177) with organometallic CVD-precursors: inclusion compounds of the type $[L_nM]_a$@MOF-177 and the formation of Cu and Pd nanoparticles inside MOF-177, J. Mater. Chem. 18 (2008) 5274–5281.
[112] M. Müller, S. Hermes, K. Kähler, M.W.E. van den Berg, M. Muhler, R.A. Fischer, Loading of MOF-5 with Cu and ZnO nanoparticles by gas-phase infiltration with organometallic precursors: properties of Cu/ZnO@MOF-5 as catalyst for methanol synthesis, Chem. Mater. 20 (2008) 4576–4587.
[113] D. Esken, H. Noei, Y. Wang, C. Wiktor, S. Turner, G.V. Tendeloo, R.A. Fischer, ZnO@ZIF-8: stabilization of quantum confined ZnO nanoparticles by a zinc methylimidazolate framework and their surface structural characterization probed by CO_2 adsorption, J. Mater. Chem. 21 (2011) 5907–5915.
[114] M. Meilikhov, K. Yusenko, R.A. Fischer, Incorporation of metallocenes into the channel structured metal–organic frameworks MIL-53(Al) and MIL-47(V), Dalton Trans. 39 (2010) 10990–19999.
[115] N.V. Maksimchuk, K.A. Kovalenko, S.S. Arzumanov, Y.A. Chesalov, M.S. Melgunov, A.G. Stepanov, V.P. Fedin, O.A. Kholdeeva, Hybrid polyoxotungstate/MIL-101 materials: synthesis, characterization, and catalysis of H_2O_2-based alkene epoxidation, Inorg. Chem. 49 (2010) 2920–2930.
[116] D. Juli ao, A.C. Gomes, M. Pillinger, L. Cunha-Silva, B de Castro, I.S. Gonçalves, S.S. Balula, Desulfurization of model diesel by extraction/oxidation using a zinc-substituted polyoxometalate as catalyst under homogeneous and heterogeneous (MIL-101(Cr) encapsulated) conditions, Fuel Proc. Techn. 131 (2015) 78–86.
[117] F. Zhang, Y. Jin, J. Shi, Y. Zhong, W. Zhu, M.S. El-Shall, Polyoxometalates confined in the mesoporous cages of metal-organic framework MIL-100(Fe): efficient heterogeneous catalysts for esterification and acetalization reactions, Chem. Eng. J. 269 (2015) 236–244.
[118] C.T. Buru, P. Li, B.L. Mehdi, A. Dohnalkova, A.E. Platero-Prats, N.D. Browning, K.W. Chapman, J.T. Hupp, O.K. Farha, Adsorption of a catalytically accessible polyoxometalate in a mesoporous channel-type metal-organic framework, Chem. Mater. 29 (2017) 5174–5181.

[119] J. Lyu, N. Zhang, H. Liu, Z. Zeng, J. Zhang, P. Bai, X. Guo, Adsorptive removal of boron by zeolitic imidazolate framework: kinetics, isotherms, thermodynamics, mechanism and recycling, Sep. Purif. Technol. 187 (2017) 67–75.

[120] R. Haldar, M. Inukai, S. Horike, K. Uemura, S. Kitagawa, T.K. Maji, ^{113}Cd nuclear magnetic resonance as a probe of structural dynamics in a flexible porous framework showing selective O_2/N_2 and CO_2/N_2 adsorption, Inorg. Chem. 55 (2016) 4166–4172.

[121] Y. Mito-oka, S. Horike, Y. Nishitani, T. Masumori, M. Inukai, Y. Hijikata, S. Kitagawa, Siloxane D4 capture by hydrophobic microporous materials, J. Mater. Chem. A 1 (2013) 7885–7888.

[122] M.G. Lope, P. Canepa, T. Thonhauser, NMR study of small molecule adsorption in MOF-74-Mg, J. Chem. Phys. 138 (2013) 154704.

[123] T. Wittmann, A. Mondal, C.B.L. Tschense, J.J. Wittmann, O. Klimm, R. Siegel, B. Corzilius, B. Weber, M. Kaupp, J. Senker, Probing interactions of Ndonor molecules with open metal sites within paramagnetic Cr-MIL-101: a solid-state NMR spectroscopic and density functional theory study, J. Am. Chem. Soc. 140 (2018) 2135–2144.

[124] T. Gullion, J. Schaefer, Rotational-echo double-resonance NMR, J. Magn. Reson. 81 (1989) 196–200.

[125] M. Wehring, P.C.M.M. Magusin, S. Amirjalayer, R. Schmid, F. Stallmach, Ferrocene in the metalorganic framework MOF-5 studied by homo- and heteronuclear correlation NMR and MD simulation, Microporous Mesoporous Mater. 186 (2014) 130–136.

[126] M. Peksa, S. Burraekaew, R. Schmid, J. Lang, F. Stallmach, Rotational and translational dynamics of CO_2 adsorbed in MOF $Zn_2(bdc)_2(dabco)$, Microporous Mesoporous Mater. 216 (2015) 75–81.

[127] V.J. Witherspoon, L.M. Yu, S. Jawahery, E. Braun, S.M. Moosavi, S.K. Schnell, B. Smit, J.A. Reimer, Translational and rotational motion of C8 aromatics adsorbed in isotropic porous media (MOF-5): NMR studies and md simulations, J. Phys. Chem. C 121 (2017) 15456–15462.

[128] V. Bon, J. Pallmann, E. Eisbein, H.C. Hoffmann, I. Senkovska, I. Schwedler, A. Schneemann, S. Henke, D. Wallacher, R.A. Fischer, G. Seifert, E. Brunner, S. Kaskel, Characteristics of flexibility in metal-organic framework solid solutions of composition $[Zn_2(BME\text{-}bdc)_x(DB\text{-}bdc)_{2-x}dabco]_n$: *In situ* powder X-ray diffraction, *in situ* NMR spectroscopy, and molecular dynamics simulations, Microporous Mesoporous Mater. 216 (2016) 64–74.

[129] A.J. Rossini, A. Zagdoun, M. Lelli, J. Canivet, S. Aguado, O. Ouari, P. Tordo, M. Rosay, W.E. Maas, C. Copéret, D. Farrusseng, L. Emsley, A. Lesage, Dynamic nuclear polarization enhanced solid-state NMR spectroscopy of functionalized metal-organic frameworks, Angew. Chem. Int. Ed. 51 (2012) 123–127.

[130] Z. Guo, T. Kobayashi, L.-L. Wang, T.W. Goh, C. Xiao, M.A. Caporini, M. Rosay, D.D. Johnson, M. Pruski, W. Huang, Selective host-guest interaction between metal ions and metal-organic frameworks using dynamic nuclear polarization enhanced solid-state NMR spectroscopy, Chem. Eur. J. 20 (2014) 16308–16313.

[131] F. Pourpoint, A.S.L. Thankamony, C. Volkringer, T. Loiseau, J. Trébosc, F. Aussenac, D. Carnevale, G. Bodenhausen, H. Vezin, O. Lafon, J.-P. Amoureux, Probing ^{27}Al-^{13}X proximities in metal-organic frameworks using dynamic nuclear polarization enhanced NMR spectroscopy, Chem. Commun. 50 (2014) 933–935.

[132] T. Kobayashi, F.A. Perras, T.W. Goh, T.L. Metz, W. Huang, M. Pruski, DNP-enhanced ultrawideline solid-state NMR spectroscopy: studies of platinum in metal-organic frameworks, J. Phys. Chem. Lett. 7 (2016) 2322–2327.

CHAPTER TWO

Acidity characterization of solid acid catalysts by solid-state ^{31}P NMR of adsorbed phosphorus-containing probe molecules: An update

Xianfeng Yi[a], Anmin Zheng[a,*], and Shang-Bin Liu[b,*]
[a]State Key Laboratory of Magnetic Resonance and Atomic and Molecular Physics, National Center for Magnetic Resonance in Wuhan, Key Laboratory of Magnetic Resonance in Biological System, Wuhan Institute of Physics and Mathematics, Innovation Academy for Precision Measurement Science and Technology, The Chinese Academy of Sciences, Wuhan, PR China
[b]Institute of Atomic and Molecular Sciences, Academia Sinica, Taipei, Taiwan
*Corresponding authors: e-mail address: zhenganm@wipm.ac.cn; sbliu.iams@gmail.com

Contents

Abstract

Recent developments on acidity characterization of solid acid catalysts, specifically those invoking ^{31}P solid-state nuclear magnetic resonance (SSNMR) spectroscopy using phosphorus-containing molecules as probes has been summarized. In particular, various ^{31}P SSNMR approaches using trimethylphosphine (TMP), phosphines, diphosphines, and trialkylphosphine oxides (R_3PO) will be introduced and their practical applications for characterization of important qualitative and quantitative features, viz. type, distribution, accessibility (location/proximity), concentration (amount), and strength of acid

Annual Reports on NMR Spectroscopy, Volume 101
ISSN 0066-4103
https://doi.org/10.1016/bs.arnmr.2020.07.004

sites in various solid acids are illustrated. This is an updated version of the original review (Annu. Rep. NMR Spectr. 81 (2014) 47–108).

Keywords: Solid acid catalysts, Phosphorus-containing molecules, Solid-state NMR, Acidity, ^{31}P chemical shift

1. Introduction

While homogeneous catalysis has been exploited for industrial petrochemical processes in the past several decades, it is handicapped by the extensive use of toxic and corrosive liquid acids catalysts such as H_2SO_4 and HF, and hence also suffers from inherent high risks and costs during waste acid treatments. Thus, there has been an urgent demand in the development of heterogeneous catalysis processes, particularly those related to R&D of environmental-friendly solid acid catalysts. Over the years, a wide variety of solid acids have been developed and exploited in industrial processes, including zeolites, complex oxides, sulphated metal oxides, heteropoly acids, etc., as depicted in Table 1. It is well-known that detailed features of acid sites, namely their types (Brønsted vs. Lewis acid), distributions/locations (intra- vs. extracrystalline), concentrations (i.e. amounts), and strengths, normally dictate the catalytic performances (e.g. catalytic activity and product selectivity) of solid acid catalysts. Thus, a comprehensive understanding of these detailed properties of acid sites is most crucial for the design, modification, and practical applications of solid acid catalysts. For this reason, a variety of different spectroscopic and analytical techniques have been developed to unravel the nature and characteristics of acid sites in solid acids.

Common acidity characterization methods for solid acid catalysts normally utilize the direct observation of the acidic hydroxyl OH groups either by Fourier-transform infrared (FT-IR) [2] or ^1H nuclear magnetic resonance (NMR) [3] spectroscopy or through the analysis and/or detection of adsorbed basic probe molecules containing unpaired electrons (N, P, etc.), such as ammonia (NH_3), pyridine (C_5H_5N), methylamine (CH_3NH_2), or trimethylphosphine ($(CH_3)_3P$; TMP) [4]. For the latter, analytical or spectroscopic techniques such as titration, microcalorimetry, temperature-programmed-desorption (TPD), IR, and NMR are invoked [5]. In general, conventional analytical methods are more useful for quantitative description (acid amount and strength) of the *overall* acidity. On the other hand, conventional spectroscopic techniques are more suitable for

Table 1 List of assorted solid acid catalysts and their practical industrial applications [1].

Catalysts	Formula/example	Acid group and/or catalytic active species	Catalytic applications	Reaction type	Temperature range (°C)
Alumina (silicated)	Al_2O_3; $Al_{2-x}Si_xO_{3+x/2}$	Al-OH, Al-OH-Al, $(SiOH)Al^{3+}$	Olefin skeletal isomerization Alcohol dehydration	Gas/solid	>400
Silicalite	SiO_2	SiOH	Beckmann rearrangement of cyclohexanone-oxime	Gas/solid	300
Chloride alumina	$Al_2O_{3-x}Cl_{2x}$	Cl_xAlOH Al^{3+}	Paraffin isomerization Aliphatic alkylation	Gas/solid Liquid/solid	120–200
Acid-treated montmorillonite clays	$Na_x[Al_{2-x}Mg_xSi^{4-}O_{10}(OH)_2]\cdot nH_2O$	SiOH Al^{3+}	Cracking	Gas/solid Liquid/solid	200–550
Silica-alumina/aluminated silica	$H_ySi_{1-x}Al_xO_{2-x/2+2y}$	SiOH	Cracking, dehydrochlorination, alkylation	Gas/solid Liquid/solid	200–550 200
H-zeolites	$H_xSi_{1-x}Al_xO_2$	Si-OH-Al	Aromatics alkylation Paraffin and olefin isomerization Cracking	Gas/solid Liquid/solid	200–550 150–250
SAPO	$H_{x-y}Si_{1-x-y}P_yAl_xO_2$	Si-OH-Al	Methanol to olefins	Gas/solid	400–450
Sulphated zirconia	SO_4^{2-}/ZrO_2	SOH	Paraffin isomerization	Gas/solid	170–230
Tungstated zirconia	WO_3/ZrO_2	WOH	Paraffin isomerization	Gas/solid	200–270

Continued

Table 1 List of assorted solid acid catalysts and their practical industrial applications [1].—cont'd

Catalysts	Formula/example	Acid group and/or catalytic active species	Catalytic applications	Reaction type	Temperature range (°C)
Solid phosphoric acid	H_3PO_4/SiO_2 (kiselghur)	POH $[H(H_2O)_n]^+$	Olefin oligomerization and hydration Aromatics alkylation	Gas/solid	150–300
Heteropolyacid	$Cs_xH_{3-x}PW_{12}O_{40}$	W-OH-W $[H(H_2O)_n]^+$	Ethyl acetate synthesis	Gas/solid Liquid/solid	140–250 60
Niobic acid	$Nb_2O_5 \cdot nH_2O$	NbOH $[H(H_2O)_n]^+$	Ethylene hydration Fructose dehydration	Gas/solid Water/solid	200 100
Sulphonated polystyrene-poly(divinylbenzene) resins	Sulphonated PPS-PDVB	$-SO_3H$ $[H(H_2O)_n]^+$ $[H(ROH)]^+$	Ether synthesis Olefin oligomerization	Liquid/solid	40–100

Reproduced from G. Busca, Acid catalysts in industrial hydrocarbon chemistry, Chem. Rev. 107 (2007) 5366–5410. Copyright 2007 American Chemical Society.

qualitative characterization of acid sites. However, they are also drawback by the strong correlation between experimental parameters and surrounding environment of the solid acids. Taking HY zeolite as an example, the hydrogen-bonding interactions between the acidic proton and its adjacent framework oxygen atom are found to be much stronger in small sodalite cages than that in large supercages, leading to a considerably higher ^{1}H NMR chemical shift (CS; $\delta^{1}H$) value (i.e. notable shift of ^{1}H resonance towards downfield) [6]. Nonetheless, this does not mean that acid sites in sodalite cages possess higher acidic strength than those in supercages. By the same token, indirect detection methods such as those through the adsorption of basic probes are also limited by the ambiguous interpretation of experimental results. For example, there have been some debates on the assignments of desorption peaks obtained from the commonly used NH_3-TPD method [5]. The most common conception is that the multiple desorption peaks centred at low and high temperatures may be inferred to Brønsted acid sites with relatively weak and strong acidic strength, respectively. However, ab initio calculations revealed that, for NH_3 adsorbed in zeolites, the dual low- and high-temperature desorption peak may be arising from interactions among encapsulated NH_3 molecules and interactions between NH_3 with the zeolite pore-walls, respectively. Provided that the shape and size of the pores are similar, the above finding prevailed even in the absence of strong acid centres in zeolite [7]. Apparently, the notion whether NH_3-TPD is a reliable technique for acidity characterization of solid acid catalysts remains questionable.

2. Acidity characterization by ^{31}P solid-state NMR spectroscopy

Among various experimental methods available for acidity characterization of solid acid catalysts, solid-state nuclear magnetic resonance (SSNMR) spectroscopy is a desirable and most versatile technique [8–10]. Among them, ^{1}H magic-angle-spinning (MAS) NMR represents a direct approach to identify structures of hydroxyl groups in a catalyst. Typically, protons residing at Brønsted acid sites, for examples, bridging hydroxyls (Si-OH-Al) in zeolites or strongly acidic terminal hydroxyls (–OH) in heterpolyacids possess $\delta^{1}H$ in the range of 3.6–5.6 ppm, whereas weakly acidic terminal hydroxyls (e.g. Al-OH or Si-OH) tend to appear at 1.5–2.0 ppm [11,12]. Nonetheless, the shortcoming of this technique is its capability to explore the type, strength, and distribution of acid sites.

Alternatively, MAS NMR of various adsorbed basic probe molecules, such as pyridine, acetone, and trimethylphosphine, exploiting specific nucleus of interest have been developed. For example, for the ^{1}H-pridine-d_5 NMR approach, it has been found that when the pyridine-d_5 probe molecule is adsorbed on the weakly acidic silanol groups (Si-OH) in zeolites, the occurrence of hydrogen-bonding interaction normally leads to resonance with δ^{1}H within the range of ca. 2–10 ppm. Whereas for pyridine-d_5 adsorbs on bridging hydroxyl groups (Al-OH-Si) in zeolites or strongly acidic terminal –OH groups in heteropolyacids, the formation of pyridinium ion ($C_5D_5NH^+$) complexes typically result in δ^{1}H in the range of ca. 12–20 ppm [12,13]. The ^{13}C-acetone NMR approach, which utilizes isotope-labelled 2-^{13}C-acetone as the probe molecule, is another well-established acidity characterization technique for determining relative Brønsted acidic strengths in solid acids. Upon adsorbing the probe molecule on Brønsted acid sites, couplings between the acidic proton and the carbonyl oxygen of 2-^{13}C-acetone normally cause the ^{13}C resonance to shift towards downfield direction (i.e. towards a higher δ^{13}C value) [14,15]. Indeed, these aforediscussed ^{1}H-pyridine and ^{13}C-acetone NMR approaches are capable of determining the *relative* acidic strength of solid acid catalysts. However, in terms of the two key ingredients of spectroscopic techniques, namely sensitivity and resolution, these two approaches are limited by the characteristics of their respective probing nuclei. While ^{1}H nucleus ($I=1/2$) has a nearly 100% natural abundance, it is handicapped by its rather narrow chemical shift range (ca. 20 ppm), thus, an inferior spectral resolution. On the other hand, although the ^{13}C nucleus ($I=1/2$) has a desirable resolution (chemical shift range ca. 300 ppm), it has a rather low natural abundance (1.07%), thus, relies on ^{13}C-enriched reagents and hence is less favourable in terms of experimental convenience and cost. Moreover, acidity characterization invoking the ^{1}H-pyridine and ^{13}C-acetone NMR approaches are also disadvantaged by their non-uniform correlations between acidic strengths (of the solid catalyst) and NMR chemical shifts (of the adsorbed probe molecule; vide infra).

Following the above arguments, phosphorus-containing molecules, such as trimethylphosphine (TMP) and trialkylphosphine oxides (R_3PO), should be more desirable as NMR probes. This is owing to the fact that, like ^{1}H and ^{13}C, ^{31}P nucleus also possesses spin $I=1/2$ but with a 100% natural abundance and a much larger chemical shift range (>650 ppm). The main objectives of this review are two folds: (1) to provide background introduction of ^{31}P SSNMR acidity characterization technique using phosphorus-containing molecules as probes and (2) to illustrate practical applications of such technique for characterization of acid properties (such as type, distribution, and strength) in various solid acid catalyst systems.

Typically, the ^{31}P NMR experiments are performed using a 4mm MAS probe operating at a spinning rate of 12kHz. It is noteworthy that the ^{31}P spectral resolution could be improved with fast spinning rates. In order to prevent contamination of the probe molecule-loaded sample by moisture or O_2, prior to the NMR experiments, one should transfer the adsorbate-loaded sample (in a sealed vessel) into a 4mm ZrO_2 rotor and seal with a gas-tight Kel-F cap under inert environment (in a N_2 bag or glovebox).

2.1 ^{31}P-TMP approach

Pioneered by Lunsford and co-workers, TMP molecule was first adopted as an NMR probe to characterize the acidity of HY zeolite [16]. Thereafter, ^{31}P MAS NMR of adsorbed TMP has been widely utilized for probing Brønsted and Lewis acidities of various solid acid catalysts, such as microporous zeolites, metal oxides, mixed metal oxides, sulphated metal oxides and heteropolyacids. However, as will be shown latter that while the ^{31}P resonances responsible for TMP adsorbed on Lewis acid sites span over a considerable chemical shift range of ca. $-20 \sim -60$ ppm, signals correspond to TMP bound to Brønsted acid sites occur in a rather narrow range (ca. $-2 \sim -5$ ppm) [16,17]. As a consequence, the ^{31}P-TMP NMR approach is more sensitive for identifying the strength of Lewis acid sites, and less sensitive for Brønsted acidity in solid acids. In this regards, as will be shown later that the ^{31}P-R_3PO NMR approach appears to be more superior in terms of identifying variations of Brønsted acid strengths in solid acid catalysts (vide infra).

As illustrated in Fig. 1, three typical interaction models may be envisaged for the adsorbed TMP with the surfaces of solid acid catalysts, namely (i) chemisorbed on Lewis acid centres, (ii) physisorbed on weakly acidic hydroxyls (i.e. weak H-bonding interaction), and (iii) chemisorbed on Brønsted acid sites. For the latter, upon adsorbing TMP onto Brønsted acid sites, effective proton transfers from the acid site to the probe molecule result in formation of $TMPH^+$ complexes, which typically give rise to ^{31}P resonances with $\delta^{31}P$ in the range of ca. $-2 \sim -5$ ppm. Table 2 depicted $\delta^{31}P$ of TMP adsorbed on Brønsted and Lewis acid sites of assorted solid acids. As revealed by density functional theory (DFT) calculations, an

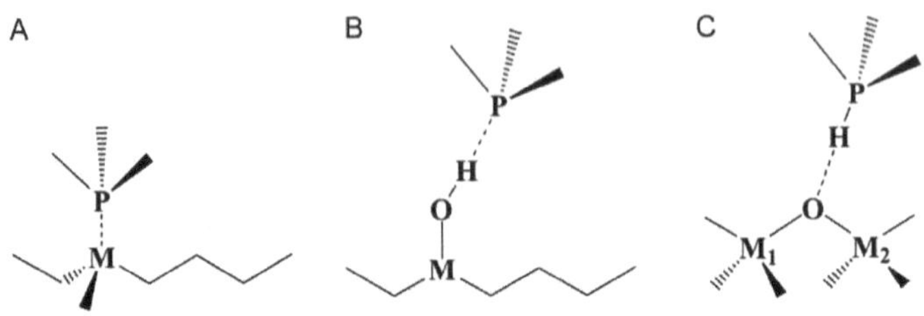

Fig. 1 Scenarios of adsorbed trimethylphosphine (TMP) on surfaces of solid catalysts: (A) chemisorbed on Lewis acid centres, (B) physisorbed on hydroxyls, and (C) chemisorbed on Brønsted acid sites [18]. *Reproduced from Y. Chu, Z. Yu, A. Zheng, H. Fang, H. Zhang, S.J. Huang, S.B. Liu, F. Deng, Acidic strengths of Brønsted and Lewis acid sites in solid acids scaled by* ^{31}*P NMR chemical shifts of adsorbed trimethylphosphine, J. Phys. Chem. C 115 (2011) 7660–7667. Copyright 2011 American Chemical Society.*

Table 2 Experimental ^{31}P NMR chemical shifts of TMP adsorbed on Brønsted and Lewis acid sites of various solid acid catalysts [18,19].

	δ^{31}P (ppm)		
Catalyst	**Brønsted acid**	**Lewis acid**	**Lewis acid centre (M)**
Al_2O_3	–	−51	Al
$CHCl_3/Al_2O_3$	−3.5	−44	Al
$AlCl_3/Al_2O_3$	−3.4	−41; −53	Al
H-ZSM-5	−3.5	−45	Al
H-Beta	−4.5	−32; −47	Al
HY (T_{cal} = 700 °C)	−4.2	−32; −43.5; −46.5; −50.5; −54.5	Al
BF_3/Al_2O_3	−4.0	−26	B
TiO_2	–	−35	Ti
SO_4^{2-}/TiO_2	−4.0	−24	Ti
TS zeolite	−4.8	−34.2; −32	Ti
ZrO_2	–	−30.3; −42.2; −50.1	Zr
SO_4^{2-}/ZrO_2	−3.8	−33.7; −39.7; −44.4	Zr
WO_3/ZrO_2 (W = 5.3%)	−4.0	−35	Zr or W
CeO_2	–	−19.3	Ce

Reproduced from Y. Chu, Z. Yu, A. Zheng, H. Fang, H. Zhang, S.J. Huang, S.B. Liu, F. Deng, Acidic strengths of Brønsted and Lewis acid sites in solid acids scaled by ^{31}P NMR chemical shifts of adsorbed trimethylphosphine, J. Phys. Chem. C 115 (2011) 7660–7667. Copyright 2011 American Chemical Society.

increasing extent of proton transfer (from Brønsted acid site to TMP) would correspond to a decrease in deprotonation energy (DPE; i.e. increase in acidic strength), thus an increase in the observed $\delta^{31}P$, as shown in Fig. 2 [18]. In view of the fact that typical solid acid catalysts such as zeolites possess mostly Brønsted acid sites with moderate acidic strengths, corresponding to a typical DPE value of ca. 270–310 kcal/mol [18]. Together with the well-accepted threshold DPE value for superacidity (250 kcal/mol) [21], it is noteworthy that only marginal variations in $\delta^{31}P$ was observed for the $TMPH^{+}$ adduct as DPE decreases from ca. 320 (weak acidic strength) to 250 kcal/mol (super strong acidic strength) [21]. As such, it is obvious that the ^{31}P-TMP NMR approach is inferior for differentiating Brønsted acidic strengths in solid acid catalysts.

In the case of TMP physisorbed on weakly acidic hydroxyl groups residing on the surfaces of solid acids (Fig. 1B), the formation of hydrogen-bonded complexes prevail, leading to ^{31}P resonance with $\delta^{31}P$ at ca. −60 ppm (Fig. 2). On the other hand, for TMP chemisorbed on Lewis acid sites

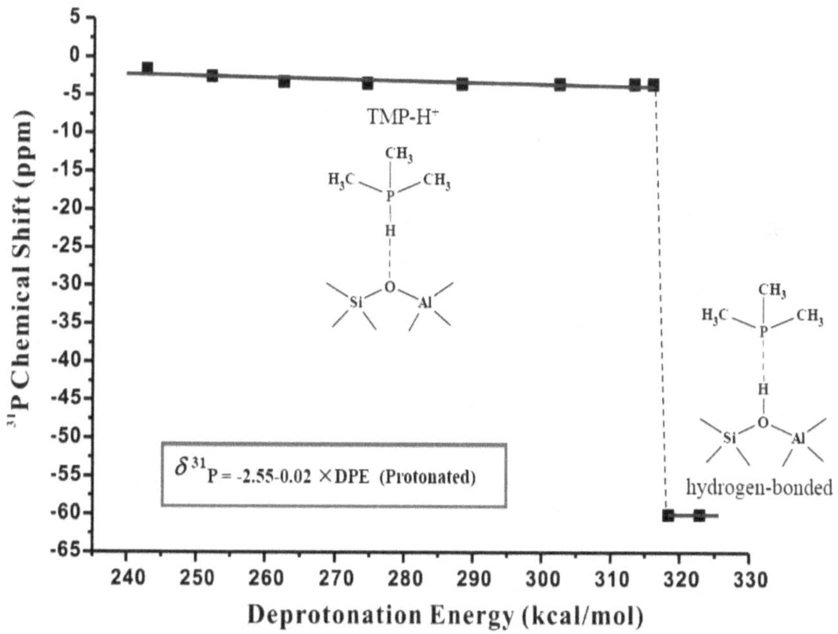

Fig. 2 Correlations of ^{31}P chemical shift of adsorbed TMP with deprotonation energy of Brønsted acid sites in solid acids [20]. *Reproduced from A. Zheng, S.B. Liu, F. Deng, Acidity characterization of heterogeneous catalysts by solid-state NMR spectroscopy using probe molecules, Solid State Nucl. Magn. Reson. 55–56 (2013) 12–27. Copyright 2013 Elsevier.*

(Fig. 1A), the lowest unoccupied molecular orbital (LUMO) of the Lewis acid centre (M) tends to share electron pairs with the phosphorus (P) atom on the TMP, the formation of such M-P chemical bonds typically give rise to ^{31}P signals in a considerable range of δ^{31}P within ca. $-20\sim-60$ ppm (Table 2). As an illustration, ^{31}P spectra obtained from mixed metal oxides by the ^{31}P-TMP NMR approach will be discussed. As shown in Fig. 3, three distinct ^{31}P resonances at -30.3, -42.2, and -50.1 ppm were observed for TMP adsorbed on zirconia (ZrO_2), indicating the presences of three types of Lewis sites with varied acidic strengths [22]. That the signal responsible for $TMPH^+$ complex was not observed, indicating the absence of Brønsted acidic protons in the ZrO_2 catalyst. In contrast to pure ZrO_2, a major ^{31}P peaks at -3.3 and a weak shoulder around -48 ppm were observed for TMP adsorbed on MoO_x/ZrO_2 (Fig. 3B) and WO_x/ZrO_2 (Fig. 3C) mixed oxide systems. The ^{31}P signal occurred at -3.3 ppm is clearly due to the presence of Brønsted acid sites, which are anticipated to exhibit much stronger acid strengths than hydroxyl groups. Whereas the signal located -48 ppm

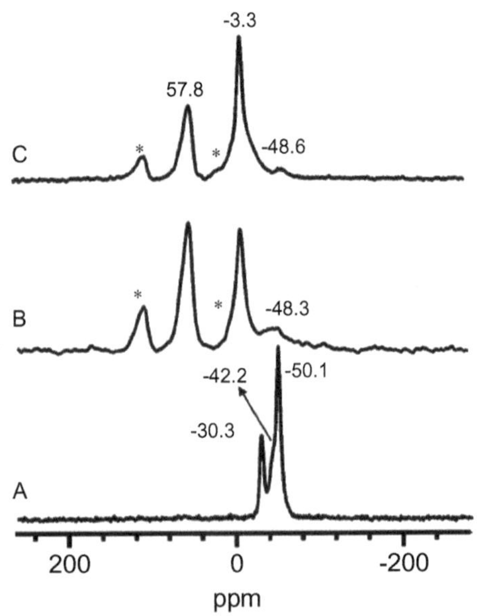

Fig. 3 ^{31}P MAS NMR spectra (with ^{1}H decoupling) of TMP adsorbed on mesoporous materials: (A) ZrO_2, (B) MoO_x/ZrO_2, and (C) WO_x/ZrO_2. Asterisks denote spinning sidebands [22]. *Reproduced from J. Xu, A. Zheng, J. Yang, Y. Su, J. Wang, D. Zeng, M. Zhang, C. Ye, F. Deng, Acidity of mesoporous MoO_x/ZrO_2 and WO_x/ZrO_2 materials: a combined solid-state NMR and theoretical calculation study, J. Phys. Chem. B 110 (2006) 10662–10671. Copyright 2006 American Chemical Society.*

was attributed to TMP adsorbed on Lewis acid sites. It is noteworthy that TMP tends to oxidize easily in these mixed metal oxides to form trimethylphosphine oxide (TMPO) during sample treatment, which attributed the ^{31}P resonance at 57.8 ppm, as can be seen in Fig. 3B and C [22].

The same ^{31}P-TMP NMR approach has also been employed for the characterization of sulphated metal oxides [23–25]. For TMP adsorbed on pristine TiO_2, the symmetrical ^{31}P resonance peak centred at $\delta^{31}P \sim -35$ ppm (Fig. 4A) may be attributed to TMP adsorbed on Lewis acid sites. Upon sulphonation treatment, a notable shift of this resonance towards downfield ($\delta^{31}P \sim -27$ ppm; Fig. 4B) was observed for the SO_4^{2-}/TiO_2 catalyst [25]. In addition, Similar to that observed for MoO_x/ZrO_2 and WO_x/ZrO_2 in Fig. 3, an additional ^{31}P peak at ca. −4 ppm was also evident [22], which is unambiguously due to TMP adsorbed on Brønsted acid sites. Similar conclusions may be inferred for TMP adsorbed on pristine Al_2O_3 ($\delta^{31}P \sim -51$ ppm; Fig. 4C) and SO_4^{2-}/Al_2O_3 ($\delta^{31}P \sim -49$ and −3 ppm; Fig. 4D). Likewise, the additional weak peak appearing at $\delta^{31}P \sim 34$ ppm (Fig. 4C) has been ascribed due to the presence of TMPO. On the basis of above results obtained from the ^{31}P-TMP NMR approach, it is indicative while metal oxides such as ZrO_2, TiO_2, and Al_2O_3 possess mostly Lewis acidity, further sulphonation treatment of these metal oxide catalysts tends to not only increase the Lewis acidic

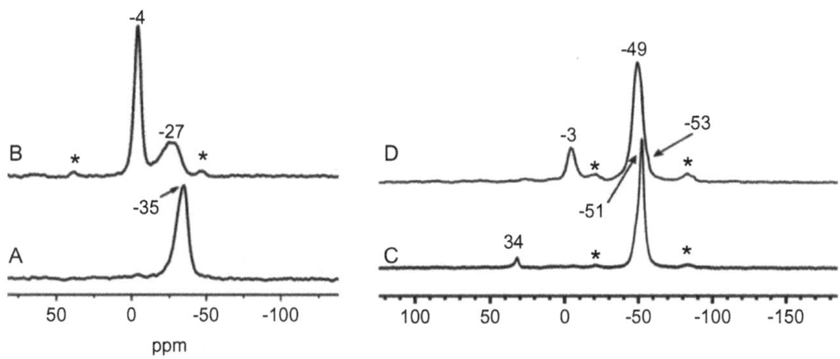

Fig. 4 ^{31}P MAS NMR spectra of TMP adsorbed on (A) TiO_2, (B) SO_4^{2-}/TiO_2, (C) Al_2O_3, and (D) SO_4^{2-}/Al_2O_3. Asterisks denote spinning sidebands. The weak signal at 34 ppm in (C) is due to the presence of trimethylphosphine oxide (TMPO) [24,25]. *Reproduced from J. Yang, M. Zhang, F. Deng, Q. Luo, D. Yi, C. Ye, Solid state NMR study of acid sites formed by adsorption of SO_3 onto γ-Al_2O_3, Chem. Commun. 7 (2003) 884–885. Copyright 2003 The Royal Society of Chemistry; H. Zhang, H. Yu, A. Zheng, S. Li, W. Shen, F. Deng, Reactivity enhancement of 2-propanol photocatalysis on SO_4^{2-} /TiO_2: insights from solid-state NMR spectroscopy, Environ. Sci. Technol. 42 (2008) 5316–5321. Copyright 2008 American Chemical Society.*

strength (as indicated by the downfield shift of the ^{31}P resonance) but also promote formation of new Brønsted acid sites.

As shown above the ^{31}P-TMP approach appears to be insensitive in terms of discernment of acidic strength of Brønsted acid sites due to the limited range of $\delta^{31}P$ ($-2 \sim -5$ ppm). In this context, the same approach is more desirable for characterization of Lewis acidity, for which a much wider $\delta^{31}P$ range (ca. $-20 \sim -55$ ppm) is available. This is exemplified by a series of theoretical Lewis acid models with different metallic centres, viz. boron (BCl_nF_{3-n}; $n=0$–3), aluminium ($AlCl_nF_{3-n}$; $n=0$–3), and titanium ($TiCl_nF_{4-n}$; $n=0$–4). The optimized structures of assorted TMP-Lewis acid complexes are depicted in Fig. 5. The variations of calculated $\delta^{31}P$ with binding energy (BE) for the aforementioned theoretical Lewis acid models are shown in Fig. 6 [18]. Here, the BE value deduced from TMP adsorbed

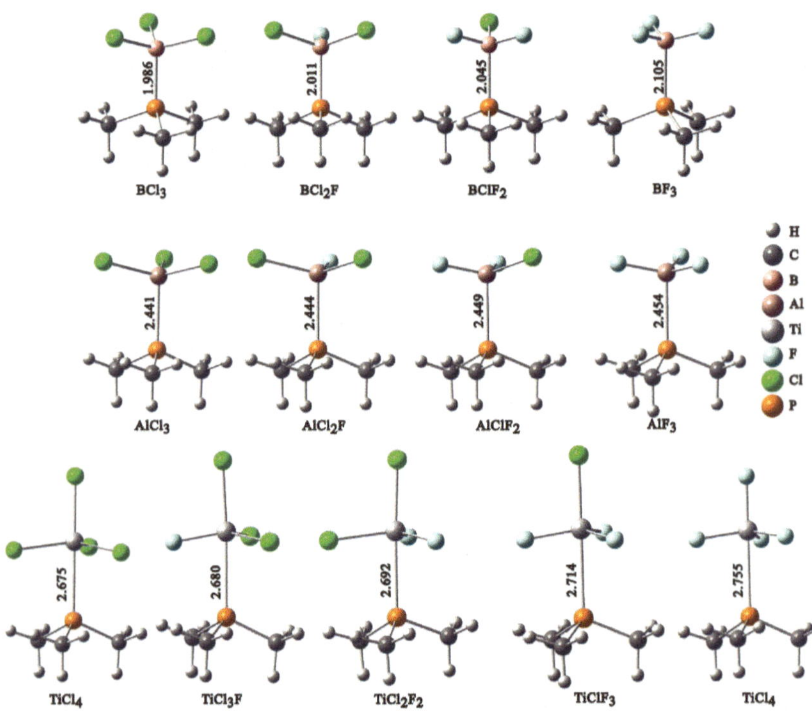

Fig. 5 Optimized geometries of various TMP-Lewis acid complexes (see text) at the HF/6-311++G(2d,2p) level. Selective bond distances (in Å) are indicated [18]. *Reproduced from Y. Chu, Z. Yu, A. Zheng, H. Fang, H. Zhang, S.J. Huang, S.B. Liu, F. Deng, Acidic strengths of Brønsted and Lewis acid sites in solid acids scaled by ^{31}P NMR chemical shifts of adsorbed trimethylphosphine, J. Phys. Chem. C 115 (2011) 7660–7667. Copyright 2011 American Chemical Society.*

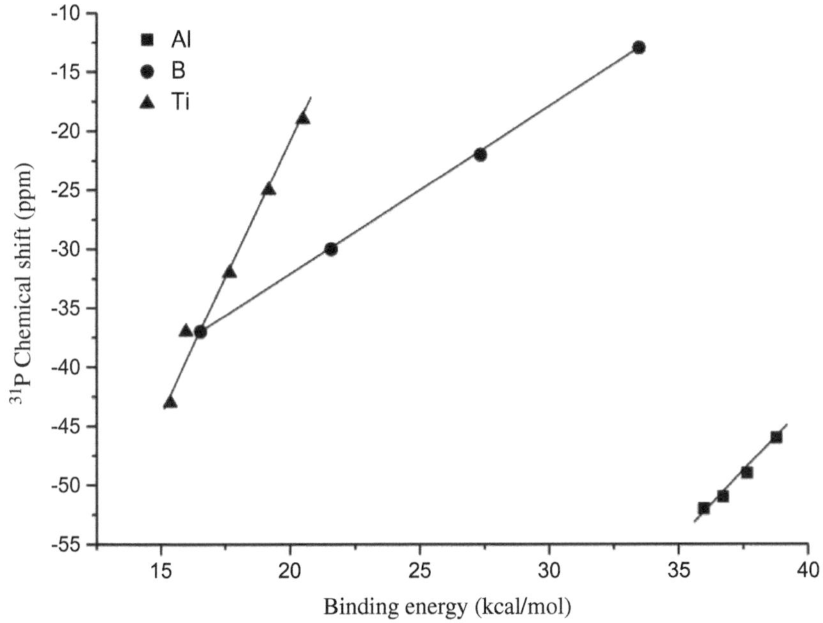

Fig. 6 Correlations of calculated ^{31}P chemical shift with corresponding binding energy of TMP adsorbed on various Lewis acid systems with B, Al, and Ti metal centres [18]. *Reproduced from Y. Chu, Z. Yu, A. Zheng, H. Fang, H. Zhang, S.J. Huang, S.B. Liu, F. Deng, Acidic strengths of Brønsted and Lewis acid sites in solid acids scaled by ^{31}P NMR chemical shifts of adsorbed trimethylphosphine, J. Phys. Chem. C 115 (2011) 7660–7667. Copyright 2011 American Chemical Society.*

on the Lewis acid site is used as a criterion for evaluating the strength of Lewis acidity. It is obvious that a linear correlation between the theoretical δ^{31}P and BE may be respectively inferred for various Lewis acid systems with varied B-, Al-, and Ti-centres [18]:

$$\begin{aligned}&BCl_nF_{3-n}\,(n=0-3):\\&\quad\delta^{31}P=1.41(\pm0.01)-60.50(\pm0.33)\times BE;R^2=1.00\end{aligned}\tag{1}$$

$$\begin{aligned}&AlCl_nF_{3-n}\,(n=0-3):\\&\quad\delta^{31}P=2.15(\pm0.20)-129.64(\pm7.46)\times BE;R^2=0.99\end{aligned}\tag{2}$$

$$\begin{aligned}&TiCl_nF_{4-n}\,(n=0-4):\\&\quad\delta^{31}P=3.37(\pm0.86)-93.28(\pm16.00)\times BE;R^2=0.91\end{aligned}\tag{3}$$

Accordingly, such linear correlations facilitate utilization of the ^{31}P-TMP approach for quantitative characterization of Lewis acidic strength. That the $\delta^{31}P$ observed for the TMP-Lewis acid complexes indeed span over a wider range compared to that obtained from the ^{31}P-TMPO and ^{13}C-acetone NMR approaches, indicating that the ^{31}P-TMP approach is indeed a more sensitive technique for characterization of Lewis acidity [18]. Unfortunately, most metallic centres in a wide variety of solid Lewis acid systems tend to interact strongly with the adsorbed TMP molecule. Such complex interactions, in turn, make a universal correlation between the observed $\delta^{31}P$ and Lewis acidic strength rather impossible for various metallic Lewis centres. This, together with the fact that volatile TMP is more cumbersome to handle in terms of sample preparation represent the major shortcomings of the ^{31}P-TMP NMR approach.

2.2 ^{31}P-R_3PO approach

As mentioned earlier, the ^{31}P-TMP NMR approach is more useful for differentiating the acid types (i.e. Brønsted and/or Lewis) and subtle differences in Lewis acidic strength in solid acids, but, less sensitive for Brønsted acidity. It is well-known that the distribution and acidic strength of Brønsted acid sites are inhomogeneous in most solid acids, however, they normally dictate the catalytic performance of the catalyst. In this context, trialkylphosphine oxide (R_3PO; $R=C_nH_{2n+1}$, $n=1$, 2, 4, 8) molecules, including trimethylphosphine oxide (TMPO) and its homologues viz. triethylphosphine oxide (TEPO), tributylphosphine oxide (TBPO), and trioctylphosphine oxide (TOPO), are highly desirable for probing the strength and distribution of acid sites in solid catalysts [26,27]. Comparing to the ^{31}P-TMP approach, the ^{31}P-R_3PO approach has been shown to be superior in terms of identifying variations of Brønsted acid strengths. This is due to the fact that phosphorus oxide probe molecules tend to interact with the Brønsted acid proton to form R_3POH^+ complexes, which exhibit a wide range of $\delta^{31}P$ spanning over ca. 50–98 ppm with the varied acidities. Another unique advantage of the ^{31}P-R_3PO NMR approach is that the observed $\delta^{31}P$ is directly proportional to the acidic strength of the catalyst [27].

Previously, comprehensive theoretical studies have been made to explore the correlations between the Brønsted acidic strength and $\delta^{31}P$ of adsorbed R_3PO by DFT calculations [21,28]. Accordingly, the adsorption structures and $\delta^{31}P$ of R_3PO adsorbed on Brønsted acid protons with varied acid strengths (from weak, medium, strong, to super acid) were carried out on a 8 T zeolite cluster model by varying the Si–H bond distance (Fig. 7A). Interestingly, linear correlations between the $\delta^{31}P$ of homologous R_3PO probe molecules and the proton affinity (PA; in kcal/mol) were obtained,

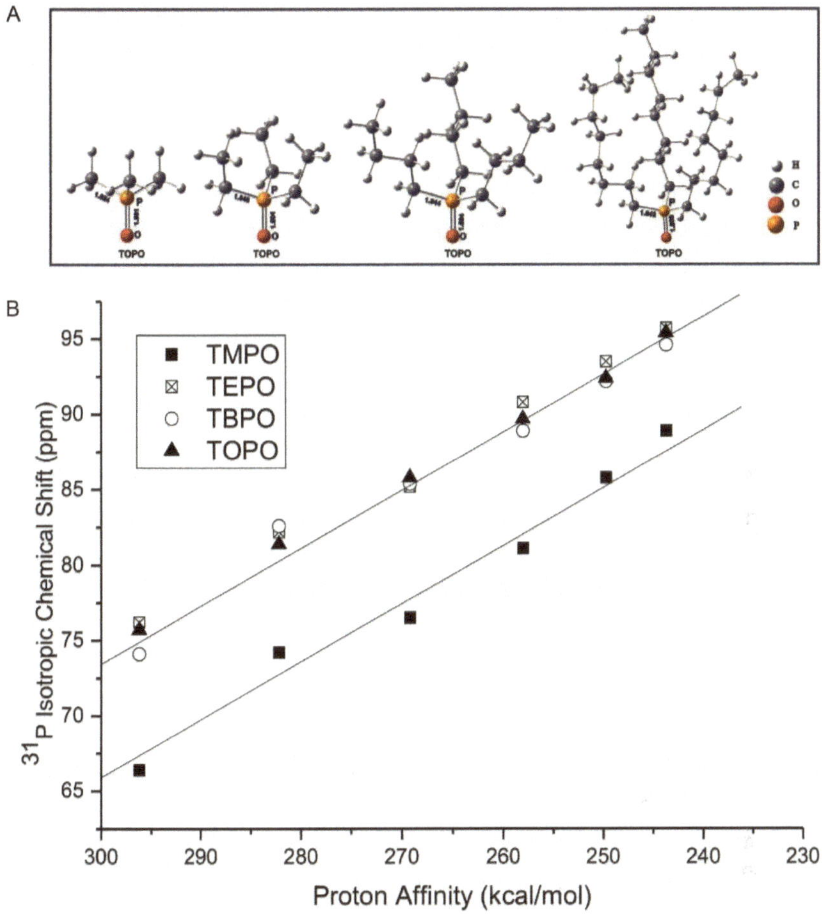

Fig. 7 (A) Optimized equilibrium configurations of free TMPO, TEPO, TBPO, and TOPO probe molecules. Selected interatomic distances (in Å) are indicated and (B) correlations of calculated ^{31}P chemical shift of adsorbed R_3POH^+ complexes and proton affinity (PA) predicted based on the 8T zeolite cluster models [28]. *Reproduced from A. Zheng, S.J. Huang, W.H. Chen, P.H. Wu, H. Zhang, H.K. Lee, L.C. De Ménorval, F. Deng, S.B. Liu, ^{31}P Chemical shift of adsorbed trialkylphosphine oxides for acidity characterization of solid acids catalysts, J. Phys. Chem. A 112 (2008) 7349–7356. Copyright 2008 American Chemical Society.*

as shown in Fig. 7B. The PA value is equivalent to the deprotonation energy (DPE), that is, the energy required to rip-off the proton from a Brønsted acid site. Taking TMPO as an example, the correlation can be expressed as:

$$\text{TMPO}: \quad \delta^{31}\text{P} = 182.866(\pm 5.314) - 0.3902(\pm 0.020) \times \text{PA}; R^2 = 0.9913. \tag{4}$$

It is obvious that the linear dependences observed for TEPO, TBPO, and TOPO not only nearly overlap with one another, they also exhibit a similar slope compared to that observed for TMPO (Eq. 4). Thus, with the exception of TMPO, variations in alkyl chain length of the R_3PO ($R = C_nH_{2n+1}$, $n = 1, 2, 4, 8$) probe molecules appear to have negligible effect on the $\delta^{31}P$ (within an experimental error of ca. ± 2 ppm) observed for corresponding R_3POH^+ adsorption complexes. Note that an average offset of 8 ± 2 ppm was observed for δ_{31P} of adsorbed $(C_nH_{2n+1})_3PO$ with $n \geq 2$ (i.e. TEPO, TBPO, and TOPO) relative to TMPO ($n = 1$). On the bases of theoretical predictions and experimental results obtained from the ^{31}P-TMPO NMR approach, a threshold $\delta^{31}P$ value of ca. 86 ppm was derived for superacidity, corresponding to a threshold PA value of ca. 250 kcal/mol [21]. Consequently, a threshold $\delta^{31}P$ value of ca. 92–94 ppm may be inferred for R_3PO probes with $n \geq 2$ [28]. As such, quantitative information on acidic strength of Brønsted acid sites in solid acids may readily be attained by combining the results obtain from the ^{31}P-R_3PO NMR approach. Moreover, by incorporating spectral deconvolution with elemental analysis, quantitative information such as distribution and concentration of acid sites may also be determined [27,29].

As illustrated in Fig. 8, a standard operation procedure invoking the ^{31}P-R_3PO NMR experimental approach, including adsorption of guest molecules (such as TMPO or TBPO) on solid catalysts and relevant experimental considerations, has been explored [30,31]. Taking the adsorption of TMPO on typical 10-membered ring (10-MR; typical pore diameter ca. 0.6 nm) H-ZSM-5 zeolite as an example, the solid catalyst was first subjected to a

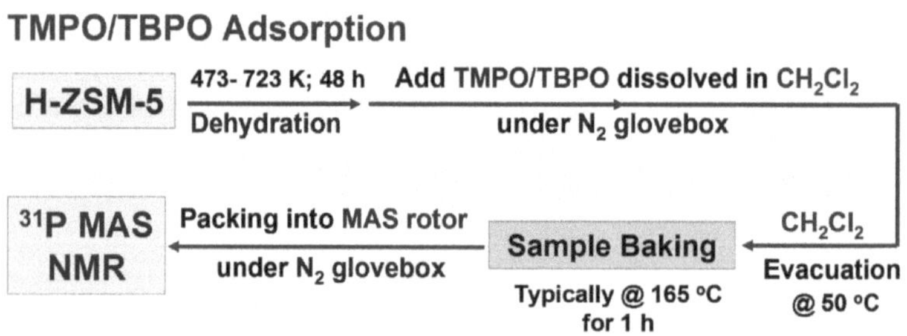

Fig. 8 Standard operation procedure invoked for the ^{31}P-R_3PO NMR approach [30,31]. *Reproduced from A. Zheng, S.J. Huang, S.B. Liu, F. Deng, Acid properties of solid acid catalysts characterized by solid-state ^{31}P NMR of adsorbed phosphorous probe molecules, Phys. Chem. Chem. Phys. 13 (2011) 14889–14901. Copyright 2011 The Royal Society of Chemistry.*

thorough dehydration treatment under desirable temperature and duration (for zeolites, typically 673–723 K for at least 12 h) in a vacuum manifold ($<10^{-5}$ Torr) prior to the adsorption of guest molecule [17]. Subsequently, a known amount of crystalline TMPO adsorbate dissolved in anhydrous CH_2Cl_2 was introduced into a vessel containing the dehydrated solid sample under inert environment (e.g. in a N_2 bag or glovebox), followed by removal of the CH_2Cl_2 solvent by evacuation at ca. 323–343 K. To ensure a uniform adsorption of adsorbate probe molecules in the pores/channels of the porous adsorbents, the sealed sample vessel was further subjected to thermal treatment at ca. 438 K (note that the melting point of TMPO is ca. 413 K) for at least 1 h. It is noteworthy that an excessive loading of adsorbate molecules is desirable in ensuring a complete coverage of all acid sites and that adequate thermal pre-treatment of the adsorbate-loaded sample is an inevitable procedure to warrant a homogeneous distribution of probe molecule on acid sites. Finally, the TMPO-loaded sample (in sealed vessel) was transferred into a MAS rotor, which is sealed by a gas-tight (Kel-F) cap, under inert environment (e.g. N_2 bag/glovebox). Regarding to the subsequent solid-state ^{31}P MAS NMR experiment, typically ca. 800–1000 free-induction-decay (FID) signals were accumulated using a single pulse sequence with a recycle delay of ca. 10 s and a sample spinning frequency of ca. 12 kHz. The above experimental parameters are of course depending on the estimated numbers of Brønsted acid sites (or Al content in zeolite) present in the intra- and/or extracrystalline surfaces of the detecting solid catalyst.

As shown in Fig. 9, an extensive investigation has been made invoking the ^{31}P-TMPO approach for acidity characterization of various solid acids, including microporous zeolites (H-ZSM-5, H-Mordenite, H-Beta, HUSY, and MCM-22), mesoporous molecular sieves (MCM-41), and metal oxides such as sulphated ZrO_2 [32]. Except for the latter, porous catalyst samples with a similar Al content (Si/Al $\sim$ 15) were chosen for comparison purpose. It is indicative that, the ^{31}P resonances and corresponding observed δ^{31}P readily render comparison of relative acid strengths in various solid catalysts. In addition, other qualitative information such as relative distribution and concentration of acid sites may be derived from the peak areas of the deconvoluted spectra. Furthermore, by combing the spectral analysis data with results obtained from elemental analysis (i.e. by ICP-MS method), quantitative information of acid sites in solid acids may also be attained, affording detailed understanding of acid features, viz. type, distribution, amount, and strength of acid sites [32].

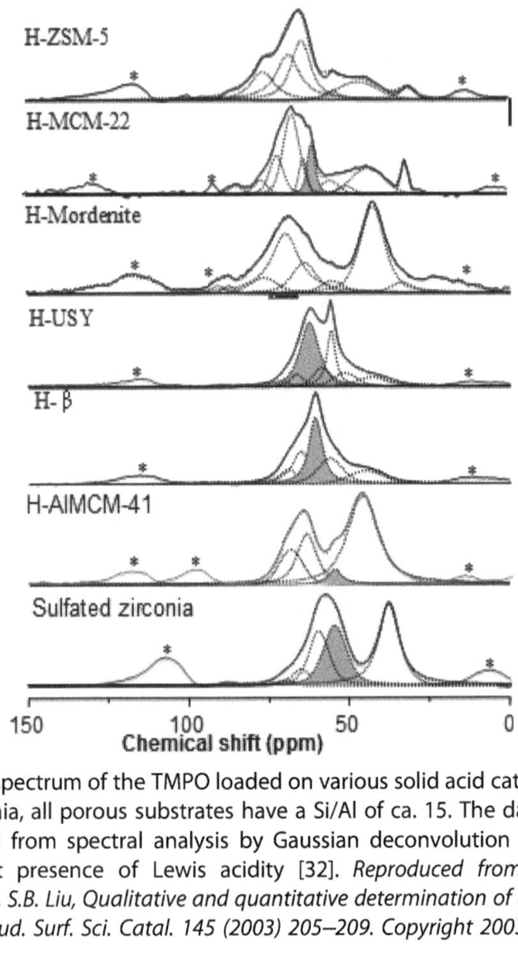

Fig. 9 ^{31}P MAS spectrum of the TMPO loaded on various solid acid catalysts. Except for sulphated zirconia, all porous substrates have a Si/Al of ca. 15. The dashed curves are results obtained from spectral analysis by Gaussian deconvolution and the shaded peaks represent presence of Lewis acidity [32]. *Reproduced from Q. Zhao, W.H. Chen, S.J. Huang, S.B. Liu, Qualitative and quantitative determination of acid sites on solid acid catalysts, Stud. Surf. Sci. Catal. 145 (2003) 205–209. Copyright 2003 Elsevier.*

2.3 Other ^{31}P SSNMR approaches using phosphorus-containing probes

In addition to the aforediscussed ^{31}P-TMP and R_3PO NMR approaches, some other phosphorus-containing probe molecules have been exploited for acidity characterization in solid acid catalysts. For example, triphenylphosphine (PPh_3) molecule has been used for probing the distribution of Brønsted acid sites in solid acids [33]. Owing to its size (ca. 1.17×0.71 nm), the bulky PPh_3 molecule is desirable for detecting acid sites presiding on the external surfaces of zeolitic catalysts with 8- and 10-MR pore channels. Related applications of the ^{31}P-PPh_3 NMR approach will be illustrated in a later section. Apart from the properties of isolated acid sites, proximities of acid centres also play a crucial role during catalytic reactions that involve multiple acid sites. In this context, the homologous series of diphenyldiphosphines ($Ph_2P(CH_2)_nPPh_2$; $n=$integer) whose physical size may be tuned by varying the $-CH_2-$ group in the axial position may also be used to determine proximities of Brønsted acid centres in solid acid catalysts. For examples, a P–P distance of 0.30, 0.56, and 0.94 nm was found for $Ph_2P(CH_2)_nPPh_2$ with $n=1$, 3, and 6, respectively. Accordingly, Grey and co-workers utilized ^{31}P–^{31}P DQ NMR [34] of adsorbed diphenyldiphosphine (Ph_2P-CH_2-PPh_2) to determine distance between adjacent Brønsted acid sites in solid catalysts [35]. The applicability of the above technique is, of course, relying on the fix distance between two phosphorus atoms in the probe molecule. The above notion was further justified by the 2D DQ-SQ (double quantum-single quantum) correlation NMR technique [34], which is capable of probing the spatial proximity of two like-spins. Accordingly, by comparing the ^{31}P NMR spectra obtained from different diphenyldiphosphines probe molecules with varied P–P distances and loadings, the distance and concentration of acidic sites in zeolitic catalysts may readily be determined (vide infra) [35].

3. Applications of ^{31}P SSNMR for acidity characterization

3.1 Discernment of acid types

As mentioned earlier, the ^{31}P-TMP NMR approach is a reliable technique for probing Lewis acidity in solid acid catalysts [9,10,18]. As shown in Fig. 10, the ^{31}P MAS NMR spectra of TMP adsorbed on parent and dealuminated H-form Y zeolites revealed a series of characteristic $\delta^{31}P$ peaks at 28, −5, −33, −47, and −57 ppm. For example, the parent HY with Si/Al = 2.8 exhibited a singlet resonance at −5 ppm, which may be unambiguously assigned due to the formation of $TMPH^+$ ionic complex associated with Brønsted acid sites. On the other hand, multiple resonance peaks at 28, −33, −47, and −57 ppm were observed for TMP adsorbed on dealuminated HY-d450 zeolite sample. Among them, the weak signal at ca. 28 ppm has been assigned to $P(CH_3)_4^+$ species formed during the disproportionation of TMP probe molecules [37], whereas peaks at −33 and −47 ppm were attributed to TMP bound to Lewis acid sites with different acid strengths [18]. That the signal at −33 ppm whose $\delta^{31}P$ exceeds that of superacidic $AlCl_3$ (−41 ppm) [37] indicates the presence of ultra-strong Lewis acid site in dealuminated HY-450 zeolite. Likewise, similar peak assignments may also be inferred for the parent HUSY (Si/Al = 3.5) and dealuminated

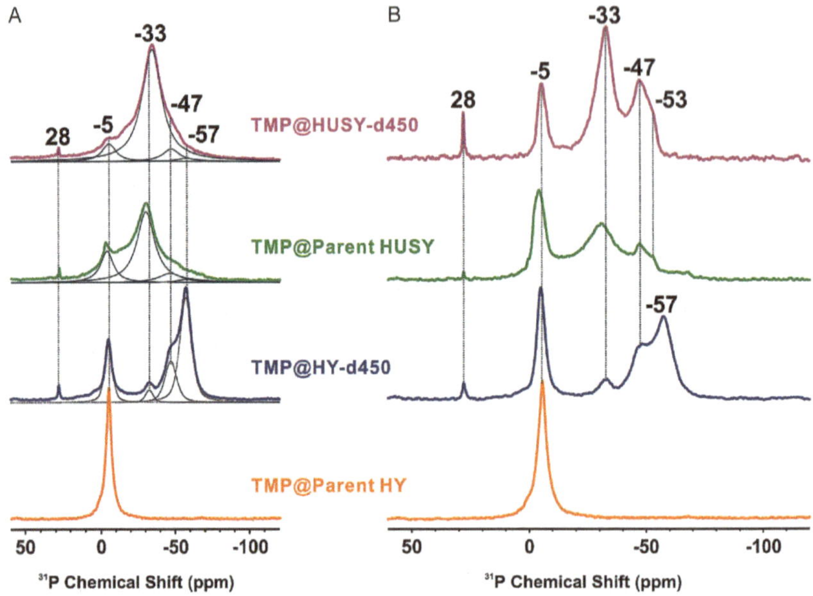

Fig. 10 ^{31}P MAS NMR spectra of TMP adsorbed on parent HY (Si/Al = 2.8) and HUSY (Si/Al = 3.5) zeolites and their dealuminated counterparts (which were calcined in air at 450 °C) obtained (A) without and (B) with cross-polarization (CP; contact time = 4.0 ms). The dashed lines in (A) represent Gaussian deconvolution results [36]. *Reproduced from X. Yi, K. Liu, W. Chen, J. Li, S. Xu, C. Li, Y. Xiao, H. Liu, X. Guo, S.B. Liu, A. Zheng, Origin and structural characteristics of tri-coordinated extra-framework aluminum species in dealuminated zeolites, J. Am. Chem. Soc. 140 (2018) 10764–10774. Copyright 2018 American Chemical Society.*

HUSY-450 zeolites. It is noteworthy that the assignment of Lewis acidity may further be confirmed by partial hydration of the sample, as illustrated in Fig. 11 for the dealuminated HUSY-d450. Owing to their very high affinity for water, Lewis acid sites tend to transform into Brønsted acidity upon hydration, leading to diminishing of ^{31}P resonances associated with Lewis acid sites, i.e. peaks with $\delta^{31}P$ of −33 and −47 ppm. Thus, it is indicative that dealumination treatment indeed provokes formation of extraframework Al (EFAL) species in zeolites with characteristics of Lewis acidity [36–39].

The assignments of the ^{31}P NMR resonances observed for the parent and dealuminated HY and HUSY zeolites in Fig. 10 may further be verified by using the 2D ^{1}H–^{31}P heteronuclear correlation (HETCOR) NMR technique [40]. As shown in Fig. 12A, the HETCOR spectrum of the dealuminated HY-d450 recorded with a short contact time (τ_c = 0.1 ms)

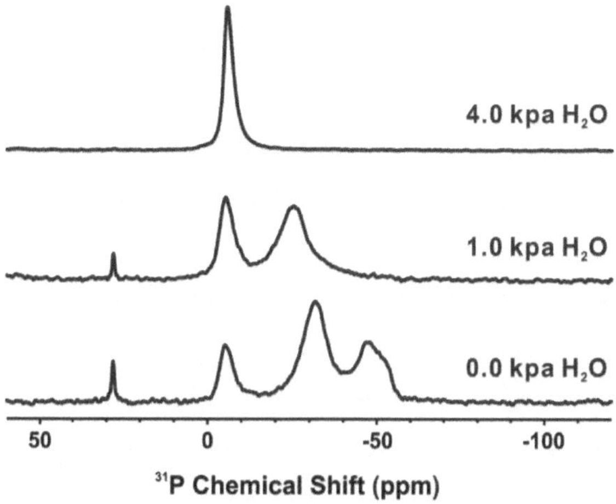

Fig. 11 ^{31}P MAS NMR spectra of TMP adsorbed on dealuminated HUSY-d450 zeolite sample with varied H_2O loadings [36]. *Reproduced from X. Yi, K. Liu, W. Chen, J. Li, S. Xu, C. Li, Y. Xiao, H. Liu, X. Guo, S.B. Liu, A. Zheng, Origin and structural characteristics of tri-coordinated extra-framework aluminum species in dealuminated zeolites, J. Am. Chem. Soc. 140 (2018) 10764–10774. Copyright 2018 American Chemical Society.*

revealed two correlation peaks at (−5, 6.0) and (−5, 7.1) ppm. These correlation peaks are clearly associated with TMP adsorbed on bridge hydroxyl (Si-OH-Al) protons (i.e. Brønsted acid sites) to form $TMPH^+$ complexes. Upon increasing contact time, progressive increases in correlation peaks were observed. The spectrum recorded with a prolonged contact time ($\tau_c = 4.0\,ms$; Fig. 12D) revealed additional correlation peaks. The additional peak at (−5, 1.7) ppm was attributed to correlation of $TMPH^+$ with ^1H resonance of methyl groups ($\delta^1H = 1.7$ ppm). While correlation peaks at (−33, 1.7) and (−47 ~ − 53, 1.4) ppm were attributed to TMP adsorbed on Lewis acid sites associated with the methyl protons and that at (−56 ~ − 60, 1.4) ppm was due to TMP adsorbed on weakly acidic Si-OH groups. On the other hand, the correlation peak at (−56, 6.9) ppm was ascribed due to hydrogen-bonding interactions [36]. As such, three possible interaction schemes may be envisaged for the adsorption of TMP on dealuminated zeolites, namely chemisorbed on Brønsted acid sites (i.e. formation of $TMPH^+$ complexes), chemisorbed on Lewis acid sites, and physisorbed on hydroxyl groups (i.e. hydrogen-bonding interactions), as illustrated in Fig. 13.

Although the ^{31}P-TMP NMR approach is useful for differentiating acidic strength of Lewis acid sites, the ^{31}P-R_3PO NMR approach is more desirable for probing Brønsted acidity. In other words, in terms of validity of

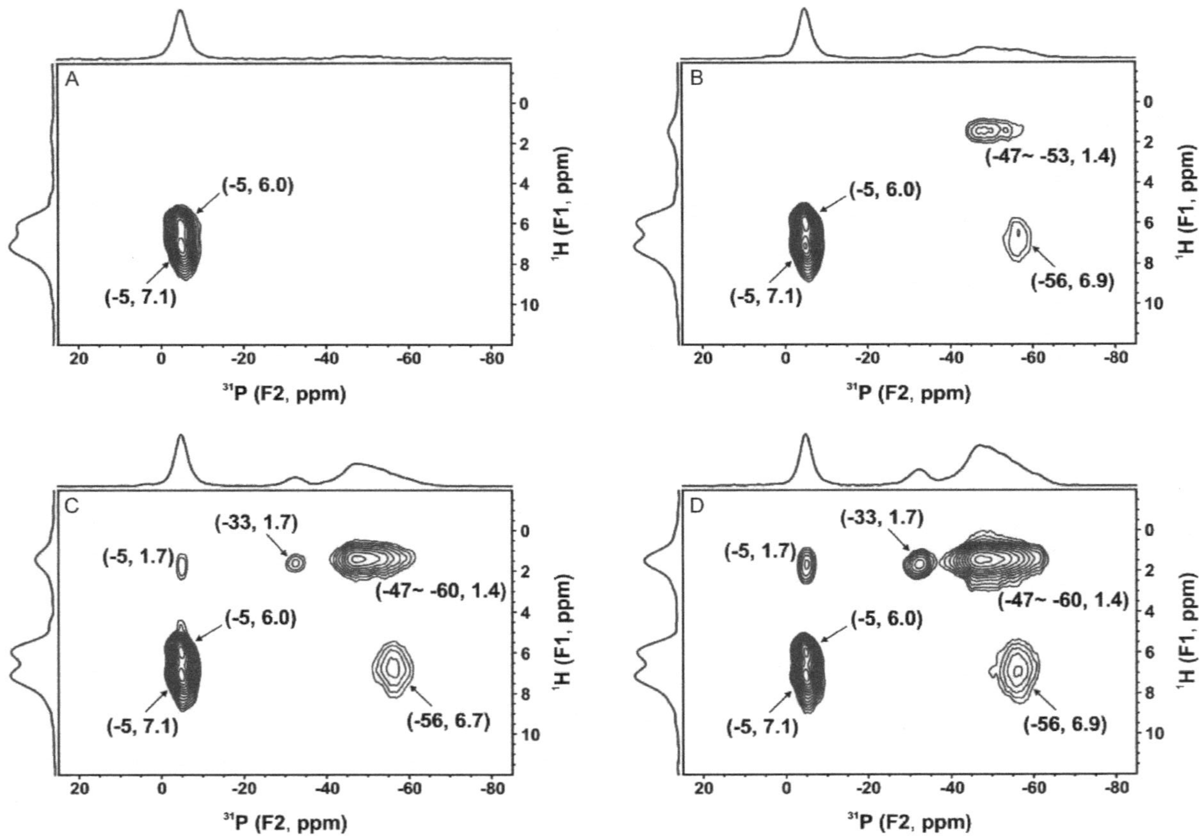

Fig. 12 2D ^{1}H–^{31}P HETCOR NMR spectra of TMP adsorbed on dealuminated HY-d450 recorded with a contact time of (A) 0.1 ms, (B) 1.0 ms, (C) 2.0 ms, and (D) 4.0 ms [36]. *Reproduced from X. Yi, K. Liu, W. Chen, J. Li, S. Xu, C. Li, Y. Xiao, H. Liu, X. Guo, S.B. Liu, A. Zheng, Origin and structural characteristics of tri-coordinated extra-framework aluminum species in dealuminated zeolites, J. Am. Chem. Soc. 140 (2018) 10764–10774. Copyright 2018 American Chemical Society.*

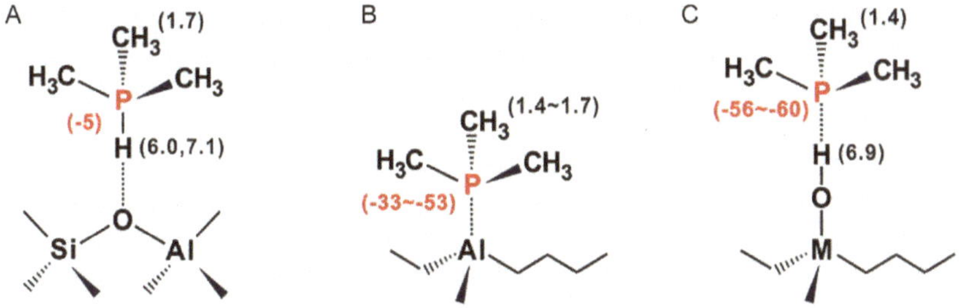

Fig. 13 Possible schemes for adsorption of TMP on dealuminated zeolites: (A) chemisorbed on Brønsted acid sites, (B) chemisorbed on Lewis acid centres, and (C) physisorbed on hydroxyls (hydrogen-bonding interactions; M = Al or Si). The corresponding ^{1}H and ^{31}P NMR chemical shifts (in ppm) obtained from the NMR experiments are also shown [36]. *Reproduced from X. Yi, K. Liu, W. Chen, J. Li, S. Xu, C. Li, Y. Xiao, H. Liu, X. Guo, S.B. Liu, A. Zheng, Origin and structural characteristics of tri-coordinated extra-framework aluminum species in dealuminated zeolites, J. Am. Chem. Soc. 140 (2018) 10764–10774. Copyright 2018 American Chemical Society.*

probe molecule, TMP is less sensitive for probing Brønsted acid sites while R_3PO molecules are inferior for detecting Lewis acidity. Since volatile TMP is more cumbersome to handle in terms of sample preparation, the ^{31}P-R_3PO approach is more preferable in this context. Among them, TMPO represents the most popular probe molecule chosen owing to its suitable dimension (vide supra). In practice, differentiation of Brønsted vs. Lewis acidity invoking the ^{31}P-TMPO NMR approach is normally carried out by comparing results obtained from experiments performed on fully dehydrated as well as partially hydrated samples [29]. It is known that Lewis acid sites (tri-coordinated Al species in zeolites), which interact weakly with TMPO, may easily react with H_2O to form weak Brønsted acid sites. However, the strong hydrogen bonding between TMPO and the Brønsted acid sites is unlikely to be dissociated while in the presence of H_2O. Thus, upon progressive exposure of the TMPO-loaded samples to humidity, ^{31}P resonances associated with Lewis acid sites will be diminished eventually, while those bound to Brønsted acid sites will remain intact. As such, this offers an indirect method to identify the presence of Brønsted and/or Lewis acidity in the solid catalyst. For illustration, Fig. 14 displays the ^{31}P MAS NMR spectra of TMPO adsorbed on dehydrated ZrO_2 catalyst before (Fig. 14A) and after (Fig. 14B and C) partial hydration treatments [29]. Four distinct resonances at 62, 53, 41, and 34 ppm were observed for fully dehydrated pure ZrO_2. While the latter two peaks with δ^{31}P lower than 50 ppm are unambiguously due to physisorbed TMPO, the

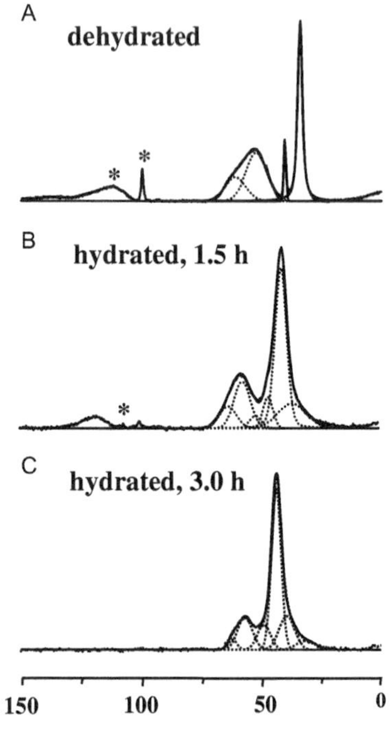

Fig. 14 ^{31}P MAS NMR spectra of TMPO adsorbed on (A) fully dehydrated and (B and C) partially hydrated ZrO_2. The dashed curves represent spectral simulation results and the asterisks denote spinning sidebands (spinning rate 12kHz) [29]. *Reproduced from W.H. Chen, H.H. Ko, A. Sakthivel, S.J. Huang, S.H. Liu, A.Y. Lo, T.C. Tsai, S.B. Liu, A solid-state NMR, FT-IR and TPD study on acid properties of sulfated and metal-promoted zirconia: Influence of promoter and sulfation treatment, Catal. Today 116 (2006) 111–120. Copyright 2006 Elsevier.*

other two peaks at 62 and 53 ppm may be, a priori, assigned due to TMPO adsorbed on Brønsted and/or Lewis acid sites. Upon progressive exposure of the TMPO-loaded ZrO_2 sample to humidity, the peak intensities at 62 and 53 ppm decrease with increasing severity of hydration (i.e. H_2O content) while the peaks responsible for physisorbed TMPO (at 41 and 34 ppm) are broadened and shifted towards downfield (i.e. higher $\delta^{31}P$) direction. Accordingly, the two peaks at 62 and 53 ppm observed for the parent ZrO_2 sample can thus be ascribed due to interaction of TMPO with Lewis sites with different acidic strengths, in excellent agreement with earlier reports by IR spectroscopy of adsorbed CO [41–43]. Accordingly, detailed Brønsted and Lewis acid properties of heteroatom doped zeolite catalysts, such as the H-[B]-ZSM-5 borosilicate, have also been clearly identified [44].

More recently, the ^{31}P-TMPO NMR approach has also been adopted to explore detailed Brønsted and Lewis acid properties of dehydrated H-ZSM-5 zeolites (Si/Al=25) [45], which revealed up to six characteristic ^{31}P peaks at 88, 85, 76, 69, 65, and 51 ppm (Fig. 15). For dehydrated H-ZSM-5 zeolite with a relative low TMPO loading (P/Al=0.18), the corresponding 2D $^{31}P\{^{1}H\}$ HETCOR spectrum (Fig. 15E) exhibited two correlation peaks centred at (76, 12.2) and (88, 6.4) ppm, which may be assigned due to distinct protonated $TMPOH^+$ ionic pair complexes. In this context, since a δ^{31}P value of ca. 86 ppm has been identified as the threshold of superacidity (i.e. acid strength comparable to that of pure H_2SO_4) [21], the ^{31}P resonance at 88 ppm confirmed the presence of super-acidic acid sites in dehydrated ZSM-5 zeolites. However, since no correlation peak was observed for the ^{31}P signals around 65 ppm with Brønsted acidic protons (i.e. $TMPOH^+$ complexes), the ^{31}P resonance observed at 65 ppm therefore may not be accountable for Brønsted acid sites. Additional study performed on rehydrated TMPO-loaded ZSM-5 sample revealed a notable decline in relative intensity of the signals at 65 and 69 ppm (Fig. 15B and D). In particular, both signals at 88 and 85 ppm originally observed for dehydrated samples with different TMPO loadings (Fig. 15A and C, respectively) disappeared completely after sample rehydration. Meanwhile, a notable increase in peak intensity of the resonance with $\delta^{31}P=51$ ppm due to TMPO adsorbed on weak acid sites was observed. Thus, both ^{31}P signals at 65 and 85–88 ppm should either be directly or indirectly associated with Lewis acid sites. Since the peak at ca. 88 ppm has been previously ascribed to TMPO adsorbed on Brønsted acid sites, it is anticipated that the presence of such ultra-strong Brønsted acidity may be due to synergistic effect with neighbouring Lewis acid sites.

To further verify correlations between the adsorbed TMPO and various Al species in dehydrated H-ZSM-5 zeolites, additional experiments invoking 2D ^{31}P–^{27}Al population-transfer dipolar-mediated heteronuclear multiple quantum coherence (PT-D-HMQC) spectroscopy were conducted [46–48]. As shown in Fig. 16B, the 2D $^{31}P\{^{27}Al\}$ PT-D-HMQC MAS NMR spectrum recorded at 11.7 T for the TMPO-adsorbed H-ZSM-5 zeolite (P/Al=0.42) showed two notable correlation regions. The correlation peak centred at (76, 54) ppm should be associated with TMPO adsorbed on Brønsted acid sites. Whereas correlation features associated with $\delta^{31}P=65$–69 ppm and the broad ^{27}Al resonances (which originated from distorted four-coordinated aluminium species) indicate the presences of two types of Lewis acid sites with different acidic strengths, as further

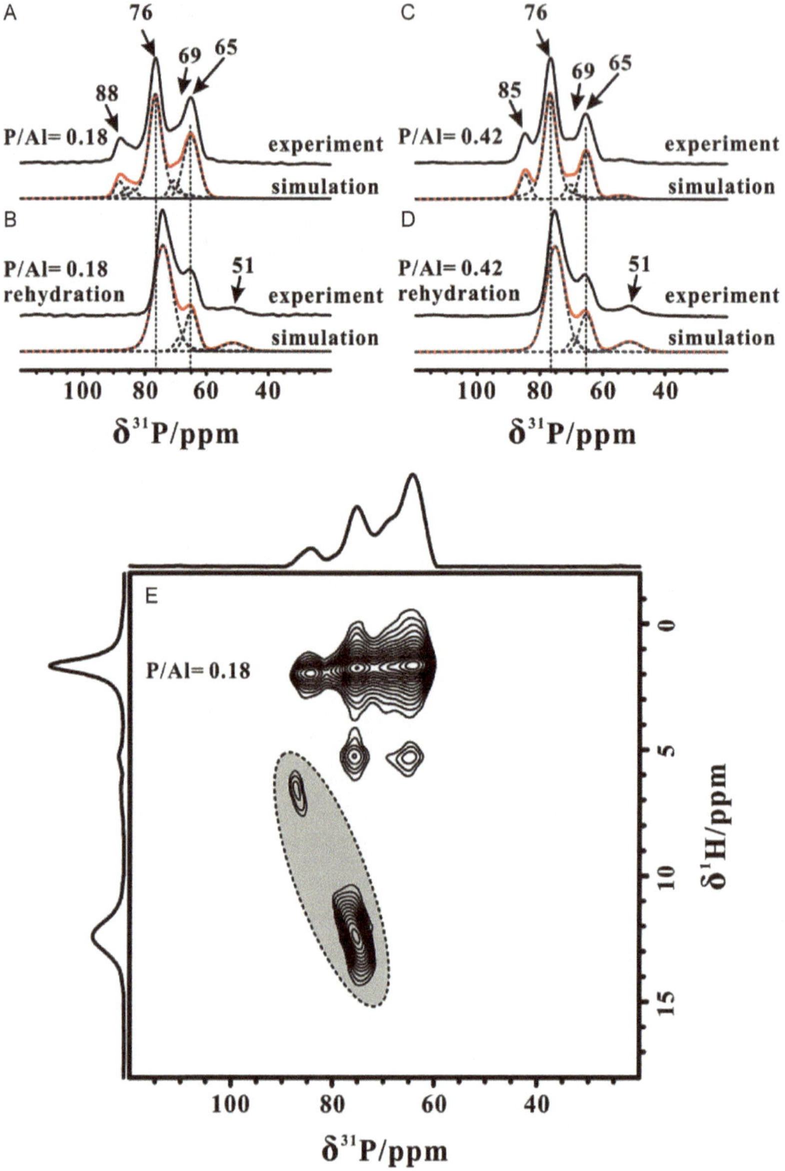

Fig. 15 1D one-pulse ^{31}P MAS spectra of dehydrated H-ZSM-5 zeolite with (A, C) different TMPO loadings and (B, D) subsequent rehydration, and (E) corresponding 2D $^{31}P\{^1H\}$ CP HETCOR MAS NMR spectrum observed for sample with a TMPO loading of P/Al = 0.18 [45]. *Reproduced from S. Xin, Q. Wang, J. Xu, Y. Chu, P. Wang, N. Feng, G. Qi, J. Trébosc, O. Lafon, W. Fan, F. Deng, The acidic nature of "NMR-invisible" tricoordinated framework aluminum species in zeolites, Chem. Sci. 10 (2019) 10159–10169. Copyright 2019 The Royal Society of Chemistry.*

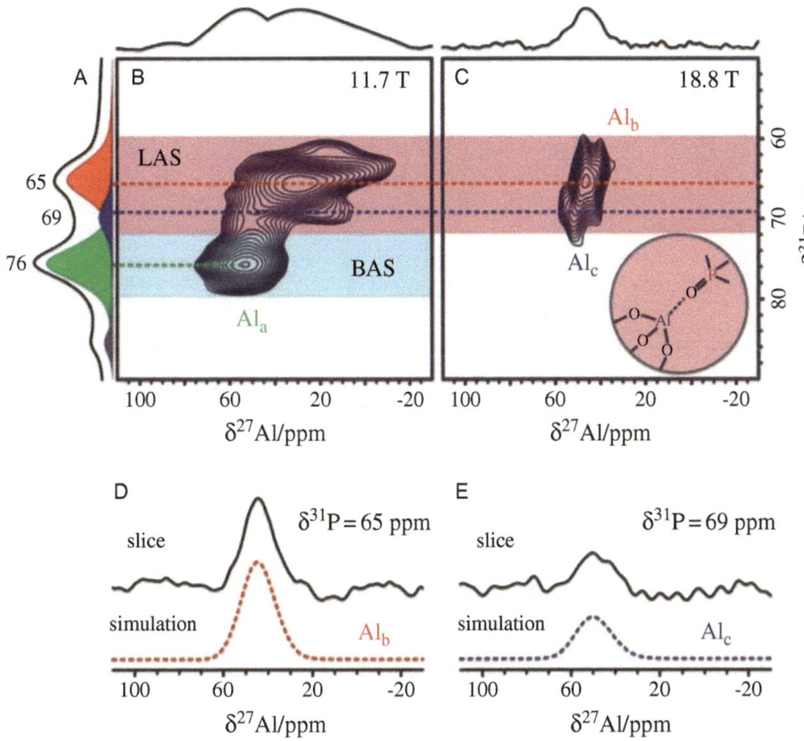

Fig. 16 (A) 1D one-pulse ^{31}P and (B) 2D $^{31}P\{^{27}Al\}$ PT-D-HMQC MAS NMR spectra recorded at 11.7T, and (C) 2D $^{27}Al\{^{31}P\}$ D-HMQC MAS NMR spectrum recorded at 18.8T observed for the dehydrated H-ZSM-5 zeolite (Si/Al = 25) with adsorbed TMPO (P/Al = 0.42). (D and E) Selected slices are also shown with corresponding simulated spectra [45]. *Reproduced from S. Xin, Q. Wang, J. Xu, Y. Chu, P. Wang, N. Feng, G. Qi, J. Trébosc, O. Lafon, W. Fan, F. Deng, The acidic nature of "NMR-invisible" tricoordinated framework aluminum species in zeolites, Chem. Sci. 10 (2019) 10159–10169. Copyright 2019 The Royal Society of Chemistry.*

supported by the spectrum in Fig. 16C, which was recorded by using 2D $^{27}Al\{^{31}P\}$ D-HMQC NMR method at a higher magnetic field (18.8 T).

In addition, symmetry-based resonance-echo saturation-pulse double-resonance (S-RESPDOR) technique [49] was employed to estimate the ^{31}P–^{27}Al internuclear distance through heteronuclear dipolar interactions. This was accomplished by performing the experiments on a sample with relatively low TMPO loading and by analysing the signal fraction $\Delta S/S_0 = (S_0 - S')/S_0$ as a function of recoupling time (Fig. 17A and B), where S' and S_0 represent the signal intensity with and without dipolar dephasing, respectively. As a result, a P–Al distance of 2.9 ± 0.2 Å was derived, which

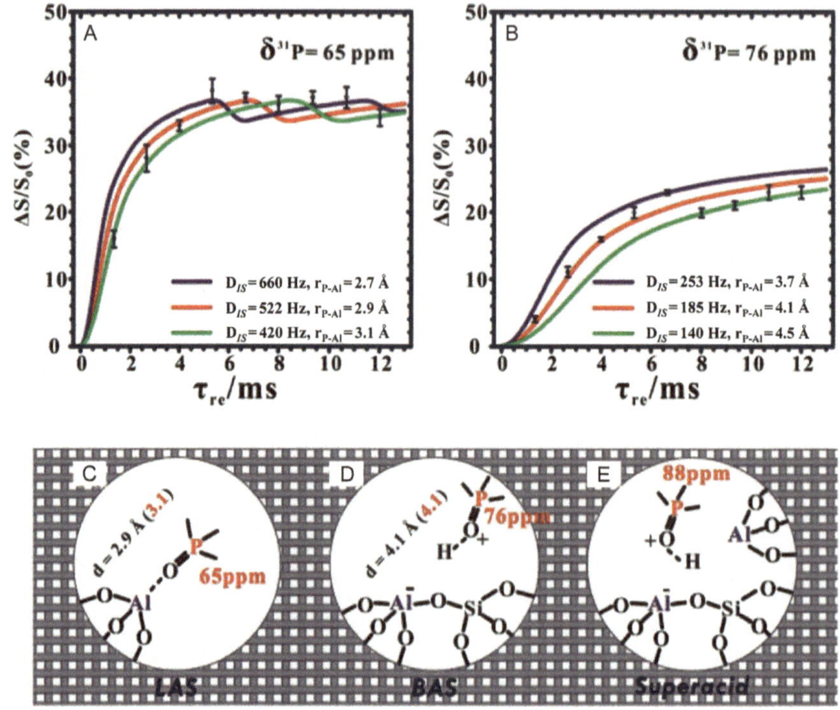

Fig. 17 $^{31}P\{^{27}Al\}$ S-RESPDOR built-up curves of TMPO (loading P/Al = 0.18) adsorbed on (A) Lewis (tri-coordinated FAL) and (B) Brønsted acid sites of dehydrated H-ZSM-5 zeolite. D_{IS} and $r_{P–Al}$ represent the dipolar interaction constant and internuclear distance of the $^{31}P–^{27}Al$ spin pair, respectively. Plausible adsorption structure models for TMPO adsorbed on (C) LAS (tri-coordinated FAL), (D) BAS (bridging hydroxyls), and (E) a BAS with a tri-coordinated FAL in close proximity. The numbers in parentheses were predicted P–Al distances obtained from DFT calculations [45]. *Reproduced from S. Xin, Q. Wang, J. Xu, Y. Chu, P. Wang, N. Feng, G. Qi, J. Trébosc, O. Lafon, W. Fan, F. Deng, The acidic nature of "NMR-invisible" tricoordinated framework aluminum species in zeolites, Chem. Sci. 10 (2019) 10159–10169. Copyright 2019 The Royal Society of Chemistry.*

was rationalized in terms of a structure model of TMPO adsorbed on Lewis acid sites (LAS) associated with tri-coordinated framework Al (FAL) species (i.e. peak with $\delta^{31}P = 65$ ppm; Fig. 17A). Likewise, a P–Al distance of 4.1 ± 0.4 Å was determined for TMPO adsorbed on Brønsted acid sites (BAS, i.e. peak with $\delta^{31}P = 76$ ppm; Fig. 17B). In the former scenario, formation of distorted tetrahedral FAL species are highly anticipated upon adsorption of the TMPO probe molecule on tri-coordinated FAL. The $^{31}P–^{27}Al$ distances so extracted were in good agreement with those by theoretical DFT calculations, in which a P–Al distance predicted for the

TMPO-Lewis acid [$(CH_3)_3PO\ldots Al\equiv$] and TMPO-Brønsted acid [$(CH_3)_3PO\ldots H\text{–}OAl\equiv$] complexes were 3.1 and 4.1 Å, respectively, as illustrated in Fig. 17C and D.

Aside from the more sophisticated ^{31}P SSNMR approaches available for indirect detection of acid types in solid acid catalysts (vide supra), direct identification of Brønsted and Lewis acid sites would be highly desirable. In this regard, Huang et al. developed a novel SSNMR technique [50] for direct discernment of Brønsted and Lewis acid sites in solid acids through simple spectral editing. This is accomplished by incorporating a selective ^{1}H excitation pulse on non-acidic protons while observing the ^{31}P resonance in conjunction with the Lee-Goldburg cross-polarization (LG-CP) method [51,52], which effectively suppress homonuclear dipole-dipole couplings. As shown in Fig. 18A, five distinct ^{31}P resonances at 41, 45, 55, 65, and 75 ppm were observed for TMPO loaded on HY zeolite. Based on the results obtained from additional experiments using the $^{1}H/^{31}P/^{27}Al$ TRAPDOR technique [53], the authors were able to assign most of the ^{31}P resonance signals in Fig. 18A. Peaks V (75 ppm) and I (41 ppm) are ascribed due to TMPO adsorbed on strong Brønsted acid sites and physisorbed TMPO, respectively [54]. However, the assignment of peaks II (45 ppm), III (55 ppm), and IV (65 ppm) remained ambiguous. In the selective LG-CP experiment, a Gaussian-shaped pulse was used to selectively suppress the ^{1}H signal at 2.1 ppm (i.e. the anticipated $\delta^{1}H$ for non-acidic Si-OH groups in the HY zeolite), while keeping the ^{1}H resonance at 7.2 ppm (arising from Brønsted acid sites) unperturbed. Accordingly, the $^{31}P\{^{1}H\}$ selective LG-CP spectrum can thus be compared with the spectrum obtained from conventional LG-CP sequence in the absence of ^{1}H selective excitation to render further assignment of the ^{31}P resonances, as shown in Fig. 18B. Note that since the spectrum obtained from selective LG-CP exhibited much weaker signal sensitivity compared to that of regular LG-CP, it was scaled up intentionally so that the most intense signal at 75 ppm (peak V) may be diminished in the difference spectrum. As a result, three ^{31}P resonance peaks at 65, 55, and 41 ppm were observed in the difference spectrum (Fig. 18B). A priori, it is anticipated that only ^{31}P signals arising from TMPO adsorbed on Lewis acid sites, silanol sites, or Brønsted acid sites which have different polarization transfer dynamics would remain in the difference spectrum. Thus, the absence of signal at 45 ppm in the difference spectrum indicates that peak II (see Fig. 18A) should be solely due to TMPO adsorbed on Brønsted acid sites (with somewhat weak acidic strength). Whereas the broader peak centred at ca. 41 ppm is attributed to

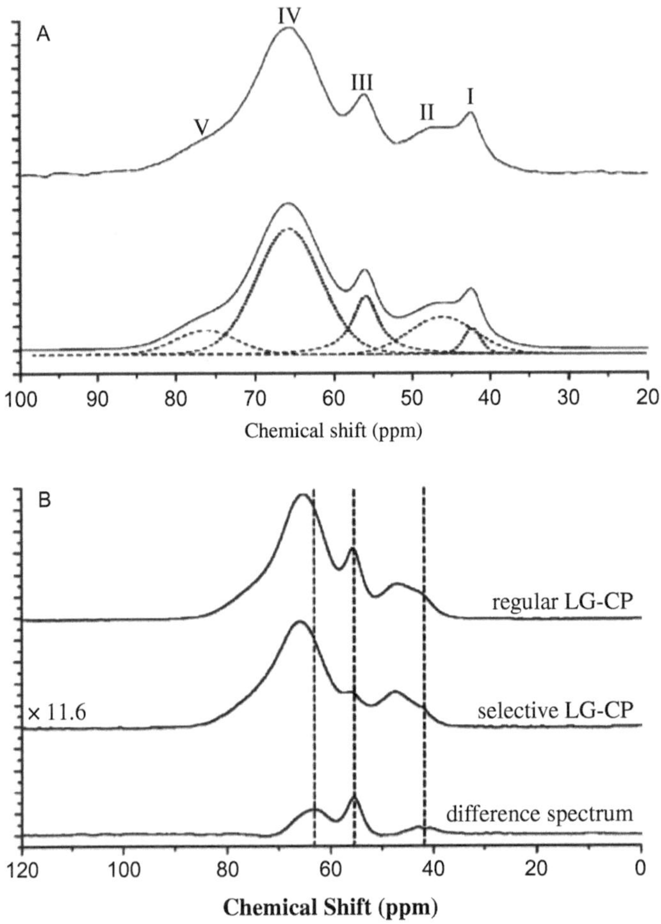

Fig. 18 (A) Top: ^{31}P MAS NMR spectrum of TMPO adsorbed on HY zeolite (Si/Al = 2.6). Bottom: deconvoluted spectrum via Gaussian simulations. (B) ^{31}P{^{1}H} LG-CP spectra with and without selective excitation of bridging hydroxyl protons [50]. *Reproduced from S.J. Huang, Y.H. Tseng, Y. Mou, S.B. Liu, S.H. Huang, C.P. Lin, J.C.C. Chan, Spectral editing based on selective excitation and Lee-Goldburg cross-polarization under magic angle spinning, Solid State Nucl. Magn. Reson. 29 (2006) 272–277. Copyright 2006 Elsevier.*

TMPO interacting with silanol groups. Thus, it is indicative that peaks III and IV, which also present in the difference spectrum, should be associated with Lewis acidity. In view of their difference in peak intensity, the authors concluded that the peak at 55 ppm should be mainly due to TMPO interacting with Lewis acid sites but partially with Brønsted acidity, and vice versa for the peak at 65 ppm. Nonetheless, this does not exclude the scenarios of

TMPO interacting with Brønsted acid sites residing in the sodalite cages (which are inaccessible to probe molecules) of HY zeolite, forming weak hydrogen bonding through the narrow windows.

Adopting the similar spectral editing scheme, Alonso et al. demonstrated that chemical environments of molecular adsorbate on surfaces of mixed metal oxide catalysts may also be attained by means of ^{31}P-TMPO NMR approach [55]. As shown in Fig. 19(a1), when TMPO is adsorbed on precipitated silica-alumina (Al_2O_3 content 5 wt%), two main signals with δ^1H of ca. 1.5 and 7.0 ppm were observed, which may be assigned due to methyl protons on TMPO and Brønsted acid protons (Si-OH-Al) bound to TMPO, respectively. On the other hand, the corresponding ^{31}P{^{1}H} CP MAS NMR spectrum exhibited four main features, which may be attributed to crystalline TMPO (42 ppm), physisorbed TMPO (41 ppm), and TMPO adsorbed on Brønsted (sites 1; 64 ppm) and Lewis (sites 2; 58 ppm) acid sites [56]. It is noteworthy that the peak intensity of the ^{31}P resonance may be manipulated by varying the contact times τ_c during ^{31}P{^{1}H} CP or Lee-Goldburg CP (CPLG) experiments, as illustrated in Fig. 19(b1) and (b2), respectively. In other words, the ^{31}P resonance signals will be predominately arising from crystalline TMPO at short contact times ($\tau_c \leq 1$ ms), while the signal of $TMPOH^+$ arising from TMPO adsorbed on Brønsted acid sites (sites 1) prevails at longer τ_c. Additional experiments by a modified 2D ^{1}H–^{31}P heteronuclear correlation (HETCOR) technique, which exploits ^{1}H phase-modulated Lee-Goldburg (PMLG) homonuclear decoupling in conjunction with CPLG heteronuclear decoupling prior to the detection period, were also performed under varied contact time (τ_c) and mixing time (τ_m). Again, at short τ_c (0.5 ms) and null τ_m, a single correlation peak corresponding to ^{31}P signal of the crystalline TMPO (referred as B in Fig. 19(c1)) and ^{1}H signal of its methyl protons was observed. Upon increasing τ_c to ca. 12 ms, additional correlation peaks associated with TMPO interacting with acid sites (referred as A) ^{31}P signals and methyl ^{1}H signals were evident (Fig. 19(c2)). A slight shift in δ^1H (ca. 0.5 ppm) was also observed between methyl ^{1}H signals associated with phase separated (crystalline) and bounded TMPO. Moreover, when allowing ^{1}H spin diffusion to take place during the mixing period ($\tau_m = 50$ ms; Fig. 19(c3)), a new correlation peak arising from ^{31}P resonance of TMPO (68 ppm) and bridging hydroxyl protons (Si-OH-Al; at ca. 7 ppm) was also observed. Accordingly, various adsorption states and interactions of the TMPO probe molecule and acid sites may be readily resolved using multinuclear NMR techniques [55].

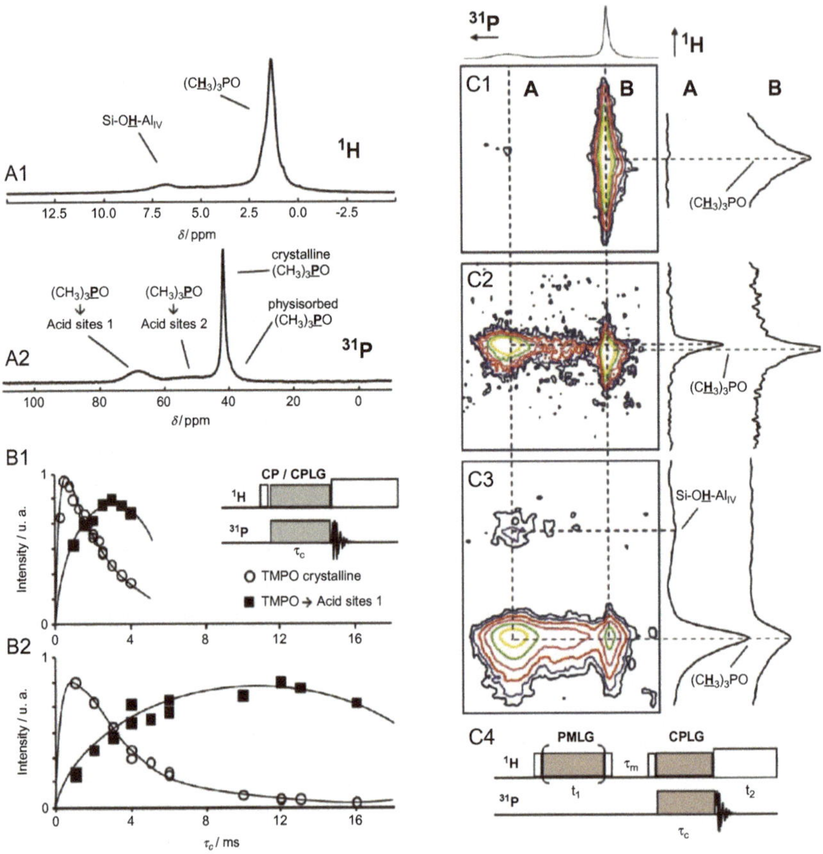

Fig. 19 (A1–A2): (A1) ^{1}H MAS and (A2) $^{31}P\{^{1}H\}$ CP MAS NMR spectra of TMPO-loaded silica-alumina sample. (B1–B2): Variations of ^{31}P signal intensities with contact time τ_c for crystalline TMPO (○) and for TMPO in interaction with acid sites 1 (■) using (B1) CP and (B2) CPLG pulse sequences. Lines are purely indicative. (C1–C4): 2D contour plots obtained from the ^{1}H–^{31}P HETCOR experiments using the pulse sequence depicted in (C4) with (C1) $\tau_c = 0.5$ ms, $\tau_m = 0$ ms, 384 scans, (C2) $\tau_c = 12$ ms, $\tau_m = 0$ ms, 384 scans, and (C3) $\tau_c = 12$ ms, $\tau_m = 50$ ms, 2048 scans. Vertical ^{1}H spectra correspond to TMPO interacting with (A) acid sites 1 and (B) crystalline TMPO, which were obtained from a multi-step fitting procedure of all rows (F1 dimension) using the four ^{31}P signals [55]. *Reproduced from B. Alonso, I. Klur, D. Massiot, Studies of surfaces through molecular probe adsorption and solid-state NMR, Chem. Commun. 8 (2002) 804–805. Copyright 2002 The Royal Society of Chemistry.*

3.2 Characterization of acidic strength

Apart from discernment of acid type, qualitative and quantitative measures of acidic strengths in solid acids are also crucial for detailed understanding the mechanism and performance of the catalyst during catalytic reactions.

Assorted illustrations for characterization of acidic strength in various solid acid catalysts invoking the aforediscussed ^{31}P SSNMR approaches will be discussed in this section.

3.2.1 Microporous zeolites and mesoporous molecular sieves

It has been shown that the ^{31}P SSNMR of adsorbed TMP and/or TMPO probe molecules are practical approaches for acidity characterization of solid acid catalysts. In particular, important acid features such as distribution and strength of acid sites may be readily attained. Fig. 20 shows the ^{31}P MAS NMR spectra of TMP adsorbed in H-form mordenite (MOR; Si/Al = 10) and beta (Si/Al = 12.5) zeolites [57], which both possess 12-MR pores [55]. In the presence of ^{1}H decoupling, the ^{31}P resonance of the adsorbed TMP revealed a singlet at −2.2 (Fig. 20A) and −5.0 (Fig. 20C) ppm for H-MOR and H-beta, respectively, revealing interaction of the probe molecule with Brønsted acid sites, forming the protonated $[(CH_3)_3PH]^+$ adduct. That no ^{31}P resonance signal was observed within

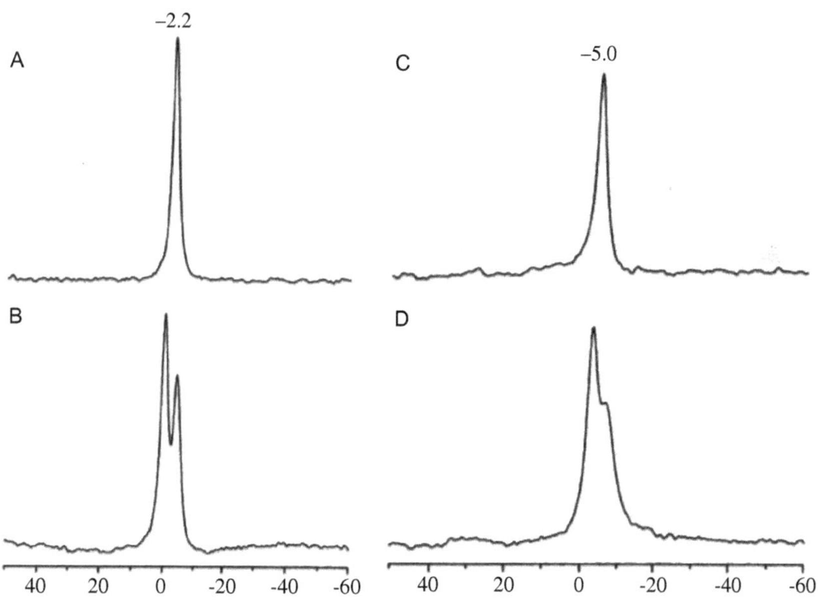

Fig. 20 Room temperature ^{31}P MAS NMR spectra of TMP adsorbed on H-MOR and H-beta acquired (A, C) with and (B, D) without proton decoupling (spinning frequency = 10 kHz) [57]. *Reproduced from H.M. Kao, C.Y. Yu, M.C. Yeh, Detection of the inhomogeneity of Brønsted acidity in H-mordenite and H-β zeolites: A comparative NMR study using trimethylphosphine and trimethylphosphine oxide as* ^{31}P *NMR probes, Micropor. Mesopor. Mater. 53 (2002) 1–12. Copyright 2002 Elsevier.*

the $\delta^{31}P$ range of -30 to -60 ppm, indicating the absence of Lewis acidity in both zeolite catalysts. For spectra recorded without ^{1}H decoupling, ^{31}P resonance responsible for the $[(CH_3)_3PH]^+$ adduct split into a well-resolved doublet, as shown in Fig. 20B and D. Accordingly, a corresponding *J*-coupling constant (J_{P-H}) of 493 and 409 Hz may be deduced for TMP adsorbed on Brønsted acid sites of H-MOR and H-beta, respectively. That a more intense downfield shift in $\delta^{31}P$ and a larger J_{P-H} constant were observed for TMP adsorbed in H-MOR than those in H-beta, suggesting that the former possesses stronger acidity than the latter, in excellent agreements with their catalytic activities. It is indicative that the variations in $\delta^{31}P$ and J_{P-H} values invoked by the ^{31}P-TMP NMR approach are useful for evaluation of Brønsted acidity and catalytic activity in solid acid catalysts such as zeolites [57].

On the other hand, the ^{1}H-decoupled ^{31}P MAS spectra observed for H-MOR (Fig. 21A) and H-beta (Fig. 21B) invoking the ^{31}P-TMPO NMR approach display multiple resonances with $\delta^{31}P$ in the range of 45–90 ppm. Accordingly, with the exception of the additional peak at 86.4 ppm observed for the TMPO/H-MOR system, both spectra revealed similar features, which could be deconvoluted into five overlapping components (peaks I–V). While peak V with $\delta^{31}P$ of ca. 45–50 ppm may be assigned due to physisorbed TMPO, the other peaks with $\delta^{31}P$ exceeding 55 ppm are attributed to TMPO adsorbed on Brønsted acid sites with

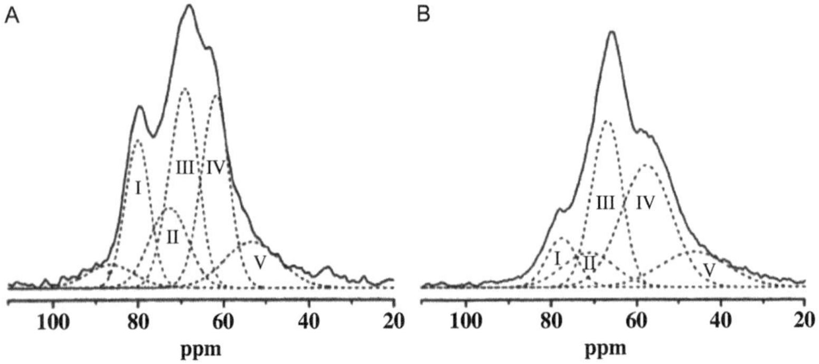

Fig. 21 Proton decoupled ^{31}P MAS NMR spectra of TMPO adsorbed on (A) H-MOR and (B) H-beta zeolites. Both spectra were recorded at room temperature with a spinning frequency of 10 kHz. The dashed curves represent spectral simulation results via Gaussian deconvolution [57]. *Reproduced from H.M. Kao, C.Y. Yu, M.C. Yeh, Detection of the inhomogeneity of Brønsted acidity in H-mordenite and H-β zeolites: a comparative NMR study using trimethylphosphine and trimethylphosphine oxide as* ^{31}P *NMR probes, Micropor. Mesopor. Mater. 53 (2002) 1–12. Copyright 2002 Elsevier.*

different acidic strengths. As revealed by theoretical DFT calculations [21], a more intense interaction between TMPO and bridging hydroxyl proton may be inferred for Brønsted acid sites with stronger acidic strength. As a result, leading to a shift of ^{31}P-TMPO resonance towards downfield direction (i.e. towards a larger δ^{31}P value). That an optimal δ^{31}P of 86.4 and 77.6 ppm was observed for TMPO adsorbed on H-MOR and H-beta, respectively, indicating that the former catalyst possesses Brønsted acidity with relatively higher acidic strength. Clearly, this notion coincides with that deduced from the aforediscussed ^{31}P-TMP approach [58]. Thus, ^{31}P SSNMR of adsorbed phosphorus-containing probes represents a unique and sensitive technique for acidity characterization of solid acids. Accordingly, detailed information of acid features (viz. type, concentration, distribution, and strength of acid sites) in solid catalysts such as zeolites may be attained.

Transition metal-incorporated zeolites have been shown to be efficient catalysts for direct conversion of methane to benzene and toluene under non-oxidative conditions [59,60]. Bao and co-workers revealed that Mo/H-MCM-22 catalysts are desirable bifunctional catalysts for methane dehydroaromatization reaction [61]. In terms of catalytic performances of Mo/H-MCM-22 with varied metal loading, catalyst with a Mo loading of ca. 6 wt% was found to exhibit the optimal benzene selectivity, suppressed naphthalene yield, and prolonged catalyst life under a moderate methane conversion. Although both Brønsted and Lewis acid sites are capable of catalysing methane conversion reaction, active sites with higher acidic strengths are anticipated to play the dominant role.

To explore effects of incorporated Mo on acid properties, Huang et al. exploited ^{31}P-TMPO NMR approach to investigate Mo/H-MCM-22 catalysts with varied Mo loading [62]. As shown in Fig. 22, as many as 10 different resonance peaks, spanning over a δ^{31}P range of 31.0–83.0 ppm, were observed for the pristine MCM-22 and various *x*MoOM22IC (*x* represents Mo loading; $x=2$–10 wt%) catalyst samples. The notation for MoOM22IC represents incorporation of molybdenum oxide (MoO) on MCM-22 zeolite (M22) by impregnation method and after a calcination treatment. The two upfield peaks with lowest δ^{31}P of 41.3 and 31.4 ppm were unambiguously assigned due to physisorbed TMPO. In addition, the peak with δ^{31}P of 60.9 ppm was attributed to TMPO adsorbed on Lewis acid sites [32], whereas the other peaks were ascribed due to TMPO adsorbed on Brønsted acid sites. Upon progressive loading of Mo (<6 wt%), slight increases (ca. 1 ppm) in δ^{31}P of resonance peaks were observed, especially

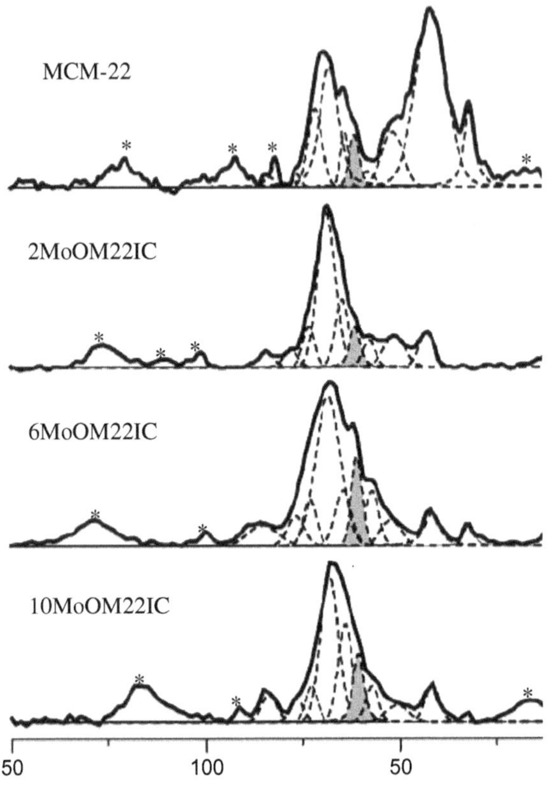

Fig. 22 Solid-state ^{31}P MAS NMR spectra of TMPO adsorbed on parent MCM-22 and various calcined xMoOM22 samples ($x=2$–10 wt%) prepared by the impregnation method. The dashed lines represent Gaussian simulation results. The shaded peaks denote resonance arising from TMPO adsorbed on Lewis acid sites. Typically, a sample spinning rate of 10–12 kHz was used [62]. *Reproduced from S.J. Huang, Q. Zhao, W.H. Chen, X. Han, X. Bao, P.S. Lo, H.K. Lee, S.B. Liu, Structure and acidity of Mo/H-MCM-22 catalysts studied by NMR spectroscopy, Catal. Today 97 (2004) 25–34. Copyright 2004 Elsevier.*

for those associated with acid sites with higher acidic strengths. As such, a mild increase in the overall acidic strength of MCM-22 with increasing Mo loading may be inferred for samples with $x<6$ wt% (Fig. 22). By combining the ^{31}P NMR results with data obtained from elemental analyses by ICP, quantitative information regarding to acid concentration may be deduced. Accordingly, a total Brønsted acidity amount of 0.470, 0.278, 0.263, and 0.203 mmol/g was derived for MCM-22, 2MoOM22IC, 6MoOM22IC, and 10MoOM22IC, respectively. A drastic decrease in both Brønsted and Lewis acidities upon immediate incorporation of Mo species onto MCM-22 may be inferred. For example, a respective decrease of ca. 41% and 94% in total Brønsted and Lewis acidities was found for the 2MoOM22IC catalyst compared to the parent MCM-22. However, only marginal decreases were observed when the Mo loading was increased from 2 to 6 wt%. Moreover, as the Mo loading was further increased to 10 wt%, considerable decrease in total acid concentration was observed.

It is noteworthy that the amounts of acid sites with highest acidic strengths (i.e. those with $\delta^{31}P$ of ca. 84 and 76 ppm) tend to increase at the expanse of weaker acidity while increasing the Mo loading, reaching an optimal value at ca. 6 wt% loading. Nonetheless, further increasing the Mo loading exceeding 6 wt% led to notable decrease in the amount of strong acid sites. Thus, it is conclusive that the ^{31}P-TMPO NMR approach is capable of providing valuable information on distribution and strength of acid sites, which in turn help to explain the peculiar catalytic performance of Mo/H-MCM-22 catalyst with a Mo loading of 6 wt% during methane dehydroaromatization reaction.

Recently, Bao and co-workers utilized the ^{31}P-TMP NMR approach to characterize the acid properties of a Zn-incorporated $Zn_{0.34}$/ZSM-5 catalyst [63]. As shown in Fig. 23, TMP adsorbed on the pristine ZSM-5 exhibited only a single ^{31}P resonance at −4.5 ppm, which may be attributed to TMP adsorbed on Brønsted acid sites. On the other hand, two additional ^{31}P peaks at −42 and −47 ppm were observed for the $Zn_{0.34}$/ZSM-5, indicating the presences of two types of Lewis acid sites with different acidic strengths. Further studies using the $^{31}P\{^{27}Al\}$ rotational echo adiabatic passage double-resonance (REAPDOR) NMR method [64] excluded the presence of correlation between these two ^{31}P signals with Al species, thus they should be solely associated with Zn sites. Given that the acidity of $[Zn\text{-}O\text{-}Zn]^{2+}$ is weaker than that of $[O^-\text{-}Zn^{2+}\text{-}O^-]$, the ^{31}P signals at −47 and −42 ppm may therefore be attributed to TMP adsorbed on $[Zn\text{-}O\text{-}Zn]^{2+}$ and $[O^-\text{-}Zn^{2+}\text{-}O^-]$, respectively. The presence of the diagonal peak at (−47, −94) ppm in the 2D ^{31}P–^{31}P DQ MAS NMR spectrum further validated the spatial proximity between the co-adsorbed TMP molecules, reflecting the spatially correlated zinc sites of $[Zn\text{-}O\text{-}Zn]^{2+}$. On the other hand, the absence of diagonal peak at (−42, −84) ppm indicated the presence of $[O^-\text{-}Zn^{2+}\text{-}O^-]$. Therefore, results obtained from the ^{31}P-TMP NMR approach confirmed the coexistence of $[Zn\text{-}O\text{-}Zn]^{2+}$ and $[O^-\text{-}Zn^{2+}\text{-}O^-]$ active sites over the $Zn_{0.34}$/ZSM-5 catalysts.

Furthermore, Lewis acidic zeolites, such as Sn- and Ti-Beta, have emerged as exceptional catalysts for the activation of oxygenated molecules

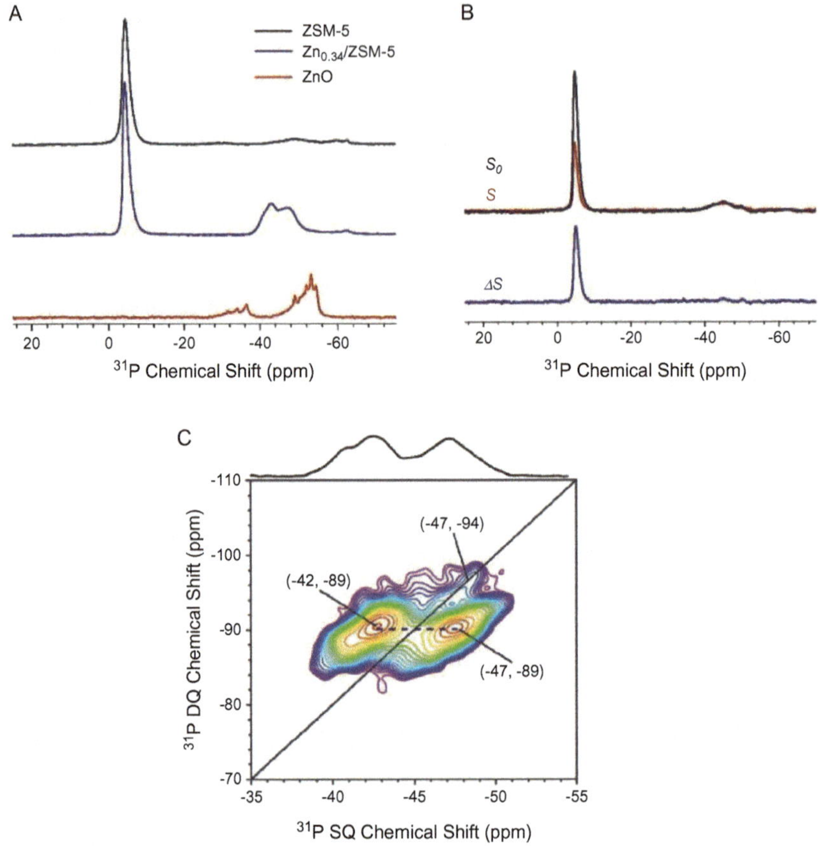

Fig. 23 (A) ^{31}P CP MAS NMR spectra of TMP adsorption on $Zn_{0.34}$/ZSM-5 in comparison to ZnO and ZSM-5. (B) ^{31}P{^{27}Al} REAPDOR NMR spectra of TMP adsorption on $Zn_{0.34}$/ZSM-5, with the full intensity spectrum shown by the black line denoted as S_0, the dipolar dephased spectrum by the red line denoted as S, and the difference spectrum shown by the blue line denoted as ΔS. (C) ^{31}P–^{31}P CP POST-C7 double quantum-single quantum (DQ-SQ) MAS NMR spectrum upon TMP adsorption on $Zn_{0.34}$/ZSM-5 [63]. *Reproduced from Y. Chen, K. Gong, F. Jiao, X. Pan, G. Hou, R. Si, X. Bao, C-C Bond formation in syngas conversion over Zn sites grafted on ZSM-5, Angew. Chem. Int. Ed., 59 (16) 2020 6529–6534. https://doi.org/10.1002/anie.201912869. Copyright 2020 Wiley.*

and have drawn considerable research interest for conversion of biomass [65–67]. Although these metal-incorporated zeolites are capable of catalysing a wide range of reactions, the nature of active sites in these catalysts remains largely unknown. On the basis of the ^{31}P-TMPO NMR approach, Lewis et al. developed a new method to characterize and quantify active sites in Lewis acidic zeolites [68]. The authors assigned the ^{31}P NMR resonances obtained from the Sn-Beta zeolite during TMPO titration with corresponding variations in the observed ^{119}Sn NMR spectrum. Moreover, it was found that the reaction rate of glucose isomerization may be directly correlated with the sum of integrated peak areas of ^{31}P resonances at 55.8 and 54.9 ppm, while the peak area of the ^{31}P signal at 58.6 ppm may be correlated with the aldol condensation activity. Dubray et al. also invoked the ^{31}P-TMPO NMR approach to verify the successful dispersion of molybdenum (Mo) atoms onto the framework of nanosized MFI zeolite [69]. As shown in Fig. 24, the ^{31}P NMR spectrum of TMPO adsorbed on atomically dispersed Mo-MFI-D zeolite sample exhibited two main ^{31}P resonances at 29 and 45 ppm, which may be ascribed due to physisorbed TMPO and TMPO interacting with framework Mo atoms, respectively. Further study by using 2D ^{31}P–^{31}P MAS radio frequency-driven recoupling (RFDR) NMR experiments [70] verified the presence of a single chemical species for framework Mo species. In contrast, the ^{31}P spectrum obtained from TMPO adsorbed on the Mo-MFI-P catalyst prepared by wet-impregnation method showed primary resonance peaks corresponding to crystalline TMPO

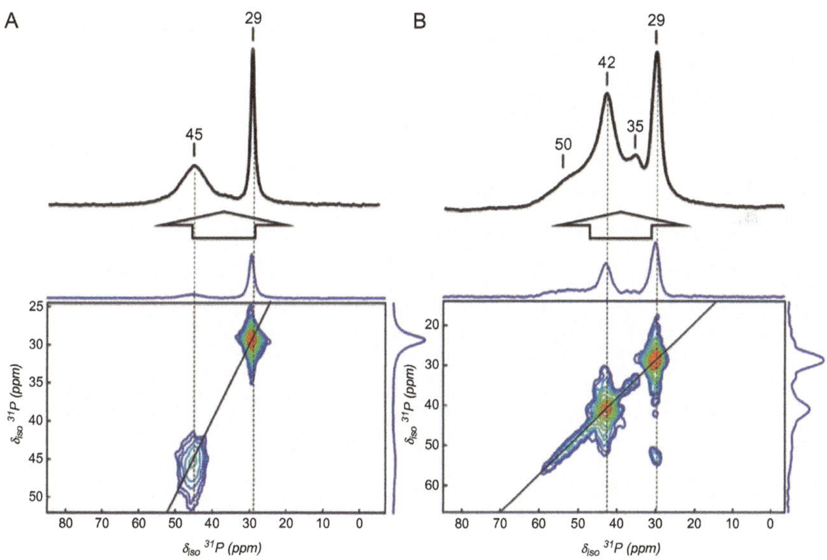

Fig. 24 2D ^{31}P–^{31}P MAS RFDR NMR spectra (top: 1D ^{31}P MAS NMR spectra) of (A) Mo-MFI-D and (B) Mo-MFI-P under ^{1}H spinal-64 decoupling [69]. *Reproduced from F. Dubray, S. Moldovan, C. Kouvatas, J. Grand, C. Aquino, N. Barrier, J.P. Gilson, N. Nesterenko, D. Minoux, S. Mintova, Direct evidence for single molybdenum atoms incorporated in the framework of MFI zeolite nanocrystals, J. Am. Chem. Soc. 141 (2019) 8689–8693. Copyright 2019 American Chemical Society.*

(42 ppm), physisorbed TMPO (29 ppm), and TMPO adsorbed on weakly acidic silanol groups (50 ppm). In this context, resonance associated with framework Mo species ($\delta^{31}P = 45$ ppm) was not observed for the Mo-MFI-P catalyst sample prepared by simple impregnation method.

3.2.2 Metal oxides and mixed metal oxides

Owing to their extensive applications in environmental and energy-related areas such as photocatalysis, dye-sensitized solar cells (DSSCs), lithium ion batteries, catalyst supports and so on, titanium dioxide (TiO_2) has become the most studied transition metal oxide. It is well-known that the size, shape, and surface structures of TiO_2 crystals were all found to have significant impact on their physical and chemical properties [71]. Recently, tailoring of surface structures, particularly in enhancing more reactive (001) facets in anatase TiO_2 has drawn considerable R&D interest. This is of course due to the fact that the crystalline surfaces of synthetic TiO_2 are mostly abundant with thermodynamically more stable (101) facets rather than (001) facets [72]. As such, there is an urgent demand in understanding detailed features of various facets on crystalline surfaces of TiO_2. By invoking the ^{31}P-TMP NMR approach, Peng et al. demonstrated that subtle changes in chemical state of titanium (Ti) provoked by surface doping of surface adsorbates such as fluorine, sulphate, or OH on anatase TiO_2 with different morphologies [73,74]. The ^{31}P MAS NMR spectrum of TMP adsorbed on powdered (PD) TiO_2 sample (Fig. 25B) revealed no significant ^{31}P signals between -2 and -5 ppm, indicating the absence of protonated $TMPH^+$ species. Whereas peaks at -36 and -29 ppm were clearly evident, which may be assigned due to interactions of TMP with surface five-coordinated Lewis acids, namely Ti^{4+} on stable (101) facet (denoted as $Ti_{5C}(101)$) and on more reactive (001) facet (denoted as $Ti_{5C}(001)$), respectively. The F^- stabilized (101) and (001) facets, denoted as F-(101) and F-(001), respectively, showed notable increase in Lewis acidic strength of Ti_{5C}, causing a notable shift in $\delta^{31}P$ to -31 and -22.5 ppm, respectively (Fig. 25D and F). Additional calcination treatment of various samples led to effective removal of surface F^- species, leading to a notable shift in $\delta^{31}P$ peak to -41 and -35 ppm for (101) and (001) facets, respectively. Similarly, variations in surface characteristics of TiO_2 modified by sample washing with 0.1 M NaOH solution and subsequent sulphation treatment may also be monitored by the ^{31}P-TMP NMR approach, as shown in Fig. 26. However, the presence of OH^- together with F^- would result in two extra ^{31}P peaks at -36.5 and -28 ppm. Upon sulphation treatment, the ^{31}P signal at -34 and -25.5 ppm

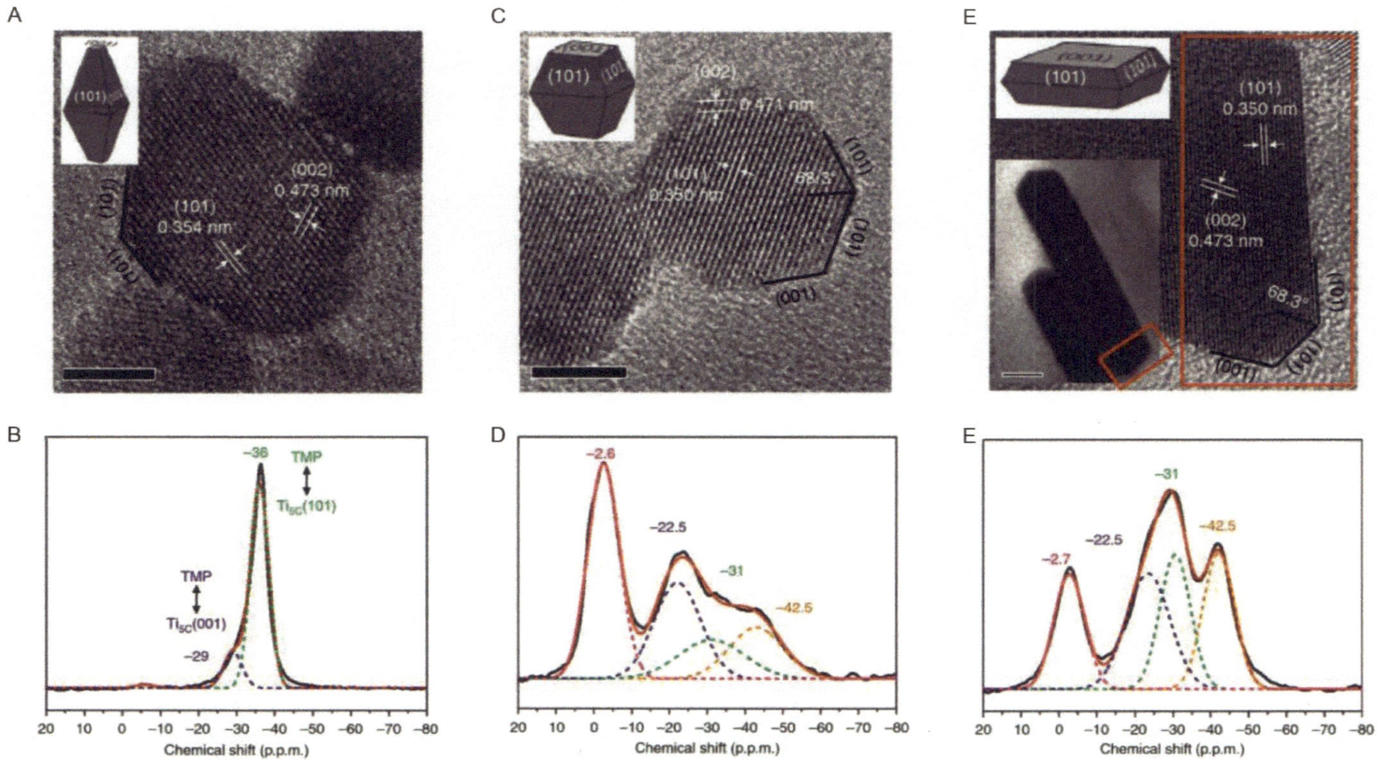

Fig. 25 TEM images of (A) powdered (PD), (C) F-(101), and (E) F-(001) TiO_2 samples and (B, D, F) their corresponding ^{31}P NMR spectra of adsorbed TMP. Four possible scenarios, namely TMP adsorbed on $Ti_{5C}(101)$ (green curve), $Ti_{5C}(001)$ (blue curve), surface Ti^{4+} containing OH and F (orange curve), and Brønsted acid site (TiO(H)-Ti) (pink curve). Among them, the F-(101) and F-(001) samples clearly possess a higher coverage of (101) and (001) facets, as indicated by the predominant ^{31}P peak at −31 (green curve) and −22.5 (blue curve) ppm, respectively. Whereas the PD sample is clearly abundant with (101) facets (green curve). The scale bar in all TEM images is 5 nm [73]. *Reproduced from Y.K. Peng, Y. Hu, H.L. Chou, Y. Fu, I.F. Teixeira, L. Zhang, H. He, S.C.E. Tsang, Mapping surface-modified titania nanoparticles with implications for activity and facet control, Nature Commun. 8 (2017) 1–13. Copyright 2017 Nature Publishing Group.*

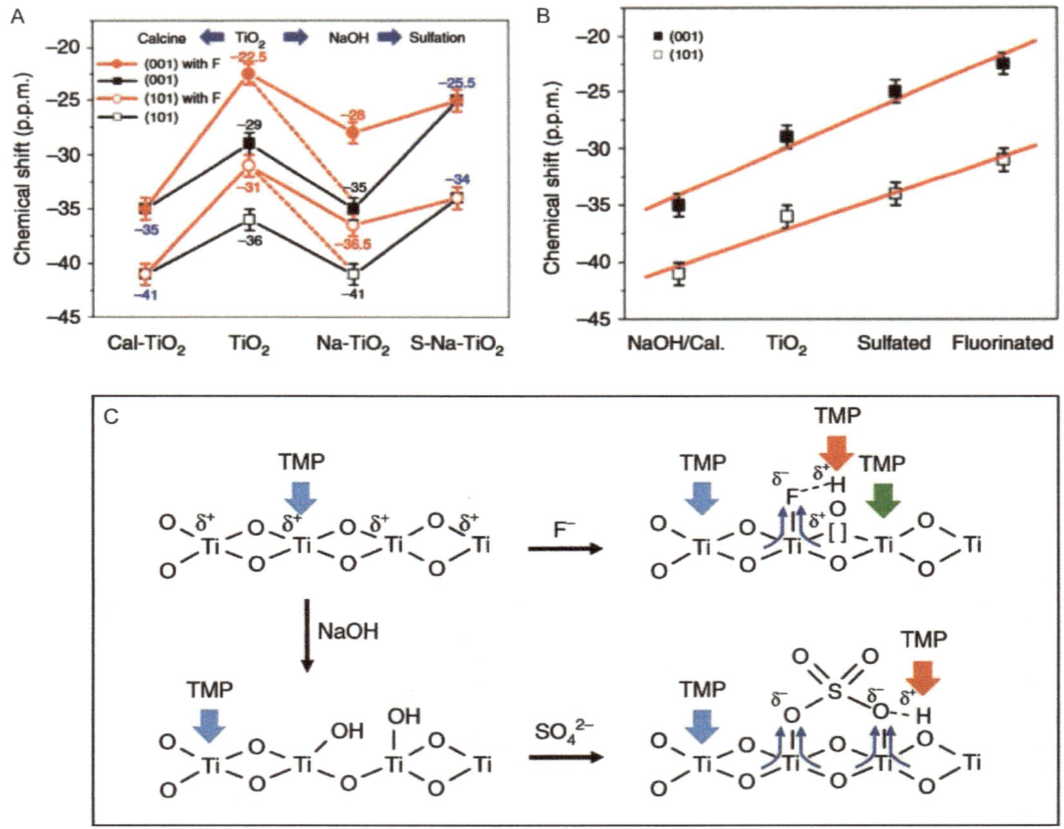

Fig. 26 Interaction between TMP and surface features on TiO_2 facet with various treatments/modifications. Variations of $\delta^{31}P$ observed for TMP adsorbed on $Ti_{5C}(001)/(101)$ facets of (A) PD, F-(001), and F-(101) with different treatment/modification. (B) Summary of $\delta^{31}P$ of TMP-adsorbed $Ti_{5C}(001)/(101)$ facets with different treatments (calcination/NaOH wash) and modification (sulphate/fluorine). The error bar for $\delta^{31}P$ in (A, B) is ± 1 ppm. (C) Illustration of interaction between TMP and surface features on TiO_2 facets under various treatment/modification [73]. *Reproduced from Y.K. Peng, Y. Hu, H.L. Chou, Y. Fu, I.F. Teixeira, L. Zhang, H. He, S.C.E. Tsang, Mapping surface-modified titania nanoparticles with implications for activity and facet control, Nature Commun. 8 (2017) 1–13. Copyright 2017 Nature Publishing Group.*

corresponding to TMP adsorbed on (101) and (001) facets, respectively, may further be manipulated. Based on such variations in the observed $\delta^{31}P$ values, the electron withdrawing ability of various surface additives to the surface Ti^{4+} was found to obey the trend: $OH^- < -O- < SO_4^{2-} < F^-$. In this context, the ^{31}P-TMP NMR approach is a feasible and reliable technique for probing both the strength and concentration of Brønsted and Lewis acid sites on the surface of TiO_2.

By invoking such probe molecule NMR technique, it has been shown that qualitative as well as quantitative information on oxygen vacancies on the specific facet of ZnO nanoparticles may also be attained [74]. Apart from TiO_2 and ZnO, the surface features of other metal oxides such as $Nb_2O_5 \cdot nH_2O$ [75], ZrO_2, Al_2O_3, MgO [76], Cu_2O [77] and CeO_2 [78] have also been explored by means of the ^{31}P NMR approach, which help to verify the structure-activity correlations of various catalytic systems at microscopic scale.

Nanostructured titanate materials, which can be fabricated with exotic morphologies (such as layered, nanosheet, nanotube, nanorod, etc.) and their applications as solid acid catalysts have drawn considerable attentions in recent years [79,80]. It has been demonstrated that titanate nanotubes exhibit excellent catalytic activity for Friedel-Crafts alkylation of toluene with benzyl chloride at ambient temperature, surpassing that of layered titanates ($H_2Ti_3O_7$) and nanosheets ($H_{0.7}Ti_{1.825}\square_{0.175}O_4 \cdot H_2O$; where $\square$ indicates a vacancy) having similar crystalline structures [81]. Again, the ^{31}P-TMPO NMR approach has been employed to elucidate correlations between catalytic reactivity and Brønsted acidity in these titanate nanomaterials [81]. For TMPO adsorbed on protonated titanate nanotubes, a distinct broad ^{31}P peak at ca. 65 ppm was observed, which may be attributed to the formation of $TMPOH^+$ species. However, this peak tends to shift towards upfield direction (at ca. 60 ppm) when TMPO is adsorbed on layered titanates or titanate nanosheets, indicating that the latter two catalysts possess relatively weaker Brønsted acid strength. It is noteworthy that the $\delta^{31}P$ observed for TMPO adsorbed on titanate nanotubes (ca. 65 ppm) corresponds to acidic strength comparable to that of niobic acid, HY zeolite, and $HTiNbO_5$ nanosheet catalysts [82–84], but weaker than those of H-ZSM-5 (ca. 86 ppm). That layered titanates and titanate nanosheets exhibit similar $\delta^{31}P$ of adsorbed TMPO indicates that the amount of defect sites in titanate is irrelevant to its Brønsted acidity. In this context, the higher acidic strength observed for titanate nanotubes should be associated with lattice distortion caused by scrolling of lamellar titanate nanosheet, as also evidenced by results from Raman spectroscopy and DFT calculations [81].

Domen and co-workers also adopted the ^{31}P-TMPO NMR approach to characterize the acid properties of nanosheet aggregates prepared from protonated layered tungstates M(*OH*)M′ (M=Nb, Ta; M′=Nb, W) [85]. As shown in Fig. 27, a broad resonance with overlapping features centring at ca. 60–75 ppm was observed for TMPO adsorbed on $HTiNbO_5$, HNb_3O_8, and $HNbWO_6$ nanosheets in addition to peaks responsible for physisorbed and crystalline TMPO ($\delta^{31}P < 50$ ppm). In the case of TMPO adsorbed on $HTaWO_6$, an additional shoulder peak at ca. 84 ppm emerged along with the two main features at ca. 72 and 66 ppm, indicating the presence of very strong acid sites. In terms of acid strengths of these nanosheet catalysts, the following order may be inferred: $HTaWO_6 > HNbWO_6 > HNb_3O_8 > HTiNbO_5$. The same research group further adopted similar approach to probe acid properties of novel mesoporous oxide materials, namely $Nb_xW_{(10-x)}$ [86]. As shown in Fig. 28, the main resonance peak of the adsorbed TMPO tends to shift towards downfield direction from 67 ppm

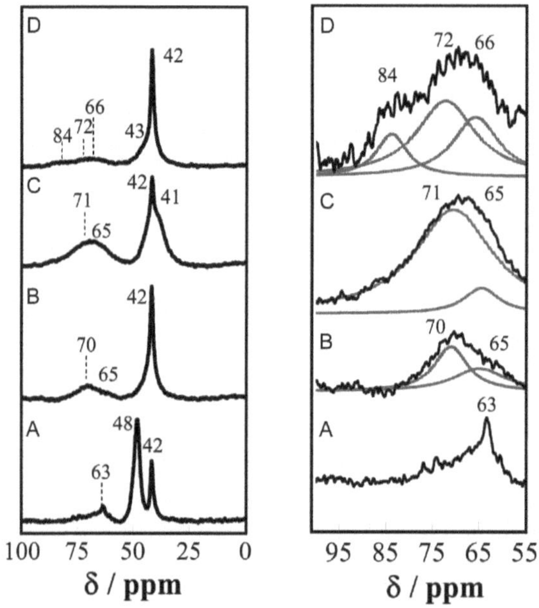

Fig. 27 Left: room temperature ^{31}P MAS NMR spectra for TMPO adsorbed on various nanosheets aggregates: (A) $HTiNbO_5$, (B) HNb_3O_8, (C) $HNbWO_6$, and (D) $HTaWO_6$. All spectra were acquired with a MAS spinning rate of 10 kHz. Right: expanded spectra [85]. *Reproduced from C. Tagusagawa, A. Takagaki, S. Hayashi, K. Domen, Characterization of $HNbWO_6$ and $HTaWO_6$ metal oxide nanosheet aggregates as solid acid catalysts, J. Phys. Chem. C 113 (2009) 7831–7837. Copyright 2009 American Chemical Society.*

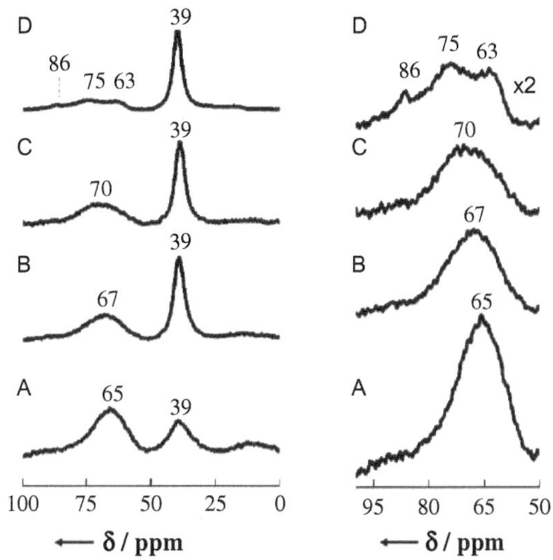

Fig. 28 Left: room temperature ^{31}P MAS NMR spectra for TMPO (loading: 0.8 mmol per gram catalyst) adsorbed on mesoporous (A) Nb, (B) Nb_7W_3, (C) Nb_5W_5, and (D) Nb_3W_7 oxides. All spectra were acquired with a MAS spinning rate of 10 kHz. Right: expanded spectra [86]. *Reproduced from C. Tagusagawa, A. Takagaki, A. Iguchi, K. Takanabe, J.N. Kondo, K. Ebitani, S. Hayashi, T. Tatsumi, K. Domen, Highly active mesoporous Nb–W oxide solid-acid catalyst, Angew. Chem. Int. Ed. Engl. 49 (2010) 1128–1132. Copyright 2010 Wiley.*

($x=7$) to 70 ppm ($x=5$), indicating an enhanced acidic strength with decreasing loading of Nb. Moreover, a distinct small peak at 86 ppm was observed for TMPO adsorbed on Nb_3W_7 ($x=3$), revealing the presence of super acidity [21,83].

Among various mixed metal oxides, titania-based binary metal oxides such as TiO_2-SiO_2 and TiO_2-ZrO_2 have received much attention due to its unique acid/base and reduction/oxidation properties suitable for heterogeneous catalysis and wide bandgap desirable for photocatalytic applications [87,88]. In terms of their acid/base properties, bare TiO_2 is known to possess exclusively Lewis acidity, while bare SiO_2 exhibit only weak Brønsted acidity owing to silanol groups abundant on the surface. Nevertheless, the nature and origins of acidic sites on TiO_2-SiO_2 mixed oxides remain as debatable issues. As mentioned earlier, acidity characterization utilizing ^{31}P SSNMR of phosphorus-containing probe molecules may have different approaches. While TMPO probe is more sensitive to Brønsted sites with varied acidic strengths in solid acid catalysts, TMP is more suitable for probing Lewis acidity and hence capable of discriminating acid types, viz. Brønsted vs. Lewis acid sites. Therefore, by combining the two different ^{31}P SSNMR approaches using TMP and TMPO probes, variations of acid sites in mesoporous TiO_2-SiO_2 mixed oxides with Ti/Si = 1 before and after hydrogenation, denoted as mTS-1.0 and H-mTS-1.0, respectively, may be evaluated.

The ^{31}P NMR spectrum of TMPO adsorbed on mTS-1.0 revealed three main resonance peaks at 62.7, 57.1, and 44.2 ppm, which may be attributed to guest molecule adsorbed on Brønsted and Lewis acid sites, and physisorbed TMPO, respectively (Fig. 29A). On the other hand, the ^{31}P spectrum of TMP adsorbed on mTS-1.0 showed two ^{31}P peaks at −5.8 and −35.5 ppm, which may be assigned due to the presence of Brønsted and Lewis acid sites, respectively. It is noteworthy that TMP adsorbed on bare SiO_2 by which interactions between the probe molecule and the weak acidic surface –OH groups tend to form hydrogen bonding, leading to ^{31}P NMR resonances at ca. 50–60 ppm. In this context, the signal at −5.8 ppm responsible for TMP adsorbed on mTS-1.0 and H-mTS-1.0, may be attributed to the presence of protonated adduct namely $TMPH^+$, indicating the presence of much stronger Brønsted acidity in the TiO_2-SiO_2 (Fig. 29B). Moreover, the intensities of the signals at −5.8 and −35.5 ppm may be used to determine the relative concentration of Brønsted and Lewis acid sites in TiO_2-SiO_2. As may be seen in Fig. 29B, a notable increase in intensity of the peak at −5.8 ppm responsible for Brønsted acidity was observed for H-mTS-1.0 compared to mTS-1.0, while no change for the peak at −35.5 ppm was found. It is indicative that hydrogenation treatment promote enhancement in concentration of Brønsted acid sites in TiO_2-SiO_2 due to partial reduction of Ti(IV) to Ti(III), while keeping Lewis acidity practically unchanged [89].

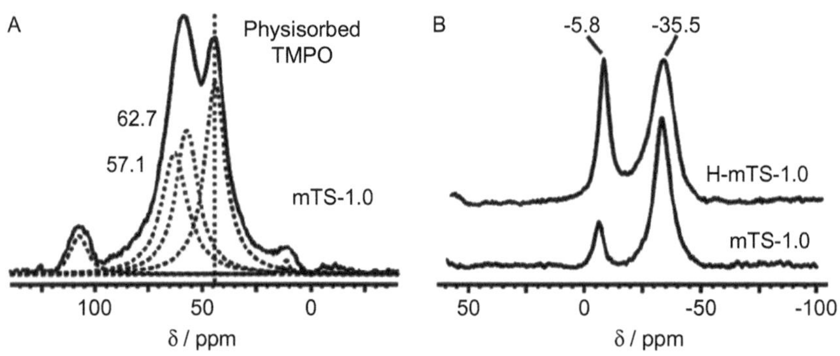

Fig. 29 ^{31}P NMR of (A) TMPO adsorbed on mTS-1.0 and (B) TMP adsorbed on mTS-1.0 and H-mTS-1.0 mesoporous TiO_2-SiO_2 catalysts [89]. *Reproduced from S. Li, H. Zhou, B. Han, F. Deng, X. Liu, L. Xiao, J. Fan, Hydrogenated mesoporous TiO_2–SiO_2 with increased moderate strong Brønsted acidic sites for Friedel-Crafts alkylation reaction, Catal. Sci. Technol. 2 (2012) 719–721. Copyright 2012 The Royal Society of Chemistry.*

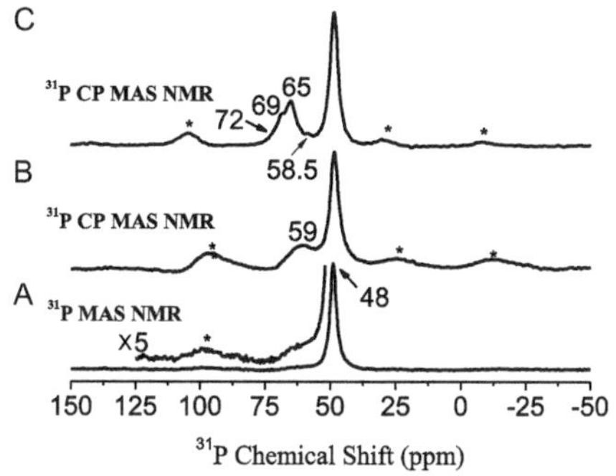

Fig. 30 ^{31}P (A) MAS and (B) CP MAS NMR spectra of TMPO adsorbed on TiO_2 and (C) MAS NMR spectrum of TMPO adsorbed on SO_4^{2-}/TiO_2. Asterisks denote spinning sidebands [25]. *Reproduced from H. Zhang, H. Yu, A. Zheng, S. Li, W. Shen, F. Deng, Reactivity enhancement of 2-propanol photocatalysis on* SO_4^{2-}/TiO_2*: Insights from solid-state NMR spectroscopy, Environ. Sci. Technol. 42 (2008) 5316–5321. Copyright 2008 American Chemical Society.*

The acid properties of sulphated TiO_2 (denoted as SO_4^{2-}/TiO_2) have also been investigated by means of the ^{31}P-TMPO NMR approach [25]. As shown in Fig. 30A, the ^{31}P MAS NMR spectrum of TMPO adsorbed on pristine TiO_2 revealed a main resonance at 48 ppm due to physisorbed TMPO and a much weaker broad signal centring at ca. 59 ppm, which may be attributed to TMPO adsorbed on Lewis acid sites. The assignment of the latter peak was further confirmed by additional ^{31}P CP MAS NMR experiment to eliminate undesirable artefacts originating from spinning sidebands. As a result, the resonance signal at 59 ppm was greatly enhanced (Fig. 30B). For TMPO adsorbed on SO_4^{2-}/TiO_2, apart from the signals of physisorbed TMPO (48 ppm) and TMPO adsorbed on Lewis acid sites (58.5 ppm), three signals at 65, 69, and 72 ppm were also observed (Fig. 30C), which may be ascribed due to TMPO adsorbed on three types of Brønsted acid sites with varied acidic strengths.

Sulphated zirconia (SO_4^{2-}/ZrO_2; SZ) is an environmental friendly and a practical solid acid catalyst that has been extensively used for low-temperature conversion of hydrocarbons. Nonetheless, SZ is also known as a strong acidic catalyst vulnerable to deactivation [90]. This drawback is normally circumvented by adding suitable amount of metal promoter, such as Pt, Al, Ga, Fe or Mn [91,92]. However, the acid properties of SZ and metal-promoted SZ (denoted as M/SZ) have not yet been clarified. Liu and co-workers adopted the ^{31}P-TMPO NMR approach to characterize the acid features (type, concentration, and strength) of various SZ and

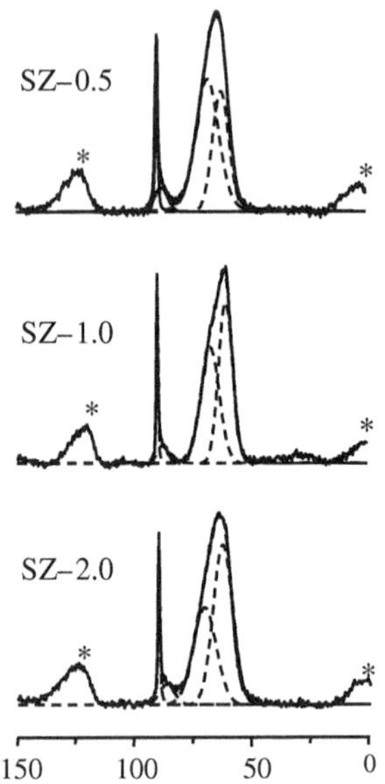

Fig. 31 ^{31}P MAS NMR spectra of TMPO adsorbed on dehydrated SZ-xN ($x=0.5$, 1.0, and 2.0) prepared with varied sulphur contents. The dashed curves represent spectral simulation results and the asterisks denote spinning sidebands (spinning rate 12 kHz) [29]. *Reproduced from W.H. Chen, H.H. Ko, A. Sakthivel, S.J. Huang, S.H. Liu, A.Y. Lo, T.C. Tsai, S.B. Liu, A solid-state NMR, FT-IR and TPD study on acid properties of sulfated and metal-promoted zirconia: Influence of promoter and sulfation treatment, Catal. Today 116 (2006) 111–120. Copyright 2006 Elsevier.*

M/SZ catalysts [29]. As shown in Fig. 31, the ^{31}P NMR spectrum of TMPO adsorbed on the dehydrated parent ZrO_2 revealed four distinct resonance peaks at 62, 53, 41, and 34 ppm. The latter two peaks with $\delta^{31}P$ lower than 50 ppm can be unambiguously assigned due to physisorbed TMPO, while the former two peaks (62 and 53 ppm) may be attributed to TMPO adsorbed on Lewis acid sites, as confirmed by additional experiments performed on partially hydrated samples (Fig. 31). Unlike the parent ZrO_2 sample, the TMPO-loaded SZ-xN samples ($x=0.5$, 1.0, and 2.0, where x represents the concentration of sulphuric acid added during the sulphation treatment) all reveal four distinct ^{31}P resonance peaks at 90, 87, 68, and 63 ppm (Fig. 31). Assorted data are depicted in Table 3. It is indicative that coexisting Lewis (at 90 and 63 ppm) and Brønsted (at 87 and 68 ppm) acid sites were present in SZ-xN samples. In particular, the emergences of the strong Lewis acid sites (at 90 ppm) as well as Brønsted acid sites with strong (87 ppm) and medium (68 ppm) acidic strengths took place mostly at the

Table 3 ^{31}P MAS NMR chemical shift assignments and distribution of acid sites for various SZ and M/SZ samples loaded with TMPO probe molecule [29].

Sample	Chemical shift (ppm)[a,b]									Acid amount (mmol/g cat.)[c]				
			87	76		68	65			P[c]	Brønsted	Lewis	Total	B/L (%)
ZrO_2	–	–	–	–	–	–	–	50.3%	49.7%	√	–	–	–	–
SZ-0.5 N	–	7.9% (0.003)	5.4% (0.002)	–	–	53.3% (0.023)	–	33.4% (0.015)	–	–	0.025	0.018	0.043	1.39
SZ-1.0 N	–	7.4% (0.012)	3.7% (0.006)	–	–	43.2% (0.071)	–	45.7% (0.075)	–	–	0.077	0.087	0.164	0.89
SZ-2.0 N	–	6.0% (0.005)	5.1% (0.004)	–	–	37.2% (0.028)	–	51.7% (0.039)	–	–	0.032	0.044	0.076	0.73
Al/SZ	–	5.3% (0.008)	4.3% (0.006)	2.8% (0.004)	3.1% (0.005)	9.7% (0.015)	15.3% (0.023)	55.8% (0.084)	3.7% (0.006)	–	0.048	0.103	0.151	0.47
Ga/SZ	–	4.1% (0.005)	5.5% (0.007)	3.5% (0.005)	7.8% (0.010)	14.4% (0.019)	41.9% (0.054)	20.0% (0.026)	2.8% (0.004)	–	0.085	0.045	0.130	1.89
Fe/SZ	4.5% (0.013)	–	6.3% (0.018)	–	–	12.9% (0.038)	13.7% (0.040)	27.0% (0.079)	35.6% (0.105)	√	0.096	0.197	0.293	0.49

[a]For comparison, simulation results (see text) obtained for various samples representing TMPO adsorbed on Brønsted and/or Lewis (denoted by chemical shift values in brackets) acid sites with practically the same chemical shift (i.e. similar acid strengths) are aligned in the same column.
[b]For SZ-xN (x = 0.5, 1.0, and 2.0) and M/SZ (M = Al, Ga, and Fe) samples, data denote relative concentration of acid sites (%); whereas data in parentheses represent acid concentration (±0.002 mmol/g cat.) of the corresponding acid site, as derived from elemental analyses by ICP-MS.
[c]Resonance peaks with chemical shift lower than 50 ppm, which arise from physisorbed TMPO, were excluded during derivations of acid amount.

expanses of the weak Lewis acidity originally existed in the pristine ZrO_2, particularly the one with weaker strength (at 53 ppm), which diminishes upon sulphation treatment. Moreover, a maximum Brønsted, Lewis, and total acidities were found on sample sulphated with a sulphuric acid concentration (x) of 1.0 N.

Upon incorporating ca. 1.0 wt% of metal (Al, Ga, and Fe) onto the SZ-1.0 N catalyst, both TMPO-loaded Al/SZ and Ga/SZ samples exhibit a total of eight ^{31}P resonances (Fig. 32A), in which four of them have

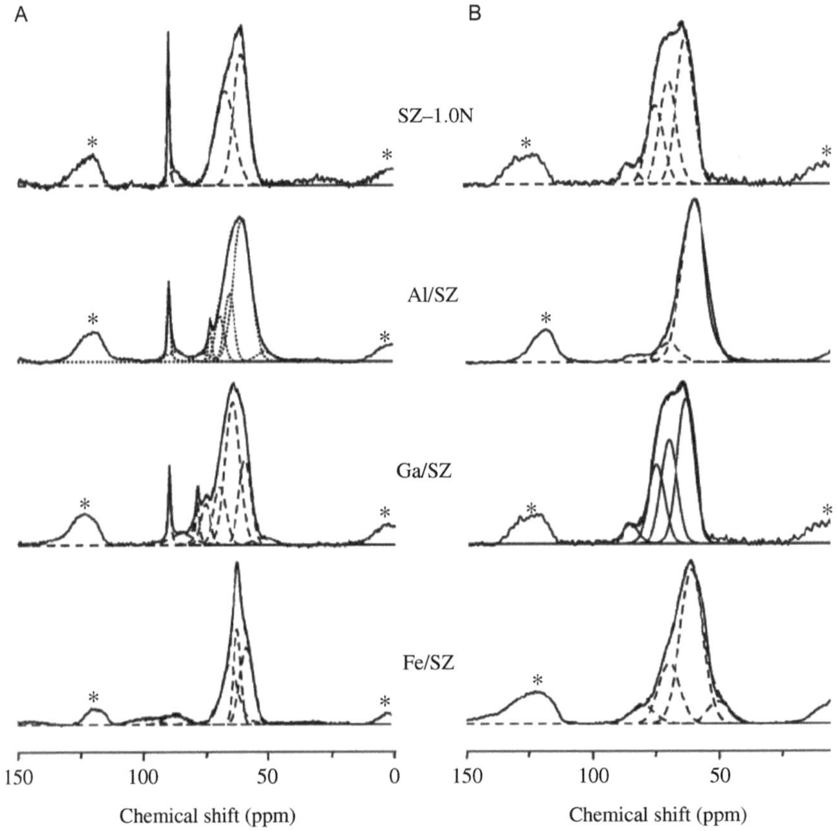

Fig. 32 ^{31}P MAS NMR spectra of TMPO adsorbed on dehydrated M/SZ (M = Al, Ga and Fe) (A) before and (B) after hydration treatment (for 1 h). The dashed curves represent spectral simulation results and the asterisks denote spinning sidebands (spinning rate 12 kHz) [29]. *Reproduced from W.H. Chen, H.H. Ko, A. Sakthivel, S.J. Huang, S.H. Liu, A.Y. Lo, T.C. Tsai, S.B. Liu, A solid-state NMR, FT-IR and TPD study on acid properties of sulfated and metal-promoted zirconia: Influence of promoter and sulfation treatment, Catal. Today 116 (2006) 111–120. Copyright 2006 Elsevier.*

identical $\delta^{31}P$ with those found for the non-promoted SZ-1.0 N. By the same token with the methodology illustrated earlier, identification of acid types may be accomplished by performing additional experiments on hydrated samples (Fig. 32B). Accordingly, the presences of extra Brønsted (76 and 65 ppm) and Lewis (73 and 53 ppm) acid sites in M/SZ may be inferred, as summarized in Table 3. Note that these additional peaks tend to span over a wide range (from weak to medium) of acidic strengths, occurred mostly at the disposals of the strongest Lewis acid sites (at 90 ppm) and medium Brønsted acid sites (at 68 ppm) in SZ-1.0 N. For the Fe/SZ catalyst, formations of very strong (99 ppm) and very weak (53 ppm) Lewis acid sites that are accompanied by the notable increase in strong (87 ppm) and additional weak (65 ppm) Brønsted acid sites were observed, mostly at the collaborative expanses of the strong (90 ppm) and weak (62 ppm) Lewis and medium Brønsted (68 ppm) acid sites of its parent SZ-1.0 N. Obviously, by incorporation of different metal species onto SZ not only resulted in formation of new Brønsted and Lewis acid sites with varied acidic strengths, but also leads to variations in concentration among different acid sites, which are anticipated to have considerable impact to their catalytic activity and selectivity.

3.2.3 Heteropolyacids

Heteropolyacids (HPAs), especially those with Keggin-type structures, have been extensively employed as solid acid catalysts and photocatalysts in various homogeneous solutions, liquid-solid, and gas-solid heterogeneous reactions due to their strong acidity [9,93]. Fig. 33 displays the ^{31}P MAS NMR spectra of varied amount of TMPO adsorbed on 12-molybdophosphoric acid ($H_3PMo_{12}O_{40}$; HPMo) with varied loading. The ^{31}P resonances in the range of -5 to -10 ppm were due to the $PM_{12}O_{40}{}^{3-}$ polyanions [95,96]. Upon adsorption of TMPO probe molecule, broad resonance peaks in the $\delta^{31}P$ range of 80–90 ppm were observed (Fig. 33) [94]. Further spectral analyses revealed that these peaks may be deconvoluted into five resonances centred at 89.9, 87.5, 86.2, 83.7, and 81.2 ppm. On the basis of DFT calculations, $\delta^{31}P$ can be predicted theoretically for various TMPO/HPMo adsorption complexes. The formal three peaks at 89.9, 87.5, and 86.2 ppm may be assigned to one TMPO adsorption complex per Keggin unit (KU), i.e. $TMPOH^+$/KU. Whereas the latter two signals at 83.7 and 81.2 ppm may be ascribed to two $TMPOH^+$ complexes in one Keggin unit, i.e. $(TMPOH^+)_2$/KU. Nevertheless, as the TMPO loading exceeding ca. 2.5 TMPO/KU, an additional broad signal centring at ca. 63 ppm was observed.

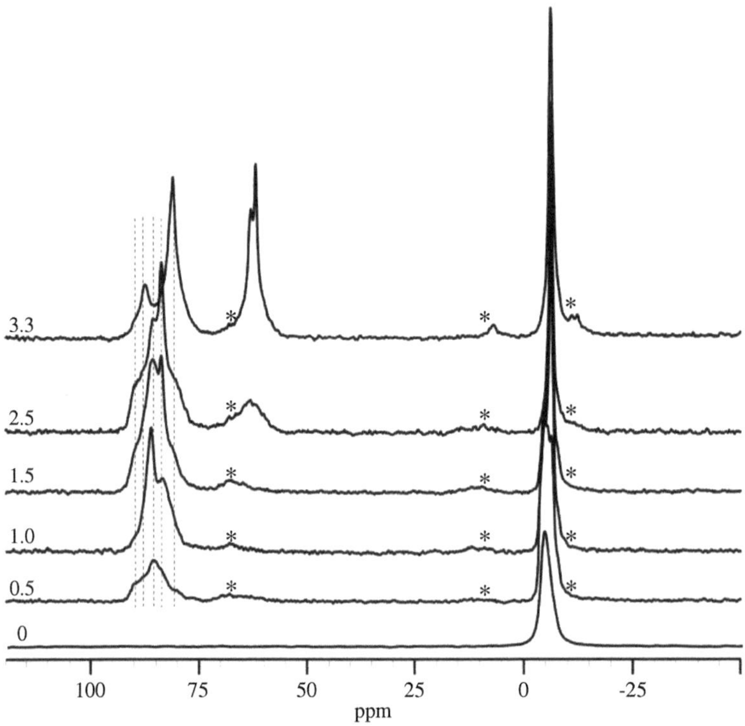

Fig. 33 ^{31}P MAS NMR spectra of bare HPMo and TMPO/HPMo samples with varied TMPO loading ranging from 0.5 to 3.3 TMPO/KU. Prior to the NMR experiments, each TMPO-loaded sample was subjected to baking treatment at 473 K for 8 h [94]. The asterisks denote spinning sidebands. *Reproduced from N. Feng, A. Zheng, S.J. Huang, H. Zhang, N. Yu, C.Y. Yang, S.B. Liu, F. Deng, Combined solid-state NMR and theoretical calculation studies of Brønsted acid properties in anhydrous 12-molybdophosphoric acid, J. Phys. Chem. C 114 (2010) 15464–15472. Copyright 2010 American Chemical Society.*

As shown in Fig. 34, additional $^{31}P\{^{1}H\}$ LG-CP HETCOR experiments [47] were used to further validate interactions between the TMPO (guest adsorbate) molecules and Brønsted acid sites in HPMo (host adsorbent). Such 2D multi-nuclei technique represents a direct approach to probe correlations between the local environments of ^{1}H and ^{31}P nuclei. The cross-peaks which correlate ^{31}P resonances at ca. 80–90 ppm and ^{1}H resonances at ca. 6–8 (Fig. 34A) and 9–13 (Fig. 34B) ppm should be originated from one TMPO adsorbed on various protonic sites (i.e. $TMPOH^{+}$) in HPMo with varied acidic strengths. Whereas, the cross-peaks that correlate the ^{31}P and ^{1}H resonances at ca. 60–65 ppm and ca. 15.5–16.5 ppm, respectively, should be due to the presence of $(TMPO)_2H^{+}$ adsorption complexes

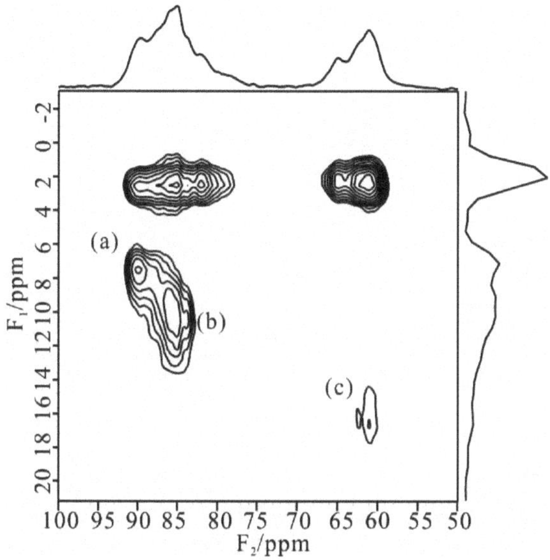

Fig. 34 $^{31}P\{^1H\}$ LG-CP HETCOR spectrum of TMPO adsorbed on HPMo sample with a loading of 2.5 TMPO/KU. Prior to the NMR experiment, the TMPO-loaded sample was subjected to baking treatment at 473K for 8h [94]. *Reproduced from N. Feng, A. Zheng, S.J. Huang, H. Zhang, N. Yu, C.Y. Yang, S.B. Liu, F. Deng, Combined solid-state NMR and theoretical calculation studies of Brønsted acid properties in anhydrous 12-molybdophosphoric acid, J. Phys. Chem. C 114 (2010) 15464–15472. Copyright 2010 American Chemical Society.*

in which the probe molecules are in close proximities with one and other (Fig. 34C). While, the cross-peaks associated with ^{31}P resonances at ca. 80–90 and 60–65 ppm and ^{1}H resonances at ca. 1.5–3.5 ppm should be arising from intra-molecular interactions between the methyl protons and the ^{31}P nucleus of the TMPO and hence are irrelevant to the distribution of protonic sites in HPMo.

It is noteworthy that the adsorption behaviour obtained from the TMPO/HPMo system is somewhat different from that of the TMPO adsorbed on 12-tungstophosphoric acid ($H_3PW_{12}O_{40}$; HPW) system. The investigation on the latter system showed that $(TMPO)_2H^+$ complexes tend to diminish when the TMPO-loaded HPW samples were subjected to baking treatment (T_b) $\geq$ 423 K, leaving only homogeneously distributed adsorption complex in the form of $(TMPOH^+)_3$/KU in HPW (see Fig. 35) [97]. This is clearly not the case for the TMPO/HPMo system, for which adsorption complexes in the form of $(TMPOH^+)_n$/KU ($n=1, 2$) and $(TMPO)_2H^+$ were also present in addition to the $(TMPOH^+)_3$/KU, even after a thorough

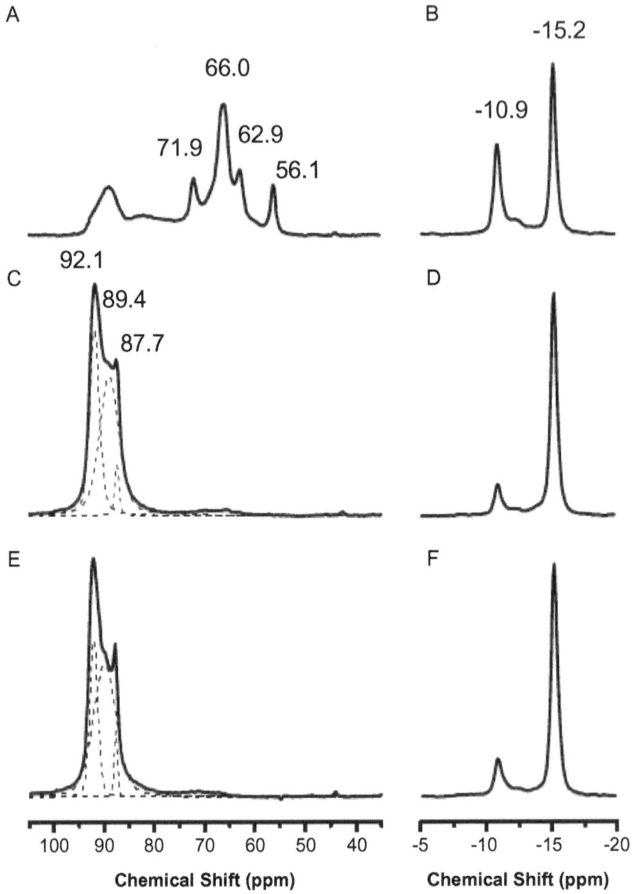

Fig. 35 Left: ^{31}P MAS NMR spectra of protonated TMPO adducts arising from TMPO adsorbed on HPW (loading 2.5 TMPO/KU). Right: ^{31}P signal corresponding to $PW_{12}O_{40}^{-3}$ polyanions. Prior to the NMR experiments, the sample was subjected to baking treatment at (A–B) 373 K, (C–D) 423 K, and (E–F) 473 K for 6 h [97]. The dashed lines represent Gaussian deconvolution results. *Reproduced from S.J. Huang, C.Y. Yang, A. Zheng, N. Feng, N. Yu, P.H. Wu, Y.C. Chang, Y.C. Lin, F. Deng, S.B. Liu, New insights into Keggin-type 12-tungstophosphoric acid from ^{31}P MAS NMR analysis of absorbed trimethylphosphine oxide and DFT calculations, Chem. Asian J. 6 (2011) 137–148. Copyright 2011 Wiley.*

sample pre-treatment at $T_b = 473\,K$ [97]. Therefore, it is hypothesized that HPMo has a relatively weaker pseudoliquid characteristic compared to that of HPW, making the TMPO probe molecules more difficult to diffuse within its secondary structure. Such discrepancy in acid properties observed for HPMo and HPW may be ascribed due to differences in the electronic configurations of Mo ($4d_5 5s_1$) and W ($5d_4 6s_2$), although these heteropolyacids have a similar crystalline structure.

Furthermore, owing to the superacidic characteristic of HPA and the "green solvent" nature of ionic liquid (IL), functionalized IL-HPA organic-inorganic composite catalysts have been developed and utilized for conversions of biomass [98–101]. For example, Hung et al. synthesized a series of HPW immobilized on imidazolium chloride ionic liquid modified mesoporous SBA-15 silica with tunable superacidity (denoted as z-HPW-IL/SBA; z=HPW/IL molar ratio), and exploited them for catalytic alkylation of benzene with 1-dodecen [101]. The acid features (i.e. amount, strength, and distribution of acid sites) of these surface-functionalized HPW-IL/SBA catalysts and their impact on catalytic performances were investigated. The acid properties of these hybrid catalysts were explored by using the ^{31}P-TMPO NMR approach. As shown in Fig. 36, ^{31}P MAS NMR spectra observed for various z-HPW-IL/SBA catalysts exhibited multiple resonances that featured at ca. 40–80 ppm and ca. 80–95 ppm. The ^{31}P peaks at 40–80 ppm may be attributed to possible presences of $(TMPO)_2H^+$ complexes, or distorted $TMPOH^+$ adducts due to spatial hindrances, whereas the ^{31}P resonances at ca. 80–95 ppm were ascribed due to $TMPOH^+$ adsorption complexes. Clearly, these HPW-immobilized catalysts also possess ultra-strong Brønsted acidity (i.e. $\delta^{31}P \geq 86$ ppm). Moreover, by monitoring the fraction of superacidity vs. relative HPW loading (z), detailed acidity-activity correlation of various z-HPW-IL/SBA catalysts may readily be manipulated.

3.2.4 other functionalized solid acid materials

Bifunctional carbon-silica composite materials (CSM), which have been shown to be efficient catalysts for conversion of sugars, are normally constructed by porous silicas (such as mesoporous MCM-41 molecular sieve) tucked with porous carbon matrix. The ^{31}P-TMPO NMR approach was used to probe the acid strength of CSM catalysts [102]. Accordingly, the ^{31}P resonances with $\delta^{31}P$ within the range of 40–70 ppm were attributed to the presence of weakly acidic oxygen-containing surface groups on the carbonized material. To avoid interference arising from physisorbed and crystalline TMPO, a TMPO loading under the estimated stoichiometric amount was adopted. As a result, no ^{31}P resonance at 41 ppm was found, indicating that crystalline TMPO were indeed absent. Consequently, two ^{31}P resonances were observed for the TMPO-loaded CSM sample post-treated at 573 K. The authors ascribed the main peak (at ca. 45 ppm) and

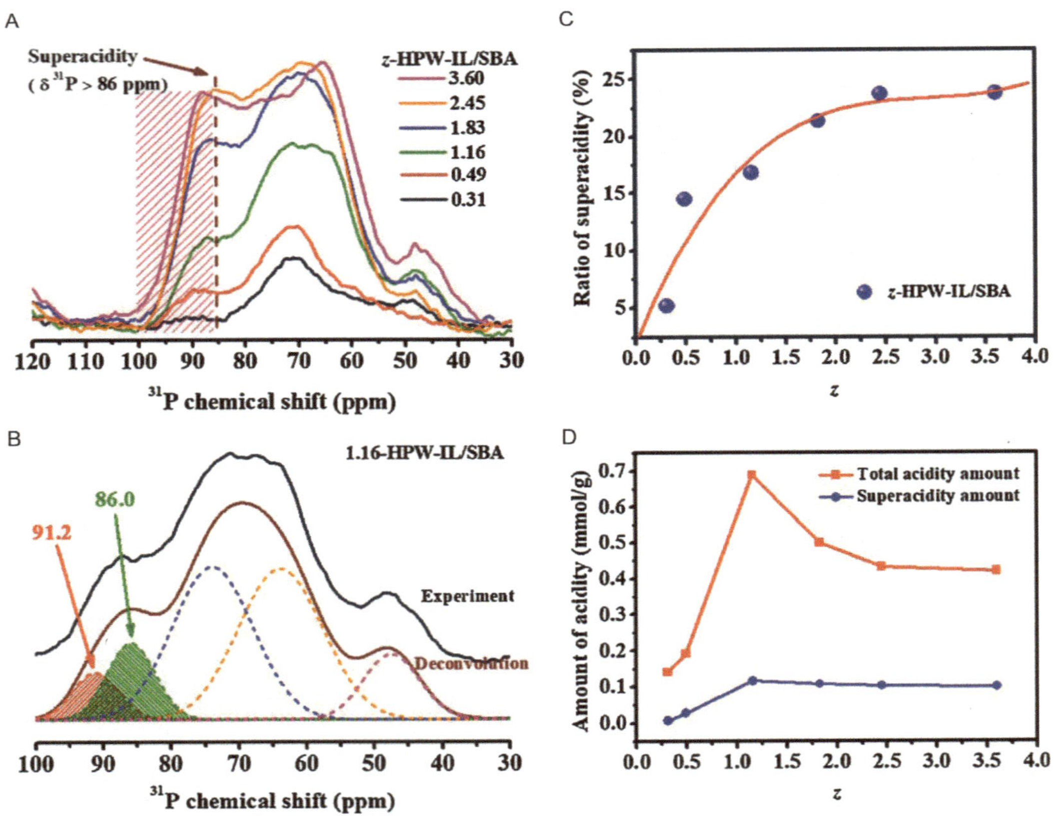

Fig. 36 (A) ^{31}P MAS NMR spectra of TMPO adsorbed on various *z*-HPW-IL/SBA catalysts ($z = 0.31–3.60$) and (B) corresponding spectral deconvolution of the 1.16-HPW-IL/SBA catalyst, (C) plot of superacidity ratio (SAR) against HPW/IL ratio (*z*), and (D) variations of total acidity amount (TAA) and superacidity amount (SAA) with *z* [101]. *Reproduced from C.T. Hung, L.L. Liu, J.J. Wang, P.H. Wu, C.B. Wang, T.C. Tsai, S.B. Liu, Acidity and alkylation activity of 12-Tungstophosphoric acid supported on ionic liquid-functionalized SBA-15, Catal. Today 327 (2019) 10–18. Copyright 2019 Elsevier.*

the weak shoulder peak (at ca. 60 ppm) to TMPO adsorbed on weak and strong acid sites in CSM, respectively. The broadening of the ^{31}P resonance observed for the post-treated samples was attributed to heterogeneous distribution of acid sites with varied acidic strengths in the catalyst. The Brønsted acidity observed for CSM mainly arises from the surface functional groups such as phenols, anhydrides, and carboxylic acids in carbon matrix, thus, exhibiting an overall acidic strength much weaker than that of typical zeolites. Nonetheless, the absence of strong Brønsted acid sites serves to avoid the occurrence of undesirable side-reactions. Moreover, the strength and concentration of Brønsted acid sites in CSM may readily be controlled by varying the carbon deposition content, pyrolysis temperature, and thermal post-treatment conditions.

Mesoporous poly(divinylbenzene) (PDVB) based solid acids, such as PDVB-SO_3H-SO_2CF_3 prepared by grafting of strong electron withdrawing group of SO_2CF_3 onto the network of mesoporous PDVB-SO_3H, have been demonstrated to exhibit excellent catalytic activities during biomass transformation towards transesterification to biodiesel and depolymerization of crystalline cellulose to sugars. To unravel the correlation between acid property and reaction activity, Liu et al. adopted the ^{31}P-TMPO NMR approach to characterize the distribution and strength of acid sites in PDVB-SO_3H and PDVB-SO_3H-SO_2CF_3 catalysts [103]. The ^{31}P MAS NMR spectrum of TMPO adsorbed on PDVB-SO_3H shows two resonances with $\delta^{31}P$ of 72 and 80 ppm with a relative peak area (concentration) of 40% and 60%, respectively. These two ^{31}P resonances may be attributed unambiguously to TMPO adsorbed on acid sites of PDVB-SO_3H with different acidic strengths, forming $TMPOH^+$ complexes with varied extents of protonation. Upon further treatment of PDVB-SO_3H by HSO_3CF_3, notable increase in Brønsted acidic strength was observed for the PDVB-SO_3H-SO_2CF_3 catalyst, leading to a more homogeneously distributed acid sites, as revealed by the singlet peak with $\delta^{31}P$ of 83 ppm. Furthermore, based on the correlation between deprotonation energy (DPE) and ^{31}P chemical shift attained for the ^{31}P-TMPO NMR approach in Eq. (4), DPE value corresponding to $\delta^{31}P$ of 72, 80, and 83 ppm should be ca. 284, 264, and 256 kcal/mol, respectively. Apparently, the acidic strengths in PDVB-SO_3H were dramatically enhanced through the treatment by the electron withdrawing CF_3 groups, thus, more favourable for enhancing the catalytic activity of the catalyst.

Similar treatment by SO_3H may also be applied to modify the surface of porous carbon materials to facilitate active functional moieties desirable for acid-catalysed hydrophobic reactions. As a result, the surface SO_3H groups may be readily accessed by hydrophobic reactants, leading to the high catalytic performance observed for such functionalized carbons. Suganuma et al. utilized the ^{31}P-TMPO NMR approach to investigate acid properties of functionalized mesoporous carbon-based solid acids, namely CAC, RF-773, and RF-973, prepared by post-synthesis sulphonation treatment [104]. Among them, CAC represents cellulose-derived bulky amorphous carbon bearing -SO_3H and phenolic –OH groups, RF-773 and RF-973 samples were amorphous carbon with different amounts of phenolic –OH groups; the former having a higher amount than the latter. The ^{31}P resonance at 83 ppm for the CAC material was attributed to TMPO adsorbed on SO_3H groups. On the other hand, the corresponding spectrum for RF-773 revealed three distinct signals at 78, 63, and 41 ppm, which were assigned due to TMPO adsorbed on acidic –SO_3H, phenolic –OH, and self-aggregated TMPO, respectively. In the case of RF-973, however, only the signal of self-aggregated TMPO (41 ppm) was observed, despite of the fact that vibrational bands responsible for –SO_3H group was evident in the IR spectrum. The authors hypothesized that this may be due to the –SO_3H and phenolic –OH groups in RF-973 were inaccessible to the TMPO probe molecule. Consequently, a null catalytic activity was observed for the RF-973 catalyst during a-methylstyrene (AMS) dimerization reaction [104].

Sulphated metal-organic framework (MOFs) materials have been shown to be potential catalysts for acid-catalysed reactions [105]. However, owing to the poor thermal stability of MOFs, conventional acidity characterization method such as NH_3-TPD is infeasible since it may lead to disintegration of MOF structure. Yaghi et al. utilized the ^{31}P-TMPO NMR approach to characterize the acid properties of Zr(IV)-based sulphated MOF-808-$x$$SO_4$ (where x represents concentration of the sulphate groups) [106]. Upon increasing the amount of the SO_4^{2-} in the sample, gradual formation of Brønsted acid sites with desirable acidic strength was found, as confirmed by the ^{31}P resonance emerged at 69 ppm. This is in excellent agreement with the superior catalytic performances observed in highly sulphated MOF-808 [106]. Further study utilizing the ^{31}P-TMPO NMR approach enable the authors to identify that the strong Brønsted acidity in the MOF-808-SO_4 catalyst is originated from a specific arrangement of adsorbed water and sulphate moieties on zirconium clusters [107]. Upon adsorbing water molecule, the zirconium atom tends to form hydrogen bonding with a sulphate moiety that is chelated to a neighbouring zirconium atom, which in turn leads to the formation of strong Brønsted acidic proton. As shown in

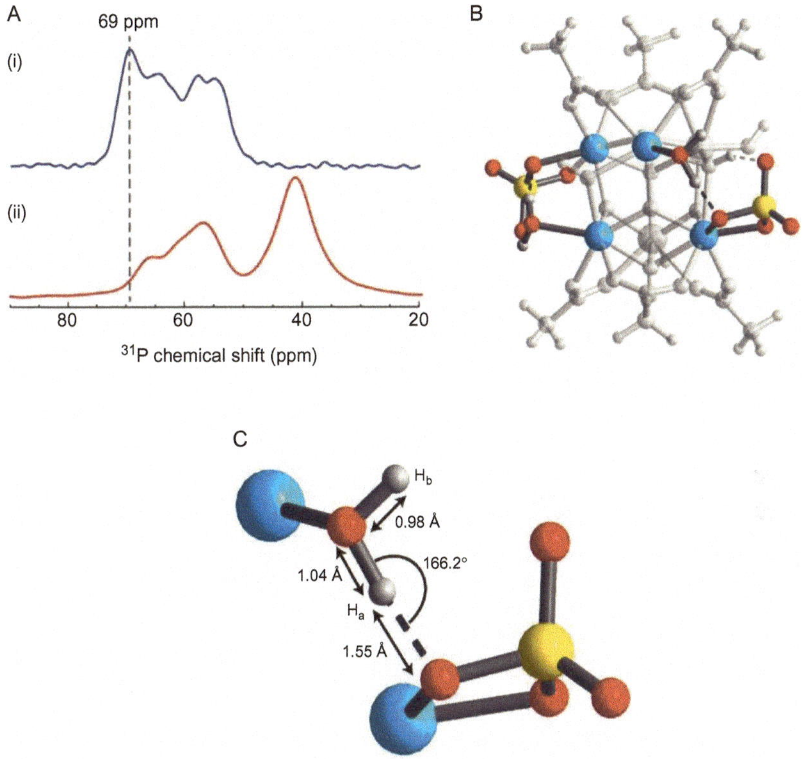

Fig. 37 (A) ^{31}P MAS solid-state NMR spectra of TMPO adsorbed on (i) hydrated and (ii) dehydrated MOF-808-SO_4. (B) Depiction of the overall cluster and (C) optimized structure of Brønsted acid site nearby to zirconium clusters in MOF-808-SO_4 [107]. *Reproduced from C.A. Trickett, T.M. Osborn Popp, J. Su, C. Yan, J. Weisberg, A. Huq, P. Urban, J.C. Jiang, M.J. Kalmutzki, Q.N. Liu, J. Baek, M.P. Head-Gordon, G.A. Somorjai, J.A. Reimer, O.M. Yaghi, Identification of the strong Brønsted acid site in a metal-organic framework solid acid catalyst, Nature Chem. 11 (2018) 170–176. Copyright 2018 Nature Publishing Group.*

Fig. 37A, the ^{31}P NMR spectrum of TMPO adsorbed on hydrated MOF-808-SO_4 showed a ^{31}P resonance at 69 ppm associated with a strong Brønsted acid site. However, this signal was invisible for TMPO adsorbed on dehydrated MOF-808-SO_4. For the latter, the peak centred at 42 ppm may be ascribed to excessive crystalline TMPO. In this context, water clearly plays a crucial role for the formation of strong Brønsted acid sites in the MOF-808-SO_4 catalyst [107]. More recently, the ^{31}P-TMPO NMR approach has also been employed to probe the acidity properties of functionalized covalent-organic frameworks (COFs) [108] and mesoporous organosilicas [109] materials.

Further improvements on mass transport and acid properties of solid acid catalysts such as zeolites remain as challenging tasks in catalysis [110]. In this regard, the emergence of two-dimensional (2D) materials provides a new opportunity in the design and fabrication of novel solid acid catalysts [111]. Recently, Liu et al. developed a series of 2D layered hybrid solid acids with outstanding mass transfer and ultra-high acidic strength by incorporating sulphonated polymers in montmorillonite (MMT) layers [112]. The authors adopted the ^{31}P-TMPO NMR approach to evaluate the acidity properties of these novel 2D catalyst materials. As shown in Fig. 38A, two distinct ^{31}P resonances at 46 and 55 ppm were observed for the bare MMT support, which may be assigned to the physisorbed TMPO and TMPO adsorbed on weak acid sites, respectively. On the other hand, the sulphonated polymer PAN-[C_3N][SO_3CF_3] exhibited the presence of Brønsted acidity with high acidic strengths, as evidenced by the observed ^{31}P resonance peak at 84 ppm when in the presence of adsorbed TMPO. It is intriguing that the intercalation of such sulphonated polymer into the MMT layer resulted in a significant increasement in the ultra-strong acid sites with δ^{31}P of 86 and 93 ppm, corresponding to two types of superacid sites in MMT@PAN-[C_3N][SO_3CF_3] (Fig. 38B) [9,10]. Further characterization by the conventional Hammett indicator method also confirmed the presence of superacidity [112], whose origin remains largely unknown. It was proposed that these superacidic sites were originated from strong interactions between the sulphonated polymer and the MMT scaffold, which acts as an efficient H^+ donor (Fig. 38C).

Petroleum cokes are one of the vast by-products in petrochemical industry. To salvage these petroleum cokes, it is common to invoke KOH activation and/or H_2SO_4 sulphonation treatment to incorporate functional groups such as –OH, –COOH, and –SO_3H onto the carbonaceous substrates. As a result, surface-functionalized petroleum cokes have been considered as novel solid acid catalysts. Zeng et al. studied the acid properties of functionalized petroleum cokes by means of the ^{31}P SSNMR of adsorbed TMP and TMPO and examined their catalytic performances during esterification of oleic acid with methanol [113]. As shown in Fig. 39A, the ^{31}P spectrum obtained for TMP adsorbed on the unmodified petroleum coke material revealed only a single peak at −60 ppm, which may be assigned due to physisorbed TMP. This indicates that neither Lewis nor Brønsted acid sites were present in the pristine petroleum cokes. Whereas two ^{31}P resonance peaks at −4 and −60 ppm were observed for TMP adsorbed on petroleum coke after the KOH/H_2SO_4 treatment (Fig. 39B). The presence

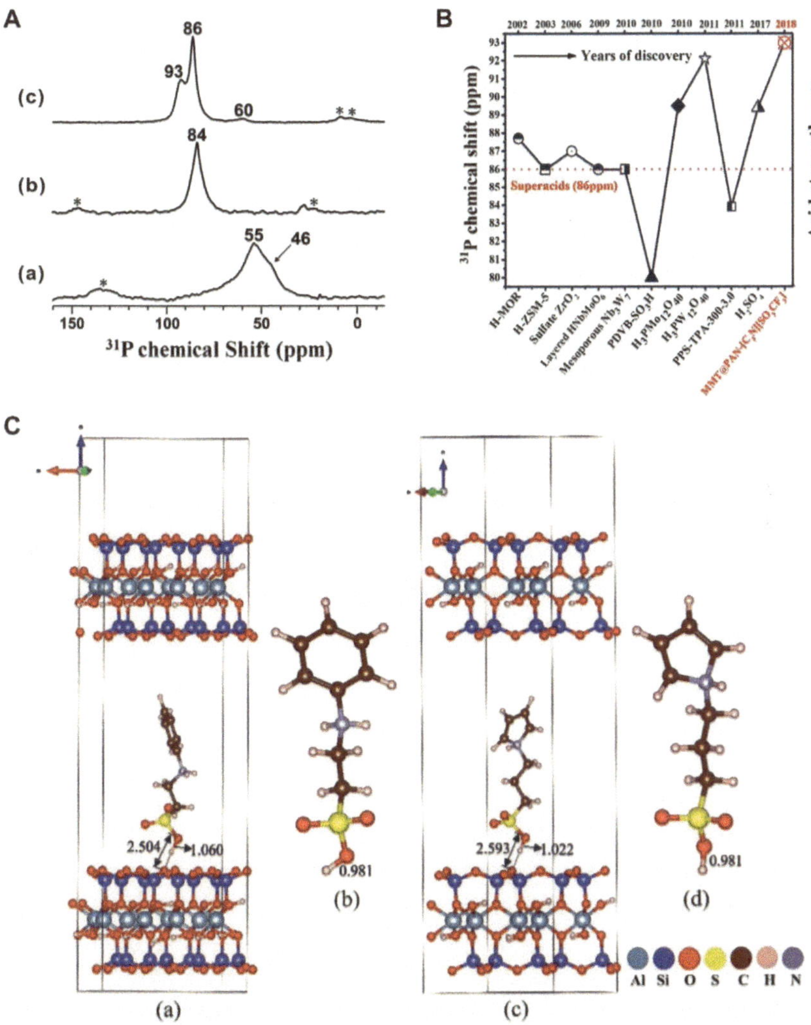

Fig. 38 (A) ^{31}P MAS NMR spectra of TMPO adsorbed on (a) MMT support, (b) PAN-[C_3N][SO_3CF_3] and (c) MMT@PAN-[C_3N][SO_3CF_3]. (B) Comparisons of acidic strength of MMT@PAN-[C_3N][SO_3CF_3] with other common solid acids. (C) Optimized structures of sulphonated aniline and pyrrole in MMT (A and C) and sulphonated aniline and pyrrole monomers (B and D). The calculated O–H interatomic distances (in Å) are also indicated [112]. *Reproduced from F. Liu, X. Yi, W. Chen, Z. Liu, W. Chen, C. Qi, Y. Song, A. Zheng, Developing two-dimensional solid superacids with enhanced mass transport, extremely high acid strength and superior catalytic performance, Chem. Sci. 10 (2019) 5875–5883. Copyright 2019 The Royal Society of Chemistry.*

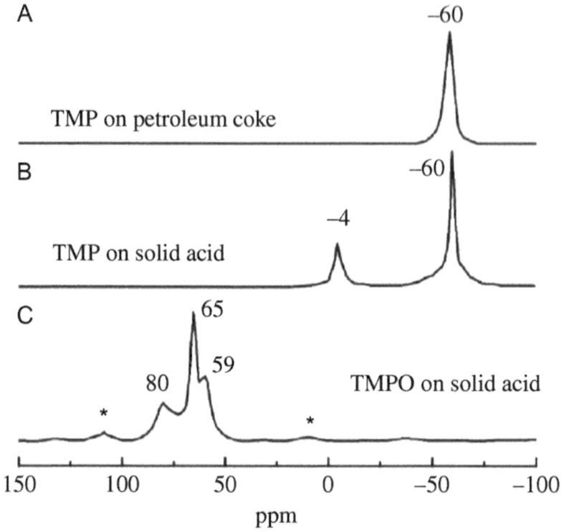

Fig. 39 ^{1}H-decoupled ^{31}P MAS NMR spectra of TMP adsorbed on (A) petroleum coke, (B) solid acids, and (C) TMPO adsorbed on solid acid. The asterisks denote spinning sidebands [113]. *Reproduced from D. Zeng, S. Liu, W. Gong, G. Wang, J. Qiu, Y. Tian, Acid properties of solid acid from petroleum coke by chemical activation and sulfonation, Catal. Commun. 40 (2013) 5–8. Copyright 2013 Elsevier.*

of the new peak at −4 ppm indicates that the chemical activation and sulphonation treatment provoked formation of Brønsted acid sites in the solid catalyst. On the other hand, three distinct ^{31}P resonances at 59, 65, and 80 ppm were observed by adopting the ^{31}P-TMPO NMR approach (Fig. 39C), which were ascribed to TMPO adsorbed on Brønsted acid sites with varied acidic strengths. Together with the results obtained from FT-IR experiments and theoretical calculations, three types of moieties on the solid acid were identified, namely weak (–OH), media (–COOH), and strong (–SO_3H) acidic functional groups. Thus, the resonances at 59, 65, and 80 ppm were assigned to TMPO adsorbed on the –OH, –COOH, and –SO_3H groups on the surfaces of the modified petroleum cokes, respectively [113].

3.2.5 Commercial catalysts

Commercial solid acid catalysts are mostly in form of granular catalyst-binder extrudates in order to sustain their physicochemical and catalytic properties over the long-term catalytic cycles. Most commercial catalysts are commonly formulated by a certain type of zeolite catalyst with a desirable amount of alumina or silica binder, and eventually extruded to form catalyst particles of various shapes and sizes [114]. Nonetheless, detailed information on acid properties of such catalyst mixture (especially in the case of commercial catalysts) are normally deficient in the literatures. Presumably, solid interactions between zeolite and binder materials may occur through intimate contact, which in turn may alter the acidity and catalytic properties of the original zeolite catalyst [115]. Moreover, the formation of carbonaceous residues (or cokes) during catalytic reaction is one of the major concerns from the industrial operation perspective. Since coking is a shape-selective reaction [116], the presence of binder, therefore, may alter the deactivation rate of the zeolite catalyst. It is well-known that while acidity plays a key role during catalytic reactions, coking is the primary cause for catalyst deactivation. It was found that carbonaceous deposits (cokes), which tend to inactivate acid sites, are mostly located in the intracrystalline voids of the zeolitic catalysts during reaction and hence should be responsible for the decline in catalytic activity [117].

Liu and co-workers studied the effects of binder, coking, and regeneration on the acid properties of a commercial H-mordenite catalyst during toluene disproportionation reaction [118]. The authors utilized ^{31}P SSNMR of adsorbed TMPO and TBPO for acidity characterization of various fresh, spent, and regenerated catalysts. As shown in Fig. 40, three resonance peaks with $\delta^{31}P$ at 68, 56, and 49 ppm were observed when TMPO was adsorbed on γ-Al_2O_3 binder material. While the peak at 49 ppm was

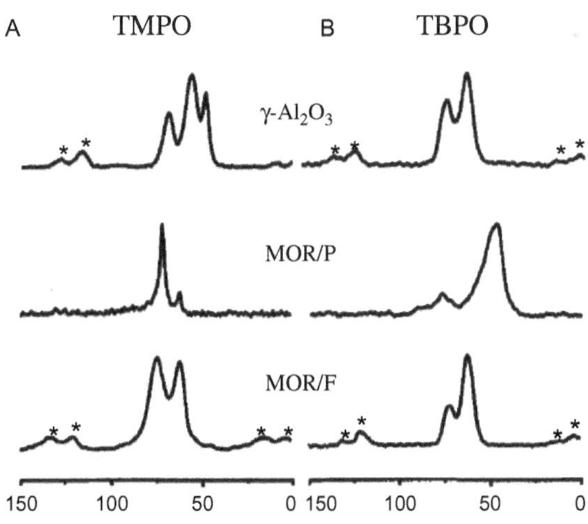

Fig. 40 ^{31}P MAS NMR spectra of (A) TMPO and (B) TBPO adsorbed on various samples. The asterisks represent spinning sidebands [118]. *Reproduced from W.H. Chen, S.J. Huang, C.S. Lai, T.C. Tsai, H.K. Lee, S.B. Liu, Effects of binder, coking and regeneration on acid properties of H-mordenite during TDP reaction, Res. Chem. Intermed. 29 (2003) 761–772. Copyright 2003 Springer.*

unambiguously assigned due to physisorbed TMPO, the other two downfield peaks were attributed to TMPO absorbed over two different Lewis acid sites. Since the pore size of the γ-Al_2O_3 binder (ca. 5 nm) is surely large enough to accommodate TBPO probe molecule (kinetic diameter 0.82 nm), three characteristic peaks at 74, 63, and 58 ppm were also observed for the TBPO/γ-Al_2O_3 system; similar to that of TMPO/γ-Al_2O_3. Beside the physisorbed TMPO, the ^{31}P spectra obtained from parent mordenite zeolite (denoted as MOR/P) and fresh commercial catalysts (denoted as MOR/F) samples revealed additional signals ($\delta^{31}P > 74$ ppm) corresponding to Brønsted acid sites in the mordenite zeolite. On the other hand, for TBPO adsorbed on MOR/P, majority of the probe molecules were adsorbed on the extracrystalline surfaces of the zeolite. Note that the presence of γ-$A1_2O_3$ binder in MOR/F slightly broaden the resonance peaks and to some extent alter the relative concentrations of the weaker acid sites. Other than that, key features of Brønsted (from MOR/P) and Lewis acid sites (from γ-$A1_2O_3$) from their respective origins were largely preserved in mixture catalyst (MOR/F).

Fig. 41 displays ^{31}P NMR spectra of TMPO and TBPO adsorbed on fresh commercial catalyst (MOR/F), various spent catalysts (MOR/C-*x*; *x* represent coke content in wt%), and a regenerated catalyst (MOR/R). As summarized in Table 4, up to six resonances with $\delta^{31}P$ at 89, 80, 74, 68, 62, and 56 ppm were observed for TMPO adsorbed on the MOR/F sample. Among them, peaks at 68 and 56 ppm were attributed to the presence of Lewis acidity originated from the γ-Al_2O_3 binder, while the other peaks were assigned to Brønsted acid sties with varied acidic strengths. It is noted that, unlike TBPO that can only probe external acid sites, TMPO (0.55 nm) may be accommodated by the 12-MR or mordenite zeolite, hence, capable of probing both internal and external acidity [118]. As such, the results depicted in Table 4 for the ^{31}P-TMPO approach are valid as far as the overall acid concentration is concerned. On the other hand, notable decreases in Brønsted acidity (especially those corresponding to peaks with $\delta^{31}P$ at 89 and 80 ppm) were observed during initial stage of reaction. This indicates that cokes tend to preferentially deposited on strongest acid sites during early time-on-stream.

As shown in Fig. 41 and Table 4, a progressive diminishing of the ^{31}P NMR resonances at downfield with increasing coke content (or duration of reaction) was evident, revealing a progressive inactivation of Brønsted acid sites. Note that such progressive decrease in Brønsted acidity is accompanied by a progressive increase in the amount (but not strength) of Lewis

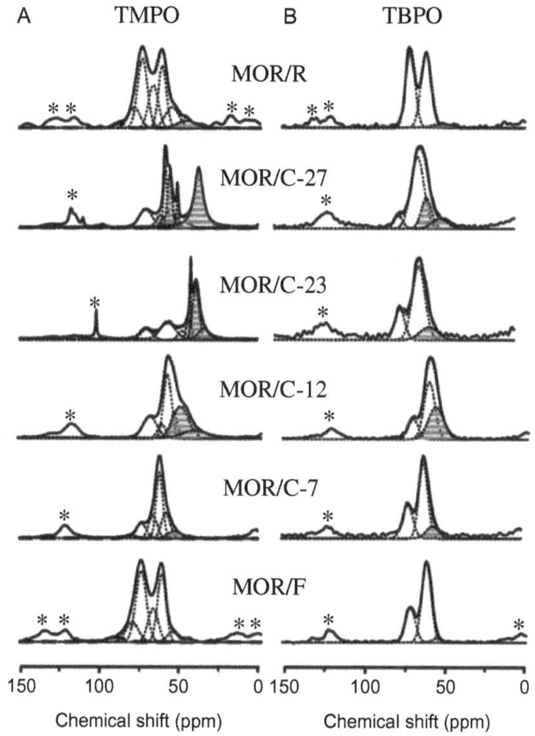

Fig. 41 ^{31}P MAS NMR spectra of (A) TMPO and (B) TBPO adsorbed on fresh (MOR/F), regenerated (MOR/R) and various coked (MOR/C) samples. The dashed curves represent simulation of resonance peaks by Gaussian deconvolution method. Shaded peaks denote contributions from physisorbed adsorbates and the spinning sidebands are indicated by asterisks [118]. *Reproduced from W.H. Chen, S.J. Huang, C.S. Lai, T.C. Tsai, H.K. Lee, S.B. Liu, Effects of binder, coking and regeneration on acid properties of H-mordenite during TDP reaction, Res. Chem. Intermed. 29 (2003) 761–772. Copyright 2003 Springer.*

acidity. It was found that, as the catalyst coke content exceeds ca. 23 wt% (which corresponding to a time-on-stream of 40 days), the ^{31}P resonances associated with Brønsted acidity were mostly deactivated, while the concentration of Lewis acidity (peaks at 68 and 56 ppm) prevailed to sustain the toluene disproportionation reaction (at least up to 400 days, in case of MOR/C-27). Moreover, it is noteworthy that majority of acid sites may be recovered upon a simple regeneration treatment, as can be seen by comparing results obtained from the MOR/R and MOR/F samples. Therefore, on the basis of such ex situ study using the ^{31}P-TMPO NMR approach, detailed variations of acid sites with reaction time can readily

Table 4 Assignments of ^{31}P NMR chemical shifts and distribution of acid sites for fresh, coked and regenerated commercial H-mordenite samples loaded with TMPO probe molecule [118].

	^{31}P NMR chemical shift vs. acid concentration (ppm)					
Sample	**89**	**80**	**74**	**68**	**62**	**56**
MOR/F	3.6 (0.014)	11.6 (0.044)	35.9 (0.135)	16.3 (0.061)	27.4 (0.103)	5.2 (0.020)
MOR/C-7	–	2.6 (0.006)	13.9 (0.031)	19.1 (0.043)	42.7 (0.096)	21.7 (0.049)
MOR/C-12	–	–	5.2 (0.016)	27.5 (0.084)	16.3 (0.050)	51.0 (0.156)
MOR/C-23	–	–	–	34.3 (0.007)	–	65.7 (0.143)
MOR/C-27	–	–	–	35.1 (0.117)	–	64.9 (0.216)
MOR/R	2.3 (0.007)	11.1 (0.035)	31.2 (0.098)	19.5 (0.061)	22.8 (0.071)	13.1 (0.041)

Values represent relative concentration of acid sites (%). Values in parentheses denote the acid amount of each site (± 0.002 mmol/g catalyst).

Reproduced from W.H. Chen, S.J. Huang, C.S. Lai, T.C. Tsai, H.K. Lee, S.B. Liu, Effects of binder, coking and regeneration on acid properties of H-mordenite during TDP reaction, Res. Chem. Intermed. 29 (2003) 761–772.

be pursued. Such approaches for acidity characterization should be valuable for assessment of commercial catalysts.

3.3 Location, distribution and spatial proximity of acid sites

The location and distribution (e.g. internal vs. external) of acid sites may also play a role during catalytic reaction, thus, are also crucial for the design and practical applications of solid acid catalysts. As mentioned earlier, the kinetic diameter (KD) of TMPO is ca. 0.55 nm, which is comparable to the pore aperture of typical 10-MR (pore size ca. 0.60 nm) zeolites such as ZSM-5. However, the sizes of its analogous such as TEPO (KD ca. 0.60 nm), TBPO (KD ca. 0.82 nm), and TOPO (KD ca. 1.10 nm) are too large to penetrate into the channels/pores of 10-MR zeolites, and hence can merely probe acid sites located on extracrystalline surfaces [27,28]. In this regard, acid sites in intracrystalline and on extracrystalline surfaces of different porous catalysts may be probed by proper choices of R_3PO probe molecules. For examples, zeolites with larger pore sizes such as 12-MR faujasite-type

(X or Y) zeolites may be probed either by TMPO, TEPO, or TBPO. As discussed above, such ^{31}P-R_3PO NMR approaches are capable of providing important information on acid features, viz. type (Brønsted vs. Lewis), strength, and distribution and concentration (i.e. relative amounts) of solid acid catalysts simultaneously. Moreover, if applied together with elemental analysis such as ICP-AES/MS, quantitative information (exact acid amounts) of acid sites may also be attained.

As shown in Fig. 42A, up to seven resonance peaks at $\delta^{31}P = 86$, 75, 67, 63, 53, 43, and 30 ppm can be identified in the ^{31}P MAS NMR spectrum of TMPO adsorbed on H-ZSM-5 zeolite. The former five downfield signals

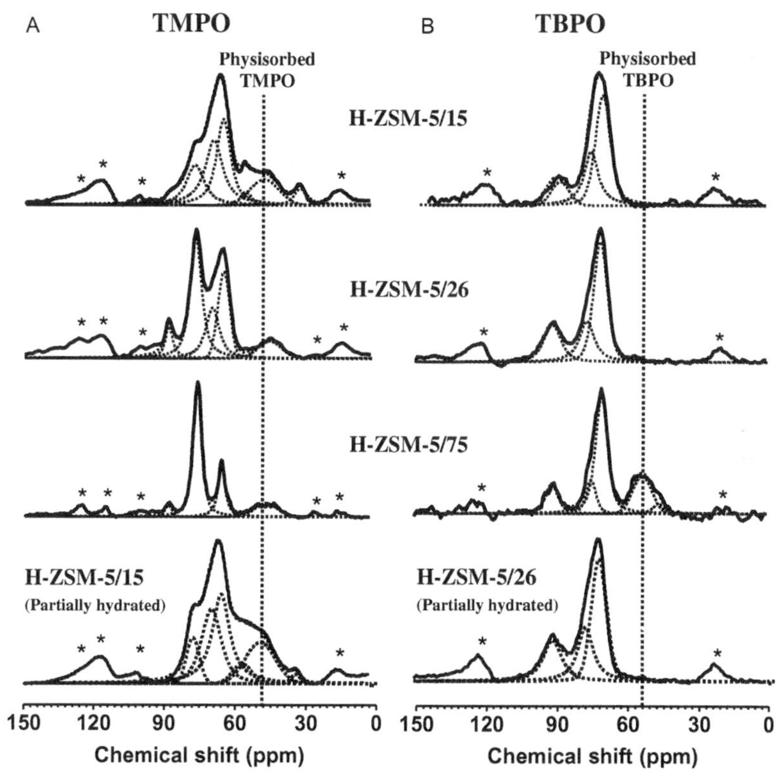

Fig. 42 ^{31}P MAS NMR spectra of (A) TMPO and (B) TBPO adsorbed on various H-ZSM-5 zeolites with different Si/Al ratios of 15, 26, and 75. The lower spectra were obtained from the H-ZSM-5/15 and H-ZSM/26 samples exposed to humidity for 1.5 h. The dashed curves indicate results of spectral analyses by Gaussian deconvolution. The sample spinning rate is 10 kHz. The asterisks denote spinning sidebands [27]. *Reproduced from Q. Zhao, W.H. Chen, S.J. Huang, Y.C. Wu, H.K. Lee, S.B. Liu, Discernment and quantification of internal and external acid sites on zeolites, J. Phys. Chem. B 106 (2002) 4462–4469. Copyright 2002 American Chemical Society.*

should be associated with TMPO adsorbed on Brønsted acid sites with varied acid strengths. The signal at 43 ppm was assigned to physisorbed TMPO; this peak remained intact even after the adsorbate-loaded sample was exposed to humidity. The signal at 30 ppm remains unchanged before and after sample silylation treatment, implying that it is likely due to "mobile" TMPO that either attached in the inter-crystalline voids or weakly adsorbed near the channel pore mouths of the zeolite. On the other hand, five characteristic peaks at $\delta^{31}P = 92$, 75, 71, 54, and 47 ppm were observed in the ^{31}P MAS NMR spectrum of TBPO adsorbed on H-ZSM-5 (Fig. 42B). The peak at $\delta^{31}P = 47$ ppm was attributed to crystalline TBPO. The resonances at $\delta^{31}P = 92$, 75, and 71 ppm should be related to TBPO adsorbed on the external acid sites, and the signal at 54 ppm was ascribed to physisorbed TBPO. By comparing the $\delta^{31}P$ values of R_3PO ($R = C_nH_{2n+1}$, $n = 1$–4) adsorbed on Brønsted acid sites with the same acidic strength, an average offset of 8 ± 2 ppm was observed based on DFT calculation results for R_3PO with $n \geq 2$ relative to TMPO ($n = 1$) [27]. Accordingly, the results obtained by using the ^{31}P-R_3PO NMR approaches with different probe molecules may be attained to provide information on detailed acid features of solid acid catalysts such as type, concentration, location, and strength of acid sites.

Ryoo and co-workers synthesized MFI zeolites with varied morphology by means of a surfactant as structure-directing agent [119]. The zeolitic catalysts so fabricated were found to exhibit superior catalytic activities during decalin cracking reactions. To correlate the observed reactivity with acid properties of the catalysts, the ^{31}P-R_3PO NMR approaches with TMPO and TBPO probe molecules were exploited. The ^{31}P NMR spectra of TMPO and TBPO adsorbed on various MFI zeolites are compared to a mesoporous Al-MCM-41 aluminosilicate, as shown in Fig. 43. The ^{31}P NMR spectrum of TMPO adsorbed on different MFI zeolites exhibited up to six resonance peaks with $\delta^{31}P$ of 30, 42, 66, 68, 76, and 86 ppm (Fig. 43A). Among them, the peak at 30 and 42 ppm were assigned to the mobile and physisorbed TMPO, respectively. Whereas the signals at downfield were attributed to TMPO chemisorbed on Brønsted acid sites. In other word, four types of Brønsted acid sites with different acidic strengths, viz. I (86 ppm), II (76 ppm), III (68 ppm), and IV (66 ppm) were present in the MFI samples. Among them, type I sites appear to possess an acidic strength close to the threshold of superacidity ($\delta^{31}P \geq 86$ ppm) [21]. It is noteworthy that, since the size of TMPO is comparable to the pore aperture of typical 10-MR zeolite, the results obtained from the ^{31}P-TMPO NMR

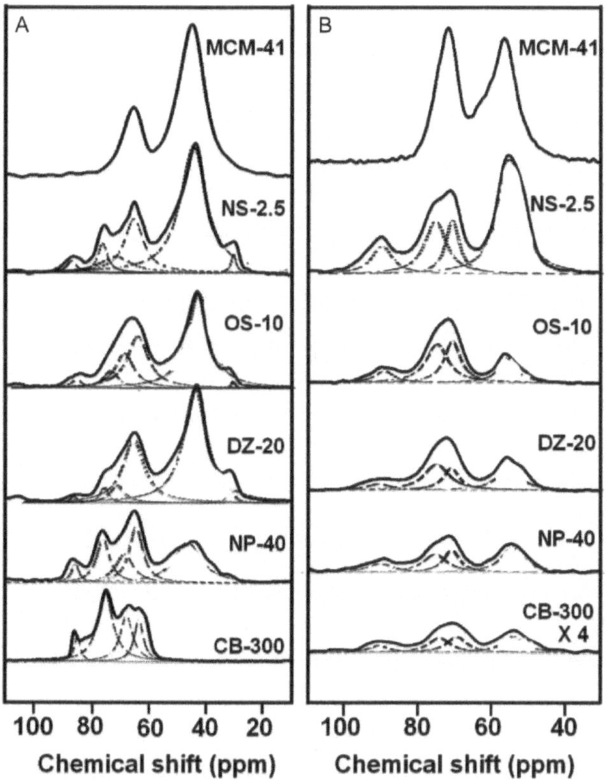

Fig. 43 ^{31}P MAS NMR spectra of (A) TMPO and (B) TBPO adsorbed on various MFI zeolites. The dashed curves indicate results of spectral analyses by Gaussian deconvolution. All spectra are plotted on the absolute intensity scale after collecting with the same number of acquisitions and being reduced to the same sample mass [119]. *Reproduced from Y. Seo, K. Cho, Y. Jung, R. Ryoo, Characterization of the surface acidity of MFI zeolite nanosheets by ^{31}P NMR of adsorbed phosphine oxides and catalytic cracking of decalin, ACS Catal. 3 (2013) 713–720. Copyright 2013 American Chemical Society.*

therefore contain information of both internal (intracrystalline) as well as external (extracrystalline) acid sites of the MFI zeolite samples as in the case of Al-MCM-41.

However, notable differences in the ^{31}P NMR spectra of MFI zeolites were observed when TBPO was used as probe, as shown in Fig. 43B. Compared to the spectra obtained by the ^{31}P-TMPO approach in Fig. 43A, the major differences of spectra in Fig. 43B arose not only from the anticipated intrinsic $\delta^{31}P$ difference (ca. 8 ± 2ppm) [28] between the TBPO and TMPO probe molecules, but also due to the fact that the size of TBPO is too bulky to enter the micropores ofMFI zeolite, hence, it can only probe acid sites located on external surfaces. In this regard, the three distinct ^{31}P resonances at 92, 74, and 72 ppm were exclusively arising from TBPO adsorbed on extracrystalline Brønsted acidic sites. Accordingly, the three peaks observed from the adsorbed TBPO may be correlated with the peaks obtained from TMPO as followed: 92 ppm (I), 74 ppm (III), and 72 ppm (IV). In this context, it is indicative that type II acid sites are exclusively located in intracrystalline voids of the MFI catalysts regardless of their differences in morphology (mainly crystalline thickness). Further quantitative information such as exact acid amounts corresponding to ^{31}P resonance peaks with different $\delta^{31}P$ values (i.e. acidic strengths) may be attained if analysed in conjunction with elemental analysis data. In practice, the amount of external acid sites may be obtained from peak areas corresponding to the adsorbed TBPO by first excluding the contribution from physisorbed TBPO. Accordingly, the amount of internal acid sites may be derived by subtracting the external amount obtained above from the total peak areas of the adsorbed TMPO (excluding contributions from mobile and physisorbed TMPO). As a result, it was found that the fraction of external acid amount increased with decreasing crystalline thickness of the zeolite. For example, the fraction of external acidity increased from 5% of the conventional CB-300 crystalline bulk zeolite to ca. 32% of the NS-2.5 zeolite nanosheet. Moreover, a linear correlation between the number of strongest external acid sites (i.e. type I from TBPO result) and catalytic activity of these MFI samples during decalin cracking may be inferred, indicating that the reaction took place exclusively on the external surfaces of the catalysts [119].

Beside trialkylphosphine oxides (R_3PO) molecules, triphenylphosphine (PPh_3, 11.7×7.1 Å) [33] has also been applied to probe acid sites in solid acid catalysts. Separate study using molecular mechanical simulations have revealed that the pyramidal shape triphenylphosphine molecule prefers to adsorb on the 12-MR external surface pockets (7.1×9.2 Å) of the MCM-22 zeolite through partial intrusion (see Fig. 44) [120], hence only capable of probing acid sites in the external surface pockets. The ^{31}P MAS NMR spectrum of PPh_3 adsorbed on H-MCM-22 zeolites ($SiO_2/Al_2O_3 = 32$) shows three distinct resonances at 14.8, 11.1, and −4.6 ppm (Fig. 45). While the latter peak may be assigned due to physisorbed PPh_3, the other peaks at downfield should be related to protonated PPh_3 bound to Brønsted acid sites with varied acidic strengths in H-MCM-22 zeolite. Further incorporating the ^{31}P NMR results with elemental analysis data confirmed that ca. 6% of total Brønsted acid sites are located at the pore openings of the external surface pockets [120].

In addition to the properties of isolated acid sites, the proximity of acid centres also plays a crucial role during catalytic reactions, particularly for

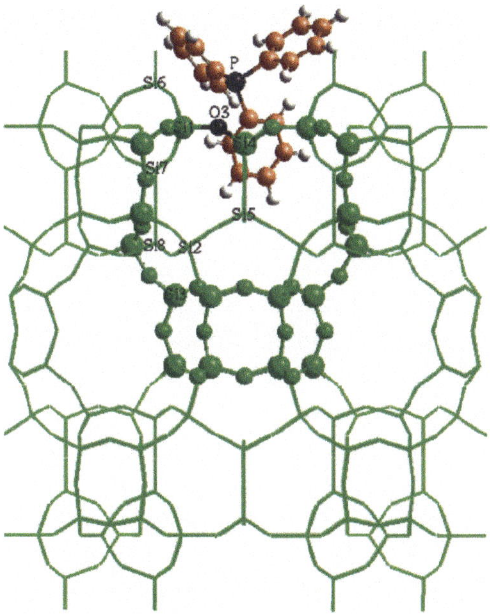

Fig. 44 Optimized structure of the PPh_3 adsorbed on unit cell of MCM-22 zeolite. The balls in green indicate external surface pocket [120]. *Reproduced from Y. Wang, J. Zhuang, G. Yang, D. Zhou, D. Ma, X. Han, X. Bao, Study on the external surface acidity of MCM-22 zeolite: theoretical calculation and* ^{31}*P MAS NMR, J. Phys. Chem. B 108 (2004) 1386–1391. Copyright 2004 American Chemical Society.*

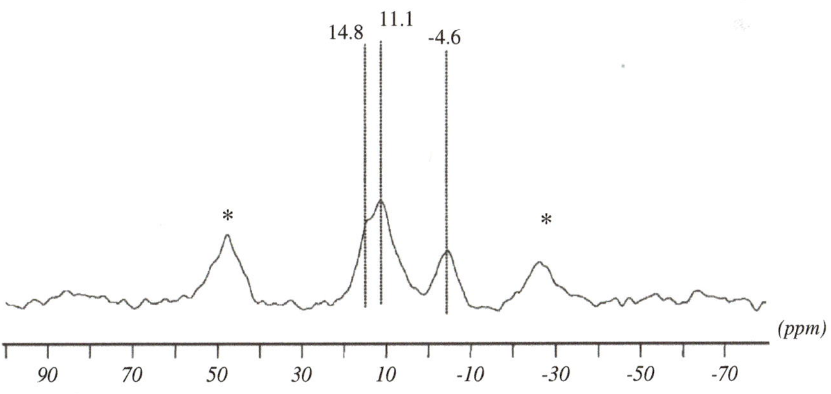

Fig. 45 ^{31}P MAS NMR spectrum of PPh_3 adsorbed on H-MCM-22 zeolite. The spinning sidebands are indicated by asterisks [120]. *Reproduced from Y. Wang, J. Zhuang, G. Yang, D. Zhou, D. Ma, X. Han, X. Bao, Study on the external surface acidity of MCM-22 zeolite: theoretical calculation and* ^{31}*P MAS NMR, J. Phys. Chem. B 108 (2004) 1386–1391. Copyright 2004 American Chemical Society.*

those involving multiple acid sites. In this context, advanced two-dimensional ^{31}P–^{31}P homonuclear correlation NMR techniques, such as the 2D ^{31}P–^{31}P PDSD correlation and DQ MAS NMR experiments with TMP probe molecule have been used to determine the interactions among various Brønsted-Lewis acid sites over the dealuminated HY zeolites [36]. As shown in Fig. 46, three off-diagonal correlation (F2, F1) peak pairs at (−5, −47), (−5, −53), (−5, −56) and (−47, −5), (−53, −5), (−56, −5) ppm were observed in 2D ^{31}P–^{31}P PDSD spectra. The presences of these crossed peaks clearly indicate that the ^{31}P resonances associated with Brønsted acid sites ($\delta^{31}P$ at −5 ppm) are in close proximity to those associated with Lewis acid sites with different acidic strengths (with $\delta^{31}P$ at −47, −53 ppm), and weakly acidic Si-OH groups ($\delta^{31}P = -56$ ppm). In other words, the spatial proximity and the synergy between the Brønsted and various active sites (including Lewis acids with different acidic strengths) in the dealuminated HY zeolite may readily be inferred. Further measurements by 2D ^{31}P–^{31}P DQ MAS NMR technique also revealed that Brønsted ($\delta^{31}P = -5$ ppm) and weak Lewis acid sites ($\delta^{31}P = -47$ ppm) are in close proximity and are spatially correlated in dealuminated zeolites [36].

On the other hand, in view of the distance (~0.30 nm) between the two phosphorus atoms in diphenyldiphosphine (Ph_2P-CH_2-PPh_2), Grey and co-workers used 1D ^{31}P MAS NMR and 2D ^{31}P–^{31}P DQ NMR of adsorbed diphenyldiphosphines ($Ph_2P(CH_2)_nPPh_2$; $n = 1$–6) to determine the distance between adjacent Brønsted acid sites in solid catalysts [35]. Since the distances between two phosphorus atoms in these diphenyldiphosphines may be tuned by varying the length of the –CH_2– group, thus, P–P distances of 0.30, 0.56, and 0.94 nm for $Ph_2P(CH_2)_nPPh_2$ were observed with $n = 1$, 3, and 6, respectively. Fig. 47 showed the ^{31}P MAS NMR spectra of zeolite HY (Si/Al = 2.6) loaded with $Ph_2P(CH_2)_nPPh_2$ ($n = 1$, 3, and 6), the resonance at −22 ppm was attributed to weakly physisorbed diphosphine molecules and the peak at −28 ppm to P atoms in more strongly physisorbed phosphine molecules that were either singly protonated or nonprotonated. Since notable $\delta^{31}P$ enhancements were observed in ^{1}H–^{31}P CP MAS experiment, the peaks at −1, 8, and 14 ppm were assigned to either singly or doubly protonated phosphines.

Additional 2D DQ-SQ (double quantum-single quantum) correlation experiment was also performed to probe the spatial proximity of the two like-spins [34]. Typically, dipolar correlation signals arising from spins with identical chemical shifts appear on the diagonal axis of the 2D NMR spectrum, whereas those with different chemical shifts appear as a pair of off-diagonal cross-peaks. As such, the chemical shift values of the peaks appear in the indirect (DQ) dimension should be the sum of chemical shifts of the

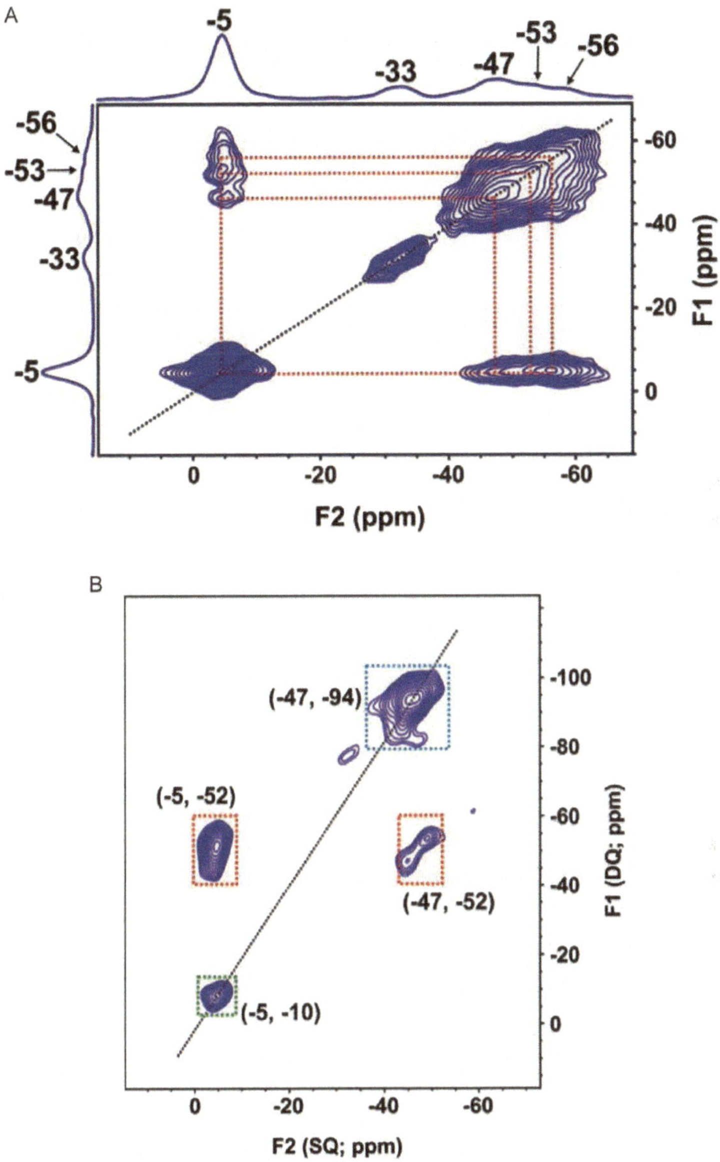

Fig. 46 2D ^{31}P–^{31}P (A) PDSD and (B) DQ MAS NMR spectra of TMP adsorbed on dealuminated HY (HY-d450) zeolites [36]. *Reproduced from X. Yi, K. Liu, W. Chen, J. Li, S. Xu, C. Li, Y. Xiao, H. Liu, X. Guo, S.B. Liu, A. Zheng, Origin and structural characteristics of tri-coordinated extra-framework aluminum species in dealuminated zeolites, J. Am. Chem. Soc. 140 (2018) 10764–10774. Copyright 2018 American Chemical Society.*

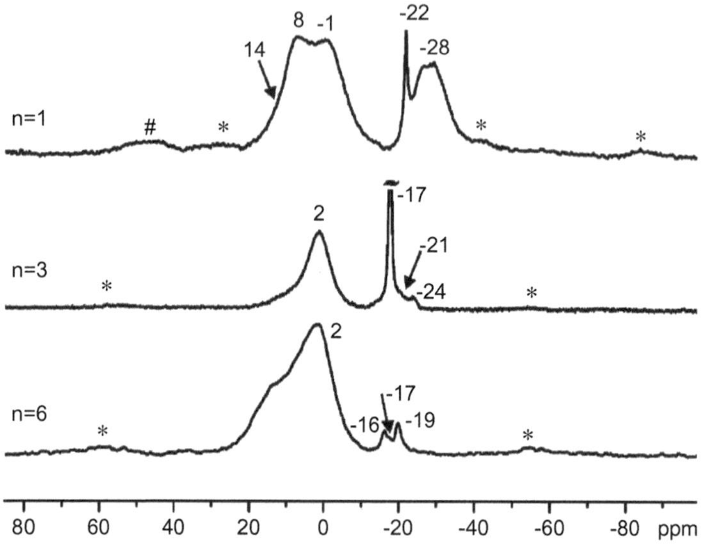

Fig. 47 ^{1}H-decoupled ^{31}P MAS NMR spectra of $Ph_2P(CH_2)nPPh_2$ adsorbed on a HY zeoilte loaded with ca. 12 molecules/unit cell. Spinning speed = 8 kHz. The asterisk denotes spinning sidebands and # represents overlapping spinning sideband and diphosphine oxide impurity [35]. *Reproduced from L. Peng, P.J. Chupas, C.P. Grey, Measuring Brønsted acid densities in zeolite HY with diphosphine molecules and solid state NMR spectroscopy, J. Am. Chem. Soc. 126 (2004) 12254–12255. Copyright 2004 American Chemical Society.*

off-diagonal peak pair in the direct (SQ) dimension. The cross-peaks at (14, −14) and (−28, −14) ppm in Fig. 48 reveal that species corresponding to δ^{31}P of −28 and 14 ppm are in close proximity. In addition, the diagonal peaks arising from the doubly protonated diphosphine at (8, 16) and (−1, −2) ppm reveal the close proximity of Brønsted acid sites in HY zeolite. Thus, by comparing ^{31}P NMR spectra for different diphenyldiphosphine probe molecules with varying P–P distance (i.e. *n* value) and loadings, the distance and concentration of acidic sites in zeolites may be determined [35].

3.4 Reaction mechanism

As mentioned earlier, the ^{31}P SSNMR approaches using various phosphorus-containing probe molecules are capable of providing detailed features of acid sites in solid acid catalysts. By monitoring variations of these acid features (i.e. type, concentration, location, and strength), information on reaction mechanism may also be obtained [121–126]. For example,

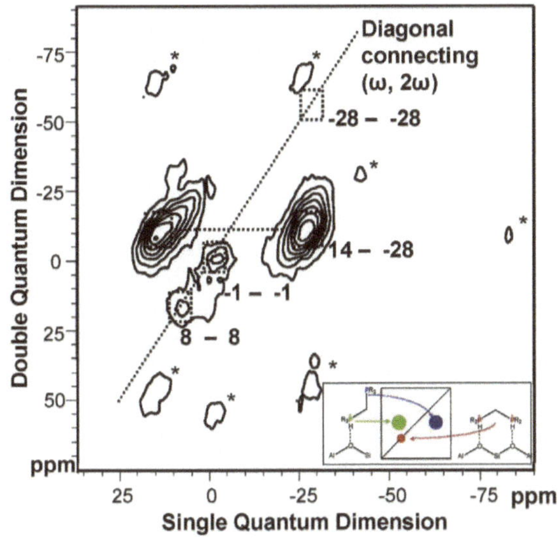

Fig. 48 2D ^{31}P DQ NMR spectra of $Ph_2P(CH_2)PPh_2$ (loading ca. 8 molecules/unit cell). Following $^{1}H–^{31}P$ CP, a post-C7 DQ sequence was adopted to prepare then to reconvert DQ coherences using a DQ excitation time of 2 ms and MAS spinning speed of 8 kHz. The asterisk denotes spinning sidebands [35]. *Reproduced from L. Peng, P.J. Chupas, C.P. Grey, Measuring Brønsted acid densities in zeolite HY with diphosphine molecules and solid state NMR spectroscopy, J. Am. Chem. Soc. 126 (2004) 12254–12255. Copyright 2004 American Chemical Society.*

TMP has been used as probe to monitor the mechanism of the titanium-substituted silicalite-1 (TS-1) during oxidation reaction. It is well-known that TS-1 zeolites are highly active solid acid catalysts for a variety of important oxidation reactions, such as phenol hydroxylation, cyclohexanone ammoximation, and propylene oxidation [88]. However, the distribution of Ti in the structural framework and consequently the active site responsible for the high catalytic activity observed for TS-1 during oxidation reactions remain largely unknown. By adopting the ^{31}P-TMP NMR approach in situ, Zhuang et al. demonstrated that the various Ti sites in TS-1 zeolite, which serve as active centres during oxidation reaction, may be identified, as shown in Fig. 49 [121]. For TS-1 before the oxidation treatment (in air), the corresponding ^{31}P NMR spectrum of the adsorbed TMP reveals as many as five resonance peaks was explicitly observed (Fig. 49A). The peak at −4.8 ppm was assigned to $[(CH_3)_3PH]^+$ ion protonated by mainly TiOH (and partly by SiOH) of the framework $Ti(OSi)_3OH$ species [122]. The signals at −62.1 and −59.8 ppm may be attributed to the physisorbed TMP. Whereas the peaks at −34.2 and −32.0 ppm were ascribed due to TMP

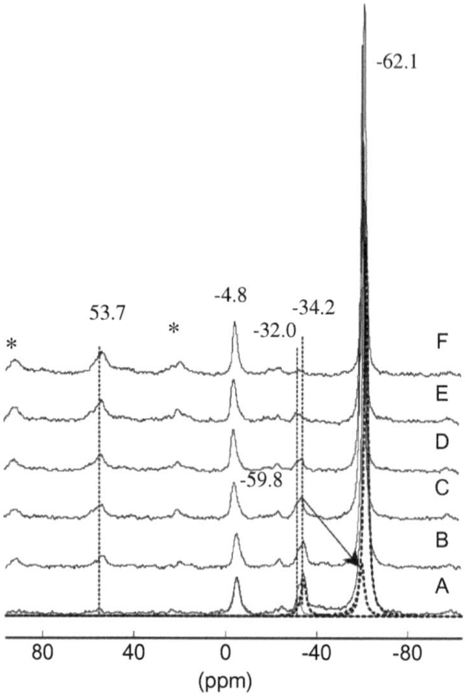

Fig. 49 Variations of ^{31}P MAS NMR spectra of TMP adsorbed on TS-1 zeolite (A) before and after oxidation treatment at 343 K in air for different exposure periods of time: (B) 10 min, (c) 20 min, (d) 30 min, (e) 40 min, and (f) 50 min. The asterisks denote spinning sidebands [121]. *Reproduced from J. Zhuang, D. Ma, Z. Yan, F. Deng, X. Liu, X. Han, X. Bao, X. Liu, X. Guo, X. Wang, Solid-state MAS NMR detection of the oxidation center in TS-1 zeolite by in situ probe reaction, J. Catal. 221 (2004) 670–673. Copyright 2004 Elsevier.*

adsorbed on Lewis acid sites with different acidic strengths [123,124]. Further investigation by theoretical calculations confirmed that the presences of $Ti(OSiO_3)_4$ and $(OSiO_3)_3Ti(OH)$ species are most likely the origins of Lewis acid sites that responsible for the peaks with $\delta^{31}P$ at −34.2 and −32.0 ppm, respectively.

Immediately after the catalyst was exposed to oxygen, notable decrease in the intensity of the ^{31}P resonance at −34.2 ppm accompanied by the appearance of a new resonance peak at 53.7 ppm were evident (Fig. 49B). The authors assigned the latter downfield peak to TMPO bound to Lewis acid sites [121]. It was estimated that after exposing the TS-1 catalyst in air for 10 min, ca. 46% of the peak area of the resonance at −34.2 ppm (for TMP adsorbed on Lewis acid sites) was converted to the Lewis-bound TMPO (at 53.7 ppm), while no significant changes in intensities of other

^{31}P resonances were observed. Moreover, a progress increase in peak intensity of the signal at 53.7 ppm at the expanse of the resonance at −34.2 ppm with increasing exposure time was observed. The peak responsible for TMP bound to Lewis acid sites with weaker acidic strength eventually diminished at an extended exposure time of about 40 min (Fig. 49E). It is noted that as the exposure time exceeded ca. 30 min, the peak intensity of the resonance at −32.0 ppm also begin to decrease, indicating the subsequent oxidation of the TMP species bound to the stronger Lewis acid sites. Almost all of the Lewis-bound TMP species were oxidized to form TMPO after exposure in air for 50 min. In contrast, the resonance corresponds to protonated $TMPH^+$ ion remained practically intact in the presence of oxidation treatment, suggesting that the inability of $Ti(OSi)_3OH$ species in TS-1 to catalyse the oxidation of TMP under such experimental conditions. Thus, it is indicative that Lewis acid sites (i.e. unsaturated tri-coordinated Ti) in TS-1 serve as the primary oxidation centres for TMP conversion. Accordingly, the results obtained from this in situ ^{31}P-TMP NMR approach readily provide information on mechanism and activity of TS-1 during oxidation reaction [121].

4. Summary

We have demonstrated that the ^{31}P SSNMR of adsorbed phosphorus-containing probe molecules (viz. TMP, R_3PO, phosphines, and diphosphines) represents a unique and practical technique for comprehensive acidity characterization of solid acid catalysts. Detailed acid features, namely type, distribution, accessibility (location/proximity), concentration (amount), and strength of acid sites in solid acids may readily be obtained both qualitatively and quantitatively. By comparison, the ^{31}P-TMP NMR approach is found to be viable for discerning acid types in the solid acids, and more sensitive for probing Lewis acidity. Whereas the ^{31}P-TMPO NMR approach is more useful for determining the Brønsted acid strength of solid acids and capable of covering the whole range from weak, medium, strong, all the way to superacidity. Moreover, by properly choosing trialkylphosphine oxide probe molecules with different sizes (viz. TMPO, TEPO, TBPO, and TOPO), the ^{31}P-R_3PO NMR approaches may be applied to distinguish acid sites located in internal or external surfaces of porous solid acid catalysts. In addition, the proximity and distance between adjacent Brønsted acid sites can be estimated by 2D ^{31}P–^{31}P DQ NMR technique by using diphenyldiphosphines ($Ph_2P(CH_2)_nPPh_2$; $n=1$–6) as the probes. These ^{31}P SSNMR approaches for acidity characterization invoking various phosphorus-containing probe molecules, if properly exploited with in situ and/or ex situ experimental techniques, are most practical not only for probing the variations of acid features in solid acids, but also for understanding synergy effect among multiple acid sites and possible reaction mechanism (activity and selectivity) in heterogeneous catalysis.

Acknowledgements

This work was supported by the National Natural Science Foundation of China (No. 91645112, 21802164, 21902180, 21991092, and U1832148), and Natural Science Foundation of Hubei Province of China (No. 2018CFA009), Key Research Program of Frontier Sciences, CAS (No. QYZDB-SSW-SLH026), Ministry of Science and Technology (MOST, Taiwan; Grants NSC 101-2113-M-001-020-MY3, and MOST 104-2113-M-001-019), and SINOPEC (No. 417012-4).

References

[1] G. Busca, Acid catalysts in industrial hydrocarbon chemistry, Chem. Rev. 107 (2007) 5366–5410.
[2] A. Zecchina, S. Bordiga, C. Lamberti, G. Spoto, L. Carnelli, A.C. Otero, Low-temperature Fourier transform infrared study of the interaction of CO with cations in alkali-metal exchanged ZSM-5 zeolites, J. Phys. Chem. 98 (1994) 9577–9582.
[3] P. Sarv, T. Tuherm, E. Lippmaa, E. Keskinen, A. Root, Mobility of the acidic proton in Broensted sites of H-Y, H-mordenite, and H-ZSM-5 zeolites, studied by high-temperature 1H MAS NMR, J. Phys. Chem. 99 (1995) 13763–13768.
[4] H.M. Kao, C.P. Grey, Probing the Brønsted and Lewis acidity of zeolite HY: a $^{1}H/^{27}Al$ and $^{15}N/^{27}Al$ TRAPDOR NMR study of monomethylamine adsorbed on HY, J. Phys. Chem. 100 (1996) 5105–5117.
[5] W.E. Farneth, R.J. Gorte, Methods for characterizing zeolite acidity, Chem. Rev. 95 (1995) 615–636.
[6] D. Freude, M. Hunger, H. Pfeifer, W. Schwieger, ^{1}H MAS NMR studies on the acidity of zeolites, Chem. Phys. Lett. 128 (1986) 62–66.
[7] L. Liu, L. Zhao, H. Sun, Simulation of NH_3 temperature-programmed desorption curves using an ab initio force field, J. Phys. Chem. C 113 (2009) 16051–16057.
[8] A. Zheng, F. Deng, S.-B. Liu, Acidity characterization of solid acid catalysts by solid-state ^{31}P NMR of adsorbed phosphorus-containing probe molecules, Annu. Rep. NMR Spectrosc. 81 (2014) 47–108.
[9] A. Zheng, S.B. Liu, F. Deng, ^{31}P NMR chemical shifts of phosphorous probes as reliable and practical acidity scales for solid and liquid catalysts, Chem. Rev. 117 (2017) 12475–12531.
[10] A. Zheng, S. Li, S.B. Liu, F. Deng, Acidic properties and structure-activity correlations of solid acid catalysts revealed by solid-state NMR spectroscopy, Acc. Chem. Res. 49 (2016) 655–663.
[11] H. Pfeifer, H. Ernst, NMR studies of zeolites, Annu. Rep. NMR Spectrosc. 28 (1994) 91–187.
[12] M. Hunger, Brønsted acid sites in zeolites characterized by multinuclear solid-state NMR spectroscopy, Catal. Rev. Sci. Eng. 39 (1997) 345–393.

[13] M. Xu, A. Arnold, A. Buchholz, W. Wang, M. Hunger, Low-temperature modification of mesoporous MCM-41 material with sublimated aluminum chloride in vacuum, J. Phys. Chem. B 106 (2002) 12140–12143.
[14] J.F. Haw, J.B. Nicholas, T. Xu, L.W. Beck, D.B. Ferguson, Physical organic chemistry of solid acids: lessons from in situ NMR and theoretical chemistry, Acc. Chem. Res. 29 (1996) 259–267.
[15] D.H. Barich, J.B. Nicholas, T. Xu, J.F. Haw, Theoretical and experimental study of the ^{13}C chemical shift tensors of acetone complexed with Brønsted and Lewis acids, J. Am. Chem. Soc. 120 (1998) 12342–12350.
[16] J.H. Lunsford, W.P. Rothwell, W. Shen, Acid sites in zeolite Y: a solid-state NMR and infrared study using trimethylphosphine as a probe molecule, J. Am. Chem. Soc. 107 (1985) 1540–1547.
[17] H.M. Kao, C.P. Grey, Characterization of the Lewis acid sites in zeolite HY with the probe molecule trimethylphosphine, and $^{31}P/^{27}Al$ double resonance NMR, Chem. Phys. Lett. 259 (1996) 459–464.
[18] Y. Chu, Z. Yu, A. Zheng, H. Fang, H. Zhang, S.J. Huang, S.B. Liu, F. Deng, Acidic strengths of Brønsted and Lewis acid sites in solid acids scaled by ^{31}P NMR chemical shifts of adsorbed trimethylphosphine, J. Phys. Chem. C 115 (2011) 7660–7667.
[19] Y. Wang, F. Wang, Q. Song, Q. Xin, S. Xu, J. Xu, Heterogeneous ceria catalyst with water-tolerant Lewis acidic sites for one-pot synthesis of 1,3-diols via prins condensation and hydrolysis reactions, J. Am. Chem. Soc. 135 (2013) 1506–1515.
[20] A. Zheng, S.B. Liu, F. Deng, Acidity characterization of heterogeneous catalysts by solid-state NMR spectroscopy using probe molecules, Solid State Nucl. Magn. Reson. 55–56 (2013) 12–27.
[21] A. Zheng, H. Zhang, X. Lu, S.B. Liu, F. Deng, Theoretical predictions of ^{31}P NMR chemical shift threshold of trimethylphosphine oxide absorbed on solid acid catalysts, J. Phys. Chem. B 112 (2008) 4496–4505.
[22] J. Xu, A. Zheng, J. Yang, Y. Su, J. Wang, D. Zeng, M. Zhang, C. Ye, F. Deng, Acidity of mesoporous MoO_x/ZrO_2 and WO_x/ZrO_2 materials: a combined solid-state NMR and theoretical calculation study, J. Phys. Chem. B 110 (2006) 10662–10671.
[23] T. Riemer, H. Knözinger, ^{31}P solid state NMR and ESR investigation of trimethylphosphane adsorption on zirconia and sulfated zirconia, J. Phys. Chem. 100 (1996) 6739–6742.
[24] J. Yang, M. Zhang, F. Deng, Q. Luo, D. Yi, C. Ye, Solid state NMR study of acid sites formed by adsorption of SO_3 onto γ-Al_2O_3, Chem. Commun. 7 (2003) 884–885.
[25] H. Zhang, H. Yu, A. Zheng, S. Li, W. Shen, F. Deng, Reactivity enhancement of 2-propanol photocatalysis on SO_4^{2-}/TiO_2: insights from solid-state NMR spectroscopy, Environ. Sci. Technol. 42 (2008) 5316–5321.
[26] G.E. Maciel, P.D. Ellis, NMR characterization of silica and alumina surfaces, in: A.T. Bell, A. Pines (Eds.), NMR Techniques in Catalysis, vol. 55, Marcel Dekker, New York, 1994, p. 231.
[27] Q. Zhao, W.H. Chen, S.J. Huang, Y.C. Wu, H.K. Lee, S.B. Liu, Discernment and quantification of internal and external acid sites on zeolites, J. Phys. Chem. B 106 (2002) 4462–4469.
[28] A. Zheng, S.J. Huang, W.H. Chen, P.H. Wu, H. Zhang, H.K. Lee, L.C. De Ménorval, F. Deng, S.B. Liu, ^{31}P chemical shift of adsorbed trialkylphosphine oxides for acidity characterization of solid acids catalysts, J. Phys. Chem. A 112 (2008) 7349–7356.
[29] W.H. Chen, H.H. Ko, A. Sakthivel, S.J. Huang, S.H. Liu, A.Y. Lo, T.C. Tsai, S.B. Liu, A solid-state NMR, FT-IR and TPD study on acid properties of sulfated and metal-promoted zirconia: influence of promoter and sulfation treatment, Catal. Today 116 (2006) 111–120.

[30] H.H. Ko, Preparation, Modification, and Characterization of Metal-Oxide and Nanosized Aluminosilicate Solid Acid Catalysts, MS thesis, Department of Materials Science and Nanotechnology, Chinese Culture University, Taipei, Taiwan, June 2005.
[31] A. Zheng, S.J. Huang, S.B. Liu, F. Deng, Acid properties of solid acid catalysts characterized by solid-state ^{31}P NMR of adsorbed phosphorous probe molecules, Phys. Chem. Chem. Phys. 13 (2011) 14889–14901.
[32] Q. Zhao, W.H. Chen, S.J. Huang, S.B. Liu, Qualitative and quantitative determination of acid sites on solid acid catalysts, Stud. Surf. Sci. Catal. 145 (2003) 205–209.
[33] C.S. Hsia Chen, S.E. Schramm, Type and catalytic activity of surface acid sites of medium and large pore zeolites: their deactivation with bulky organophosphorus compounds, Microporous Mater. 7 (1996) 125–132.
[34] M. Hohwy, H.J. Jakobsen, M. Eden, M.H. Levitt, N.C. Nielsen, Broadband dipolar recoupling in the nuclear magnetic resonance of rotating solids: a compensated C7 pulse sequence, J. Chem. Phys. 108 (1998) 2686–2694.
[35] L. Peng, P.J. Chupas, C.P. Grey, Measuring Brønsted acid densities in zeolite HY with diphosphine molecules and solid state NMR spectroscopy, J. Am. Chem. Soc. 126 (2004) 12254–12255.
[36] X. Yi, K. Liu, W. Chen, J. Li, S. Xu, C. Li, Y. Xiao, H. Liu, X. Guo, S.B. Liu, A. Zheng, Origin and structural characteristics of tri-coordinated extra-framework aluminum species in dealuminated zeolites, J. Am. Chem. Soc. 140 (2018) 10764–10774.
[37] J.F. Haw, J. Zhang, K. Shimizu, T.N. Venkatraman, D.P. Luigi, W. Song, D.H. Barich, J.B. Nicholas, NMR and theoretical study of acidity probes on sulfated zirconia catalysts, J. Am. Chem. Soc. 122 (2000) 12561–12570.
[38] H. Sang, H.Y. Chu, J.H. Lunsford, An NMR study of acid sites on chlorided alumina catalysts using trimethylphosphine as a probe, Catal. Lett. 26 (1994) 235–246.
[39] X. Gao, Y. Zhang, X. Wang, Z. Zhang, G. Xu, Structure and acidity changes in ultra-stable Y zeolites during hydrothermal aging: a solid-state NMR spectroscopy study, Chin. J. Magn. Reson. 37 (2020) 95–103.
[40] S.P. Brown, H.W. Spiess, Advanced solid-state NMR methods for the elucidation of structure and dynamics of molecular, macromolecular, and supramolecular systems, Chem. Rev. 101 (2001) 4125–4156.
[41] L.M. Kustov, V.B. Kazansky, F. Figueras, D. Tichit, Investigation of the acidic properties of ZrO_2 modified by SO_4^{2-} anions, J. Catal. 150 (1994) 143–149.
[42] F. Babou, G. Coudurier, J.C. Vedrine, Acidic properties of sulfated zirconia: an infrared spectroscopic study, J. Catal. 152 (1995) 341–349.
[43] A. Corma, V. Fornés, M.I. Juan-Rajadell, J.M. López Nieto, Influence of preparation conditions on the structure and catalytic properties of SO_4^{2-}/ZrO_2 superacid catalysts, Appl. Catal. A 116 (1994) 151–163.
[44] P.V. Wiper, J. Amelse, L. Mafra, Multinuclear solid-state NMR characterization of the Brønsted/Lewis acid properties in the BP HAMS-1B (H-[B]-ZSM-5) borosilicate molecular sieve using adsorbed TMPO and TBPO probe molecules, J. Catal. 316 (2014) 240–250.
[45] S. Xin, Q. Wang, J. Xu, Y. Chu, P. Wang, N. Feng, G. Qi, J. Trébosc, O. Lafon, W. Fan, F. Deng, The acidic nature of "NMR-invisible" tricoordinated framework aluminum species in zeolites, Chem. Sci. 10 (2019) 10159–10169.
[46] Q. Wang, J. Trébosc, Y. Li, J. Xu, B. Hu, N. Feng, Q. Chen, O. Lafon, J.P. Amoureux, F. Deng, Signal enhancement of *J*-HMQC experiments in solid-state NMR involving half-integer quadrupolar nuclei, Chem. Commun. 49 (2013) 6653–6655.

[47] G. Tricot, J. Trébosc, F. Pourpoint, R. Gauvin, L. Delevoye, The D-HMQC MAS-NMR technique: An efficient tool for the editing of through-space correlation spectra between quadrupolar and spin-1/2 (^{31}P, ^{29}Si, ^{1}H, ^{13}C) nuclei, Annu. Rep. NMR Spectrosc. 81 (2014) 145–184.
[48] Q. Wang, Y. Li, J. Trébosc, O. Lafon, J. Xu, B. Hu, N. Feng, Q. Chen, J.P. Amoureux, F. Deng, Population transfer HMQC for half-integer quadrupolar nuclei, J. Chem. Phys. 142 (2015), 094201.
[49] L. Chen, X.Y. Lu, Q. Wang, O. Lafon, J. Trébosc, F. Deng, J.P. Amoureux, Distance measurement between a spin-1/2 and a half-integer quadrupolar nuclei by solid-state NMR using exact analytical expressions, J. Magn. Reson. 206 (2010) 269–273.
[50] S.J. Huang, Y.H. Tseng, Y. Mou, S.B. Liu, S.H. Huang, C.P. Lin, J.C.C. Chan, Spectral editing based on selective excitation and Lee-Goldburg cross-polarization under magic angle spinning, Solid State Nucl. Magn. Reson. 29 (2006) 272–277.
[51] B.J. Van Rossum, C.P. De Groot, V. Ladizhansky, S. Vega, H.J.M. De Groot, A method for measuring heteronuclear (^{1}H-^{13}C) distances in high speed MAS NMR, J. Am. Chem. Soc. 122 (2000) 3465–3472.
[52] V. Ladizhansky, S. Vega, Polarization transfer dynamics in Lee-Goldburg cross polarization nuclear magnetic resonance experiments on rotating solids, J. Chem. Phys. 112 (2000) 7158–7168.
[53] C.P. Grey, W.S. Veeman, A.J. Vega, Rotational echo ^{14}N/^{13}C/^{1}H triple resonance solid-state nuclear magnetic resonance: a probe of ^{13}C-^{14}N internuclear distances, J. Chem. Phys. 98 (1993) 7711–7725.
[54] M.D. Karra, K.J. Stutovich, K.T. Mueller, NMR characterization of Brønsted acid sites in faujasitic zeolites with use of perdeuterated trimethylphosphine oxide, J. Am. Chem. Soc. 124 (2002) 902–903.
[55] B. Alonso, I. Klur, D. Massiot, Studies of surfaces through molecular probe adsorption and solid-state NMR, Chem. Commun. 8 (2002) 804–805.
[56] L. Baltusis, J.S. Frye, G.E. Maciel, Phosphine oxides as NMR probes for adsorption sites on surfaces, J. Am. Chem. Soc. 108 (1986) 7119–7120.
[57] H.M. Kao, C.Y. Yu, M.C. Yeh, Detection of the inhomogeneity of Brønsted acidity in H-mordenite and H-β zeolites: a comparative NMR study using trimethylphosphine and trimethylphosphine oxide as ^{31}P NMR probes, Microporous Mesoporous Mater. 53 (2002) 1–12.
[58] Catalysis Commission, International Zeolite Association. http://www.iza-online.org/catalysis/.
[59] B.M. Weckhuysen, D. Wang, M.P. Rosynek, J.H. Lunsford, Conversion of methane to benzene over transition metal ion ZSM-5 zeolites: I. Catalytic characterization, J. Catal. 175 (1998) 338–346.
[60] Y. Xu, L. Lin, Recent advances in methane dehydro-aromatization over transition metal ion-modified zeolite catalysts under non-oxidative conditions, Appl. Catal. A 188 (1999) 53–67.
[61] Y. Shu, D. Ma, L. Xu, Y. Xu, X. Bao, Methane dehydro-aromatization over Mo/MCM-22 catalysts: a highly selective catalyst for the formation of benzene, Catal. Lett. 70 (2000) 67–73.
[62] S.J. Huang, Q. Zhao, W.H. Chen, X. Han, X. Bao, P.S. Lo, H.K. Lee, S.B. Liu, Structure and acidity of Mo/H-MCM-22 catalysts studied by NMR spectroscopy, Catal. Today 97 (2004) 25–34.
[63] Y. Chen, K. Gong, F. Jiao, X. Pan, G. Hou, R. Si, X. Bao, C-C bond formation in syngas conversion over Zn sites grafted on ZSM-5, Angew. Chem. Int. Ed. 59 (16) (2020) 6529–6534.
[64] T. Gullion, J. Schaefer, Rotational-echo double-resonance NMR, J. Magn. Reson. 81 (1989) 196–200.

[65] T. Ennaert, J. Van Aelst, J. Dijkmans, R. De Clercq, W. Schutyser, M. Dusselier, D. Verboekend, B.F. Sels, Potential and challenges of zeolite chemistry in the catalytic conversion of biomass, Chem. Soc. Rev. 45 (2016) 584–611.
[66] A. Corma, L.T. Nemeth, M. Renz, S. Valencia, Sn-zeolite beta as a heterogeneous chemoselective catalyst for Baeyer-Villiger oxidations, Nature 412 (2001) 423–425.
[67] M. Moliner, State of the art of Lewis acid-containing zeolites: lessons from fine chemistry to new biomass transformation processes, Dalton Trans. 43 (2014) 4197–4208.
[68] J.D. Lewis, M. Ha, H.L. Luo, A. Faucher, V.K. Michaelis, Y. Román-Leshkov, Distinguishing active site identity in Sn-beta zeolites using ^{31}P MAS NMR of adsorbed trimethylphosphine oxide, ACS Catal. 8 (2018) 3076–3086.
[69] F. Dubray, S. Moldovan, C. Kouvatas, J. Grand, C. Aquino, N. Barrier, J.P. Gilson, N. Nesterenko, D. Minoux, S. Mintova, Direct evidence for single molybdenum atoms incorporated in the framework of MFI zeolite nanocrystals, J. Am. Chem. Soc. 141 (2019) 8689–8693.
[70] D. Aucoin, D. Camenares, X. Zhao, J. Jung, T. Sato, S.O. Smith, High-resolution ^{1}H MAS RFDR NMR of biological membranes, J. Magn. Reson. 197 (2009) 77–86.
[71] J. Schneider, M. Matsuoka, M. Takeuchi, J.L. Zhang, Y. Horiuchi, M. Anpo, D.W. Bahnemann, Understanding TiO_2 photocatalysis: mechanisms and materials, Chem. Rev. 114 (2014) 9919–9986.
[72] G. Liu, H.G. Yang, J. Pan, Y.Q. Yang, G.Q. Lu, H.M. Cheng, Titanium dioxide crystals with tailored facets, Chem. Rev. 114 (2014) 9559–9612.
[73] Y.K. Peng, Y. Hu, H.L. Chou, Y. Fu, I.F. Teixeira, L. Zhang, H. He, S.C.E. Tsang, Mapping surface-modified titania nanoparticles with implications for activity and facet control, Nat. Commun. 8 (2017) 1–13.
[74] Y. Peng, L. Ye, J. Qu, L. Zhang, Y. Fu, I.F. Teixeira, I.J. McPherson, H. He, S.C.E. Tsang, Trimethylphosphine-assisted surface fingerprinting of metal oxide nanoparticle by ^{31}P solid-state NMR: a zinc oxide case study, J. Am. Chem. Soc. 138 (2016) 2225–2234.
[75] H.T. Kreissl, M.M.J. Li, Y.K. Peng, K. Nakagawa, T.J.N. Hooper, J.V. Hanna, A. Shepherd, T.S. Wu, Y.L. Soo, S.C.E. Tsang, Structural studies of bulk to nano-size niobium oxides with correlation to their acidity, J. Am. Chem. Soc. 139 (2017) 12670–12680.
[76] Y. Fu, L. Zhang, B. Yue, X. Chen, H. He, Simultaneous characterization of solid acidity and basicity of metal oxide catalysts via solid state NMR technique, J. Phys. Chem. C 122 (2018) 24094–24102.
[77] Y. Wu, D. Huang, Y. Fu, L. Zhang, S. Liu, G. Tang, Y. Ren, L. Ye, X. Chen, B. Yue, H. He, The effects of exposed specific facets and sulfation on the surface acidity of Cu_2O solids, Chem. Eur. J. 25 (2019) 14771–14774.
[78] Z. Tan, G. Li, H.L. Chou, Y. Li, X. Yi, A.H. Mahadi, A. Zheng, S.C.E. Tsang, Y.K. Peng, Differentiating surface Ce species among CeO_2 facets by solid-state NMR for catalytic correlation, ACS Catal. 10 (2020) 4003–4011.
[79] E.L. Tae, K.E. Lee, J.S. Jeong, K.B. Yoon, Synthesis of diamond-shape titanate molecular sheets with different sizes and realization of quantum confinement effect during dimensionality reduction from two to zero, J. Am. Chem. Soc. 130 (2008) 6534–6543.
[80] D.V. Bavykin, F.C. Walsh, Elongated titanate nanostructures and their applications, Eur. J. Inorg. Chem. 8 (2009) 977–997.
[81] M. Kitano, E. Wada, K. Nakajima, S. Hayashi, S. Miyazaki, H. Kobayashi, M. Hara, Protonated titanate nanotubes with Lewis and Brønsted acidity: relationship between nanotube structure and catalytic activity, Chem. Mater. 25 (2013) 385–393.
[82] E.F. Rakiewicz, A.W. Peters, R.F. Wormsbecher, K.J. Sutovich, K.T. Mueller, Characterization of acid sites in zeolitic and other inorganic systems using solid-state

^{31}P NMR of the probe molecule trimethylphosphine oxide, J. Phys. Chem. B 102 (1998) 2890–2896.
[83] C. Tagusagawa, A. Takagaki, S. Hayashi, K. Domen, Evaluation of strong acid properties of layered $HNbMoO_6$ and catalytic activity for Friedel-Crafts alkylation, Catal. Today 142 (2009) 267–271.
[84] A. Takagaki, C. Tagusagawa, S. Hayashi, M. Hara, K. Domen, Nanosheets as highly active solid acid catalysts for green chemical syntheses, Energ. Environ. Sci. 3 (2010) 82–93.
[85] C. Tagusagawa, A. Takagaki, S. Hayashi, K. Domen, Characterization of $HNbWO_6$ and $HTaWO_6$ metal oxide nanosheet aggregates as solid acid catalysts, J. Phys. Chem. C 113 (2009) 7831–7837.
[86] C. Tagusagawa, A. Takagaki, A. Iguchi, K. Takanabe, J.N. Kondo, K. Ebitani, S. Hayashi, T. Tatsumi, K. Domen, Highly active mesoporous Nb-W oxide solid-acid catalyst, Angew. Chem. Int. Ed. 49 (2010) 1128–1132.
[87] J. Fierro, Metal Oxides: Chemistry and Applications, CRC Press, Boca Raton, FL, 2006.
[88] C. Perego, A. Carati, P. Ingallina, M.A. Mantegazza, G. Bellussi, Production of titanium containing molecular sieves and their application in catalysis, Appl. Catal. A 221 (2001) 63–72.
[89] S. Li, H. Zhou, B. Han, F. Deng, X. Liu, L. Xiao, J. Fan, Hydrogenated mesoporous TiO_2–SiO_2 with increased moderate strong Brønsted acidic sites for Friedel-Crafts alkylation reaction, Catal. Sci. Technol. 2 (2012) 719–721.
[90] A. Corma, Inorganic solid acids and their use in acid-catalyzed hydrocarbon reactions, Chem. Rev. 95 (1995) 559–614.
[91] J.A. Moreno, G. Poncelet, Isomerization of n-butane over sulfated Al- and Ga-promoted zirconium oxide catalysts. Influence of promoter and preparation method, J. Catal. 203 (2001) 453–465.
[92] K. Föttinger, K. Zorn, H. Vinek, Influence of the sulfate content on the activity of Pt containing sulfated zirconia, Appl. Catal. A 284 (2005) 69–75.
[93] C. Xu, Z. Cai, Q. Wang, X.Y. Mei, Y.S. Zhou, Y.M. Xu, J.H. Duan-Mu, S.T. Wang, X.X. Han, Preparation of biodiesel using silver-modified phosphotungstic acid as catalyst, Chin. J. Magn. Reson. 37 (2020) 86–94.
[94] N. Feng, A. Zheng, S.J. Huang, H. Zhang, N. Yu, C.Y. Yang, S.B. Liu, F. Deng, Combined solid-state NMR and theoretical calculation studies of Brønsted acid properties in anhydrous 12-molybdophosphoric acid, J. Phys. Chem. C 114 (2010) 15464–15472.
[95] T. Okuhara, H. Watanabe, T. Nishimura, K. Inumaru, M. Misono, Microstructure of cesium hydrogen salts of 12-tungstophosphoric acid relevant to novel acid catalysis, Chem. Mater. 12 (2000) 2230–2238.
[96] U. Filek, A. Bressel, B. Sulikowski, M. Hunger, Structural stability and Brønsted acidity of thermally treated $AlPW_{12}O_{40}$ in comparison with $H_3PW_{12}O_{40}$, J. Phys. Chem. C 112 (2008) 19470–19476.
[97] S.J. Huang, C.Y. Yang, A. Zheng, N. Feng, N. Yu, P.H. Wu, Y.C. Chang, Y.C. Lin, F. Deng, S.B. Liu, New insights into Keggin-type 12-tungstophosphoric acid from ^{31}P MAS NMR analysis of absorbed trimethylphosphine oxide and DFT calculations, Chem. Asian J. 6 (2011) 137–148.
[98] X.X. Han, K.K. Chen, W. Yan, C.T. Hung, L.L. Liu, P.H. Wu, K.C. Lin, S.B. Liu, Amino acid-functionalized heteropolyacids as efficient and recyclable catalysts for esterification of palmitic acid to biodiesel, Fuel 165 (2016) 115–122.
[99] M.Y. Huang, X. Han, C.T. Hung, J.C. Lin, P.H. Wu, J.C. Wu, S.B. Liu, Heteropolyacid-based ionic liquids as efficient homogeneous catalysts for acetylation of glycerol, J. Catal. 320 (2014) 42–51.

[100] X.X. Han, X. Zhang, G. Zhu, J. Liang, X. Cao, R. Kan, C.T. Hung, L.L. Liu, S.B. Liu, Ionic liquid-silicotungstic acid composites as efficient and recyclable catalysts for the selective esterification of glycerol with lauric acid to monolaurin, ChemCatChem 9 (2017) 2727–2738.

[101] C.T. Hung, L.L. Liu, J.J. Wang, P.H. Wu, C.B. Wang, T.C. Tsai, S.B. Liu, Acidity and alkylation activity of 12-Tungstophosphoric acid supported on ionic liquid-functionalized SBA-15, Catal. Today 327 (2019) 10–18.

[102] F. De Clippel, M. Dusselier, R. Van Rompaey, P. Vanelderen, J. Dijkmans, E. Makshina, L. Giebeler, S. Oswald, G.V. Baron, J.F. Denayer, P.P. Pescarmona, P.A. Jacobs, B.F. Sels, Fast and selective sugar conversion to alkyl lactate and lactic acid with bifunctional carbon-silica catalysts, J. Am. Chem. Soc. 134 (2012) 10089–10101.

[103] F. Liu, A. Zheng, I. Noshadi, F.S. Xiao, Design and synthesis of hydrophobic and stable mesoporous polymeric solid acid with ultra strong acid strength and excellent catalytic activities for biomass transformation, Appl. Catal. B 136–137 (2013) 193–201.

[104] S. Suganuma, K. Nakajima, M. Kitano, H. Kato, A. Tamura, H. Kondo, S. Yanagawa, S. Hayashi, M. Hara, SO_3H-bearing mesoporous carbon with highly selective catalysis, Microporous Mesoporous Mater. 143 (2011) 443–450.

[105] J. Jiang, O.M. Yaghi, Brønsted acidity in metal-organic frameworks, Chem. Rev. 115 (2015) 6966–6997.

[106] J. Jiang, F. Gandara, Y.B. Zhang, K. Na, O.M. Yaghi, W.G. Klemperer, Superacidity in sulfated metal-organic framework-808, J. Am. Chem. Soc. 136 (2014) 12844–12847.

[107] C.A. Trickett, T.M. Osborn Popp, J. Su, C. Yan, J. Weisberg, A. Huq, P. Urban, J.C. Jiang, M.J. Kalmutzki, Q.N. Liu, J. Baek, M.P. Head-Gordon, G.A. Somorjai, J.A. Reimer, O.M. Yaghi, Identification of the strong Brønsted acid site in a metal-organic framework solid acid catalyst, Nat. Chem. 11 (2018) 170–176.

[108] Q. Sun, Y.Q. Tang, B. Aguila, S. Wang, F.S. Xiao, P.K. Thallapally, A.M. Alenizi, A. Nafady, S.Q. Ma, Reaction environment modification in covalent organic frameworks for catalytic performance enhancement, Angew. Chem. 131 (2019) 8762–8767.

[109] Q. Wu, F. Liu, X. Yi, Y. Zou, L. Jiang, A solvent-free, one-step to the synthesis of sulfonic group functionalized mesoporous organosilica with ultra-high acid concentrations and excellent catalytic activities, Green Chem. 20 (2018) 1020–1030.

[110] X. Zhang, Y. Zhao, S. Xu, Y. Yang, J. Liu, Y. Wei, Q. Yang, Polystyrene sulphonic acid resins with enhanced acid strength via macromolecular self-assembly within confined nanospace, Nat. Commun. 5 (2014) 3170.

[111] C. Tan, X. Cao, X.J. Wu, Q. He, J. Yang, X. Zhang, J. Chen, W. Zhao, S. Han, G.H. Nam, M. Sindoro, H. Zhang, Recent advances in ultrathin two-dimensional nanomaterials, Chem. Rev. 117 (2017) 6225–6331.

[112] F. Liu, X. Yi, W. Chen, Z. Liu, W. Chen, C. Qi, Y. Song, A. Zheng, Developing two-dimensional solid superacids with enhanced mass transport, extremely high acid strength and superior catalytic performance, Chem. Sci. 10 (2019) 5875–5883.

[113] D. Zeng, S. Liu, W. Gong, G. Wang, J. Qiu, Y. Tian, Acid properties of solid acid from petroleum coke by chemical activation and sulfonation, Catal. Commun. 40 (2013) 5–8.

[114] Y.Y. Li, S.P. Perera, B.D. Crittenden, J. Bridgwater, The effect of the binder on the manufacture of a 5A zeolite monolith, Powder Technol. 116 (2001) 85–96.

[115] M. Raimondo, G. Perez, A. DeStefanis, A.A.G. Tomlinson, O. Ursini, PLS vs. zeolites as sorbents and catalysts 4. Effects of acidity and porosity on alkylation of benzene by primary alcohols, Appl. Catal. A 164 (1997) 119–126.

[116] L.D. Rollmann, D.E. Walsh, Shape selectivity and carbon formation in zeolites, J. Catal. 56 (1979) 139–140.

[117] A.R. Pradhan, J.F. Wu, S.J. Jong, W.H. Chen, T.C. Tsai, S.B. Liu, Influences of zeolite structure on formation and location of coke: a ^{129}Xe and ^{13}C CP-MAS NMR study, Appl. Catal. A 159 (1997) 187–209.
[118] W.H. Chen, S.J. Huang, C.S. Lai, T.C. Tsai, H.K. Lee, S.B. Liu, Effects of binder, coking and regeneration on acid properties of H-mordenite during TDP reaction, Res. Chem. Intermed. 29 (2003) 761–772.
[119] Y. Seo, K. Cho, Y. Jung, R. Ryoo, Characterization of the surface acidity of MFI zeolite nanosheets by ^{31}P NMR of adsorbed phosphine oxides and catalytic cracking of decalin, ACS Catal. 3 (2013) 713–720.
[120] Y. Wang, J. Zhuang, G. Yang, D. Zhou, D. Ma, X. Han, X. Bao, Study on the external surface acidity of MCM-22 zeolite: theoretical calculation and ^{31}P MAS NMR, J. Phys. Chem. B 108 (2004) 1386–1391.
[121] J. Zhuang, D. Ma, Z. Yan, F. Deng, X. Liu, X. Han, X. Bao, X. Liu, X. Guo, X. Wang, Solid-state MAS NMR detection of the oxidation center in TS-1 zeolite by *in situ* probe reaction, J. Catal. 221 (2004) 670–673.
[122] J. Zhuang, Z. Yan, X. Liu, X. Liu, X. Han, X. Bao, U. Müeller, NMR study on the acidity of TS-1 zeolite, Catal. Lett. 83 (2002) 87–91.
[123] G. Yang, J. Zhuang, D. Ma, X. Lan, L. Zhou, X. Liu, X. Han, X. Bao, A joint experimental-theoretical study on trimethylphosphine adsorption on the Lewis acidic sites present in TS-1 zeolite, J. Mol. Struct. 882 (2008) 24–29.
[124] G. Yang, X. Lan, J. Zhuang, D. Ma, L. Zhou, X. Liu, X. Han, X. Bao, Acidity and defect sites in titanium silicalite catalyst, Appl. Catal. A 337 (2008) 58–65.
[125] J. Wang, Y. Su, J. Xu, C. Ye, F. Deng, Acid sites and oxidation center in molybdena supported on tin oxide as studied by solid-state NMR spectroscopy and theoretical calculation, Phys. Chem. Chem. Phys. 8 (2006) 2378–2384.
[126] L. Zhou, S. Li, J. Li, Q. Wang, F. Deng, Valence state alternation of copper species doped in HY zeolite as revealed by paramagnetic relaxation enhancement NMR spectroscopy, Solid State Nucl. Magn. Reson. 74-75 (2016) 10–15.

CHAPTER THREE

^{15}N NMR coordination shifts in transition metal complexes and organometallics with heterocycles containing nitrogen—Update for 2012–20

Leszek Pazderski*
Faculty of Chemistry, Nicolaus Copernicus University, PL, Toruń, Poland
*Corresponding author: e-mail address: leszekp@chem.umk.pl

Contents

Annual Reports on NMR Spectroscopy, Volume 101
ISSN 0066-4103
https://doi.org/10.1016/bs.arnmr.2020.07.001

Abstract

An update of a review paper published in 2013 (L. Pazderski, ^{15}N and ^{31}P NMR coordination shifts in transition metal complexes with nitrogen- and phosphorus-containing heterocycles, *Annu. Rep. NMR Spectr.* 80 (2013) 33–180), devoted to ^{15}N NMR and covering years 2012–20 (till June), with 184 references. It collects ^{15}N NMR data for 407 complexes or organometallics of such transition metal ions as Ni(II), Pd(II), Pt(II), Pt(IV), Rh(I), Ir(I), Co(III), Rh(III), Ir(III), Fe(II), Ru(II), Os(II), Cu(I), Ag(I), Au(I), Au(III), Zn(II), Cd(II) and Hg(II) with nitrogen-containing heterocycles (azines, azoles, purines and 1,2,4-triazolo-[1,5*a*]-pyrimidines). ^{15}N NMR coordination shifts, i.e., differences between ^{15}N chemical shifts of the same nitrogen in the molecules of the complex or organometallic and the ligand ($\Delta^{15N}{}_{coord} = \delta^{15N}{}_{compl/organomet} - \delta^{15N}{}_{lig}$) are discussed in relation to some structural features of the reviewed coordination compounds, mainly to the type of the metal central ion and the donor atom in *trans*-position in respect to the concerned N atom.

Keywords: ^{15}N NMR transition metal complexes, Azine complexes, Azole complexes, Coordination shift

The term *NMR coordination shift* has two principal meanings: the phenomenon of NMR signal shift caused by coordination of a ligand with a metal ion, and the numerical parameter describing the magnitude of this effect, defined as a difference between NMR chemical shifts of the same atom in the molecules of a given complex or organometallic (δ_{compl} or $\delta_{organomet}$; further denoted simply as δ_{compl}) and the parent ligand (δ_{lig}), measured preferably in the same medium: $\Delta_{coord} = \delta_{compl} - \delta_{lig}$.

NMR coordination shifts can be defined for any nuclei in the coordinated molecule; however, the most interesting are those concerning donor atoms. The present review is focused at transition metal coordination compounds with such nitrogen-containing heterocycles as azines, azoles, purines and 1,2,4-triazolo-[1,5*a*]-pyrimidines, involving both complexes (N-donor ligand(s) bound to metal only by heteroatom(s)) and organometallics (N-donor ligand(s) bound to metal also by carbon(s)); thus, the ^{15}N coordination shifts ($\Delta^{15N}{}_{coord}$) are discussed. As it is an update of the analogous review published in 2013 (further: "Review 2013") [1], this work covers mainly the period 2014–19; however, also some papers published in 2012–13 (if they were not quoted in Review 2013) and the ones printed in early 2020 (till June) have been cited.

The reviewed coordination compounds contain the following diamagnetic metal ions: Ni(II), Pd(II), Pt(II), Rh(I), Ir(I) and Au(III) (square-planar, d^8); Pt(IV), Co(III), Rh(III), Ir(III), Fe(II), Ru(II) and Os(II) (octahedral, d^6, low-spin); Cu(I), Ag(I), Au(I), Zn(II), Cd(II) and Hg(II) (various geometries, d^{10}).

In this review, N-donor ligands have been chosen from variously substituted derivatives of simple azines (pyridine—**1**, 2,2′-bipyridine—**2**, 1,10-phenanthroline—**3**, 2,2′:6′,2″-terpyridine—**4**, quinoline—**5**, isoquinoline—**6**, 7,8-benzoquinoline—**7**, 2,2′-biquinoline—**8**, 1,1′-biisoquinoline—**9**, pyridazine—**10**, pyrimidine—**11**, pyrazine—**12**, 1,3,5-triazine—**13**, quinazoline—**14**, phthalazine—**15**, 4,5-diazafluorene—**16**, 1,5-naphthyridine—**17**, 1,8-naphthyridine—**18**, 4-azaindole—**19**, 7-azaindole—**20**; Scheme 1), azoles (pyrazole—**21**, imidazole—**22**, indazole—**23**, benzoimidazole—**24**, 1,2,3-triazole—**25**, 1,2,4-triazole—**26**, benzotriazole—**27**, 4,5,6,7-tetrahydrobenzotriazole—**28**; Scheme 2) and such azoloazines like purine—**29** and 1,2,4-triazolo-[1,5*a*]-pyrimidine—**30** (Scheme 3). However, some complicated coordination compounds with the above transition metals and ligands, for which the analysis of $\Delta^{15N}{}_{coord}$ parameters with respect to the molecular structures would be difficult, have been skipped.

Unfortunately, a number of reports had to be neglected, due to their incomplete or evidently erroneous content. There were several reasons for that. In many papers the authors did not specify the reference compound, assuming it was obvious for a reader—while it was not, because two different ones, i.e., neat nitromethane and liquid ammonia, were widely used by various research groups. Of course, in some cases it could be easily deduced whether the ^{15}N chemical shifts were referenced to CH_3NO_2 or to NH_3—owing to the large 381.7 ppm relative shift of both species. Sometimes, however, the published δ^{15N} values were not typical and did not certainly correspond to any common reference.

Furthermore, in some articles the ^{15}N chemical shifts, also of the well-known heterocycles, were significantly different from many literature reports—which suggested that their determination may have been based on an unspecified reference standard. The same difficulty concerned a few coordination compounds, for which the δ^{15N} values were totally inconsistent with the majority of known NMR data—indicating systematic or gross errors during their determination.

Another problem was that sometimes ^{15}N chemical shifts of the studied complexes or organometallics were nearly identical with those for free

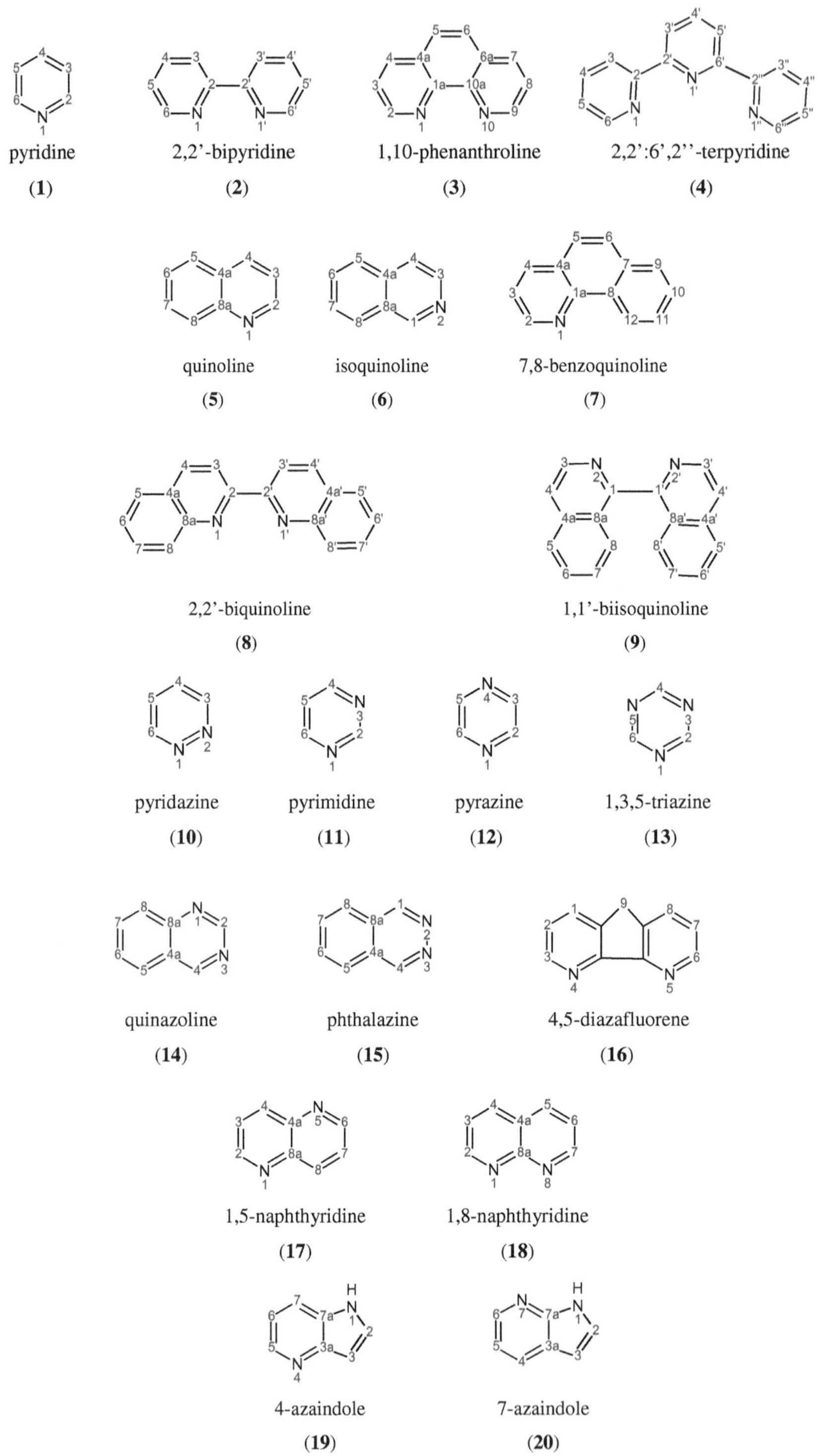

Scheme 1 Azine ligands.

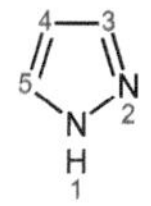

1H-pyrazole

(21)

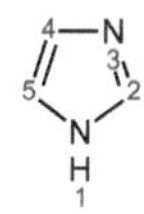

1H-imidazole

(22)

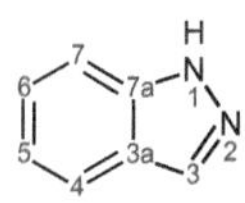

1H-indazole

(23)

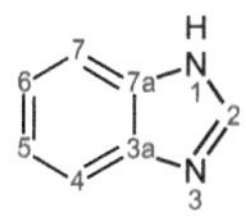

1H-benzoimidazole

(24)

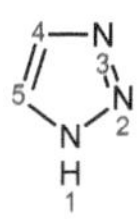

1H-1,2,3-triazole

(25)

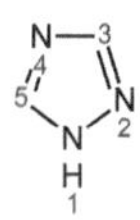

1H-1,2,4-triazole

(26)

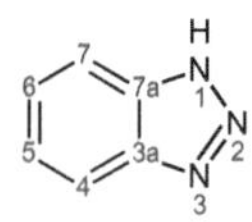

1H-benzotriazole

(27)

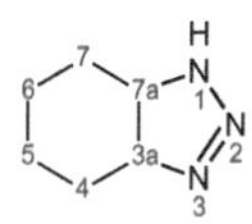

1H-tetrahydrobenzotriazole

(28)

Scheme 2 Azole ligands.

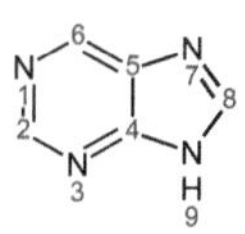

purine

(29)

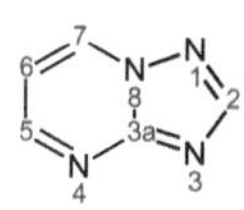

1,2,4-triazolo-[1,5a]-pyrimidine

(30)

Scheme 3 Azoloazine ligands.

ligands (described in the same paper or elsewhere). Such a situation often occurred when NMR measurements were performed in solvents having their own ligating ability (e.g., DMSO-d_6), which suggested decomposition of the studied coordination compounds. The authors were often not conscious of this possibility, mistakenly treating the measured δ^{15N} values as parameters characterizing given complexes or organometallics—whereas, in reality, they concerned parent ligands in situ. In such cases, the determined ^{15}N coordination shifts would be nearly equal to zero, having, in fact, no physical sense; this is why it has not been reasonable to quote these data in the present review.

The ^{15}N chemical and coordination shifts for the reviewed compounds are presented in Tables 1–15, which have the following construction.

Table 1 Square-planar Ni(II), Pd(II) and Pt(II) (d^8), and octahedral Pt(IV) (d^6, low-spin) complexes and organometallics with azines (**1–3**, **5**, **16**, **19–20** ring systems), ^{15}N NMR studied in 2012–20.

Compound	Parent ring system	^{15}N chemical shift (δ^{15N}_{compl}, ppm)	^{15}N coordination shift (Δ^{15N}_{coord}, ppm)	Nitrogen *trans* to
Ni(II) compounds	No interesting species			
Pd(II) compounds				
trans(Cl,$C^{ph,C2}$)-[Pd^{II}(2-phenyl**pyridine***-κ^2-N,$C^{ph,C2}$)(**pyridine**)Cl]^{CDCl3} [2][a,b1]	**1**	N^{py}: −133.7	−65.0[aaa]	(N^{2ppy*})[b1]
		N^{2ppy*}: −145.1	−70.2[aab]	(N^{py})[b1]
trans(Cl,$C^{ph,C2}$)-[Pd^{II}(2-phenyl**pyridine***-κ^2-N,$C^{ph,C2}$)(2-methyl**pyridine**)Cl]^{CDCl3} [2][a,b1]	**1**	N^{py}: −130.6	−59.0[aac]	(N^{2ppy*})[b1]
		N^{2ppy*}: −145.0	−70.1[aab]	(N^{py})[b1]
trans(Cl,$C^{ph,C2}$)-[Pd^{II}(2-phenyl**pyridine***-κ^2-N,$C^{ph,C2}$)(3-methyl**pyridine**)Cl]^{CDCl3} [2][a,b1]	**1**	N^{py}: −135.0	−59.2[aad]	(N^{2ppy*})[b1]
		N^{2ppy*}: −144.7	−69.8[aab]	(N^{py})[b1]
trans(Cl,$C^{ph,C2}$)-[Pd^{II}(2-phenyl**pyridine***-κ^2-N,$C^{ph,C2}$)(4-methyl**pyridine**)Cl]^{CDCl3} [2][a,b1]	**1**	N^{py}: −143.1	−59.0[aae]	(N^{2ppy*})[b1]
		N^{2ppy*}: −144.7	−69.8[aab]	(N^{py})[b1]
trans(Cl,$C^{ph,C2}$)-[Pd^{II}(2-phenyl**pyridine***-κ^2-N,$C^{ph,C2}$)(2,3-dimethyl**pyridine**)Cl]^{CDCl3} [2][a,b1]	**1**	N^{py}: −129.8	−58.4[aaf]	(N^{2ppy*})[b1]
		N^{2ppy*}: −144.2	−69.3[aab]	(N^{py})[b1]
trans(Cl,$C^{ph,C2}$)-[Pd^{II}(2-phenyl**pyridine***-κ^2-N,$C^{ph,C2}$)(2,4-dimethyl**pyridine**)Cl]^{CDCl3} [2][a,b1]	**1**	N^{py}: −139.2	−59.2[aag]	(N^{2ppy*})[b1]
		N^{2ppy*}: −144.3	−69.4[aab]	(N^{py})[b1]
trans(Cl,$C^{ph,C2}$)-[Pd^{II}(2-phenyl**pyridine***-κ^2-N,$C^{ph,C2}$)(2,6-dimethyl**pyridine**)Cl]^{CDCl3} [2][a,b1]	**1**	N^{py}: −125.4	−53.4[aah]	(N^{2ppy*})[b1]
		N^{2ppy*}: −143.6	−68.7[aab]	(N^{py})[b1]

trans(Cl,$C^{ph,C2}$)-[Pd^{II}(2-phenyl**pyridine***-κ^2-N,$C^{ph,C2}$)(3,5-dimethyl**pyridine**)Cl]^{CDCl3} [2]^{a,b1}	**1**	N^{py}: −135.5	−64.5^{aai}	(N^{2ppy*})^{b1}
		N^{2ppy*}: −143.8	−68.9^{aab}	(N^{py})^{b1}
trans(Cl,$C^{ph,C2}$)-[Pd^{II}(2-phenyl**pyridine***-κ^2-N,$C^{ph,C2}$)(2,4,6-trimethyl**pyridine**)Cl]^{CDCl3} [2]^{a,b2,A}	**1**	N^{py}: −133.5	−53.4^{aaj}	N^{2ppy*} $^{X\text{-}ray}$
		N^{2ppy*}: −143.5	−68.6^{aab}	N^{py} $^{X\text{-}ray}$
trans(Cl,$C^{ph,C2}$)-[Pd^{II}(2-phenyl**pyridine***-κ^2-N,$C^{ph,C2}$)(NH_3)Cl]$^{DMSO\text{-}d6}$ [2]^{a,b1}	**1**	N^{2ppy*}: −142.0	−70.1^{aak}	(NH_3)^{b1}
[Pd^{II}(2-phenyl**pyridine***-κ^2-N,$C^{ph,C2}$)(CD_3SOCD_3)$_2$]$^+$	**1**	N^{2ppy*} (cation): −141	−69^{aal}	(S)^{aaaa}
[Pd^{II}(2-phenyl**pyridine***-κ^2-N,$C^{ph,C2}$)Cl_2]$^-$ $_{DMF\text{-}d7+DMSO\text{-}d6\ (3:1),243K}$ [3]^a	**1**	N^{2ppy*} (anion): −145	−73^{aal}	(Cl)^{aaaa}
[Pd^{II}(2-phenyl**pyridine***-κ^2-N,$C^{ph,C2}$)(μ-Cl)]$_2^{DMF\text{-}d7}$ [3]^{a,B}	**1**	N^{2ppy*}: −146	−74^{aam}	Cl $^{X\text{-}ray}$
[Pd^{II}(2-phenyl**pyridine***-κ^2-N,$C^{ph,C2}$)(μ-Cl)]$_2^{solid}$ [3]^{a,B}	**1**	N^{2ppy*}: −147	−74^{aan}	Cl $^{X\text{-}ray}$
trans-[Pd^{II}(4-methyl**pyridine**)(1-methyl-3-(2-(phtalimide-κ^1-N)-ethyl)imidazol-2-ylidene-κ^1-C^2)I_2]$^{DMSO\text{-}d6}$ [4]^C	**1**	N^{py}: −143	−65^{aao}	$C^{ylidene,C2\ X\text{-}ray}$
trans-[Pd^{II}(4-methyl**pyridine**)(1-methyl-3-(3-triethoxysilyl-*n*-propyl)imidazol-2-ylidene-κ^1-C^2)I_2]^{CDCl3} [5]^D	**1**	N^{py}: −143	−59^{aae}	$C^{ylidene,C2\ X\text{-}ray}$
trans-[Pd^{II}(4-*tert*-butyl**pyridine**)$_2$(*E*-(N-methoxy-2,6-difluorobenzamidate-κ^1-N))$_2$]^{CDCl3} [6]^c	**1**	N^{py}: −165.2^{aa1}	−95.2^{aap}	(N^{py})^c
trans-[Pd^{II}(4-*tert*-butyl**pyridine**)$_2$(*Z*-(N-methoxy-2,6-difluorobenzamidate-κ^1-N))$_2$]^{CDCl3} [6]^c	**1**	N^{py}: −165.0^{aa2}	−95.0^{aap}	(N^{py})^c
trans-[Pd^{II}(4-*tert*-butyl**pyridine**)$_2$(*E*-(N-methoxy-2,6-difluorobenzamidate-κ^1-N))(*Z*-(N-methoxy-2,6-difluorobenzamidate-κ^1-N))]^{CDCl3} [6]^c	**1**	N^{py}: −165.5^{aa3}	−95.5^{aap}	(N^{py})^c

Continued

Table 1 Square-planar Ni(II), Pd(II) and Pt(II) (d^8), and octahedral Pt(IV) (d^6, low-spin) complexes and organometallics with azines (**1–3**, **5**, **16**, **19–20** ring systems), ^{15}N NMR studied in 2012–20.—cont'd

Compound	Parent ring system	^{15}N chemical shift (δ^{15N}_{compl}, ppm)	^{15}N coordination shift (Δ^{15N}_{coord}, ppm)	Nitrogen *trans* to
trans-[Pd^{II}(4-*tert*-butyl**pyridine**)$_2$(*E*-(N-methoxy-2,6-difluorobenzamidate-κ^1-N))(CH_3COO)]^{CDCl3} [6][c]	**1**	N^{py}: −166.5[aa4]	−96.5[aap]	(N^{py})[c]
trans-[Pd^{II}(4-*tert*-butyl**pyridine**)$_2$(*Z*-(N-methoxy-2,6-difluorobenzamidate-κ^1-N))(CH_3COO)]^{CDCl3} [6][c]	**1**	N^{py}: −166.3[aa5]	−96.3[aap]	(N^{py})[c]
trans-[Pd^{II}(4-*tert*-butyl**pyridine**)$_2$(*E*-(N-methoxy-2,6-difluorobenzamidate-κ^1-N))(C_6H_5COO)]^{CDCl3} [6][c]	**1**	N^{py}: −166.7[aa6]	−96.7[aap]	(N^{py})[c]
trans-[Pd^{II}(4-*tert*-butyl**pyridine**)$_2$(*Z*-(N-methoxy-2,6-difluorobenzamidate-κ^1-N))(C_6H_5COO)]^{CDCl3} [6][c]	**1**	N^{py}: −166.7[aa7]	−96.7[aap]	(N^{py})[c]
trans-[Pd^{II}(3-chloro**pyridine**)(1-(4-methylphenyl)-3-methyl-4-phenyl-1,2,3-triazolium-5-yl-κ^1-C^5)Cl_2]^{CDCl3} [7][c]	**1**	N^{py}: −127[ab]	−69[aaq]	($C^{tz,C5}$)[c]
[Pd^{II}(2-phenylmethyl(amino)**pyridine**-κ^2-N^{py},N^{NH2})Cl_2]$^{DMSO\text{-}d6}$ [8]	**1**	N^{py}: −155	−88[aar]	(Cl)[aaaa]
[Pd^{II}(2-phenylmethylaminomethyl**pyridine**-κ^2-N^{py},N^{NH})Cl_2]$^{DMSO\text{-}d6}$ [8]	**1**	N^{py}: −154	−89[aas]	(Cl)[aaaa]
[Pd^{II}(2-phenyl(1-ethyl)aminomethyl**pyridine**-κ^2-N^{py},N^{NH})Cl_2]$^{DMSO\text{-}d6}$ [8] diastereomer A	**1**	N^{py}: −155	−89[aat]	(Cl)[aaaa]
[Pd^{II}(2-phenyl(1-ethyl)aminomethyl**pyridine**-κ^2-N^{py},N^{NH})Cl_2]$^{DMSO\text{-}d6}$ [8] diastereomer B	**1**	N^{py}: −153	−87[aat]	(Cl)[aaaa]
[Pd^{II}(2-phenylmethylamino(methyl)methyl(phenyl)**pyridine**-κ^2-N^{py},N^N)Cl_2]$^{DMSO\text{-}d6}$ [8]	**1**	N^{py}: −156	−93[aau]	(Cl)[aaaa]

$[Pd^{II}$(2-phenyl(1-ethyl)amino(methyl)methyl(phenyl)**pyridine**-κ^2-$N^{py},N^{N})Cl_2]^{DMSO\text{-}d6}$ [8]	**1**	N^{py}: −156	−93[aav]	(Cl)[aaaa]
trans(Cl,N^{py})-$[Pd^{II}$(2-phenyliminomethyl**pyridine**-κ^2-$N^{py}, N^{N=})(CH_3)Cl]^{CD2Cl2}$ [9][d2,E]	**1**	N^{py}: −144.8	−81.8[aaw]	Cl [X-ray]
trans(CH_3,N^{py})-$[Pd^{II}$(2-phenyliminomethyl**pyridine**-κ^2-$N^{py}, N^{N=})(CH_3)Cl]^{CD2Cl2}$ [9][d1]	**1**	N^{py}: −111.4	−48.4[aaw]	(CH_3)[d1]
trans(CH_3,N^{py})-$[Pd^{II}$(2-(2,6-dimethylphenyl) iminomethyl**pyridine**-κ^2-$N^{py},N^{N=})(CH_3)Cl]^{CD2Cl2}$ [9][F]	**1**	N^{py}: −111.7	Unknown[aax1]	CH_3 [X-ray]
trans(CH_3,N^{py})-$[Pd^{II}$(2-phenylimino(1-ethyl)**pyridine**-κ^2-$N^{py}, N^{N=})(CH_3)Cl]^{CD2Cl2}$ [9][G]	**1**	N^{py}: −112.5	Unknown[aax2]	CH_3 [X-ray]
trans(CH_3,N^{py})-$[Pd^{II}$(2-(2,6-dimethylphenyl)imino(1-ethyl) **pyridine**-κ^2-$N^{py},N^{N=})(CH_3)Cl]^{CD2Cl2}$ [9][H]	**1**	N^{py}: Unknown[al]	−54[aay]	(Cl)[aaaa]
$[Pd^{II}$(2-(*1H*-benzotriazol-1-yl)**pyridine**-κ^2-$N^{py},N^{btz,N2})$ $Cl_2]^{DMSO\text{-}d6}$ [10][#]	**1**	N^{py}: Unknown[al]	−54[aay]	(Cl)[aaaa]
$[Pd^{II}$(2-(1-(pyrid-2-yl)-5-(4-methoxyphenyl)-*1H*-pyrazol-3-yl) **pyridine**-κ^2-$N^{pyA},N^{pz,N2})Cl_2]^{DMSO\text{-}d6}$ [11][e,I,#]	**1**	N^{pyA}: −154.7	−75.1[aaz]	Cl [X-ray]
		N^{pyB}: undetected[am]	Not applicable[aba]	None
$[Pd^{II}$(2-(1-(pyrid-2-yl)-5-(4-methoxyphenyl)-*1H*-pyrazol-3-yl) **pyridine**-κ^2-$N^{pyA},N^{pz,N2})Cl_2]^{solid}$ [11][e,I,#]	**1**	N^{pyA}: −159.7	−91.9[abb]	Cl [X-ray]
		N^{pyB}: −65.0/−62.0[am]	Not applicable[abc]	None
$[Pd^{II}$(2-(1-(**pyrid-2-yl**)-5-(4-methoxyphenyl)-*1H*-pyrazol-3-yl)pyridine-κ^2-$N^{pyB},N^{pz,N2})Cl_2]^{DMSO\text{-}d6}$ [11][f]	**1**	N^{pyA}: undetected[am]	Not applicable[aaz]	None
		N^{pyB}: −169.7	−93.5[aba]	(Cl)[aaaa]

Continued

Table 1 Square-planar Ni(II), Pd(II) and Pt(II) (d^8), and octahedral Pt(IV) (d^6, low-spin) complexes and organometallics with azines (**1–3**, **5**, **16**, **19–20** ring systems), ^{15}N NMR studied in 2012–20.—cont'd

Compound	Parent ring system	^{15}N chemical shift (δ^{15N}_{compl}, ppm)	^{15}N coordination shift (Δ^{15N}_{coord}, ppm)	Nitrogen *trans* to
$[Pd^{II}$(**2,2′-bipyridine**)(2-(2,3-dimethylphenylamino)benzoate-κ^1-O$)_2]^{CDCl3}$ [12][g]	**2**	N^{bpy}: -165^{ac}	-90^{abd}	$(O)^{aaaa}$
$[Pd^{II}$(**2,2′-bipyridine**)$(NH_2CSNH_2)_2]Cl_2^{CD3OH,260K}$ [13]	**2**	N^{bpy}: -148^{ad1}	-66^{abe}	$(S)^{aaaa}$
$[Pd^{II}$(**2,2′-bipyridine**)$(CH_3NHCSNH_2)_2]Cl_2^{CD3OH,260K}$ [13]	**2**	N^{bpy}: -147^{ad2}	-65^{abe}	$(S)^{aaaa}$
$[Pd^{II}$(**2,2′-bipyridine**)$(CH_3(CH_2)_3NHCSNH_2)_2]Cl_2^{CD3OH,260K}$ [13]	**2**	N^{bpy}: -147^{ad3}	-65^{abe}	$(S)^{aaaa}$
$[Pd^{II}$(**2,2′-bipyridine**)$(CH_3CH_2NHCSNHCH_2CH_3)_2]Cl_2^{CD3OH,260K}$ [13]	**2**	N^{bpy}: -147^{ad4}	-65^{abe}	$(S)^{aaaa}$
$[Pd^{II}$(**2,2′-bipyridine**)$(CH_3(CH_2)_3NHCSNH(CH_2)_3CH_3)_2]Cl_2^{CD3OH,260K}$ [13]	**2**	N^{bpy}: -149^{ad5}	-67^{abe}	$(S)^{aaaa}$
$[Pd^{II}$(bis(*tert*-butyl) **2,2′-bipyridine-**3,3′-dicarboxylate$)_2](B(3,5\text{-bis(trifluoromethyl)phenyl})_4)_2^{CD3OD,258K}$ [14]	**2**	N^{bpy}: -160	-84^{abf}	$(N^{bpy})^{aaaa}$
$[Pd^{II}$(**1,10-phenanthroline**)$(NH_2CSNH_2)_2]Cl_2^{CD3OH,260K}$ [13][J]	**3**	N^{phen}: -150^{ad6}	-66^{abg}	$(S)^{aaaa}$
$[Pd^{II}$(**1,10-phenanthroline**)$(CH_3NHCSNH_2)_2]Cl_2^{CD3OH,260K}$ [13]	**3**	N^{phen}: -150^{ad7}	-66^{abg}	$(S)^{aaaa}$
$[Pd^{II}$(**1,10-phenanthroline**)$(CH_3(CH_2)_3NHCSNH_2)_2]Cl_2^{CD3OH,260K}$ [13]	**3**	N^{phen}: -149^{ad8}	-65^{abg}	$(S)^{aaaa}$
$[Pd^{II}$(**1,10-phenanthroline**)$(CH_3CH_2NHCSNHCH_2CH_3)_2]Cl_2^{CD3OH,260K}$ [13]	**3**	N^{phen}: -150^{ad9}	-66^{abg}	$(S)^{aaaa}$

$[Pd^{II}($**1,10-phenanthroline**$)(CH_3(CH_2)_3NHCSNH(CH_2)_3CH_3)_2)_2]Cl_2{}^{CD3OD,260K}$ [13]	**3**	N^{phen}: -149^{ad10}	-65^{abg}	$(S)^{aaaa}$
trans(P,N^{quin})-$[Pd^{II}$(8-methyl**quinoline***-κ^2-$N,C^{CH2})(P(C_6H_5)_3)(NO_3)]^{CDCl3}$ [15][h,K]	**5**	N^{quin*}: -127^{ae1}	-51^{abh}	P $^{X\text{-ray}}$
trans(N^{NO2},N^{quin})-$[Pd^{II}$(8-methyl**quinoline***-κ^2-$N,C^{CH2})(CH_3CN)(NO_2)]^{CDCl3}$ [15][h,L]	**5**	N^{quin*}: -149^{ae2}	-73^{abh}	NO_2 $^{X\text{-ray}}$
$[Pd_2^{II}$(8-methyl**quinoline***-κ^2-N, $C^{CH2})_2(\mu\text{-}CH_3COO)_2]^{CDCl3}$ [15][h,M]	**5**	N^{quin*}: -149^{ae3}	-73^{abh}	O $^{X\text{-ray}}$
$[Pd_2^{II}$(**quinoline**-8-carboxylate-κ^2-N, $O)_2(\mu\text{-}CH_3COO)_2]^{CDCl3}$ [15][N]	**5**	N^{quin}: -202^{ae4}	-95^{abi}	O $^{X\text{-ray}}$
$[Pd^{II}$(8-(ethoxycarbonylmethylaminoimino methyl)**quinoline**-κ^2-$N^{quin},N^{N=})Cl_2]^{DMSO\text{-}d6}$ [16]	**5**	N^{quin}: -146.2^{af}	Unknown^{abj}	$(Cl)^{aaaa}$
trans-$[Pd^{II}$(4-(2-bromophenoxy) **quinoline**$)_2(CH_3COO)_2]^{CDCl3}$ [17] [i2,O]	**5**	N^{quin}: -185.0^{ag1}	-95.0^{abk}	N $^{X\text{-ray}}$
trans-$[Pd^{II}$(4-(2-bromophenoxy)-7-bromo**quinoline**$)_2(CH_3COO)_2]^{CDCl3}$ [17][i1]	**5**	N^{quin}: -187.5^{ag2}	-96.3^{abl}	$(N)^{i1}$
$[Pd^{II}$(9,9-dimethyl-**4,5-diaza**-*9H*-**fluorene**-κ^2-$N^4,N^5)(CH_3COO)_2]^{CDCl3}$ [18][P]	**16**	$N^{dafl,N4,N5}$: -172.0^{ah1}	Unknown^{abm1}	$2 \times$ O $^{X\text{-ray}}$
$[Pd_2^{II}$(μ-9,9-dimethyl-**4,5-diaza**-*9H*-**fluorene**-κ^2-$N^4,N^5)(\mu\text{-}CH_3COO)_2(CH_3COO)_2]^{CDCl3}$ [18][Q]	**16**	$N^{dafl,N4,N5}$: -182.3^{ah2}	Unknown^{abm1}	$2 \times$ O $^{X\text{-ray}}$
$[Pd^{II}$(**4,5-diaza**-*9H*-**fluoren**-9-one-κ^2-$N^4,N^5)(CH_3COO)_2]^{CDCl3,228K}$ [18]	**16**	$N^{dafl,N4,N5}$: -168.4^{ah3}	Unknown^{abm2}	$2 \times (O)^{aaaa}$

Continued

Table 1 Square-planar Ni(II), Pd(II) and Pt(II) (d^8), and octahedral Pt(IV) (d^6, low-spin) complexes and organometallics with azines (**1–3**, **5**, **16**, **19–20** ring systems), ^{15}N NMR studied in 2012–20.—cont'd

Compound	Parent ring system	^{15}N chemical shift (δ^{15N}_{compl}, ppm)	^{15}N coordination shift (Δ^{15N}_{coord}, ppm)	Nitrogen *trans* to
trans/anti-[Pd^{II}(**4,5-diaza**-*9H*-**fluoren**-9-one-κ^1-N^4)$_2$ $(CH_3COO)_2]^{CDCl3,228K}$ [18][R]	**16**	$N^{dafl,N4}$: −174.6[ah4]	Unknown[abm2]	O [X-ray]
		$N^{dafl,N5}$: −75.0 [ah5,am]	Not applicable	None
trans/syn-[Pd^{II}(**4,5-diaza**-*9H*-**fluoren**-9-one-κ^1-N^4)$_2$ $(CH_3COO)_2^{CDCl3,228K}$ [18]	**16**	$N^{dafl,N4}$: −175.4[ah6]	Unknown[abm2]	(O)[aaaa]
		$N^{dafl,N5}$: −72.4[ah7,am]	Not applicable	None
cis/anti-[Pd^{II}(**4,5-diaza**-*9H*-**fluoren**-9-one-κ^1-N^4)$_2$$(CH_3COO)_2]^{CDCl3,228K}$ [18]	**16**	$N^{dafl,N4}$: −178.0[ah8]	Unknown[abm2]	(O)[aaaa]
		$N^{dafl,N5}$: −77.4[ah9,am]	Not applicable	None
[Pd^{II}(**4,5-diaza**-*9H*-**fluoren**-9-one-κ^2-N^4,N^5) $(CF_3COO)_2]^{CDCl3}$ [18][S]	**16**	$N^{dafl,N4,N5}$: −183.0[ah10]	Unknown[abm2]	2 × O [X-ray]
[Pd_2^{II}(**4,5-diaza**-*9H*-**fluoren**-9-one-κ^1-N^4)$_2$(μ-CH_3COO)$_2$ $(CH_3COO)_2]^{CDCl3,228K}$ [18][T]	**16**	$N^{dafl,N4}$: −192.4[ah11]	Unknown[abm2]	O [X-ray]
		$N^{dafl,N5}$: −77.1[ah12,am]	Not applicable	None
[Pd_2^{II}(**4,5-diaza**-*9H*-**fluoren**-9-one-μ-N^4,N^5) (μ-CH_3COO)$_2$$(CH_3COO)_2]^{CDCl3,228K}$ [18][U]	**16**	$N^{dafl,N4,N5}$: −178.0[ah13]	Unknown[abm2]	2 × O [X-ray]
[Pd_2^{II}(**4,5-diaza**-*9H*-**fluoren**-9-one-μ-N^4,N^5) (μ-CF_3COO)$_2$$(CF_3COO)_2]^{CDCl3}$ [18][V]	**16**	$N^{dafl,N4,N5}$: −184.4[ah14]	Unknown[abm2]	2 × O [X-ray]
[Pd_2^{II}(**4,5-diaza**-*9H*-**fluoren**-9-one-μ-N^4,N^5)$_2$ $((CH_3)_3CCOO)_4]^{CDCl3}$ [18][W]	**16**	$N^{dafl,N4,N5}$: −165.4[ah15]	Unknown[abm2]	2 × $N^{dafl,N4,N7}$ [X-ray]
[Pd^{II}(**4,5-diaza**-*9H*-**fluoren**-9-one-κ^1-N^4) $(CH_3COO)_3]^{1,4\text{-dioxane}/CD3COOD\ (3:1),278K}$ [19]	**16**	$N^{dafl,N4}$: −181.8[ah16]	Unknown[abm2]	(O)[aaaa]
		$N^{dafl,N5}$: −75.8[ah17,am]	Not applicable	None

Pt(II) compounds				
[Pt^{II}(2-phenyl**pyridine***-κ^2-N,C^{ph,C2})(μ-Cl)]$_2^{DMF-d7}$ [3][a]	**1**	N^{2ppy*}: −164	−92[aam]	(Cl)[aaaa]
[Pt^{II}(2-phenyl**pyridine***-κ^2-N,C^{ph,C2})(μ-Cl)]$_2^{solid}$ [3][a]		N^{2ppy*}: −165	−92[aan]	(Cl)[aaaa]
[Pt^{II}(2-iminomethyl**pyridine**-κ^2-N^{py},N$^{NH=}$)Cl$_2$]$^{DMF-d7}$ [20][X]	**1**	N^{py}: −171.9[ai]	Unknown[abn]	Cl [X-ray]
cis-[Pt^{II}(N-(*n*-dodecyl)-N′-(**pyrid-3-yl**)urea-κ^1-N^{py})$_2$Cl$_2$]$^{DMF-d7}$ [21][j]	**1**	N^{py}: Unknown	−138.4[abo]	(Cl)[i]
cis-[Pt^{II}(N-(*n*-dodecyl)-N′-(**pyrid-3-yl**methyl)urea-κ^1-N^{py})$_2$Cl$_2$]$^{DMF-d7}$ [21][j]	**1**	N^{py}: Unknown	−132.3[abo]	(Cl)[i]
cis-[Pt^{II}(N-(*n*-dodecyl)-N′-(**pyrid-3-yl**ethyl)urea-κ^1-N^{py})$_2$Cl$_2$]$^{DMF-d7}$ [21][j]	**1**	N^{py}: Unknown	−132.2[abo]	(Cl)[i]
[Pt^{II}(2-phenyl-6-(1-(12-hydroxo-*n*-dodecyl)-*1H*-1,2,3-triazol-4-yl)**pyridine***-κ^3-N^{py},C^{ph,C2},N^{tz,N3})(**pyridine**)]PF$_6^{acetone-d6}$ [22][k,#]	**1**	N^{py}: −168[aj1]	−106[abp]	(N$^{py\ in\ 2p6tzpy*}$)[aaaa]
		N$^{py\ in\ 2p6tzpy*}$: −167[aj2]	Unknown[abq1]	(N^{py})[aaaa]
[Pt^{II}(2-phenyl-6-(1-(12-trimethylammonioethoxy-*n*-dodecyl)-*1H*-1,2,3-triazol-4-yl)**pyridine***-κ^3-N^{py},C^{ph,C2},N^{tz,N3})(**pyridine**)](PF$_6$)$_2^{acetone-d6}$ [22][l,#]	**1**	N^{py}: −168[aj3]	−106[abp]	(N$^{py\ in\ 2p6tzpy*}$)[aaaa]
		N$^{py\ in\ 2p6tzpy*}$: −167[aj4]	Unknown[abq2]	(N^{py})[aaaa]
[Pt^{II}(1-β-D-ribofuranosyl-4-(**pyrid-2-yl**)-*1H*-1,2,3-triazole-κ^2-N^{py},N^{tz,N3})Cl$_2$]$^{DMF-d7}$ [23][#]	**1**	N^{py}: −175.5	−101.7[abr]	(Cl)[aaaa]
[Pt^{II}(**2,2′-bipyridine**)(NH$_2$CSNH$_2$)$_2$]Cl$_2^{CD3OH,260K}$ [13]	**2**	N^{bpy}: −159[ad11]	−77[abe]	(S)[aaaa]
[Pt^{II}(**2,2′-bipyridine**)(CH$_3$NHCSNH$_2$)$_2$]Cl$_2^{CD3OH,260K}$ [13]	**2**	N^{bpy}: −159[ad12]	−77[abe]	(S)[aaaa]

Continued

Table 1 Square-planar Ni(II), Pd(II) and Pt(II) (d^8), and octahedral Pt(IV) (d^6, low-spin) complexes and organometallics with azines (**1–3**, **5**, **16**, **19–20** ring systems), ^{15}N NMR studied in 2012–20.—cont'd

Compound	Parent ring system	^{15}N chemical shift (δ^{15N}_{compl}, ppm)	^{15}N coordination shift (Δ^{15N}_{coord}, ppm)	Nitrogen *trans* to
$[Pt^{II}$(**2,2′-bipyridine**)$(CH_3(CH_2)_3NHCSNH_2)_2]Cl_2^{CD3OH,260K}$ [13]	**2**	N^{bpy}: -160^{ad13}	-78^{abe}	$(S)^{aaaa}$
$[Pt^{II}$(**2,2′-bipyridine**)$(CH_3CH_2NHCSNHCH_2CH_3)_2]Cl_2^{CD3OH,260K}$ [13]	**2**	N^{bpy}: -160^{ad14}	-78^{abe}	$(S)^{aaaa}$
$[Pt^{II}$(**2,2′-bipyridine**)$(CH_3(CH_2)_3NHCSNH(CH_2)_3CH_3)_2]Cl_2^{CD3OH,260K}$ [13]	**2**	N^{bpy}: -160^{ad15}	-78^{abe}	$(S)^{aaaa}$
$[Pt^{II}$(**1,10-phenanthroline**)$(NH_2CSNH_2)_2]Cl_2^{CD3OH,260K}$ [13]	**3**	N^{phen}: -163^{ad16}	-79^{abg}	$(S)^{aaaa}$
$[Pt^{II}$(**1,10-phenanthroline**)$(CH_3NHCSNH_2)_2]Cl_2^{CD3OH,260K}$ [13]	**3**	N^{phen}: -163^{ad17}	-79^{abg}	$(S)^{aaaa}$
$[Pt^{II}$(**1,10-phenanthroline**)$(CH_3(CH_2)_3NHCSNH_2)_2]Cl_2^{CD3OH,260K}$ [13]	**3**	N^{phen}: -162^{ad18}	-78^{abg}	$(S)^{aaaa}$
$[Pt^{II}$(**1,10-phenanthroline**)$(CH_3CH_2NHCSNHCH_2CH_3)_2]Cl_2^{CD3OH,260K}$ [13]	**3**	N^{phen}: -162^{ad19}	-78^{abg}	$(S)^{aaaa}$
$[Pt^{II}$(**1,10-phenanthroline**)$(CH_3(CH_2)_3NHCSNH(CH_2)_3CH_3)_2]Cl_2^{CD3OH,260K}$ [13]	**3**	N^{phen}: -162^{ad20}	-78^{abg}	$(S)^{aaaa}$
cis-$[Pt^{II}$(**4-azaindole**-κ^1-$N^4)_2I_2]^{DMF-d7}$ $[24]^{j}$	**19**	$N^{azaind,N4}$: -181.1^{ak1}	-92.6^{abs1}	$(I)^{i}$
cis-$[Pt^{II}$(**4-azaindole**-κ^1-$N^4)(NH_3)I_2]^{DMF-d7}$ $[24]^{j}$	**19**	$N^{azaind,N4}$: -179.4^{ak2}	-90.9^{abs1}	$(I)^{i}$
cis-$[Pt^{II}$(3-chloro-**7-azaindole**-κ^1-$N^7)_2Cl_2]^{DMF-d7}$ $[25]^{m2,n2,Y}$	**20**	$N^{azaind,N7}$: -208.3^{ak3}	-101.5^{abs2}	Cl^{X-ray}
trans-$[Pt^{II}$(3-chloro-**7-azaindole**-κ^1-$N^7)_2Cl_2]^{DMF-d7}$ $[26]^{n1}$	**20**	$N^{azaind,N7}$: -213.5^{ak4}	-106.7^{abt1}	$(N^{azaind,N7})^{m1}$
$[Pt^{II}$(3-chloro-**7-azaindole**-κ^1-$N^7)_2(C_2O_4)]^{DMF-d7}$ [25]	**20**	$N^{azaind,N7}$: -221.8^{ak5}	-115.0^{abs2}	$(O)^{aaaa}$

cis-[Pt^{II}(4-chloro-**7-azaindole**-κ^1-N^7)$_2$Cl$_2$]$^{DMF\text{-}d7}$ [27][m1]	**20**	N^{azaind,N7}: −213.7[ak6]	−101.2[abs3]	(Cl)
[Pt^{II}(4-chloro-**7-azaindole**-κ^1-N^7)$_2$(C$_2$O$_4$)]$^{DMF\text{-}d7}$ [28]	**20**	N^{azaind,N7}: −227.1[ak7]	−114.6[abs3]	(O)[aaaa]
cis-[Pt^{II}(3-bromo-**7-azaindole**-κ^1-N^7)$_2$Cl$_2$]$^{DMF\text{-}d7}$ [27][m1]	**20**	N^{azaind,N7}: −208.9[ak8]	−101.7[abs4]	(Cl)
[Pt^{II}(3-bromo-**7-azaindole**-κ^1-N^7)$_2$(C$_2$O$_4$)]$^{DMF\text{-}d7}$ [28]	**20**	N^{azaind,N7}: −222.3[ak9]	−115.1[abs4]	(O)[aaaa]
cis-[Pt^{II}(4-bromo-**7-azaindole**-κ^1-N^7)$_2$Cl$_2$]$^{DMF\text{-}d7}$ [27][m2,Z]	**20**	N^{azaind,N7}: −213.3[ak10]	−101.0[abs5]	Cl [X-ray]
[Pt^{II}(4-bromo-**7-azaindole**-κ^1-N^7)$_2$(C$_2$O$_4$)]$^{DMF\text{-}d7}$ [28]	**20**	N^{azaind,N7}: −226.3[ak11]	−114.0[abs5]	(O)[aaaa]
cis-[Pt^{II}(5-bromo-**7-azaindole**-κ^1-N^7)$_2$Cl$_2$]$^{DMF\text{-}d7}$ [25][m1,n2]	**20**	N^{azaind,N7}: −206.8[ak12]	−102.7[abs6]	(Cl)
trans-[Pt^{II}(5-bromo-**7-azaindole**-κ^1-N^7)$_2$Cl$_2$]$^{DMF\text{-}d7}$ [26][n1]	**20**	N^{azaind,N7}: −213.6[ak13]	−109.5[abt2]	(N^{azaind,N7})[m1]
[Pt^{II}(5-bromo-**7-azaindole**-κ^1-N^7)$_2$(C$_2$O$_4$)]$^{DMF\text{-}d7}$ [25]	**20**	N^{azaind,N7}: −220.9[ak14]	−116.8[abs6]	(O)[aaaa]
cis-[Pt^{II}(3-iodo-**7-azaindole**-κ^1-N^7)$_2$Cl$_2$]$^{DMF\text{-}d7}$ [25][m1,n2]	**20**	N^{azaind,N7}: −209.1[ak15]	−102.4[abs7]	(Cl)
trans-[Pt^{II}(3-iodo-**7-azaindole**-κ^1-N^7)$_2$Cl$_2$]$^{DMF\text{-}d7}$ [26][n1]	**20**	N^{azaind,N7}: −210.8[ak16]	−104.1[abt3]	(N^{azaind,N7})[m1]
[Pt^{II}(3-iodo-**7-azaindole**-κ^1-N^7)$_2$(C$_2$O$_4$)]$^{DMF\text{-}d7}$ [25]	**20**	N^{azaind,N7}: −222.3[ak17]	−115.6[abs7]	(O)[aaaa]
cis-[Pt^{II}(3-chloro-5-bromo-**7-azaindole**-κ^1-N^7)$_2$Cl$_2$]$^{DMF\text{-}d7}$ [27][m1]	**20**	N^{azaind,N7}: −215.8[ak18]	−114.6[abs8]	(Cl)
cis-[Pt^{II}(3-iodo-5-bromo-**7-azaindole**-κ^1-N^7)$_2$Cl$_2$]$^{DMF\text{-}d7}$ [27][m1]	**20**	N^{azaind,N7}: −204.3[ak19]	−102.7[abs9]	(Cl)
Pt(IV) compounds		No interesting species		

[#]See also Table 8 for pyrazole-,1,2,3-triazole- and benzotriazole-derived ligand(s).
[a]2-Phenylpyridine* is deprotonated (at the phenyl C(2) atom), anionic (charge −1) form of 2-phenylpyridine, acting as bidentate (**κ**^{2}-N(1)^{py},C(2)^{ph}) ligand.
[b1-b2]The assumed *trans*(Cl,C^{ph,C2}) isomer ([b1]) is analogous to the X-ray studied *trans*(Cl,C^{ph,C2})-[Pd^{II}(2-phenylpyridine*)(2,4,6-trimethylpyridine)Cl] ([b2]) [2].
[c]This is probably *trans* isomer.
[d1-d2]The assumed *trans*(CH$_3$,N^{py}) form ([d1]) is isomeric to the X-ray studied *trans*(Cl,N^{py}) species ([d2]) [9].
[e]In this complex the pyridine ring at pyrazolic C(3) atom (pyridine A) is bound to Pd(II), while that at pyrazolic N(1) atom (pyridine B) remains uncoordinated.

Continued

Table 1 Square-planar Ni(II), Pd(II) and Pt(II) (d^8), and octahedral Pt(IV) (d^6, low-spin) complexes and organometallics with azines (**1–3**, **5**, **16**, **19–20** ring systems), ^{15}N NMR studied in 2012–20.—cont'd

[f]In this complex the pyridine ring at pyrazolic N(1) atom (pyridine B) is bound to Pd(II), while that at pyrazolic C(3) atom (pyridine A) remains uncoordinated.

[g]Parent acid of 2-((2,3-dimethylphenyl)amino)benzoate anion, i.e., 2-((2,3-dimethylphenyl)amino)benzoic acid, is commonly known as "mefenamic acid."

[h]8-Methylquinoline* is deprotonated (at the methyl C atom), anionic (charge −1) form of 8-methylquinoline, acting as bidentate (κ^2-$N(1)^{quin}$,C^{CH2}) ligand.

[i1-i2]The assumed *trans* isomer ([i1]) is analogous to the X-ray studied *trans*-[Pd^{II}(4-(2-bromophenoxy)quinoline)$_2$(CH_3COO)$_2$] ([i2]) [17].

[j]This is probably *cis* isomer.

[k]2-Phenyl-6-(1-(12-hydroxo-*n*-dodecyl)-*1H*-1,2,3-triazol-4-yl)pyridine* is deprotonated (at the phenyl C(2) atom), anionic (charge −1) form of 2-phenyl-6-(1-(12-hydroxo-*n*-dodecyl)-*1H*-1,2,3-triazol-4-yl)pyridine, acting as tridentate (κ^3-$N(1)^{py}$,$C(2)^{ph}$,$N(3)^{tz}$) ligand.

[l]2-Phenyl-6-(1-(12-trimethylammonioethoxy-*n*-dodecyl)-*1H*-1,2,3-triazol-4-yl)pyridine* is zwitterion, having (+1) charge at quarternary nitrogen in the trimethylammonio-moiety and (−1) charge at the deprotonated phenyl C(2) atom, acting as tridentate (κ^3-$N(1)^{py}$,$C(2)^{ph}$,$N(3)^{tz}$) ligand.

[m1-m2]The assumed *cis* isomer ([m1]) is analogous to the X-ray studied *cis*-[Pt^{II}(3-chloro-7-azaindole)$_2Cl_2$] and *cis*-[Pt^{II}(4-bromo-7-azaindole)$_2Cl_2$] ([m2]) [25,67].

[n1-n2]The assumed *trans* form ([n1]) is isomeric to the assumed or X-ray studied *cis* species ([n2]) [25,67].

[aa1-aa7]Original $\delta^{15N}{}_{compl}$ values (−163.3 ppm ([aa1]), −163.1 ppm ([aa2]), −163.6 ppm ([aa3]), −164.6 ppm ([aa4]), −164.4 ppm ([aa5]), −164.8 ppm ([aa6]), −164.8 ppm ([aa7])), primarily referenced to CH_3NO_2 in $CDCl_3$ by Tereniak et al. [6], were recalculated in respect to neat CH_3NO_2 by subtracting 1.9 ppm relative shift [37].

[ab-ak]Original $\delta^{15N}{}_{compl}$ values, primarily referenced to liquid NH_3 by the following authors, were recalculated in respect to neat CH_3NO_2 by subtracting 381.7 ppm (or 382 ppm) relative shift [37]:

[ab]255 ppm by Virant et al. [7].

[ac]217 ppm by Carvalho et al. [12].

[ad1-ad20]234 ppm ([ad1]), 235 ppm ([ad2]), 235 ppm ([ad3]), 235 ppm ([ad4]), 233 ppm ([ad5]), 232 ppm ([ad6]), 232 ppm ([ad7]), 233 ppm ([ad8]), 232 ppm ([ad9]), 233 ppm ([ad10]), 223 ppm ([ad11]), 223 ppm ([ad12]), 222 ppm ([ad13]), 222 ppm ([ad14]), 222 ppm ([ad15]), 219 ppm ([ad16]), 219 ppm ([ad17]), 220 ppm ([ad18]), 220 ppm ([ad19]), 220 ppm ([ad20]) by Rotondo et al. [13].

[ae1-ae4]55 ppm ([ae1]), 233 ppm ([ae2]), 233 ppm ([ae3]), 180 ppm ([ae4]) by Wenzel et al. [15].

[af]235.5 ppm by Filipovic et al. [16].

[ag1-ag2]196.7 ppm ([ag1]), 194.2 ppm ([ag2]) by Shanahan et al. [17].

[ah1-ah17]209.7 ppm ([ah1]), 199.4 ppm ([ah2]), 213.3 ppm ([ah3]), 207.1 ppm ([ah4]), 306.7 ppm ([ah5]), 206.3 ppm ([ah6]), 309.3 ppm ([ah7]), 203.7 ppm ([ah8]), 304.3 ppm ([ah9]), 198.7 ppm ([ah10]), 189.3 ppm ([ah11]), 304.6 ppm ([ah12]), 203.7 ppm ([ah13]), 197.3 ppm ([ah14]), 216.3 ppm ([ah15]), 199.9 ppm ([ah16]), 305.9 ppm ([ah17]) by White et al. [18] and Jaworski et al. [19].

[ai]209.8 ppm by Pinter et al. [20].

[aj1-aj4]214 ppm ([aj1]), 215 ppm ([aj2]), 214 ppm ([aj3]), 215 ppm ([aj4]) by Hebenbrock et al. [22].

[ak1-ak19]200.6 ppm ([ak1]), 202.3 ppm ([ak2]), 173.4 ppm ([ak3]), 168.2 ppm ([ak4]), 159.9 ppm ([ak5]), 168.0 ppm ([ak6]), 154.6 ppm ([ak7]), 172.8 ppm ([ak8]), 159.4 ppm ([ak9]), 168.4 ppm ([ak10]), 155.4 ppm ([ak11]), 174.9 ppm ([ak12]), 168.1 ppm ([ak13]), 160.8 ppm ([ak14]), 172.6 ppm ([ak15]), 170.9 ppm ([ak16]), 159.4 ppm ([ak17]), 165.9 ppm ([ak18]), 177.4 ppm ([ak19]) by Starha et al. [24–28].

[al]Original $\delta^{15N}{}_{compl}$ value (203 ppm) [10] cannot be recalculated in respect to neat CH_3NO_2 due to the unknown ^{15}N NMR reference.

[am]This $\delta^{15N}{}_{compl}$ concerns uncoordinated nitrogen.

[aaa]vs pyridine in $CDCl_3$: $\delta^{15N}{}_{lig} = -68.7$ ppm [68–70].

[aab]vs 2-phenylpyridine in $CDCl_3$: $\delta^{15N}{}_{lig} = -74.9$ ppm [71].

[aac]vs 2-methylpyridine in $CDCl_3$: $\delta^{15N}{}_{lig} = -71.6$ ppm [72].

[aad]vs 3-methylpyridine in $CDCl_3$: $\delta^{15N}{}_{lig} = -75.8$ ppm [72].

[aae]vs 4-methylpyridine in $CDCl_3$: $\delta^{15N}{}_{lig} = -84.1$ ppm [72].

[aaf]vs 2,3-dimethylpyridine in $CDCl_3$: $\delta^{15N}{}_{lig} = -71.4$ ppm [73].

[aag]vs 2,4-dimethylpyridine in $CDCl_3$: $\delta^{15N}{}_{lig} = -80.0$ ppm [73].

[aah]vs 2,6-dimethylpyridine in $CDCl_3$: $\delta^{15N}{}_{lig} = -72.0$ ppm [73].
[aai]vs 3,5-dimethylpyridine in $CDCl_3$: $\delta^{15N}{}_{lig} = -71.0$ ppm [73].
[aaj]vs 2,4,6-trimethylpyridine in $CDCl_3$: $\delta^{15N}{}_{lig} = -80.1$ ppm [73].
[aak]vs 2-phenylpyridine in DMSO-d_6: $\delta^{15N}{}_{lig} = -71.9$ ppm [71,74].
[aal]vs weighted average for 2-phenylpyridine in DMSO-d_6 and DMF-d_7 (3:1): −72 ppm [3,71,74].
[aam]vs 2-phenylpyridine in DMF-d_7: $\delta^{15N}{}_{lig} = -72$ ppm [3].
[aan]vs average for 2-phenylpyridine in $CDCl_3$ and DMSO-d_6 (^{15}N NMR data for neat or solid 2-phenylpyridine being unavailable): $\delta^{15N}{}_{lig} = -73$ ppm [71,74].
[aao]vs 4-methylpyridine in DMSO-d_6: $\delta^{15N}{}_{lig} = -78.3$ ppm (303.4 ppm in respect to liquid NH_3) [75].
[aap]vs 4-*tert*-butylpyridine in $CDCl_3$: $\delta^{15N}{}_{lig} = -70.0$ ppm (311.7 ppm in respect to liquid NH_3) [76].
[aaq]vs 3-chloropyridine in DMSO-d_6 (^{15}N NMR data in $CDCl_3$ being unavailable): $\delta^{15N}{}_{lig} = -57.8$ ppm [77].
[aar]vs 2-phenylmethyl(amino)pyridine in DMSO-d_6: $\delta^{15N}{}_{lig} = -67$ ppm [8].
[aas]vs 2-phenylmethylaminomethylpyridine in DMSO-d_6: $\delta^{15N}{}_{lig} = -65$ ppm [8].
[aat]vs 2-phenyl(1-ethyl)aminomethylpyridine in DMSO-d_6: $\delta^{15N}{}_{lig} = -66$ ppm [8].
[aau]vs 2-phenylmethylamino(methyl)methyl(phenyl)pyridine in DMSO-d_6: $\delta^{15N}{}_{lig} = -63$ ppm [8].
[aav]vs 2-phenyl(1-ethyl)amino(methyl)methyl(phenyl)pyridine in DMSO-d_6: $\delta^{15N}{}_{lig} = -63$ ppm [8].
[aaw]vs 2-phenyliminomethylpyridine in CD_2Cl_2: $\delta^{15N}{}_{lig} = -63.0$ ppm [9].
[aax1-aax3]$\Delta^{15N}{}_{coord}$ cannot be determined due to the lack of $\delta^{15N}{}_{lig}$ for 2-(2,6-dimethylphenyl)iminomethylpyridine ([aax1]), 2-phenylimino(1-ethyl)pyridine ([aax2]) and 2-(2,6-dimethylphenyl)imino(1-ethyl)pyridine ([aax3]).
[aay]$\Delta^{15N}{}_{coord}$ determined as difference of ca. 149 ppm (for N^{py} in [Pd^{II}(2-(*1H*-benzotriazol-1-yl)pyridine)Cl_2]) and ca. 203 ppm (for N^{py} in 2-(*1H*-benzotriazol-1-yl)pyridine), these values being published in respect to the unknown ^{15}N NMR reference standard [10].
[aaz]vs pyridine-A ring (at pyrazolic C(3) atom) in 2-(1-(pyrid-2-yl)-5-(4-methoxyphenyl)-*1H*-pyrazol-3-yl)pyridine in DMSO-d_6: $\delta^{15N}{}_{lig} = -79.6$ ppm [11].
[aba]vs pyridine-B ring (at pyrazolic N(1) atom) in 2-(1-(pyrid-2-yl)-5-(4-methoxyphenyl)-*1H*-pyrazol-3-yl)pyridine in DMSO-d_6: $\delta^{15N}{}_{lig} = -76.2$ ppm [11].
[abb]vs pyridine-A ring (at pyrazolic C(3) atom) in solid 2-(1-(pyrid-2-yl)-5-(4-methoxyphenyl)-*1H*-pyrazol-3-yl)pyridine: $\delta^{15N}{}_{lig} = -67.8$ ppm [11].
[abc]vs pyridine-B ring (at pyrazolic N(1) atom) in solid 2-(1-(pyrid-2-yl)-5-(4-methoxyphenyl)-*1H*-pyrazol-3-yl)pyridine: $\delta^{15N}{}_{lig} = -81.0$ ppm [11].
[abd]vs 2,2′-bipyridine in $CDCl_3$: $\delta^{15N}{}_{lig} = -75.1$ ppm (own data).
[abe]vs 2,2′-bipyridine in CD_3OH at 260 K: $\delta^{15N}{}_{lig} = -82$ ppm (300 ppm in respect to liquid NH_3) [13].
[abf]vs bis(*tert*-butyl) 2,2′-bipyridine-3,3′-dicarboxylate in CD_3OD: $\delta^{15N}{}_{lig} = -75.6$ ppm [14].
[abg]vs 1,10-phenanthroline in CD_3OH at 260 K: $\delta^{15N}{}_{lig} = -84$ ppm (298 ppm in respect to liquid NH_3) [13].
[abh]vs 8-methylquinoline in $CDCl_3$: $\delta^{15N}{}_{lig} = -76$ ppm (306 ppm in respect to liquid NH_3) [15].
[abi]vs quinoline-8-carboxylic acid in $CDCl_3$: $\delta^{15N}{}_{lig} = -107$ ppm (275 ppm in respect to liquid NH_3) [15].
[abj]$\Delta^{15N}{}_{coord}$ cannot be determined due to the lack of $\delta^{15N}{}_{lig}$ for 8-(ethoxycarbonylmethylaminoiminomethyl)quinoline.
[abk]vs 4-(2-bromophenoxy)quinoline in $CDCl_3$: $\delta^{15N}{}_{lig} = -90.0$ ppm (291.7 ppm in respect to liquid NH_3) [17].
[abl]vs 4-(2-bromophenoxy)-7-bromoquinoline in $CDCl_3$: $\delta^{15N}{}_{lig} = -91.2$ ppm (290.5 ppm in respect to liquid NH_3) [17].
[abm1-abm2]$\Delta^{15N}{}_{coord}$ cannot be determined due to the lack of $\delta^{15N}{}_{lig}$ for 9,9-dimethyl-4,5-diaza-*9H*-fluorene ([abm1]) and 4,5-diaza-*9H*-fluoren-9-one ([abm2]).
[abn]$\Delta^{15N}{}_{coord}$ cannot be determined due to the lack of $\delta^{15N}{}_{lig}$ for 2-iminomethylpyridine.

Continued

Table 1 Square-planar Ni(II), Pd(II) and Pt(II) (d^8), and octahedral Pt(IV) (d^6, low-spin) complexes and organometallics with azines (**1–3**, **5**, **16**, **19–20** ring systems), ^{15}N NMR studied in 2012–20.—cont'd

[abo] Δ^{15N}_{coord} were published directly, with no δ^{15N}_{compl} and δ^{15N}_{lig}.

[abp] vs pyridine in acetone-d_6: $\delta^{15N}_{lig} = -61.8$ ppm [78].

[abq1-abq2] Δ^{15N}_{coord} cannot be determined due to the lack of δ^{15N}_{lig} for 2-phenyl-6-(1-(12-hydroxo-*n*-dodecyl)-*1H*-1,2,3-triazol-4-yl)pyridine ([abq1]) and (2-phenyl-6-(1-(12-(trimethylammonioethoxy)-*n*-dodecyl)-*1H*-1,2,3-triazol-4-yl)pyridine)$^+$ cation ([abq2]).

[abr] vs pyridine ring in 1-β-D-ribofuranosyl-4-(pyrid-2-yl)-*1H*-1,2,3-triazole in DMF-d_7: $\delta^{15N}_{lig} = -73.8$ ppm [23].

[abs1-abs9] Δ^{15N}_{coord} were published directly (with no δ^{15N}_{lig}, which can be determined as $\delta^{15N}_{compl} - \Delta^{15N}_{coord}$ differences: N(4) in 4-azaindole −88.5 ppm ([abs1]), N(7) in 3-chloro-7-azaindole −106.8 ppm ([abs2]), N(7) in 4-chloro-7-azaindole −112.5 ppm ([abs3]), N(7) in 3-bromo-7-azaindole −107.2 ppm ([abs4]), N(7) in 4-bromo-7-azaindole −112.3 ppm ([abs5]), N(7) in 5-bromo-7-azaindole −104.1 ppm ([abs6]), N(7) in 3-iodo-7-azaindole −106.7 ppm ([abs7]), N(7) in 3-chloro-5-bromo-7-azaindole −101.2 ppm ([abs8]), N(7) in 3-iodo-5-bromo-7-azaindole −101.6 ppm ([abs9]), all in DMF-d_7) [24–28].

[abt1-abt3] vs N(7) in 3-chloro-7-azaindole −106.8 ppm ([abt1]), 5-bromo-7-azaindole −104.1 ppm ([abt2]), 3-iodo-7-azaindole −106.7 ppm ([abt3]), for which δ^{15N}_{lig} were determined basing on data in paper [25] (see footnotes ([abs2]), ([abs6]), ([abs7])).

[aaaa] The only stereomeric possibility, resulting from the molecular formula.

[A] X-ray structure: DUBLAP [2].
[B] X-ray structure: SOHDUO (SOHDUO 01–03) [79–82].
[C] X-ray structure: BUNQAE [4].
[D] X-ray structure: UQUNUR [5].
[E] X-ray structure of CH_2Cl_2 solvate: UVIKAN [9].
[F] X-ray structure of CH_2Cl_2 solvate: UVIKER [9].
[G] X-ray structure: UVIKIV [9].
[H] X-ray structure: UVIKOB [9].
[I] X-ray structure of CH_2Cl_2 solvate: GOCRUN [11].
[J] X-ray structure of the respective nitrate analogue ([Pd^{II}(1,10-phenanthroline)(NH_2CSNH_2)$_2$](NO_3)$_2$: LIXPOZ [83].
[K] X-ray structure: VOQYAE [15].
[L] X-ray structure: VOQYEI [15].
[M] X-ray structure: VASYOG [15].
[N] X-ray structure of CH_3OH solvate: VOQYOS [18]. X-ray structure of H_2O solvate: NUKYUQ [17].
[P] X-ray structure of $CHCl_3$ solvate: IMIYUA [18].
[Q] X-ray structure of $CHCl_3$ solvate: IMIYOU [18].
[R] X-ray structure of $CHCl_3$ solvate: IMIYIO [18].
[S] X-ray structure of CF_3COOH solvate: IMIZEL [18].
[T] X-ray structure of $CHCl_3 \bullet CH_3(CH_2)_4(CH_3)$ solvate: IMIYEK [18].
[U] X-ray structure of $CHCl_3$ solvate: IMIYAG [18].
[V] X-ray structure of CF_3COOH solvate: IMIZAH [18].
[W] X-ray structure of $CHCl_3$ solvate: IMIZIP [18].
[X] X-ray structure: NINKOM [20].
[Y] X-ray structure of DMF solvate: YEFJAW [25].
[Z] X-ray structure of DMF solvate: POTNAQ [67].

Table 2 Square-planar Rh(I) and Ir(I) (d^8) complexes and organometallics with azines (**1**, **5**, **10**, **15**, **17** ring systems), ^{15}N NMR studied in 2012–20.

Compound	Parent ring system	^{15}N chemical shift (δ^{15N}_{compl}, ppm)	^{15}N coordination shift (Δ^{15N}_{coord}, ppm)	Nitrogen *trans* to
Rh(I) *compounds*				
trans(Cl,C=C^{cyclooct})-[Rh^{I}(**pyridine**)(1,3-bis(2,6-diisopropylphenyl)imidazol-2-ylidene-κ^1-C^2)(cyclooctene)Cl]$^{toluene\text{-}d8,223K}$ [55][a]	**1**	N^{py}: −108.5[aa]	−45.1[aaa]	(C^{ylidene,C2})[a]
trans(Cl,C=C^{vinyl})-[Rh^{I}(**pyridine**)(1,3-bis(2,6-diisopropylphenyl)imidazol-2-ylidene-κ^1-C^2)(1-vinyl-*1H*-pyrazole-κ^1-C$^{C=C}$)Cl]$^{toluene\text{-}d8}$ [56][b1]	**1**	N^{py}: −114.9[ab1]	−54.3[aab]	(C^{ylidene,C2})[b1]
trans(Cl,C=C^{vinyl})-[Rh^{I}(**pyridine**)(1,3-bis(2,6-diisopropylphenyl)imidazol-2-ylidene-κ^1-C^2)(1-vinyl-3,5-dimethyl-*1H*-pyrazole-κ^1-C$^{C=C}$)Cl]$^{toluene\text{-}d8,243K}$ [56][b2,A]	**1**	N^{py}: −118.0[ab2]	−57.4[aab]	C$^{ylidene,C2\ X\text{-}ray}$
trans(Cl,C=C^{vinyl})-[Rh^{I}(2-vinyl**pyridine**-κ^2-N,C$^{C=C}$)(1,3-bis(2,6-diisopropylphenyl)imidazol-2-ylidene-κ^1-C^2)Cl]$^{toluene\text{-}d8,243K}$ [57][B]	**1**	N^{py}: −132.5[ab3]	−62.5[aac]	C$^{ylidene,C2\ X\text{-}ray}$
trans(Cl,C=C^{cyclooct})-[Rh^{I}(**pyrid-2-yl**acetonitrile-κ^1-N^{py})(1,3-bis(2,6-diisopropylphenyl)imidazol-2-ylidene-κ^1-C^2)(cyclooctene)Cl]$^{toluene\text{-}d8}$ [58][C]	**1**	N^{py}: −111.2[ac1]	Unknown[aad1]	C$^{ylidene,C2\ X\text{-}ray}$
trans(CO,O)-[Rh^{I}(**pyrid-2-yl**methanolate-κ^2-N,O)(1,3-bis(2,6-diisopropylphenyl)imidazol-2-ylidene-κ^1-C^2)(CO)]$^{toluene\text{-}d8,243K}$ [59][D]	**1**	N^{py}: −124.4[ac2]	Unknown[aad2]	C$^{ylidene,C2\ X\text{-}ray}$
trans(CO,S)-[Rh^{I}(**pyrid-2-yl**methanol-κ^1-N)(1,3-bis(2,6-diisopropylphenyl)imidazol-2-ylidene-κ^1-C^2)(CO)($C_6H_5CH_2S$)]$^{toluene\text{-}d8,243K}$ [59][c]	**1**	N^{py}: −127.7[ac3]	Unknown[aad2]	(C^{ylidene,C2})[c]

Continued

Table 2 Square-planar Rh(I) and Ir(I) (d^8) complexes and organometallics with azines (**1**, **5**, **10**, **15**, **17** ring systems), ^{15}N NMR studied in 2012–20.—cont'd

Compound	Parent ring system	^{15}N chemical shift (δ^{15N}_{compl}, ppm)	^{15}N coordination shift (Δ^{15N}_{coord}, ppm)	Nitrogen *trans* to
trans(O,$C{=}C^{CH2CH2}$)-[Rh^I(**quinolin**-8-olate-κ^2-N,O)(1,3-bis(2,6-diisopropylphenyl)imidazol-2-ylidene-κ^1-C^2)(CH_2CH_2)]$^{toluene\text{-}d8}$ [60][d]	**5**	N^{quin}: −128.4[ac4]	−32.7[aae]	($C^{ylidene,C2}$)[d]
trans(O,$C{=}C^{cyclooct}$)-[Rh^I(**quinolin**-8-olate-κ^2-N,O)(1,3-bis(2,6-diisopropylphenyl)imidazol-2-ylidene-κ^1-C^2)(cyclooctene)]$^{toluene\text{-}d8,243K}$ [60][e1]	**5**	N^{quin}: −133.4[ac5]	−37.7[aae]	($C^{ylidene,C2}$)[e1]
trans(N^{quin},$C{=}C^{cyclooct}$)-[Rh^I(**quinolin**-8-olate-κ^2-N,O)(1,3-bis(2,6-diisopropylphenyl)imidazol-2-ylidene-κ^1-C^2)(cyclooctene)]$^{toluene\text{-}d8,243K}$ [60][e2,E]	**5**	N^{quin}: −128.9[ac6]	−33.2[aae]	$C{=}C^{cyclooctene}$ [X-ray]
trans(Cl,$C{=}C^{ind}$)-[Rh^I(4-(*1H*-inden-3-yl)-**1,5-naphthyridine**-κ^2-N^5,$C^{ind,C2=C3}$)(CO)Cl]^{C6D6} [61][F]	**17**	$N^{naptht,N1}$: −72.8[ad1,af]	Not applicable[aaf]	None
		$N^{naptht,N5}$: −136.8[ad2]	−67.7[aag]	CO [X-ray]
Ir(I) compounds				
[Ir^I(**quinolin**-8-olate-κ^2-N,O)(1,5-cyclooctadiene)]^{CDCl3} [62][f,G]	**5**	N^{quin}: −148.2	−52.5[aae]	$C{=}C^{cyclooctadiene}$ [X-ray]
[Ir^I(2-methyl**quinolin**-8-olate-κ^2-N,O)(1,5-cyclooctadiene)]^{CDCl3} [62][f,H]	**5**	N^{quin}: −145.6	Unknown[aah]	$C{=}C^{cyclooctadiene}$ [X-ray]
[Ir^I(**pyridazine**-κ^1-N^1)(1,3-bis(2,4,6-trimethylphenyl)imidazol-2-ylidene-κ^1-C^2)(1,5-cyclooctadiene)]Cl^{CD3OD} [63][f]	**10**	$N^{pd,N1}$: −69.9[ae1]	−37.5	($C{=}C^{cyclooctadiene}$)
		$N^{pd,N2}$: 4.4[ae2,af]	Not applicable[aai]	None
[Ir^I(**phthalazine**-κ^1-N^2)(1,3-bis(2,4,6-trimethylphenyl)imidazol-2-ylidene-κ^1-C^2)(1,5-cyclooctadiene)]Cl^{CD3OD} [63][f]	**15**	$N^{phth,N2}$: −70.2[ae3]	−38.1[aaj]	($C{=}C^{cyclooctadiene}$)[aaaa]
		$N^{phth,N3}$: 4.3[ae4,af]	Not applicable[aaj]	None

[a]This is probably *trans*(Cl,C=C^{cyclooct}) isomer.
[b1-b2]The assumed *trans*(Cl,C=C^{vinyl}) isomer ([b1]) is analogous to the X-ray studied *trans*(Cl,C=C^{vinyl})-[Rh^{I}(pyridine)(1,3-bis(2,6-diisopropylphenyl)imidazol-2-ylidene)(1-vinyl-3,5-dimethyl-*1H*-pyrazole)Cl] ([b2]) [56].
[c]This is probably *trans*(CO,S) isomer.
[d]This is probably *trans*(O,C=C^{CH2CH2}) isomer.
[e1-e2]The assumed *trans*(O,C=C^{cyclooct}) form ([e1]) is isomeric to the X-ray studied *trans*(N^{quin},C=C^{cyclooct}) species ([e2]) [60].
[f]1,5-Cyclooctadiene ligand occupies two coordination sites, within square-planar geometry.
[aa-ae]Original $\delta^{15N}{}_{compl}$ values, primarily referenced to liquid NH_3 by the following authors, were recalculated in respect to neat CH_3NO_2 by subtracting 381.7 ppm (or 382 ppm) relative shift [37]:
[aa]273.2 ppm by Rubio-Perez et al. [55].
[ab1-ab3]266.8 ppm ([ab1]), 263.7 ppm ([ab2]), 249.2 ppm ([ab3]) by Azpiroz et al. [56,57].
[ac1-ac6]270.5 ppm ([ac1]), 257.3 ppm ([ac2]), 254.0 ppm ([ac3]), 253.3 ppm ([ac4]), 248.3 ppm ([ac5]), 252.8 ppm ([ac6]) by Palacios et al. [58–60].
[ad1-ad2]308.9 ppm ([ad1]), 244.9 ppm ([ad2]) by Sieb et al. [61].
[ae1-ae4]311.8 ppm ([ae1]), 386.1 ppm ([ae2]), 311.5 ppm ([ae3]), 386.0 ppm ([ae4]) by Appleby et al. [63].
[af]This $\delta^{15N}{}_{compl}$ concerns uncoordinated nitrogen.
[aaa]vs free pyridine in situ: $\delta^{15N}{}_{lig} = -63.4$ ppm (318.3 ppm in respect to liquid NH_3) [55].
[aab]vs pyridine in toluene-d_8: $\delta^{15N}{}_{lig} = -60.6$ ppm [64].
[aac]vs 2-vinylpyridine in C_6D_6 (^{15}N NMR data in toluene-d_8 being unavailable): $\delta^{15N}{}_{lig} = -70.0$ ppm [65].
[aad1-aad2]$\Delta^{15N}{}_{coord}$ cannot be determined due to the lack of $\delta^{15N}{}_{lig}$ for pyrid-2-ylacetonitrile ([aad1]) and pyrid-2-ylmethanol ([aad2]).
[aae]vs quinoline-8-ol in $CDCl_3$ (^{15}N NMR data in toluene-d_8 being unavailable): $\delta^{15N}{}_{lig} = -95.7$ ppm [66].
[aaf]vs N(1) in 4-(*1H*-inden-3-yl)-1,5-naphthyridine in $CDCl_3$: $\delta^{15N}{}_{lig} = -71.2$ ppm (310.5 ppm in respect to liquid NH_3) [61].
[aag]vs N(5) in 4-(*1H*-inden-3-yl)-1,5-naphthyridine in $CDCl_3$: $\delta^{15N}{}_{lig} = -69.1$ ppm (312.6 ppm in respect to liquid NH_3) [61].
[aah]$\Delta^{15N}{}_{coord}$ cannot be determined due to the lack of $\delta^{15N}{}_{lig}$ for 2-methylquinolin-8-ol.
[aai]vs pyridazine in CD_3OD: $\delta^{15N}{}_{lig} = -32.4$ ppm (349.3 ppm in respect to liquid NH_3) [63].
[aaj]vs phthalazine in CD_3OD: $\delta^{15N}{}_{lig} = -32.1$ ppm (349.6 ppm in respect to liquid NH_3) [63].
[aaaa]The only stereomeric possibility, resulting from the molecular formula.
[A]X-ray structure: GUVQOF [56].
[B]X-ray structure: MEYXEW [57].
[C]X-ray structure of $CH_3(CH_2)_4CH_3$ solvate: COTTOW [58].
[D]X-ray structure of C_6H_6 solvate: RAWKAE [59].
[E]X-ray structure: AZIBES [60].
[F]X-ray structure of CH_2Cl_2 solvate: MASKEY [61].
[G]X-ray structure: TUKVOM [62].
[H]X-ray structure: TUKVIG [62].

Table 3 Octahedral Co(III), Rh(III) and Ir(III) (d^6, low-spin) complexes and organometallics with azines (**1–3**, **5**, **10**, **14**, **15**, **20** ring systems), ^{15}N NMR studied in 2012–20.

Compound	Parent ring system	^{15}N chemical shift (δ^{15N}_{compl}, ppm)	^{15}N coordination shift (Δ^{15N}_{coord}, ppm)	Nitrogen *trans* to
Co(III) compounds				
mer-[Co^{III}(2-(2-(trifluoromethyl)vinyl-2-olate) **pyridine**-κ^2-N,O)$_3$]$^{acetone\text{-}d6}$ [84][A]	**1**	N^{py3}: −202^{aa1}	Unknown^{aaa}	O $^{X\text{-}ray}$
		$N^{py1,py2}$: −187^{aa2}, −194^{aa3}	Unknown^{aaa}	2 × N^{py} $^{X\text{-}ray}$
[Co^{III}(2,5-bis(**pyrid-2-yl**)-3,4-dimethylpyrrolate-κ^3-N^{py},$N^{py'}$, N^{pr})$_2$]$PF_6$$^{acetone-d6}$ [85]	**1**	N^{py}: −191.7^{ab}	−109.0^{aab}	(N^{py})^{aaaa}
trans($N^{bpy,N1}$,$NH^{NHCOC(CN)NO}$),*trans*($N^{bpy,N1'}$,$NO^{NHC(OCH2CH3)C(CN)NO}$)-[Co^{III}(**2,2′-bipyridine**)(NHCOC(*CN*)NO-κ^2-N^{NH},N^{NO})(NHC(OCH_2CH_3)C(*CN*)NO-κ^2-N^{NH}, N^{N})]$^{DMSO\text{-}d6}$ [86][a,b,B]	**2**	$N^{bpy,N1}$: −154.5	−81.9^{aac}	N^{NH} $^{X\text{-}ray}$
		$N^{bpy,N1'}$: −140.2	−67.6^{aac}	N^{NO} $^{X\text{-}ray}$
Rh(III) compounds				
trans(H,N^{py}),*trans*(Cl,CCSi)-[Rh^{III}(**pyridine**)$_2$(1,3-bis(2,6-diisopropylphenyl)imidazol-2-ylidene-κ^1-C^2)($(CH_3)_3SiCC$)(H)Cl]$^{toluene\text{-}d8,223K}$ [55][c]	**1**	N^{py1}: −103.2^{ac1}	−39.8^{aad}	(H)^{c}
		N^{py2}: −125.2^{ac2}	−61.8^{aad}	($C^{ylidene,C2}$)^{c}
[Rh^{III}(5,5′-bis(pyren-1-yl)-**2,2′-bipyridine**)(pentamethylcyclopentadienyl)Cl]Cl^{CD3OD} [87][d1]	**2**	N^{bpy}: −134.0	Unknown^{aae}	(C=C^{dienyl})^{aaaa}

Ir(III) compounds				
fac-[Ir^{III}(**pyridine**)$_3$(1,3-dimethylimidazol-2-ylidene-κ^1-C^2)(H)$_2$]Cl^{CD3OD,253K} [88]^e	**1**	2 × N^{py1,py2}: −130.1^{ad1}	−51.0^{aaf}	2 × (H)^e
		N^{py3}: −145.0^{ad2}	−65.9^{aaf}	(C^{ylidene,C2})^e
fac-[Ir^{III}(**pyridine**)$_3$(1,3-diisopropyl-4,5-dimethylimidazol-2-ylidene-κ^1-C^2)(H)$_2$]Cl^{CD3OD,253K} [88]^e	**1**	2 × N^{py1,py2}: −127.9^{ad3}	−48.8^{aaf}	2 × (H)^e
		N^{py3}: −139.0^{ad4}	−59.9^{aaf}	(C^{ylidene,C2})^e
fac-[Ir^{III}(**pyridine**)$_3$(1,3-bis(cyclohexyl)imidazol-2-ylidene-κ^1-C^2)(H)$_2$]Cl^{CD3OD,253K} [88]^e	**1**	2 × N^{py1,py2}: −127.7^{ad5}	−48.6^{aaf}	2 × (H)^e
		N^{py3}: −145.4^{ad6}	−66.3^{aaf}	(C^{ylidene,C2})^e
fac-[Ir^{III}(**pyridine**)$_3$(1,3-bis(2,6-diisopropylphenyl)imidazol-2-ylidene-κ^1-C^2)(H)$_2$]Cl^{CD3OD,253K} [88]^e	**1**	2 × N^{py1,py2}: −126.5^{ad7}	−47.4^{aaf}	2 × (H)^e
		N^{py3}: −142.8^{ad8}	−63.7^{aaf}	(C^{ylidene,C2})^e
fac-[Ir^{III}(**pyridine**)$_3$(1-methyl-3-phenylimidazol-2-ylidene-κ^1-C^2)(H)$_2$]Cl^{CDCl3} [89]^e	**1**	2 × N^{py1,py2}: −128.3^{ad9}	−59.6^{aag}	2 × (H)^e
		N^{py3}: −145.6^{ad10}	−76.9^{aag}	(C^{ylidene,C2})^e
fac-[Ir^{III}(**pyridine**)$_3$(1,3-bis(2,4,6-trimethylphenyl)imidazol-2-ylidene-κ^1-C^2)(H)$_2$]Cl^{CD3OD,253K} [88]^e	**1**	2 × N^{py1,py2}: −126.1^{ad11}	−47.0^{aaf}	2 × (H)^e
		N^{py3}: −142.6^{ad12}	−63.5^{aaf}	(C^{ylidene,C2})^e
fac-[Ir^{III}(**pyridine**)$_3$(1,3-bis(2,6-diisopropylphenyl)-4,5-dihydroimidazol-2-ylidene-κ^1-C^2)(H)$_2$]Cl^{CD3OD,253K} [88]^e	**1**	2 × N^{py1,py2}: −127.3^{ad13}	−48.2^{aaf}	2 × (H)^e
		N^{py3}: −142.5^{ad14}	−63.4^{aaf}	(C^{ylidene,C2})^e
fac-[Ir^{III}(**pyridine**)$_3$(1,3-bis(2,4,6-trimethylphenyl)-4,5-dihydroimidazol-2-ylidene-κ^1-C^2)(H)$_2$]Cl^{CD3OD,253K} [88]^e	**1**	2 × N^{py1,py2}: −127.0^{ad15}	−47.9^{aaf}	2 × (H)^e
		N^{py3}: −141.6^{ad16}	−62.5^{aaf}	(C^{ylidene,C2})^e
trans(H,N^{py}),*trans*(H,N^{CH3CN})-[Ir^{III}(**pyridine**)(1,3-bis(2,4,6-trimethylphenyl)imidazol-2-ylidene-κ^1-C^2)(P(C$_6$H$_5$)$_3$)(CH$_3$CN)(H)$_2$]BF$_4$^{CDCl3,263K} [90]^C	**1**	N^{py}: −143.6^{ae}	−74.9^{aag}	H $^{X-ray}$

Continued

Table 3 Octahedral Co(III), Rh(III) and Ir(III) (d^6, low-spin) complexes and organometallics with azines (**1–3**, **5**, **10**, **14**, **15**, **20** ring systems), ^{15}N NMR studied in 2012–20.—cont'd

Compound	Parent ring system	^{15}N chemical shift (δ^{15N}_{compl}, ppm)	^{15}N coordination shift (Δ^{15N}_{coord}, ppm)	Nitrogen *trans* to
trans(H,N^{py}),*trans*(H,Cl)-[Ir^{III}(**pyridine**)$_2$(cyclooctene)(H)$_2$Cl]$^{THF\text{-}d8}$ [91][D]	**1**	N^{py1}: −127.0[af1]	−58.3[aag,aah]	H [X-ray]
		N^{py2}: −157.0[af2]	−88.3[aag,aah]	C=C^{cyclooctene} [X-ray]
trans(H,N^{py})-[Ir^{III}(**pyridine**)(2,6-bis(di-*tert*-butylphosphinomethyl)phenyl-**κ**^{3}-$C^{ph,C1}$,P,P′)(H)Cl]^{CD2Cl2,260K} [92,93][f1]	**1**	N^{py}: −138.7	−73.4[aai]	(H)[f1]
trans(H,Cl)-[Ir^{III}(**pyridine**)(2,6-bis(di-*tert*-butylphosphinomethyl)phenyl-**κ**^{3}$C^{ph,C1}$,P,P′)(H)Cl]^{CD2Cl2,260K} [92,93][f2,E]	**1**	N^{py}: −149.5	−84.2[aai]	$C^{ph,C1}$ [X-ray]
trans(H,N^{py})-[Ir^{III}(**pyridine**)(2,6-bis(di-*tert*-butylphosphinite)phenyl-**κ**^{3}-$C^{ph,C1}$,P,P′)(H)Cl]^{CD2Cl2,200K} [93][g]	**1**	N^{py}: −136.0	−70.7[aai]	(H)[g]
trans(H,N^{py})-[Ir^{III}(**pyridine**)(2,6-bis(di-*tert*-butylphosphinite)-4-(ethoxycarbonyl)phenyl-**κ**^{3}-$C^{ph,C1}$,P,P′)(H)Cl]^{CD2Cl2,200K} [93][g]	**1**	N^{py}: −138.1	−72.8[aai]	(H)[g]
[Ir^{III}(2-((3-methyl-1,3-dihydro-2*H*-imidazol-2-ylidene)-1-yl)-4,6-di-*tert*-butyl**pyridine**-**κ**^{2}-N^{py},$C^{im,2}$)(pentamethylcyclopentadienyl)Cl]PF_6^{CD2Cl2,233K} [94][d1]	**1**	N^{py}: −181.4	−76.9[aaj]	(C=C^{dienyl})[aaaa]
trans(H,$N^{bpy,N1}$),*trans*(H,$N^{bpy,N1'}$)-[Ir^{III}(**2,2′-bipyridine**)(**pyridine**-3-carbonamide-N^{py})(1,3-bis(2,4,6-trimethylphenyl)imidazol-2-ylidene-**κ**^{1}-C^2)(H)$_2$]BF_4^{CD3OD,253K} [95][h]	**1, 2**	2 × N^{bpy}: −114[ag1]	−39[aak]	2 × (H)[h]
		N^{py}: −141[ag2]	−67[aal]	($C^{ylidene,C2}$)[h]
trans(H,$N^{phen,N1}$),*trans*(H,$N^{phen,N1'}$)-[Ir^{III}(**1,10-phenanthroline**)(**pyridine**)(1,3-bis(2,4,6-trimethylphenyl)imidazol-2-ylidene-**κ**^{1}-C^2)(H)$_2$]BF_4^{CD3OD} [95][h]	**1, 3**	2 × N^{phen}: −116[ag3]	−36[aam]	2 × (H)[h]
		N^{py}: −143[ag4]	−64[aaf]	($C^{ylidene,C2}$)[h]

trans(H,O)-[Ir^{III}(2-methyl**quinolin**-8-olate-κ^2-N,O)(1,5-cyclooctadiene)(Si(C_6H_5)$_3$)(H)]^{C6D6} [62][i,F]	**5**	N^{quin}: −153.2	Unknown[aan]	C=C^{cyclooctadiene} X-ray
trans(H,O)-[Ir^{III}(2-methyl**quinolin**-8-olate-κ^2-N,O)(1,5-cyclooctadiene)(Sn(C_6H_5)$_3$)(H)]^{CDCl3} [62][i,G]	**5**	N^{quin}: −163.0	Unknown[aan]	C=C^{cyclooctadiene} X-ray
[Ir^{III}(1-(**pyrid-2-yl**)-**7-azaindole**-κ^2-N^{py},N^{ind,N7})(pentamethylcyclopentadienyl)Cl]PF$_6$$^{DMSO-d6}$ [96][d1,H]	**1, 20**	N^{azaind,N7}: −208.7[ah1]	−94.2[aao]	C=C^{dienyl} X-ray
		N^{py}: −193.3[ah2]	−93.8[aap]	C=C^{dienyl} X-ray
[Ir^{III}(1-(**pyrid-2-yl**)-**7-azaindole**-κ^2-N^{py},N^{ind,N7})(tetramethylphenylcyclopentadienyl)Cl]PF$_6$$^{DMSO-d6}$ [96][d2,I]	**1, 20**	N^{azaind,N7}: −212.0[ah3]	−97.5[aao]	C=C^{dienyl} X-ray
		N^{py}: −195.9[ah4]	−96.4[aap]	C=C^{dienyl} X-ray
[Ir^{III}(1-phenyl-**7-azaindole***-κ^2-N^{ind,N7},C^{ph,C2})(pentamethylcyclopentadienyl)Cl]$^{DMSO-d6}$ [96][d1,j,J]	**20**	N^{azaind,N7}: −220.7[ah5]	−105.2[aaq]	C=C^{dienyl} X-ray
[Ir^{III}(1-phenyl-**7-azaindole***-κ^2-N^{ind,N7},C^{ph,C2})(tetramethylphenylcyclopentadienyl)Cl]$^{DMSO-d6}$ [96][d2,j]	**20**	N^{azaind,N7}: −220.0[ah6]	−104.5[aaq]	(C=C^{dienyl})[aaaa]
[Ir^{III}(1-(thiophen-2-yl)-**7-azaindole***-κ^2-N^{ind,N7},C^{th,C3})(pentamethylcyclopentadienyl)Cl]$^{DMSO-d6}$ [96][d1,k]	**20**	N^{azaind,N7}: −225.0[ah7]	−110.8[aar]	(C=C^{dienyl})[aaaa]
[Ir^{III}(1-(thiophen-2-yl)-**7-azaindole***-κ^2-N^{ind,N7},C^{th,C3})(tetramethylphenylcyclopentadienyl)Cl]$^{DMSO-d6}$ [96][d2,k]	**20**	N^{azaind,N7}: −227.3[ah8]	−113.1[aar]	(C=C^{dienyl})[aaaa]
trans(H,N^{pd})-[Ir^{III}(**pyridazine**-κ^1-N^1)(1,3-bis(2,4,6-trimethylphenyl)imidazol-2-ylidene-κ^1-C^2)(1,5-cyclooctadiene)(H)$_2$]Cl^{CD3OD} [63][i,l]	**10**	N^{pd,N1}: −87.4[ai1]	−55.0[aas]	(H)[l]
		N^{pd,N2}: 18.4[ai2,ak]	Not applicable[aas]	None

Continued

Table 3 Octahedral Co(III), Rh(III) and Ir(III) (d^6, low-spin) complexes and organometallics with azines (**1–3**, **5**, **10**, **14**, **15**, **20** ring systems), ^{15}N NMR studied in 2012–20.—cont'd

Compound	Parent ring system	^{15}N chemical shift (δ^{15N}_{compl}, ppm)	^{15}N coordination shift (Δ^{15N}_{coord}, ppm)	Nitrogen *trans* to
fac-$[Ir^{III}$(**pyridazine**-κ^1-$N^1)_3$(1,3-bis(2,4,6-trimethylphenyl)imidazol-2-ylidene-κ^1-C^2)$(H)_2]Cl^{CD3OD,253K}$ [63][e]	**10**	2 × $N^{pd1,pd2,N1}$: −55.5[ai3]	−23.1[aas]	2 × (H)[e]
		2 × $N^{pd1,pd2,N2}$: 8.3[ai4,ak]	Not applicable[aas]	None
		$N^{pd3,N1}$: −75.9[ai5]	−43.5[aas]	$(C^{ylidene,C2})$[e]
		$N^{pd3,N2}$: 10.5[ai6,ak]	Not applicable[aas]	None
trans(H,Cl),*trans*(H,N^{qnz})-$[Ir^{III}$(**quinazoline**-κ^1-$N^3)_2$(1,3-bis(2,4,6-trimethylphenyl)imidazol-2-ylidene-κ^1-C^2)$(H)_2Cl]^{CD2Cl2}$ [97][m]	**14**	$N^{qnz1,N1}$: −98.1[aj1,ak]	Not applicable[aat]	None
		$N^{qnz1,N3}$: −135.2[aj2]	−44.8[aau]	(H)[m]
		$N^{qnz2,N1}$: −98.8[aj3,ak]	Not applicable[aat]	None
		$N^{qnz2,N3}$: −155.5[aj4]	−65.1[aau]	$(C^{ylidene,C2})$[m]
fac-$[Ir^{III}$(**quinazoline**-κ^1-$N^3)_3$(1,3-bis(2,4,6-trimethylphenyl)imidazol-2-ylidene-κ^1-C^2)$(H)_2]Cl^{CD3OD}$ [97][e]	**14**	2 × $N^{qnz1,qnz2,N1}$: −99.2[aj5,ak]	Not applicable[aat]	None
		2 × $N^{qnz1,qnz2N3}$: −143.9[aj6]	−53.5[aau]	2 × (H)[e]
		$N^{qnz3,N1}$: −96.6[aj7,ak]	Not applicable[aat]	None
		$N^{qnz3,N3}$: −156.4[aj8]	−66.0[aau]	$(C^{ylidene,C2})$[e]
trans(H,N^{pht})-$[Ir^{III}$(**phthalazine**-κ^1-N^2)(1,3-bis(2,4,6-trimethylphenyl)imidazol-2-ylidene-κ^1-C^2)(1,5-cyclooctadiene)$(H)_2]Cl^{CD3OD}$ [63][i,l]	**15**	$N^{phth,N2}$: −115.1[ai7]	−83.0[aav]	(H)[l]
		$N^{phth,N3}$: −12.8[ai8,ak]	Not applicable[aav]	None

fac-[Ir^{III}(**phthalazine**-κ^1-N^2)$_3$(1,3-bis(2,4,6-trimethylphenyl) imidazol-2-ylidene-κ^1-C^2)$(H)_2$]$Cl^{CD3OD,253K}$ [63][e]	**15**	2 × $N^{phth1,phth2,N2}$: −82.0[ai9]	−49.9[aav]	2 × (H)[e]
		2 × $N^{phth1,phth2,N3}$: −20.8[ai10,ak]	Not applicable[aav]	None
		$N^{phth3,N2}$: −102.9[ai11]	−70.8[aav]	($C^{ylidene,C2}$)[e]
		$N^{phth3,N3}$: −18.9[ai12,ak]	Not applicable[aav]	None

[a]NHCOC(*CN*)NO (i.e., NH—CO—C(CN)=N—O) is anionic (charge −2) form of NH_2—CO—C(CN)=N—OH, doubly deprotonated at amino and hydroxo group, acting as bidentate (κ^2-N^{NH},N^{NO}) ligand.
[b]NHC(OCH_2CH_3)C(*CN*)NO (i.e., NH=C(OCH_2CH_3)—C(CN)=N—O) is anionic (charge −1) form of NH=C(OCH_2CH_3)—C(CN)=N—OH, deprotonated at hydroxo group, acting as bidentate (κ^2-N^{NH},N^{NO}) ligand.
[c]This is probably *trans*(H,N^{PY}),*trans*(Cl,CCSi) isomer.
[d1-d2]Pentamethylcyclopentadienyl ([d1]) and tetramethylphenylcyclopentadienyl ([d2]) ligands occupy three coordination sites, within octahedral geometry.
[e]This is probably *fac* isomer.
[f1-f2]The assumed *trans*(H,N^{PY}) form ([f1]) is isomeric to the X-ray studied *trans*(H,Cl) species ([f2]) [92].
[g]This is probably *trans*(H,N^{PY}) isomer.
[h]This is probably *trans*(H,$N^{LL,N1}$),*trans*(H,$N^{LL,N1'}$) isomer.
[i]1,5-Cyclooctadiene ligand occupies two coordination sites, within octahedral geometry.
[j]1-Phenyl-7-azaindole* is deprotonated (at the phenyl C(2) atom), anionic (charge −1) form of 1-phenyl-7-azaindole, acting as bidentate (κ^2-$N(7)^{ind}$,$C(2)^{ph}$) ligand.
[k]1-(Thiophen-2-yl)-7-azaindole* is deprotonated (at the thiophene C(3) atom), anionic (charge −1) form of 1-(thiophen-2-yl)-7-azaindole, acting as bidentate (κ^2-$N(7)^{ind}$,$C(3)^{th}$) ligand.
[l]This is probably *trans*(*H*,N^L) isomer.
[m]This is probably *trans*(H,Cl),*trans*(H,N^L) isomer.
[aa-aj]Original $\delta^{15}N_{compl}$ values, primarily referenced to liquid NH_3 by the following authors, were recalculated in respect to neat CH_3NO_2 by subtracting 381.7 ppm (or 382 ppm) relative shift [37]:
[aa1-aa3]180 ppm ([aa1]), 195 ppm ([aa2]), 188 ppm ([aa3]) by Buyukyazi et al. [84].
[ab]190.0 ppm by McPherson et al. [85].
[ac1-ac2]278.5 ppm ([ac1]), 256.5 ppm ([ac2]) by Rubio-Perez et al. [55].
[ad1-ad16]251.6 ppm ([ad1]), 236.7 ppm ([ad2]), 253.8 ppm ([ad3]), 242.7 ppm ([ad4]), 254.0 ppm ([ad5]), 236.3 ppm ([ad6]), 255.2 ppm ([ad7]), 238.9 ppm ([ad8]), (253.4 ppm ([ad9]), 236.1 ppm ([ad10]), 255.6 ppm ([ad11]), 239.1 ppm ([ad12]), 254.4 ppm ([ad13]), 239.2 ppm ([ad14]), 254.7 ppm ([ad15]), 240.1 ppm ([ad16]) by Lloyd et al. [88] and Wong et al. [89].
[ae]238.1 ppm by Fekete et al. [90].
[af1-af2]254.7 ppm ([af1]), 224.7 ppm ([af2]) by Iali et al. [91].
[ag1-ag4]268 ppm ([ag1]), 241 ppm ([ag2]), 266 ppm ([ag3]), 239 ppm ([ag4]) by Mewis et al. [95].
[ah1-ah8]173.0 ppm ([ah1]), 188.4 ppm ([ah2]), 169.7 ppm ([ah3]), 185.8 ppm ([ah4]), 161.0 ppm ([ah5]), 161.7 ppm ([ah6]), 156.7 ppm ([ah7]), 154.4 ppm ([ah8]) by Starha et al. [96].
[ai1-ai12]294.3 ppm ([ai1]), 400.1 ppm ([ai2]), 326.2 ppm ([ai3]), 390.0 ppm ([ai4]), 305.8 ppm ([ai5]), 392.2 ppm ([ai6]), 266.6 ppm ([ai7]), 368.9 ppm ([ai8]), 299.7 ppm ([ai9]), 360.9 ppm ([ai10]), 278.8 ppm ([ai11]), 362.8 ppm ([ai12]) by Appleby et al. [63].

Continued

Table 3 Octahedral Co(III), Rh(III) and Ir(III) (d^6, low-spin) complexes and organometallics with azines (**1–3, 5, 10, 14, 15, 20** ring systems), ^{15}N NMR studied in 2012–20.—cont'd

[aj1–aj8]283.6 ppm ([aj1]), 246.5 ppm ([aj2]), 282.9 ppm ([aj3]), 226.2 ppm ([aj4]), 282.5 ppm ([aj5]), 237.8 ppm ([aj6]), 285.1 ppm ([aj7]), 225.3 ppm ([aj8]) by Richards et al. [97].
[ak]This $\delta^{15N}{}_{compl}$ concerns uncoordinated nitrogen.
[aaa]$\Delta^{15N}{}_{coord}$ cannot be determined due to the lack of $\delta^{15N}{}_{lig}$ for 2-(2-(trifluoromethyl)vinyl-2-ol)pyridine.
[aab]vs pyridyl substituents in 2,5-bis(pyrid-2-yl)-3,4-dimethylpyrrole in $CDCl_3$: $\delta^{15N}{}_{lig} = -82.7$ ppm (299.0 ppm in respect to liquid NH_3) [85].
[aac]vs 2,2′-bipyridine in DMSO-d_6: $\delta^{15N}{}_{lig} = -72.6$ ppm [68,69,98].
[aad]vs free pyridine in situ: $\delta^{15N}{}_{lig} = -63.4$ ppm (318.3 ppm in respect to liquid NH_3) [55].
[aae]Δ^{15N}_{coord} cannot be determined due to the lack of $\delta^{15N}{}_{lig}$ for 5,5′-bis(pyren-1-yl)-2,2′-bipyridine.
[aaf]vs pyridine in CD_3OD: $\delta^{15N}{}_{lig} = -79.1$ ppm [99].
[aag]vs pyridine in $CDCl_3$: $\delta^{15N}{}_{lig} = -68.7$ ppm [68–70].
[aah]^{15}N NMR data for pyridine in THF-d_8 are unavailable.
[aai]vs pyridine in CD_2Cl_2: $\delta^{15N}{}_{lig} = -65.3$ ppm (own data).
[aaj]vs pyridine ring in (2-(3-methylimidazol-1-ium-1-yl)-4,6-di-*tert*-butylpyridine)$^+$ (PF_6^- salt in $CDCl_3$): $\delta^{15N}{}_{lig} = -104.5$ ppm [94].
[aak]vs 2,2′-bipyridine in $CDCl_3$ (^{15}N NMR data in CD_3OD being unavailable): $\delta^{15N}{}_{lig} = -75.1$ ppm (own data).
[aal]vs pyridine-3-carbonamide in CD_3CD_2OD (^{15}N NMR data in CD_3OD being unavailable): $\delta^{15N}{}_{lig} = -74.3$ ppm [101].
[aam]vs 1,10-phenanthroline in CD_3CD_2OD (^{15}N NMR data in CD_3OD being unavailable): $\delta^{15N}{}_{lig} = -79.9$ ppm (301.8 ppm in respect to liquid NH_3) [100].
[aan]$\Delta^{15N}{}_{coord}$ cannot be determined due to the lack of $\delta^{15N}{}_{lig}$ for 2-methylquinolin-8-ol.
[aao]vs N(7) in 7-azaindole moiety in 1-(pyrid-2-yl)-7-azaindole in DMSO-d_6: $\delta^{15N}{}_{lig} = -114.5$ ppm (267.2 ppm in respect to liquid NH_3) [96].
[aap]vs pyridyl substituent in 1-(pyrid-2-yl)-7-azaindole in DMSO-d_6: $\delta^{15N}{}_{lig} = -99.5$ ppm (282.2 ppm in respect to liquid NH_3) [96].
[aaq]vs N(7) in 1-phenyl-7-azaindole in DMSO-d_6: $\delta^{15N}{}_{lig} = -115.5$ ppm (266.2 ppm in respect to liquid NH_3) [96].
[aar]vs N(7) in 1-(thiophen-2-yl)-7-azaindole in DMSO-d_6: $\delta^{15N}{}_{lig} = -114.2$ ppm (267.5 ppm in respect to liquid NH_3) [96].
[aas]vs pyridazine in CD_3OD: $\delta^{15N}{}_{lig} = -32.4$ ppm (349.3 ppm in respect to liquid NH_3) [63].
[aat]vs N(1) in quinazoline in $CDCl_3$ (^{15}N NMR data in CD_2Cl_2 or CD_3OD being unavailable): $\delta^{15N}{}_{lig} = -100.9$ ppm (280.8 ppm in respect to liquid NH_3) [75].
[aau]vs N(3) in quinazoline in $CDCl_3$ (^{15}N NMR data in CD_2Cl_2 or CD_3OD being unavailable): $\delta^{15N}{}_{lig} = -90.4$ ppm (291.3 ppm in respect to liquid NH_3) [75].
[aav]vs phthalazine in CD_3OD: $\delta^{15N}{}_{lig} = -32.1$ ppm (349.6 ppm in respect to liquid NH_3) [63].
[aaaa]The only stereomeric possibility, resulting from the molecular formula.
[A]X-ray structure: VOZPUX [84].
[B]X-ray structure of H_2O solvate: XUDNIW [86].
[C]X-ray structure of C_6H_6 solvate: PIVPAN [90].
[D]X-ray structure: UBILUP [91].
[E]X-ray structure of CH_2Cl_2 solvate: CUTCUR [92].
[F]X-ray structure: TUKVUS [62].
[G]X-ray structure: TUKTUQ (TUKTUQ 01) [62].
[H]X-ray structure: SIKXIW [96].
[I]X-ray structure: SIKXOC [96].
[J]X-ray structure: SIKXES [96].

Table 4 Octahedral Fe(II), Ru(II) and Os(II) (d^6, low-spin) complexes and organometallics with azines (**1–6**, **9**, **11** ring systems), ^{15}N NMR studied in 2012–20. #

Compound	Parent ring system	^{15}N chemical shift (δ^{15N}_{compl}, ppm)	^{15}N coordination shift (Δ^{15N}_{coord}, ppm)	Nitrogen *trans* to
Fe(II) compounds				
[Fe^{II}(2,6-bis(**pyrid-2-yl**carbonyl)**pyridine**-κ^3-N^{pyA},N^{pyB1},N^{pyB2})$_2$]$(PF_6)_2$^{CD3CN} [102][A]	**1**	N^{pyA}: −125.5[aa1]	Unknown[aaa1]	N^{pyA} [X-ray]
		2 × N^{pyB}: −133.8[aa2]	Unknown[aaa1]	2 × N^{pyB} [X-ray]
[Fe^{II}(2,6-bis(**pyrid-2-yl**amino(methyl))**pyridine**-κ^3-N^{pyC},N^{pyD1},N^{pyD2})$_2$]$(PF_6)_2$^{CD3CN} [102][B]	**1**	N^{pyC}: −158.8[aa3]	Unknown[aaa2]	N^{pyC} [X-ray]
		2 × N^{pyD}: −152.5[aa4]	Unknown[aaa2]	2 × N^{pyD} [X-ray]
[Fe^{II}(2,6-bis(**pyrid-2-yl**carbonyl)**pyridine**-κ^3-N^{pyA},N^{pyB1},N^{pyB2})(2,6-bis(**pyrid-2-yl**amino(methyl))**pyridine**-κ^3-N^{pyC},N^{pyD1},N^{pyD2})]$(PF_6)_2$$^{DMSO-d6}$ [102][C]	**1**	N^{pyA}: −109.5[aa5]	Unknown[aaa1, aaa2]	N^{pyC} [X-ray]
		2 × N^{pyB}: −120.2[aa6]	Unknown[aaa1, aaa2]	2 × N^{pyB} [X-ray]
		N^{pyC}: −170.8[aa7]	Unknown[aaa1, aaa2]	N^{pyA} [X-ray]
		2 × N^{pyD}: −166.9[aa8]	Unknown[aaa1, aaa2]	2 × N^{pyD} [X-ray]
[Fe^{II}(2-((**pyrid-2-yl**)methyl(hydroxo)iminemethyl)**pyridine**-κ^3-N^{pyA},$N^{N=}$,N^{pyB})$_2$]$(ClO_4)_2$^{CD3CN} [103]	**1**	N^{pyA} or N^{pyB} (unassigned): −145[ab1], −134[ab2]	Unknown[aab]	(N^{pyB} or N^{pyA})[aaaa]
trans(CO,H),*trans*(CO,N^{py})-[Fe^{II}(2-(1-(2,4,6-trimethylphenylimino)ethyl)-6-(di-*tert*-butylphosphinomethyl)**pyridine-**κ^3-N^{py},$N^{N=}$,P)$(CO)_2$(H)]BF_4^{CDCl3} [104][D]	**1**	N^{py}: −123.1[ac1]	−52.9[aac]	CO [X-ray]
trans(CO,F),*trans*(CO,N^{py})-[Fe^{II}(2-(1-(2,4,6-trimethylphenylimino)ethyl)-6-(di-*tert*-butylphosphinomethyl)**pyridine-**κ^3-N^{py},$N^{N=}$,P)$(CO)_2$F]BF_4^{CD2Cl2} [104][E]	**1**	N^{py}: −124.1[ac2]	−53.9[aac,aad]	CO [X-ray]

Continued

Table 4 Octahedral Fe(II), Ru(II) and Os(II) (d^6, low-spin) complexes and organometallics with azines (**1–6**, **9**, **11** ring systems), ^{15}N NMR studied in 2012–20. #—cont'd

Compound	Parent ring system	^{15}N chemical shift (δ^{15N}_{compl}, ppm)	^{15}N coordination shift (Δ^{15N}_{coord}, ppm)	Nitrogen *trans* to
Ru(II) compounds				
trans(Cl,Cl)-[Ru^{II}(2-(**pyrid-2-yl**)**pyrimidine**-4-carboxylic acid-κ^2-N^{py},$N^{pm,N1}$)$(CO)_2Cl_2$]$^{DMSO\text{-}d6}$ [105][a1,F]	**1, 11**	N^{py}: −126.5	Unknown[aae]	CO [X-ray]
		$N^{pm,N1}$: −130.2	Unknown[aae]	CO [X-ray]
		$N^{pm,N3}$: −108.9[am]	Not applicable	None
trans(Cl,Cl)-[Ru^{II}(2-(**pyrid-2-yl**)**pyrimidine**-4-carboxylic acid-κ^2-N^{py},$N^{pm,N3}$)$(CO)_2Cl_2$]$^{DMSO\text{-}d6}$ [105][a2]	**1, 11**	N^{py}: −126.5	Unknown[aae]	(CO)[a2]
		$N^{pm,N1}$: −86.2[am]	Not applicable	None
		$N^{pm,N3}$: −143.6	Unknown[aae]	(CO)[a2]
trans(*S*,*S*)-[Ru^{II}(2-(**pyrid-2-yl**)-4-methyl**pyrimidine**-κ^2-N^{py}, $N^{pm,N1}$)$(CH_3SOCH_3)_2Cl_2$]^{CDCl3} [105][b2,G]	**1, 11**	N^{py}: −145.0	−73.8[aaf]	Cl [X-ray]
		$N^{pm,N1}$: −160.4	−59.2[aag]	Cl [X-ray]
		$N^{pm,N3}$: −90.2[am]	Not applicable	None
trans(Cl,S),*trans*(Cl,$N^{pm,N1}$)-[Ru^{II}(2-(**pyrid-2-yl**)-4-methyl**pyrimidine**-κ^2-N^{py},$N^{pm,N1}$)$(CH_3SOCH_3)_2Cl_2$]^{CDCl3} [105][b2,H]	**1, 11**	N^{py}: −130.7	−59.5[aaf]	S [X-ray]
		$N^{pm,N1}$: −153.4	−52.2[aag]	Cl [X-ray]
		$N^{pm,N3}$: −87.7[am]	Not applicable	None
trans(Cl,S),*trans*(Cl,N^{py})-[Ru^{II}(2-(**pyrid-2-yl**)-4-methyl**pyrimidine**-κ^2-N^{py},$N^{pm,N1}$)$(CH_3SOCH_3)_2Cl_2$]^{CDCl3} [105][b1]	**1, 11**	N^{py}: −135.9	−64.7[aaf]	(Cl)[b1]
		$N^{pm,N1}$: −147.6	−46.4[aag]	(S)[b1]
		$N^{pm,N3}$: −88.7[am]	Not applicable	None

trans(Cl,Cl)-[Ru^{II}(2-(**pyrid-2-yl**)-4-methyl**pyrimidine**-κ^2-N^{py}, N^{pm,N1})(CH_3SOCH_3)$_2$$Cl_2$]^{CDCl3} [105][b2,I]	**1, 11**	N^{py}: −124.6	−53.4[aaf]	S $^{X\text{-ray}}$
		N^{pm,N1}: −143.3	−42.1[aag]	S $^{X\text{-ray}}$
		N^{pm,N3}: −89.5[am]	Not applicable	None
trans(P,P),*trans*(P,N^{py})-[Ru^{II}(5-(**pyrid-2-yl**)pyrazolate-κ^2-N^{py},N^{pz,N1})(P(CH_3)$_3$)$_3$Cl]^{CDCl3} [106][c1,#]	**1**	N^{py}: −152.8	−47.1[aah]	(P)[c1]
trans(P,P),*trans*(P,N^{py})-[Ru^{II}(3-phenyl-5-(**pyrid-2-yl**)pyrazolate-κ^2-N^{py},N^{pz,N1})(P(CH_3)$_3$)$_3$Cl]^{CDCl3} [106][c2,J,#]	**1**	N^{py}: −172.5	Unknown[aai1]	P $^{X\text{-ray}}$
trans(P,P),*trans*(P,N^{py})-[Ru^{II}(4-bromo-5-(**pyrid-2-yl**)pyrazolate-κ^2-N^{py},N^{pz,N1})(P(CH_3)$_3$)$_3$Cl]^{CDCl3} [106][c2,K,#]	**1**	N^{py}: −151.5	−49.8[aaj]	P $^{X\text{-ray}}$
trans(P,P),*trans*(P,N^{py})-[Ru^{II}(4-nitro-5-(**pyrid-2-yl**)pyrazolate-κ^2-N^{py},N^{pz,N1})(P(CH_3)$_3$)$_3$Cl]^{CDCl3} [106][c2,L,#]	**1**	N^{py}: −137.0	Unknown[aai2]	P $^{X\text{-ray}}$
trans(P,P),*trans*(P,N^{py})-[Ru^{II}(5-(**pyrid-2-yl**)-1,2,4-triazolate-κ^2-N^{py}, N^{tz,N1})(P(CH_3)$_3$)$_3$Cl]^{CDCl3} [106][M,#]	**1**	N^{py}: −150.4	Unknown[aai3]	P $^{X\text{-ray}}$
trans(P,P),*trans*(P,N^{py})-[Ru^{II}(3-methyl-5-(**pyrid-2-yl**)-1,2,4-triazolate-κ^2-N^{py},N^{tz,N1})(P(CH_3)$_3$)$_3$Cl]^{CDCl3} [106][N,#]	**1**	N^{py}: −151.3	−48.5[aak]	P $^{X\text{-ray}}$
[Ru^{II}(2-(3-methyl-4-phenyl-*1H*-1,2,3-triazolium-1-yl)**pyridine***-κ^2-N^{py},C^{tz,C5})(1-isopropyl-4-methylbenzene)Cl]CF_3SO_3^{CDCl3} [107][d,e,O]	**1**	N^{py}: −156[ad1]	−62[aal]	C=C$^{bz\ X\text{-ray}}$
[Ru^{II}(2-(3-methyl-4-(4-methylphenyl)-*1H*-1,2,3-triazolium-1-yl) **pyridine***-κ^2-N^{py},C^{tz,C5})(1-isopropyl-4-methylbenzene) Cl]CF_3SO_3^{CDCl3} [107][d,e]	**1**	N^{py}: −156[ad2]	−61[aam]	(C=C^{bz})[aaaa]
[Ru^{II}(2-(3-methyl-4-(4-methoxyphenyl)-*1H*-1,2,3-triazolium-1-yl) **pyridine***-κ^2-N^{py},C^{tz,C5})(1-isopropyl-4-methylbenzene) Cl]CF_3SO_3^{CDCl3} [107][d,e]	**1**	N^{py}: −156[ad3]	−60[aan]	(C=C^{bz})[aaaa]

Continued

Table 4 Octahedral Fe(II), Ru(II) and Os(II) (d^6, low-spin) complexes and organometallics with azines (**1–6**, **9**, **11** ring systems), ^{15}N NMR studied in 2012–20. #—cont'd

Compound	Parent ring system	^{15}N chemical shift (δ^{15N}_{compl}, ppm)	^{15}N coordination shift (Δ^{15N}_{coord}, ppm)	Nitrogen *trans* to
[Ru^{II}(2-(3-methyl-4-(4-cyanophenyl)-*1H*-1,2,3-triazolium-1-yl) **pyridine***-κ^2-N^{py},$C^{tz,C5}$)(1-isopropyl-4-methylbenzene) Cl]CF_3SO_3^{CDCl3} [107][d,e]	**1**	N^{py}: −157[ad4]	−62[aao]	(C=C^{bz})[aaaa]
[Ru^{II}(2-(3-methyl-4-(4-trifluoromethylphenyl)-*1H*-1,2,3-triazolium-1-yl)**pyridine***-κ^2-N^{py},$C^{tz,C5}$)(1-isopropyl-4-methylbenzene)Cl]CF_3SO_3^{CDCl3} [107][d,e]	**1**	N^{py}: −156[ad5]	−61[aap]	(C=C^{bz})[aaaa]
[Ru^{II}(2-(3-methyl-4-(4-bromophenyl)-*1H*-1,2,3-triazolium-1-yl) **pyridine***-κ^2-N^{py},$C^{tz,C5}$)(1-isopropyl-4-methylbenzene) Cl]CF_3SO_3^{CDCl3} [107][d,e]	**1**	N^{py}: −156[ad6]	−61[aaq]	(C=C^{bz})[aaaa]
[Ru^{II}(2-(3-methyl-4-(4-nitrophenyl)-*1H*-1,2,3-triazolium-1-yl) **pyridine***-κ^2-N^{py},$C^{tz,C5}$)(1-isopropyl-4-methylbenzene) Cl]CF_3SO_3^{CDCl3} [107][d,e]	**1**	N^{py}: −157[ad7]	−62[aar]	(C=C^{bz})[aaaa]
[Ru^{II}(2-(1-methyl-3-phenyl-*1H*-1,2,3-triazolium-5-yl)**pyridine***-κ^2-N^{py},$C^{tz,C4}$)(1-isopropyl-4-methylbenzene)Cl]BF_4^{CDCl3} [107][d,f]	**1**	N^{py}: −139[ad8]	−66[aas]	(C=C^{bz})[aaaa]
[Ru^{II}(2-(1-methyl-3-(4-methylphenyl)-*1H*-1,2,3-triazolium-5-yl) **pyridine***-κ^2-N^{py},$C^{tz,C4}$)(1-isopropyl-4-methylbenzene)Cl]BF_4^{CDCl3} [107][d,f]	**1**	N^{py}: −140[ad9]	−67[aat]	(C=C^{bz})[aaaa]
[Ru^{II}(2-(1-methyl-3-(4-methoxyphenyl)-*1H*-1,2,3-triazolium-5-yl) **pyridine***-κ^2-N^{py},$C^{tz,C4}$)(1-isopropyl-4-methylbenzene)Cl]BF_4^{CDCl3} [107][d,f]	**1**	N^{py}: −139[ad10]	−66[aau]	(C=C^{bz})[aaaa]

[Ru^{II}(2-(1-methyl-3-(4-cyanophenyl)-*1H*-1,2,3-triazolium-5-yl)**pyridine***-κ^2-N^{py},C^{tz,C4})(1-isopropyl-4-methylbenzene)Cl]BF_4^{CDCl3} [107][d,f]	**1**	N^{py}: −140[ad11]	−66[aav]	(C=C^{bz})[aaaa]
[Ru^{II}(2-(1-methyl-3-(4-trifluoromethylphenyl)-*1H*-1,2,3-triazolium-5-yl)**pyridine***-κ^2-N^{py},C^{tz,C4})(1-isopropyl-4-methylbenzene)Cl]BF_4^{CDCl3} [107][d,f]	**1**	N^{py}: −139[ad12]	−66[aaw]	(C=C^{bz})[aaaa]
[Ru^{II}(2-(1-methyl-3-(4-bromophenyl)-*1H*-1,2,3-triazolium-5-yl)**pyridine***-κ^2-N^{py},C^{tz,C4})(1-isopropyl-4-methylbenzene)Cl]BF_4^{CDCl3} [107][d,f,P]	**1**	N^{py}: −139[ad13]	−66[aax]	C=C^{bz} [X-ray]
[Ru^{II}(2-(3-methyl-4-phenyl-*1H*-1,2,3-triazolium-1-yl)methyl**pyridine***-κ^2-N^{py},C^{tz,C5})(1-isopropyl-4-methylbenzene)Cl]PF_6^{CDCl3} [107][d,g,Q]	**1**	N^{py}: −147[ad14]	−76[aay]	C=C^{bz} [X-ray]
[Ru^{II}(2-(3-methyl-4-(4-methylphenyl)-*1H*-1,2,3-triazolium-1-yl)methyl**pyridine***-κ^2-N^{py},C^{tz,C5})(1-isopropyl-4-methylbenzene)Cl]PF_6^{CDCl3} [107][d,g]	**1**	N^{py}: −147[ad15]	−76[aaz]	(C=C^{bz})[aaaa]
[Ru^{II}(2-(3-methyl-4-(4-methoxyphenyl)-*1H*-1,2,3-triazolium-1-yl)methyl**pyridine***-κ^2-N^{py},C^{tz,C5})(1-isopropyl-4-methylbenzene)Cl]PF_6^{CDCl3} [107][d,g]	**1**	N^{py}: −146[ad16]	−76[aba]	(C=C^{bz})[aaaa]
[Ru^{II}(2-(3-methyl-4-(4-cyanophenyl)-*1H*-1,2,3-triazolium-1-yl)methyl**pyridine***-κ^2-N^{py},C^{tz,C5})(1-isopropyl-4-methylbenzene)Cl]PF_6^{CDCl3} [107][d,g]	**1**	N^{py}: −147[ad17]	−77[abb]	(C=C^{bz})[aaaa]
[Ru^{II}(2-(3-methyl-4-(4-trifluoromethylphenyl)-*1H*-1,2,3-triazolium-1-yl)methyl**pyridine***-κ^2-N^{py},C^{tz,C5})(1-isopropyl-4-methylbenzene)Cl]PF_6^{CDCl3} [107][d,g]	**1**	N^{py}: −147[ad18]	−77[abc]	(C=C^{bz})[aaaa]

Continued

Table 4 Octahedral Fe(II), Ru(II) and Os(II) (d^6, low-spin) complexes and organometallics with azines (**1–6**, **9**, **11** ring systems), ^{15}N NMR studied in 2012–20. #—cont'd

Compound	Parent ring system	^{15}N chemical shift (δ^{15N}_{compl}, ppm)	^{15}N coordination shift (Δ^{15N}_{coord}, ppm)	Nitrogen *trans* to
$[Ru^{II}$(2-(3-methyl-4-(4-chlorophenyl)-*1H*-1,2,3-triazolium-1-yl) methyl**pyridine***-κ^2-N^{py},$C^{tz,C5}$)(1-isopropyl-4-methylbenzene) Cl]PF_6^{CDCl3} [107]^{d,g}	**1**	N^{py}: -147^{ad19}	-77^{abd}	$(C{=}C^{bz})^{aaaa}$
$[Ru^{II}$(**2,2′-bipyridine**)$_2$(4-cyano-4′-methyl-**2,2′-bipyridine**)]$(PF_6)_2$^{CD3CN} [108]^R	**2**	$N^{bpyA,N1}$: -132.1^{ae1}	$-57.0^{abe,abf}$	$N^{bpyB,N1\ X\text{-}ray}$
		$N^{bpyA,N1'}$: -132.1^{ae1}	$-57.0^{abe,abf}$	$N^{RR'bpy,N1'\ X\text{-}ray}$
		$N^{bpyB,N1}$: -132.1^{ae1}	$-57.0^{abe,abf}$	$N^{bpyA,N1\ X\text{-}ray}$
		$N^{bpyB,N1'}$: -132.1^{ae1}	$-57.0^{abe,abf}$	$N^{RR'bpy,N1\ X\text{-}ray}$
		$N^{RR'bpy,N1}$: -122.6^{ae2}	Unknown^{abg}	$N^{bpyA,N1'\ X\text{-}ray}$
		$N^{RR'bpy,N1'}$: -141.1^{ae3}	Unknown^{abg}	$N^{bpyB,N1'\ X\text{-}ray}$
$[Ru^{II}$(**2,2′-bipyridine**)(**2,2′:6′,2″-terpyridine**)Cl] Cl$^{acetone\text{-}d6+H2O,85:15}$ [109]^S	**2, 4**	$N^{bpy,N1}$: -119.8	$-47.2^{abe,abh}$	$N^{tpy,N1'\ X\text{-}ray}$
		$N^{bpy,N1'}$: -130.7	$-58.1^{abe,abh}$	Cl $^{X\text{-}ray}$
		2 × $N^{tpy,N1,N1''}$: -132.5	$-59.5^{abi,abj}$	2 × $N^{tpy,N1,N1''\ X\text{-}ray}$
		$N^{tpy,N1'}$: -94.9	$-12.1^{abi,abk}$	$N^{bpy,N1\ X\text{-}ray}$
$[Ru^{II}$(**2,2′-bipyridine**)(**2,2′:6′,2″-terpyridine**)Cl]Cl$^{DMSO\text{-}d6}$ [110]^S	**2, 4**	$N^{bpy,N1}$: -117.2	-44.6^{abh}	$N^{tpy,N1'\ X\text{-}ray}$
		$N^{bpy,N1'}$: -127.6	-55.0^{abh}	Cl $^{X\text{-}ray}$
		2 × $N^{tpy,N1,N1''}$: -129.6	-56.6^{abj}	2 × $N^{tpy,N1,N1''\ X\text{-}ray}$
		$N^{tpy,N1'}$: -91.9	-9.1^{abk}	$N^{bpy,N1\ X\text{-}ray}$

[Ru^{II}(**2,2′-bipyridine**)(**2,2′:6′,2″-terpyridine**)(H_2O)](CF_3SO_3)$_2$$^{acetone-d6+H2O,85:15}$ [109][T]	**2, 4**	$N^{bpy,N1}$: −121.7	−49.1[abe,abh]	$N^{tpy,N1'}$ X-ray
		$N^{bpy,N1'}$: −137.9	−65.3[abe,abh]	O X-ray
		2 × $N^{tpy,N1,N1''}$: −139.1	−66.1[abi,abj]	2 × $N^{tpy,N1,N1''}$ X-ray
		$N^{tpy,N1'}$: −101.1	−18.3[abi,abk]	$N^{bpy,N1}$ X-ray
[Ru^{II}(4,4′-dihydroxo-**2,2′-bipyridine**)(**2,2′:6′,2″-terpyridine**)Cl]Cl$^{acetone-d6+H2O,85:15}$ [109]	**2, 4**	$N^{bpy,N1}$: −147.9	Unknown[abl1]	($N^{tpy,N1'}$)[aaaa]
		$N^{bpy,N1'}$: −163.8	Unknown[abl1]	(Cl)[aaaa]
		2 × $N^{tpy,N1,N1''}$: −129.7	−56.7[abi,abj]	(2 × $^{tpy,N1,N1''}$)[aaaa]
		$N^{tpy,N1'}$: −89.5	−6.7[abi,abk]	($N^{bpy,N1}$)[aaaa]
[Ru^{II}(4,4′-dihydroxo-**2,2′-bipyridine**)(**2,2′:6′,2″-terpyridine**)(H_2O)](CF_3SO_3)$_2$$^{acetone-d6+H2O,85:15}$ [109]	**2, 4**	$N^{bpy,N1}$: −150.1	Unknown[abl1]	($N^{tpy,N1'}$)[aaaa]
		$N^{bpy,N1'}$: −172.5	Unknown[abl1]	(O)[aaaa]
		2 × $N^{tpy,N1,N1''}$: −136.8	−63.8[abi,abj]	(2 × $N^{tpy,N1,N1''}$)[aaaa]
		$N^{tpy,N1'}$: −96.6	−13.8[abi,abk]	($N^{bpy,N1}$)[aaaa]
[Ru^{II}(6,6′-dihydroxo-**2,2′-bipyridine**)(**2,2′:6′,2″-terpyridine**)Cl]Cl$^{acetone-d6+H2O,85:15}$ [109][U,V]	**2, 4**	$N^{bpy,N1}$: −161.4	Unknown[abl2]	$N^{tpy,N1'}$ X-ray
		$N^{bpy,N1'}$: −178.3	Unknown[abl2]	Cl X-ray
		2 × $N^{tpy,N1,N1''}$: −133.4	−60.4[abi,abj]	2 × $N^{tpy,N1,N1''}$ X-ray
		$N^{tpy,N1'}$: −89.0	−6.2[abi,abk]	$N^{bpy,N1}$ X-ray

Continued

Table 4 Octahedral Fe(II), Ru(II) and Os(II) (d^6, low-spin) complexes and organometallics with azines (**1–6**, **9**, **11** ring systems), ^{15}N NMR studied in 2012–20. #—cont'd

Compound	Parent ring system	^{15}N chemical shift (δ^{15N}_{compl}, ppm)	^{15}N coordination shift (Δ^{15N}_{coord}, ppm)	Nitrogen *trans* to
$[Ru^{II}$(6,6′-dihydroxo-**2,2′-bipyridine**)(**2,2′:6′,2″-terpyridine**)$(H_2O)](CF_3SO_3)_2$$^{acetone-d6+H2O,85:15}$ [109][W]	**2, 4**	$N^{bpy,N1}$: −161.3	Unknown[abl2]	$N^{tpy,N1'}$ X-ray
		$N^{bpy,N1'}$: −181.3	Unknown[abl2]	O X-ray
		2 × $N^{tpy,N1,N1''}$: −137.5	−64.5[abi,abj]	2 × $N^{tpy,N1,N1''}$ X-ray
		$N^{tpy,N1'}$: −93.2	−10.4[abi,abk]	$N^{bpy,N1}$ X-ray
$[Ru^{II}$(4,4′-dimethoxy-**2,2′-bipyridine**)(**2,2′:6′,2″-terpyridine**)Cl]Cl$^{acetone-d6+H2O,85:15}$ [109][X]	**2, 4**	$N^{bpy,N1}$: −143.6	Unknown[abl3]	$N^{tpy,N1'}$ X-ray
		$N^{bpy,N1'}$: −158.4	Unknown[abl3]	Cl X-ray
		2 × $N^{tpy,N1,N1''}$: −131.0	−58.0[abi,abj]	2 × $N^{tpy,N1,N1''}$ X-ray
		$N^{tpy,N1'}$: −91.8	−9.0[abi,abk]	$N^{bpy,N1}$ X-ray
$[Ru^{II}$(4,4′-dimethoxy-**2,2′-bipyridine**)(**2,2′:6′,2″-terpyridine**)$(H_2O)](CF_3SO_3)_2$$^{acetone-d6+H2O,85:15}$ [109][Y]	**2, 4**	$N^{bpy,N1}$: −145.8	Unknown[abl3]	$N^{tpy,N1'}$ X-ray
		$N^{bpy,N1'}$: −166.7	Unknown[abl3]	O X-ray
		2 × $N^{tpy,N1,N1''}$: −137.6	−64.6[abi,abj]	2 × $N^{tpy,N1,N1''}$ X-ray
		$N^{tpy,N1'}$: −98.0	−15.2[abi,abk]	$N^{bpy,N1}$ X-ray
$[Ru^{II}$(6,6′-dimethoxy-**2,2′-bipyridine**)(**2,2′:6′,2″-terpyridine**)Cl]$PF_6$$^{acetone-d6}$ [109]	**2, 4**	$N^{bpy,N1}$: −149.8	Unknown[abl4]	($N^{tpy,N1'}$)[aaaa]
		$N^{bpy,N1'}$: −160.9	Unknown[abl4]	(Cl)[aaaa]
		2 × $N^{tpy,N1,N1''}$: −128.7	−55.7[abi,abj]	(2 × $N^{tpy,N1,N1''}$)[aaaa]
		$N^{tpy,N1'}$: −83.2	−0.4[abi,abk]	($N^{bpy,N1}$)[aaaa]

$[Ru^{II}$(6,6′-dimethoxy-**2,2′-bipyridine**)(**2,2′:6′,2″-terpyridine**)$(H_2O)](CF_3SO_3)_2$$^{acetone\text{-}d6+H2O,85:15}$ [109][Z]	**2, 4**	$N^{bpy,N1}$: −154.4	Unknown[abl4]	$N^{tpy,N1'}$ X-ray
		$N^{bpy,N1'}$: −173.5	Unknown[abl4]	O X-ray
		2 × $N^{tpy,N1,N1''}$: −138.3	−65.3[abi,abj]	2 × $N^{tpy,N1,N1''}$ X-ray
		$N^{tpy,N1'}$: −94.9	−12.1[abi,abk]	$N^{bpy,N1}$ X-ray
$[Ru^{II}$(6,6′-dimethoxy-**2,2′-bipyridine**)(**2,2′:6′,2″-terpyridine**)$(H_2O)]Cl_2$$^{acetone-d6+H2O,85:15}$ [109]	**2, 4**	$N^{bpy,N1}$: −154.6	Unknown[abl4]	$(N^{tpy,N1'})$[aaaa]
		$N^{bpy,N1'}$: −173.4	Unknown[abl4]	(O)[aaaa]
		2 × $N^{tpy,N1,N1''}$: −138.3	−65.3[abi,abj]	(2 × $N^{tpy,N1,N1''}$)[aaaa]
		$N^{tpy,N1'}$: −95.0	−12.2[abi,abk]	$(N^{bpy,N1})$[aaaa]
mer- and *fac-*$[Ru^{II}$(4-methyl-4′-((pyrid-4-yl)vinyl)-**2,2′-bipyridine**$)_3](PF_6)_2$$^{acetone-d6}$ (mixture) [111][AA]	**2**	Various N^{bpy}: −140,[af1] −133[af2]	Unknown[abm]	All N^{bpy} X-ray
$K_2[Ru^{II}$(4,4′-azobis(**2,2′-bipyridine**))$(CN)_4]^{D2O}$ [112]	**2**	$N^{bpyA,N1}$, $N^{bpyA,N2}$: −116.7[ag1], −129.1[ag2]	Unknown[abn]	(2 × CN)[aaaa]
		$N^{bpyB,N1}$, $N^{bpyB,N2}$: −79.6[ag3,am], −88.6[ag4,am]	Not applicable	None
$[Ru^{II}$(**1,10-phenanthroline**)(**2,2′:6′,2″-terpyridine**)Cl]Cl$^{DMSO\text{-}d6}$ [113][BB]	**3, 4**	$N^{phen,N1}$: −120.7	−51.4[abo]	$N^{tpy,N1'}$ X-ray
		$N^{phen,N1'}$: −131.2	−61.9[abo]	Cl X-ray
		2 × $N^{tpy,N1,N1''}$: −129.6	−56.6[abj]	2 × $N^{tpy,N1,N1''}$ X-ray
		$N^{tpy,N1'}$: −92.6	−9.8[abk]	$N^{phen,N1}$ X-ray

Continued

Table 4 Octahedral Fe(II), Ru(II) and Os(II) (d^6, low-spin) complexes and organometallics with azines (**1–6**, **9**, **11** ring systems), ^{15}N NMR studied in 2012–20. #—cont'd

Compound	Parent ring system	^{15}N chemical shift (δ^{15N}_{compl}, ppm)	^{15}N coordination shift (Δ^{15N}_{coord}, ppm)	Nitrogen *trans* to
$[Ru^{II}$(4,7-diphenyl-**1,10-phenanthroline**)$_3]Cl_2$^{CD3OD,280K} [114]	**3**	N^{phen}: −139.4^{ah1}	−66.9^{abp}	$(N^{phen})^{aaaa}$
$Na_2[Ru^{II}$(4,7-diphenyl-**1,10-phenanthroline**)$_2$(4,7-bis (3-sulfonatophenyl)-**1,10-phenanthroline**)$]Cl_2$^{CD3OD,280K} [114]	**3**	2 × N^{phen}, $N^{phen(SO3)2}$: −140^{ah2} to −138^{ah3}	Unknown^{abq,abr}	(N^{phen} or $N^{phen(SO3)2}$)^{aaaa}
$Na_4[Ru^{II}$(4,7-diphenyl-**1,10-phenanthroline**)(4,7-bis (3-sulfonatophenyl)-**1,10-phenanthroline**)$_2]Cl_2$^{CD3OD,280K} [114]	**3**	N^{phen}, 2 × $N^{phen(SO3)2}$: −140^{ah4} to −139^{ah5}	Unknown^{abq,abr}	(N^{phen} or $N^{phen(SO3)2}$)^{aaaa}
$Na_6[Ru^{II}$(4,7-bis(3-sulfonatophenyl)-**1,10-phenanthroline**)$_3]Cl_2$^{CD3OD,280K} [114]	**3**	$N^{phen(SO3)2}$: −138.0^{ah6}	Unknown^{abr}	$(N^{phen(SO3)2})^{aaaa}$
$[Ru^{II}$(**2,2′:6′,2″-terpyridine**)($NH_2CH_2CH_2NH_2$)Cl]Cl$^{DMSO-d6}$ [110]^{CC}	**4**	2 × $N^{tpy,N1,N1''}$: −128.5	−55.5^{abj}	2 × $N^{tpy,N1,N1''}$ X-ray
		$N^{tpy,N1'}$: −78.7	+4.1^{abk}	N^{NH2} X-ray
$[Ru^{II}$(**2,2′:6′,2″-terpyridine**)($(CH_3)_2NCH_2CH_2N(CH_3)_2$)Cl]$ClO_4$$^{DMSO-d6}$ [113]^{DD}	**4**	2 × $N^{tpy,N1,N1''}$: −136.3	−63.3^{abj}	2 × $N^{tpy,N1,N1''}$ X-ray
		$N^{tpy,N1'}$: −89.9	−7.1^{abk}	N^{NH2} X-ray
$[Ru^{II}$(**2,2′:6′,2″-terpyridine**-4′-carboxylic acid)(2,6-bis(**pyrid-2-yl**amino(methyl))**pyridine**-κ^3-N^{pyA},N^{pyB1},N^{pyB2})]$(PF_6)_2$^{CD3CN} [115]	**1, 4**	2 × $N^{tpy,N1,N1''}$: −143.4^{ai1}	Unknown^{abs1}	(2 × $N^{tpy,N1,N1''}$)^{aaaa}
		$N^{tpy,N1'}$: −88.1^{ai2}	Unknown^{abs1}	$(N^{pyA})^{aaaa}$
		N^{pyA}: −163.3^{ai3}	Unknown^{aaa2}	$(N^{tpy,N1'})^{aaaa}$
		2 × N^{pyB}: −169.6^{ai4}	Unknown^{aaa2}	(2 × N^{pyB})^{aaaa}

[Ru^{II}(**2,2′:6′,2″-terpyridine**-4,4′,4″-tricarboxylic acid)(2,6-bis(**pyrid-2-yl**amino(methyl))**pyridine**-κ^3-N^{pyA},N^{pyB1}, N^{pyB2})](PF$_6$)$_2$^{CD3CN} [115]	**1, 4**	2 × N$^{tpy,N1,N1''}$: −132.2[ai5]	Unknown[abs2]	(2 × N$^{tpy,N1,N1''}$)[aaaa]
		N$^{tpy,N1'}$: −87.8[ai6]	Unknown[abs2]	(N^{pyA})[aaaa]
		N^{pyA}: −165.4[ai7]	Unknown[aaa2]	(N$^{tpy,N1'}$)[aaaa]
		2 × N^{pyB}: −172.7[ai8]	Unknown[aaa2]	2 × N^{pyB})[aaaa]
[Ru^{II}(**2,2′:6′,2″-terpyridine**-4,4′,4″-tricarboxylic acid)(2,6-bis(**pyrid-2-yl**amino(methyl))-4-amino**pyridine**-κ^3-N^{pyA},N^{pyB1}, N^{pyB2})](PF$_6$)$_2$^{CD3CN} [115]	**1, 4**	2 × N$^{tpy,N1,N1''}$: −131.8[ai9]	Unknown[abs2]	(2 × N$^{tpy,N1,N1''}$)[aaaa]
		N$^{tpy,N1'}$: −84.8[ai10]	Unknown[abs2]	(N^{pyA})[aaaa]
		N^{pyA}: −200.1[ai11]	Unknown[aaa3]	(N$^{tpy,N1'}$)[aaaa]
		2 × N^{pyB}: −171.7[ai12]	Unknown[aaa3]	(2 × N^{pyB})[aaaa]
[Ru^{II}(trimethyl **2,2′:6′,2″-terpyridine**-4,4′,4″-tricarboxylate)(2,6-bis(**pyrid-2-yl**amino(methyl))**pyridine**-κ^3-N^{pyA},N^{pyB1}, N^{pyB2})](PF$_6$)$_2$^{CD3CN} [116][EE]	**1, 4**	2 × N$^{tpy,N1,N1''}$: −132.6[ai13]	Unknown[abs3]	2 × N$^{tpy,N1,N1''}$ X-ray
		N$^{tpy,N1'}$: −87.3[ai14]	Unknown[abs3]	N^{pyA} X-ray
		N^{pyA}: −165.5[ai15]	Unknown[aaa2]	N$^{tpy,N1'}$ X-ray
		2 × N^{pyB}: −172.5[ai16]	Unknown[aaa2]	2 × N^{pyB} X-ray
[Ru^{II}(ethyl **2,2′:6′,2″-terpyridine**-4′-carboxylate)(2,6-bis(**pyrid-2-yl**amino(methyl))-4-amino**pyridine**-κ^3-N^{pyA},N^{pyB1}, N^{pyB2})](PF$_6$)$_2$^{CD3CN} [116][FF]	**1, 4**	2 × N$^{tpy,N1,N1''}$: −143.1[ai17]	Unknown[abs4]	2 × N$^{tpy,N1,N1''}$ X-ray
		N$^{tpy,N1'}$: −84.5[ai18]	Unknown[abs4]	N^{pyA} X-ray
		N^{pyA}: −198.6[ai19]	Unknown[aaa3]	N$^{tpy,N1'}$ X-ray
		2 × N^{pyB}: −169.0[ai20]	Unknown[aaa3]	2 × N^{pyB} X-ray

Continued

Table 4 Octahedral Fe(II), Ru(II) and Os(II) (d^6, low-spin) complexes and organometallics with azines (**1–6**, **9**, **11** ring systems), ^{15}N NMR studied in 2012–20. #—cont'd

Compound	Parent ring system	^{15}N chemical shift ($\delta^{15N}{}_{compl}$, ppm)	^{15}N coordination shift ($\Delta^{15N}{}_{coord}$, ppm)	Nitrogen *trans* to
[Ru^{II}(triethyl **2,2′:6′,2″-terpyridine**-4,4′,4″-tricarboxylate)(2,6-bis(**pyrid-2-yl**amino(methyl))-4-amino**pyridine**-κ^3-N^{pyA},N^{pyB1},N^{pyB2})]$(PF_6)_2$^{CD3CN} [116]	**1, 4**	2 × $N^{tpy,N1,N1''}$: −132.0[ai21]	Unknown[abs5]	(2 × $N^{tpy,N1,N1''}$)[aaaa]
		$N^{tpy,N1'}$: −85.1[ai22]	Unknown[abs5]	(N^{pyA})[aaaa]
		N^{py}: −200.0[ai23]	Unknown[aaa3]	($N^{tpy,N1'}$)[aaaa]
		2 × $N^{py'}$: −171.8[ai24]	Unknown[aaa3]	(2 × N^{pyB})[aaaa]
[Ru^{II}(1,3-bis(**quinolin**-8-yl)-*1H*-pyrazole-κ^3-N^{quinA},N^{quinB},$N^{pz,N2}$)$_2$]$(PF_6)_2$^{CD3CN} [117][GG,#]	**5**	4 × N^{quin}: −145,[aj1] −148,[aj2]	All	2 × N^{quin} [X-ray]
		−150,[aj3] −152[aj4]	Unknown[abt]	2 × $N^{pz,N2}$ [X-ray]
[Ru^{II}(5-chloro-7-iodo**quinolin**-8-olate-κ^2-N,O)(1-isopropyl-4-methylbenzene)Cl]^{CDCl3} [118][d,HH]	**5**	N^{quin}: −143.5[ak]	Unknown[abu]	C=C^{bz} [X-ray]
trans-[Ru^{II}(6-bromo**isoquinoline**)$_2$(1,1′-**biisoquinoline**-3,3′-dicarboxylate-κ^4-N,N′,O,O′)]^{CDCl3} [119][II]	**6, 9**	$N^{isoquin}$: −116.8[al1]	Unknown[abv]	$N^{isoquin}$ [X-ray]
		2 × $N^{biisoquin}$: −141.7[al2]	Unknown[abw]	2 × O [X-ray]
[Ru^{II}(2-(3-methyl-4-phenyl-*1H*-1,2,3-triazolium-1-yl)**pyrimidine***-κ^2-$N^{pm,N1}$,$C^{tz,C5}$)(1-isopropyl-4-methylbenzene)Cl] $CF_3SO_3$$^{DMSO-d6}$ [120][d,h]	**11**	$N^{pm,N1}$: −164[ad20]	−52[abx]	(C=C^{bz})[aaaa]
		$N^{pm,N3}$: −114[ad21,am]	Not applicable	None
Os(II) compounds				
mer- and *fac*-[Os^{II}(4-methyl-4′-((pyrid-4-yl)vinyl)-**2,2′-bipyridine**)$_3$]$(PF_6)_2$$^{acetone-d6}$ (mixture) [111][JJ,KK]	**2**	Various N^{bpy}: −161,[af3] −154[af4]	Unknown[abm]	All N^{bpy} [X-ray]
[Os^{II}(2-(3-methyl-4-phenyl-*1H*-1,2,3-triazolium-1-yl)**pyrimidine***-κ^2-$N^{pm,N1}$,$C^{tz,C5}$)(1-isopropyl-4-methylbenzene)Cl]PF_6^{CD2Cl2} [120][d,h]	**11**	$N^{pm,N1}$: −186[ad22]	−74[abx]	(C=C^{bz})[aaaa]
		$N^{pm,N3}$: undetected	Not applicable	

[#]See also Table 11 for pyrazole- and 1,2,4-triazole-derived ligand(s).

[a1-a2]These complexes have the same *trans*(Cl,Cl) geometry, differing in the pyrimidine ligand coordination mode: κ^2-N(1)^{py},N(1)^{pm} ([a1]) vs κ^2-N(1)^{py},N(3)^{pm} ([a2]).

[b1-b2]The assumed *trans*(Cl,S),*trans*(Cl,N^{py}) form ([b1]) is isomeric to the X-ray studied *trans*(S,S), *trans*(Cl,S),*trans*(Cl,N^{pm,N1}) and *trans*(Cl,Cl) species ([b2]) [105].

[c1-c2]The assumed *trans*(P,P),*trans*(P,N^{py}) isomer ([c1]) is analogous to the X-ray studied *trans*(P,P),*trans*(P,N^{py})-[Ru^{II}(3-phenyl-5-(pyrid-2-yl)pyrazolate)(P(CH$_3$)$_3$)$_3$Cl], *trans*(P,P),*trans*(P,N^{py})-[Ru^{II}(4-bromo-5-(pyrid-2-yl)pyrazolate)(P(CH$_3$)$_3$)$_3$Cl], *trans*(P,P),*trans*(P,N^{py})-[Ru^{II}(4-nitro-5-(pyrid-2-yl)pyrazolate)(P(CH$_3$)$_3$)$_3$Cl] ([c2]) [106].

[d]1-Isopropyl-4-methylbenzene ligand occupies three coordination sites, within octahedral geometry.

[e]2-(3-Methyl-4-aryl-*1H*-1,2,3-triazolium-1-yl)pyridine* are zwitterions, having both charges in the 1,2,3-triazolic ring (positive—delocalized between all three nitrogens, negative—at the deprotonated C(5) atom), acting as bidentate (κ^2-N(1)^{py},C(5)^{tz}) ligands.

[f]2-(1-Methyl-3-aryl-*1H*-1,2,3-triazolium-5-yl)pyridine* are zwitterions, having both charges in the 1,2,3-triazolic ring (positive—delocalized between all three nitrogens, negative—at the deprotonated C(4) atom), acting as bidentate (κ^2-N(1)^{py},C(4)^{tz}) ligands.

[g]2-(3-Methyl-4-aryl-*1H*-1,2,3-triazolium-1-yl)methylpyridine* are zwitterions, having both charges in the 1,2,3-triazolic ring (positive—delocalized between all three nitrogens, negative—at the deprotonated C(5) carbon), acting as bidentate (κ^2-N(1//)^{py},C(5)^{tz}) ligands.

[h]2-(3-Methyl-4-phenyl-*1H*-1,2,3-triazolium-1-yl)pyrimidine* is zwitterion, having both charges in the 1,2,3-triazolic ring (positive—delocalized between all three nitrogens, negative—at the deprotonated C(5) atom), acting as bidentate (κ^2-N(1)^{pm},C(5)^{tz}) ligand.

[aa-al]Original $\delta^{15N}{}_{compl}$ values, primarily referenced to liquid NH_3 by the following authors, were recalculated in respect to neat CH_3NO_2 by subtracting 381.7 ppm (or 382 ppm) relative shift [37]:

[aa1-aa8]256.2 ppm ([aa1]), 247.9 ppm ([aa2]), 222.9 ppm ([aa3]), 229.2 ppm ([aa4]), 272.2 ppm ([aa5]), 261.5 ppm ([aa6]), 210.9 ppm ([aa7]), 214.8 ppm ([aa8]) by Mengel et al. [102].

[ab1-ab2]237 ppm ([ab1]), 248 ppm ([ab2]) by Markiewicz et al. [103].

[ac1-ac2]258.6 ppm ([ac1]), 257.6 ppm ([ac2]) by Butschke et al. [104].

[ad1-ad22]226 ppm ([ad1]), 226 ppm ([ad2]), 226 ppm ([ad3]), 225 ppm ([ad4]), 226 ppm ([ad5]), 226 ppm ([ad6]), 225 ppm ([ad7]), 243 ppm ([ad8]), 242 ppm ([ad9]), 243 ppm ([ad10]), 242 ppm ([ad11]), 243 ppm ([ad12]), 243 ppm ([ad13]), 235 ppm ([ad14]), 235 ppm ([ad15]), 236 ppm ([ad16]), 235 ppm ([ad17]), 235 ppm ([ad18]), 235 ppm ([ad19]), 218 ppm ([ad20]), 268 ppm ([ad21]), 196 ppm ([ad22]) by Bolje et al. [107,120].

[ae1-ae3]249.6 ppm ([ae1]), 259.1 ppm ([ae2]), 240.6 ppm ([ae3]) by Ortiz et al. [108].

[af1-af4]242 ppm ([af1]), 249 ppm ([af2]), 221 ppm ([af3]), 228 ppm ([af4]) by Keisar et al. [111].

[ag1-ag4]265.0 ppm ([ag1]), 252.6 ppm ([ag2]), 302.1 ppm ([ag3]), 293.1 ppm ([ag4]) by Abate et al. [112].

[ah1-ah6]242.3 ([ah1]), 242 ([ah2]), 244 ([ah3]), 242 ([ah4]), 243 ([ah5]), 243.7 ([ah6]) by Babak et al. [114].

[ai1-ai24]238.3 ppm ([ai1]), 293.6 ppm ([ai2]), 218.4 ppm ([ai3]), 212.1 ppm ([ai4]), 249.5 ppm ([ai5]), 293.9 ppm ([ai6]), 216.3 ppm ([ai7]), 209.0 ppm ([ai8]), 249.9 ppm ([ai9]), 296.9 ppm ([ai10]), 181.6 ppm ([ai11]), 210.0 ppm ([ai12]), 249.1 ppm ([ai13]), 294.4 ppm ([ai14]), 216.2 ppm ([ai15]), 209.2 ppm ([ai16]), 238.6 ppm ([ai17]), 297.2 ppm ([ai18]), 183.1 ppm ([ai19]), 212.7 ppm ([ai20]), 249.7 ppm ([ai21]), 296.6 ppm ([ai22]), 181.7 ppm ([ai23]), 209.9 ppm ([ai24]) by Breivogel et al. [115,116].

[aj1-aj4]237 ppm ([aj1]), 234 ppm ([aj2]), 232 ppm ([aj3]), 230 ppm ([aj4]) by Jarenmark et al. [117].

[ak]238.2 ppm by Gobec et al. [118].

[al1-al2]264.9 ppm ([al1]), 240.0 ppm ([al2]) by Scherrer et al. [119].

[am]This $\delta^{15N}{}_{compl}$ concerns uncoordinated nitrogen.

[aaa1-aaa3]$\Delta^{15N}{}_{coord}$ cannot be determined due to the lack of $\delta^{15N}{}_{lig}$ for 2,6-bis(pyrid-2-ylcarbonyl)pyridine ([aaa1]), 2,6-bis(pyrid-2-ylamino(methyl))pyridine ([aaa2]) and 2,6-bis(pyrid-2-ylamino(methyl))-4-aminopyridine ([aaa3]).

[aab]$\Delta^{15N}{}_{coord}$ cannot be determined due to the lack of $\delta^{15N}{}_{lig}$ for 2-((pyrid-2-yl)methyl(hydroxo)iminemethyl)pyridine.

[aac]vs 2-(1-(2,4,6-trimethylphenylimino)ethyl)-6-(di-*tert*-butylphosphinomethyl)pyridine in $CDCl_3$: $\delta^{15N}{}_{lig} = -70.2$ ppm (311.5 ppm in respect to liquid NH_3) [104].

[aad]^{15}N NMR data for 2-(1-(2,4,6-trimethylphenylimino)ethyl)-6-(di-*tert*-butylphosphinomethyl)pyridine in CD_2Cl_2 are unavailable.

[aae]$\Delta^{15N}{}_{coord}$ cannot be determined due to the lack of $\delta^{15N}{}_{lig}$ for 2-(pyrid-2-yl)pyrimidine-4-carboxylic acid.

[aaf]vs pyridyl substituent in 2-(pyrid-2-yl)-4-methylpyrimidine in $CDCl_3$: $\delta^{15N}{}_{lig} = -71.2$ ppm [105].

[aag]vs N(1) in pyrimidine ring of 2-(pyrid-2-yl)-4-methylpyrimidine in $CDCl_3$: $\delta^{15N}{}_{lig} = -101.2$ ppm [105].

Continued

Table 4 Octahedral Fe(II), Ru(II) and Os(II) (d^6, low-spin) complexes and organometallics with azines (**1–6**, **9**, **11** ring systems), ^{15}N NMR studied in 2012–20. #—cont'd

[aah]vs pyridyl substituent in 5-(pyrid-2-yl)-*1H*/*2H*-pyrazole in $CDCl_3$: $\delta^{15N}_{lig} = -105.7$ ppm (average for two observed tautomers, differing in proton location within the pyrazole ring—*1H* (N^{py}: −109.3 ppm) and *2H* (N^{py}: −102.2 ppm)) [106].

[aai1-aai3]Δ^{15N}_{coord} cannot be determined due to the lack of δ^{15N}_{lig} for 3-phenyl-5-(pyrid-2-yl)-*1H*-pyrazole ([aai1]), 4-nitro-5-(pyrid-2-yl)-*1H*-pyrazole ([aai2]), 5-(pyrid-2-yl)-*1H*-1,2,4-triazole ([aai3]).

[aaj]vs pyridyl substituent in 4-bromo-5-(pyrid-2-yl)-*1H*-pyrazole in $CDCl_3$: $\delta^{15N}_{lig} = -101.7$ ppm [106].

[aak]vs pyridyl substituent in 3-methyl-5-(pyrid-2-yl)-*1H*-1,2,4-triazole in $CDCl_3$: $\delta^{15N}_{lig} = -102.8$ ppm [106].

[aal]vs pyridine ring in (2-(3-methyl-4-phenyl-*1H*-1,2,3-triazolium-1-yl)pyridine)$^+$ cation ($CF_3SO_3^-$ counterion) in $CDCl_3$: $\delta^{15N}_{lig} = -94$ ppm (288 ppm in respect to liquid NH_3) [107].

[aam]vs pyridine ring in (2-(3-methyl-4-(4-methylphenyl)-*1H*-1,2,3-triazolium-1-yl)pyridine)$^+$ cation ($CF_3SO_3^-$ counterion) in $CDCl_3$: $\delta^{15N}_{lig} = -95$ ppm (287 ppm in respect to liquid NH_3) [107].

[aan]vs pyridine ring in (2-(3-methyl-4-(4-methoxyphenyl)-*1H*-1,2,3-triazolium-1-yl)pyridine)$^+$ cation ($CF_3SO_3^-$ counterion) in $CDCl_3$: $\delta^{15N}_{lig} = -96$ ppm (286 ppm in respect to liquid NH_3) [107].

[aao]vs pyridine ring in (2-(3-methyl-4-(4-cyanophenyl)-*1H*-1,2,3-triazolium-1-yl)pyridine)$^+$ cation ($CF_3SO_3^-$ counterion) in DMSO-d_6 (^{15}N NMR data in $CDCl_3$ being unavailable): $\delta^{15N}_{lig} = -95$ ppm (287 ppm in respect to liquid NH_3) [107].

[aap]vs pyridine ring in (2-(3-methyl-4-(4-trifluoromethylphenyl)-*1H*-1,2,3-triazolium-1-yl)pyridine)$^+$ cation ($CF_3SO_3^-$ counterion) in DMSO-d_6 (^{15}N NMR data in $CDCl_3$ being unavailable): $\delta^{15N}_{lig} = -95$ ppm (287 ppm in respect to liquid NH_3) [107].

[aaq]vs pyridine ring in (2-(3-methyl-4-(4-bromophenyl)-*1H*-1,2,3-triazolium-1-yl)pyridine)$^+$ cation ($CF_3SO_3^-$ counterion) in DMSO-d_6 (^{15}N NMR data in $CDCl_3$ being unavailable): $\delta^{15N}_{lig} = -95$ ppm (287 ppm in respect to liquid NH_3) [107].

[aar]vs pyridine ring in (2-(3-methyl-4-(4-nitrophenyl)-*1H*-1,2,3-triazolium-1-yl)pyridine)$^+$ cation ($CF_3SO_3^-$ counterion) in DMSO-d_6 (^{15}N NMR data in $CDCl_3$ being unavailable): $\delta^{15N}_{lig} = -95$ ppm (287 ppm in respect to liquid NH_3) [107].

[aas]vs pyridine ring in (2-(1-methyl-3-phenyl-*1H*-1,2,3-triazolium-5-yl)pyridine)$^+$ cation (BF_4^- counterion) in DMSO-d_6 (^{15}N NMR data in $CDCl_3$ being unavailable): $\delta^{15N}_{lig} = -73$ ppm (309 ppm in respect to liquid NH_3) [107].

[aat]vs pyridine ring in (2-(1-methyl-3-(4-methylphenyl)-*1H*-1,2,3-triazolium-5-yl)pyridine)$^+$ cation (BF_4^- counterion) in DMSO-d_6 (^{15}N NMR data in $CDCl_3$ being unavailable): $\delta^{15N}_{lig} = -73$ ppm (309 ppm in respect to liquid NH_3) [107].

[aau]vs pyridine ring in (2-(1-methyl-3-(4-methoxyphenyl)-*1H*-1,2,3-triazolium-5-yl)pyridine)$^+$ cation (BF_4^- counterion) in DMSO-d_6 (^{15}N NMR data in $CDCl_3$ being unavailable): $\delta^{15N}_{lig} = -73$ ppm (309 ppm in respect to liquid NH_3) [107].

[aav]vs pyridine ring in (2-(1-methyl-3-(4-cyanophenyl)-*1H*-1,2,3-triazolium-5-yl)pyridine)$^+$ cation (BF_4^- counterion) in DMSO-d_6 (^{15}N NMR data in $CDCl_3$ being unavailable): $\delta^{15N}_{lig} = -74$ ppm (308 ppm in respect to liquid NH_3) [107].

[aaw]vs pyridine ring in (2-(1-methyl-3-(4-trifluoromethylphenyl)-*1H*-1,2,3-triazolium-5-yl)pyridine)$^+$ cation (BF_4^- counterion) in DMSO-d_6 (^{15}N NMR data in $CDCl_3$ being unavailable): $\delta^{15N}_{lig} = -73$ ppm (309 ppm in respect to liquid NH_3) [107].

[aax]vs pyridine ring in (2-(1-methyl-3-(4-bromophenyl)-*1H*-1,2,3-triazolium-5-yl)pyridine)$^+$ cation (BF_4^- counterion) in DMSO-d_6 (^{15}N NMR data in $CDCl_3$ being unavailable): $\delta^{15N}_{lig} = -73$ ppm (309 ppm in respect to liquid NH_3) [107].

[aay]vs pyridine ring in (2-(3-methyl-4-phenyl-*1H*-1,2,3-triazolium-1-yl)methylpyridine)$^+$ cation (PF_6^- counterion) in DMSO-d_6 (^{15}N NMR data in $CDCl_3$ being unavailable): $\delta^{15N}_{lig} = -71$ ppm (311 ppm in respect to liquid NH_3) [107].

[aaz]vs pyridine ring in (2-(3-methyl-4-(4-methylphenyl)-*1H*-1,2,3-triazolium-1-yl)methylpyridine)$^+$ cation (PF_6^- counterion) in DMSO-d_6 (^{15}N NMR data in $CDCl_3$ being unavailable): $\delta^{15N}_{lig} = -71$ ppm (311 ppm in respect to liquid NH_3) [107].

[aba]vs pyridine ring in (2-(3-methyl-4-(4-methoxyphenyl)-*1H*-1,2,3-triazolium-1-yl)methylpyridine)$^+$ cation (PF_6^- counterion) in DMSO-d_6 (^{15}N NMR data in $CDCl_3$ being unavailable): $\delta^{15N}_{lig} = -70$ ppm (312 ppm in respect to liquid NH_3) [107].

[abb]vs pyridine ring in (2-(3-methyl-4-(4-cyanophenyl)-*1H*-1,2,3-triazolium-1-yl)methylpyridine)$^+$ cation (PF_6^- counterion) in DMSO-d_6 (^{15}N NMR data in $CDCl_3$ being unavailable): $\delta^{15N}_{lig} = -70$ ppm (312 ppm in respect to liquid NH_3) [107].
[abc]vs pyridine ring in (2-(3-methyl-4-(4-trifluoromethylphenyl)-*1H*-1,2,3-triazolium-1-yl)methylpyridine)$^+$ cation (PF_6^- counterion) in DMSO-d_6 (^{15}N NMR data in $CDCl_3$ being unavailable): $\delta^{15N}_{lig} = -70$ ppm (312 ppm in respect to liquid NH_3) [107].
[abd]vs pyridine ring in (2-(3-methyl-4-(4-chlorophenyl)-*1H*-1,2,3-triazolium-1-yl)methylpyridine)$^+$ cation (PF_6^- counterion) in DMSO-d_6 (^{15}N NMR data in $CDCl_3$ being unavailable): $\delta^{15N}_{lig} = -70$ ppm (312 ppm in respect to liquid NH_3) [107].
[abe]^{15}N NMR data for 2,2′-bipyridine in CD_3CN or acetone-d_6 (or acetone-$d_6 + H_2O$) are unavailable.
[abf]vs 2,2′-bipyridine in $CDCl_3$: $\delta^{15N}_{lig} = -75.1$ ppm (own data).
[abg]Δ^{15N}_{coord} cannot be determined due to the lack of δ^{15N}_{lig} for 4-cyano-4′-methyl-2,2′-bipyridine.
[abh]vs 2,2′-bipyridine in DMSO-d_6: $\delta^{15N}_{lig} = -72.6$ ppm [68,69,98].
[abi]^{15}N NMR data for 2,2′:6′,2″-terpyridine in acetone-d_6 (or acetone-$d_6 + H_2O$) are unavailable.
[abj]vs N(1), N(1″) in 2,2′:6′,2″-terpyridine in DMSO-d_6: $\delta^{15N}_{lig} = -73.0$ ppm [129].
[abk]vs N(1′) in 2,2′:6′,2″-terpyridine in DMSO-d_6: $\delta^{15N}_{lig} = -82.8$ ppm [129].
[abl1-abl4]Δ^{15N}_{coord} cannot be determined due to the lack of δ^{15N}_{lig} for 4,4′-dihydroxo-2,2′-bipyridine ([abl1]), 6,6′-dihydroxo-2,2′-bipyridine ([abl2]), 4,4′-dimethoxy-2,2′-bipyridine ([abl3]) and 6,6′-dimethoxy-2,2′-bipyridine ([abl4]).
[abm]Δ^{15N}_{coord} cannot be determined due to the lack of δ^{15N}_{lig} for 4-methyl-4′-((pyrid-4-yl)vinyl)-2,2′-bipyridine.
[abn]Δ^{15N}_{coord} cannot be determined due to the lack of δ^{15N}_{lig} for 4,4′-azobis(2,2′-bipyridine).
[abo]vs 1,10-phenanthroline in DMSO-d_6: $\delta^{15N}_{lig} = -69.3$ ppm [68,69,98].
[abp]vs 4,7-diphenyl-1,10-phenanthroline in DMSO-d_6 (^{15}N NMR data in CD_3OD being unavailable): $\delta^{15N}_{lig} = -72.5$ ppm [130].
[abq]Δ^{15N}_{coord} cannot be determined due to the lack of clear assignment of ^{15}N signals to 4,7-diphenyl-1,10-phenanthroline and 4,7-bis(3-sulfonatophenyl)-1,10-phenanthroline ligands.
[abr]Δ^{15N}_{coord} cannot be determined due to the lack of δ^{15N}_{lig} for 4,7-bis(3-sulfonatophenyl)-1,10-phenanthroline.
[abs1-abs5]Δ^{15N}_{coord} cannot be determined due to the lack of δ^{15N}_{lig} for 2,2′:6′,2″-terpyridine-4′-carboxylic acid ([abs1]), 2,2′:6′,2″-terpyridine-4,4′,4″-tricarboxylic acid ([abs2]), trimethyl 2,2′:6′,2″-terpyridine-4,4′,4″-tricarboxylate ([abs3]), ethyl 2,2′:6′,2″-terpyridine-4′-carboxylate ([abs4]) and triethyl 2,2′:6′,2″-terpyridine-4,4′,4″-tricarboxylate ([abs5]).
[abt]Δ^{15N}_{coord} cannot be determined due to the lack of δ^{15N}_{lig} for 1,3-bis(quinolin-8-yl)-*1H*-pyrazole.
[abu]Δ^{15N}_{coord} cannot be determined due to the lack of δ^{15N}_{lig} for 5-chloro-7-iodoquinolin-8-ol.
[abv]Δ^{15N}_{coord} cannot be determined due to the lack of δ^{15N}_{lig} for 6-bromoisoquinoline.
[abw]Δ^{15N}_{coord} cannot be determined due to the lack of δ^{15N}_{lig} for 1,1′-biisoquinoline-3,3′-dicarboxylic acid.
[abx]vs N(1) in pyrimidine ring in (2-(3-methyl-4-phenyl-*1H*-1,2,3-triazolium-1-yl)pyrimidine)$^+$ cation ($CF_3SO_3^-$ counterion) in DMSO-d_6: $\delta^{15N}_{lig} = -112$ ppm (270 ppm in respect to liquid NH_3) [120].
[aaaa]The only stereomeric possibility, resulting from the molecular formula.
[A]X-ray structure of CH_3CN solvate: ZOQJOG [102].
[B]X-ray structure of CH_3CN solvate: ZOQJEW [102].
[C]X-ray structure of CH_3CN solvate: ZOQJIA [102].
[D]X-ray structure: MUHBOI [104].
[E]X-ray structure: MUHBIC [104].
[F]X-ray structure of H_2O solvate: DONTOQ [131], DONTOQ01 [105].
[G]X-ray structure: RUPGOA [105].
[H]X-ray structure: RUPGUG [105].

Continued

Table 4 Octahedral Fe(II), Ru(II) and Os(II) (d^6, low-spin) complexes and organometallics with azines (**1–6**, **9**, **11** ring systems), ^{15}N NMR studied in 2012–20. #—cont'd

[I]X-ray structure of CH_3OH solvate: RUPJIX [105].
[J]X-ray structure: TILLIL [106].
[K]X-ray structure: VIQVUO [106].
[L]X-ray structure: VIQVOI [106].
[M]X-ray structure: VIQVIC [106].
[N]X-ray structure: SARQIO [106].
[O]X-ray structure: ZOLPIB [107].
[P]X-ray structure of $CHCl_3$ solvate: ZOLPUN [107].
[Q]X-ray structure: ZOLPEX [107].
[R]X-ray structure: VIHHOL [108].
[S]X-ray structure of H_2O solvate: KAWQEH [110].
[T]X-ray structure of analogous $(PF_6^-)_2$ salt: FEPXUU [132].
[U]X-ray structure of CH_3OH solvate: VIYCUD [109].
[V]X-ray structure of H_2O solvate: ILEXEE [133].
[W]X-ray structure: VIYDAK [109].
[X]X-ray structure of CH_3OH solvate: VIYDIS [109].
[Y]X-ray structure: VIYDEO [109].
[Z]X-ray structure of CH_3CN solvate: VIYDOY [109].
[AA]X-ray structure (*mer*- and *fac*-isomers together): PIHQAB [111].
[BB]X-ray structure of H_2O solvate: SUBWAQ [113].
[CC]X-ray structure of H_2O solvate: KAWREI [110].
[DD]X-ray structure: SUBTAN [113].
[EE]X-ray structure of $CH_3CN{\cdot}CH_3CH_2OCH_2CH_3$ solvate: XIJNEL [116].
[FF]X-ray structure of CH_3CN solvate: XIJNIP [116].
[GG]X-ray structure: UJOSOD [115].
[HH]X-ray structure: DIYJUS (DIYJUS 01) [117,134].
[II]X-ray structure: ABEVIP [118].
[JJ]X-ray structure: PIHQIJ [111].
[KK]X-ray structure of $C_6H_6{\cdot}C_6H_5CH_3$ solvate: PIHQEF [111].

Table 5 Cu(I), Ag(I) and Au(I) (d^{10}) complexes and organometallics with azines (1–4 ring systems), ^{15}N NMR studied in 2012–20.

Compound	Parent ring system	^{15}N chemical shift (δ^{15N}_{compl}, ppm)	^{15}N coordination shift (Δ^{15N}_{coord}, ppm)
Cu(I) compounds			
$[Cu^{I}$(2-(2,3,4,5-tetrafluorophenyl)**pyridine***-κ^2-N,$C^{ph,C6}$)(1,3-bis(2,6-diisopropylphenyl)imidazole-2-ylidene-κ^1-C^2)$]^{THF\text{-}d8}$ [135][a,b,A]	**1**	N^{py}: −85.7	−18.5[aaa]
$[Cu^{I}$(2-(2,3,4,5-tetrafluorophenyl)**pyridine***-κ^2-N,$C^{ph,C6}$)(bis(2-(diphenylphosphino)-1-phenyl) ether-κ^2-P,P′)$]^{C6D6}$ [135][a,c1,B]	**1**	N^{py}: −100.8	−33.6[aaa]
Ag(I) compounds			
$[Ag^{I}$(**pyridine**)$_2]BF_4{}^{CD2Cl2}$ [136][d,C,D]	**1**	N^{py}: −126.5	−61.2[aab]
$[Ag^{I}$(**pyridine**)$_2]BF_4{}^{CD2Cl2}$ [137][d,C,D]	**1**	N^{py}: −122[aa1]	−57[aab]
$[Ag^{I}$(**pyridine**)$_2]ClO_4{}^{CD2Cl2}$ [136][d,E,F]	**1**	N^{py}: −124.0	−58.7[aab]
$[Ag^{I}$(**pyridine**)$_2]PF_6{}^{CD2Cl2}$ [136][d,G,H]	**1**	N^{py}: −128.4	−63.1[aab]
$[Ag^{I}$(**pyridine**)$_2]SbF_6{}^{CD2Cl2}$ [136][d,I]	**1**	N^{py}: −129.3	−64.0[aab]
$[Ag^{I}$(**pyridine**)$_2]CF_3SO_3{}^{CD2Cl2}$ [136][d,J]	**1**	N^{py}: −122.0[ad]	−56.7[aab,aac]
$[Ag^{I}$(**pyridine**)$_2](4\text{-}CH_3C_6H_4SO_3)^{CD2Cl2}$ [136][d,K]	**1**	N^{py}: −111.8	−46.5[aab]
$[Ag^{I}$(**pyridine**)$_2]NO_3{}^{CD2Cl2}$ [136][d,L]	**1**	N^{py}: −113.1	−47.8[aab]
$[Ag^{I}$(**pyridine**)$_2]CF_3COO^{CD2Cl2}$ [136][d]	**1**	N^{py}: −109.5	−44.2[aab]

Continued

Table 5 Cu(I), Ag(I) and Au(I) (d^{10}) complexes and organometallics with azines (1–4 ring systems), ^{15}N NMR studied in 2012–20.—cont'd

Compound	Parent ring system	^{15}N chemical shift (δ^{15N}_{compl}, ppm)	^{15}N coordination shift (Δ^{15N}_{coord}, ppm)
$[Ag^{I}$(2,3-dimethyl**pyridine**$)_2]NO_3^{CDCl3}$ [138][d]	**1**	N^{py}: −108.8[ab1]	−37.2[aad]
$[Ag^{I}$(2,3-dimethyl**pyridine**$)_2]NO_3^{CD3CN}$ [138][d]	**1**	N^{py}: −98.0[ab2]	−32.1[aae]
$[Ag^{I}$(2,3-dimethyl**pyridine**$)_2]NO_3^{DMSO-d6}$ [138][d]	**1**	N^{py}: −92.7[ab3]	−28.3[aaf]
$[Ag^{I}$(2,4-dimethyl**pyridine**$)_2]NO_3^{CDCl3}$ [138][d]	**1**	N^{py}: −119.8[ab4]	−39.1[aag]
$[Ag^{I}$(2,6-dimethyl**pyridine**$)_2]NO_3^{CDCl3}$ [138][d,M]	**1**	N^{py}: −104.7[ab5]	−34.2[aah]
$[Ag^{I}$(2,6-dimethyl**pyridine**$)_2]BF_4^{CD3CN}$ [137][d]	**1**	N^{py}: −98[aa2]	−28[aah,aai]
$[Ag^{I}$(2,6-dimethyl**pyridine**$)_2][Au(CF_3)_2]^{CD3CN}$ [137][d,N]	**1**	N^{py}: −94[aa3]	−24[aah,aai]
$[Ag^{I}$(3,5-dimethyl**pyridine**$)_2]NO_3^{CDCl3}$ [138][d,O]	**1**	N^{py}: −115.2[ab6]	−45.7[aaj]
$[Ag^{I}$(4-dimethylamino**pyridine**$)_2]BF_4^{CD3CN}$ [137][d,P]	**1**	N^{py}: −161[aa4,ae]	−56[aak,aal]
$[Ag^{I}$(1-(**pyrid-2-yl**)-3-methyl-4-trifluoroacetylpyrazolate-5-one-**κ**^{2}-N^{py},$N^{pz,N2}$)(2,4,6-trimethyl**pyridine**)$]^{solid}$ [139][b,e,Q,#]	**1**	N^{py}: −148.3[af1]	Unknown[aam]
		$N^{2,4.6tmpy}$: −118.9[af2]	−23.8[aan]
$[Ag^{I}$(1-(**pyrid-2-yl**)-3-methyl-4-trifluoroacetylpyrazolate-5-one-**κ**^{2}-N^{py},$N^{pz,N2}$)(2-ethyl-*1H*-imidazole)$]^{solid}$ [139][b,e,R,#]	**1**	N^{py}: −173.2[af3]	Unknown[aam]
$[Ag^{I}$(1-(**pyrid-2-yl**)-3-methyl-4-trifluoroacetylpyrazolate-5-one-**κ**^{2}-N^{py},$N^{pz,N2}$)(1-benzyl-*1H*-imidazole)$]^{solid}$ [139][b,e,#]	**1**	N^{py}: −173.2[af4]	Unknown[aam]
$[Ag^{I}$(1-(**pyrid-2-yl)-**3-methyl-4-trifluoroacetylpyrazolate-5-one)-**κ**^{2}-N^{py},$N^{pz,N2}$)(1-benzyl-2-methyl-*1H*-imidazole)$]^{solid}$ [139][b,e,#]	**1**	N^{py}: −172.7[af5]	Unknown[aam]

$[Ag^{I}$(1-(**pyrid-2-yl**)-3-methyl-4-trifluoroacetylpyrazolate-5-one-κ^2-N^{py},$N^{pz,N2}$)(2-trifluoromethyl-*1H*-benzoimidazole)$]^{solid}$ [139][b,e,#]	**1**	N^{py}: −181.6[af6]	Unknown[aam]
$[Ag^{I}{}_2$(1-(**pyrid-2-yl**)-3-methyl-4-trifluoroacetylpyrazolate-5-one-κ^2-N^{py},$N^{pz,N2}$)$_2$(1,1′-carbonyl-*1H,1H′*-diimidazole-μ^2-N^3,$N^{3'}$)$]^{solid}$ [139][b,e,#]	**1**	N^{py}: −173.9[af7]	Unknown[aam]
$[Ag^{I}$(1-(**pyrid-2-yl**)-3-methyl-4-trifluoroacetyl-*1H*-pyrazol-5-olate-μ^3-N^{py},$N^{pz,N2}$,$O^{pz,C5-O}$)$]_n^{solid}$ [139][b,e,#]	**1**	N^{py}: −147.2[af8]	Unknown[aam]
$[Ag^{I}$(**2,2′-bipyridine**)$_2]BF_4{}^{CD3CN}$ [137][c2,S,T]	**2**	N^{bpy}: −100[aa5]	−25[aao]
$[Ag^{I}$(**2,2′-bipyridine**)$_2][Au(CF_3)_2]^{CD3CN}$ [137][c2,S,T]		N^{bpy}: −98[aa6]	−23[aao]
$[Ag^{I}$(**1,10-phenanthroline**)(2-(NH_2CSNHNCH)pyridine-κ^2-$N^{N=}$,S)$]NO_3{}^{DMSO-d6}$ [140][c1]	**3**	N^{phen}: −112.0[ac]	−42.7[aap]
$[Ag^{I}$(4′-(**pyrid-2-yl**)-**2,2′:6′,2″-terpyridine**)(CD_3SOCD_3)$]CF_3SO_3{}^{DMSO-d6,353K}$ [141][c2]	**1, 4**	2 × $N^{tpy,N1,N1''}$: −95.2	−23.0[aaq]
		$N^{tpy,N1'}$: undetected	Not applicable
		N^{py}: −105.5	−23.3[aar]
$[Ag^{I}$(4′-((morpholin-4-yl)phenyl)-**2,2′:6′,2″-terpyridine**)(CD_3SOCD_3)$]CF_3SO_3{}^{DMSO-d6,353K}$ [141][c2]	**4**	2 × $N^{tpy,N1,N1''}$: −97.4	−25.6[aas]
		$N^{tpy,N1'}$: −114.8	−25.4[aat]
$[Ag^{I}$(4′-((morpholin-4-yl)phenyl)-**2,2′:6′,2″-terpyridine**)(CD_3OD)$]CF_3SO_3{}^{CD3OD,353K}$ [141][c2]	**4**	2 × $N^{tpy,N1,N1''}$: −112.1	−31.2[aau]
		$N^{tpy,N1'}$: −118.0	−28.7[aav]
Au(I) compounds			
$[Au^{I}$(**pyridine**)$_2]BF_4{}^{CD2Cl2}$ [142][U]	**1**	N^{py}: −155.6	−88.6[aaw]

Continued

Table 5 Cu(I), Ag(I) and Au(I) (d^{10}) complexes and organometallics with azines (1–4 ring systems), ^{15}N NMR studied in 2012–20.—cont'd

[#]See also Table 12 for pyrazole-ligand(s) and imidazole-derived ligand(s).

[a]2-(2,3,4,5-Tetrafluorophenyl)pyridine* is deprotonated (at the phenyl C(6) atom), anionic (charge −1) form of 2-(2,3,4,5-tetrafluorophenyl)pyridine, acting as bidentate (κ^2-N(1)^{py},C(6)^{ph}) ligand.

[b]Coordination number 3 (trigonal geometry).

[c1–c2]Coordination number 4 (tetrahedral ([c1]) or square-planar ([c2]) geometry).

[d]Coordination number 2 (linear geometry).

[e]1-(Pyrid-2-yl)-3-methyl-4-trifluoroacetylpyrazolate-5-one and 1-(pyrid-2-yl)-3-methyl-4-trifluoroacetyl-*1H*-pyrazol-5-olate are ketone and enol forms of the same anion (charge −1) formed by deprotonation of 1-(pyrid-2-yl)-3-methyl-4-trifluoroacetyl-*1H*-pyrazol-5-ol.

[aa–ac]Original $\delta^{15N}{}_{compl}$ values, primarily referenced to liquid NH_3 by the following authors, were recalculated in respect to neat CH_3NO_2 by subtracting 381.7 ppm (or 382 ppm) relative shift [37].

[aa1–aa6]260 ppm ([aa1]), 284 ppm ([aa2]), 288 ppm ([aa3]), 221 ppm ([aa4]), 282 ppm ([aa5]), 284 ppm ([aa6]) by Zopes et al. [137].

[ab1–ab6]272.9 ppm ([ab1]), 283.7 ppm ([ab2]), 289.0 ppm ([ab3]), 261.9 ppm ([ab4]), 277.0 ppm ([ab5]), 266.5 ppm ([ab6]) by Sakharov et al. [138].

[ac]269.7 ppm by Silva et al. [140].

[ad]see footnote ([aac]).

[ae]see footnote ([aal]).

[af1–af8]Original $\delta^{15N}{}_{compl}$ values (207.5 ppm ([af1]), 236.9 ppm ([af2]), 182.6 ppm ([af3]), 182.6 ppm ([af4]), 183.1 ppm ([af5]), 174.2 ppm ([af6]), 181.9 ppm ([af7]), 208.6 ppm ([af8])), primarily referenced to solid $(NH_4)_2SO_4$, by Marchetti et al. [139], were recalculated in respect to neat CH_3NO_2 by subtracting 355.8 ppm relative shift [139].

[aaa]vs 2-(2,3,4,5-Tetrafluorophenyl)pyridine in C_6D_6 (^{15}N NMR data in THF-d_8 being unavailable): $\delta^{15N}{}_{lig} = -67.2$ ppm [135].

[aab]vs pyridine in CD_2Cl_2: −65.3 ppm (own data).

[aac]Previous literature value for $[Au^{III}(pyridine)_2]^+$ ($CF_3SO_3^-$ counterion), in $CDCl_3$: $\delta^{15N}{}_{compl} = -121$ ppm, $\Delta^{15N}{}_{coord} = -52$ ppm [144].

[aad]vs 2,3-dimethylpyridine in $CDCl_3$: $\delta^{15N}{}_{lig} = -71.6$ ppm (310.1 ppm in respect to liquid NH_3) [138].

[aae]vs 2,3-dimethylpyridine in CD_3CN: $\delta^{15N}{}_{lig} = -65.9$ ppm (315.8 ppm in respect to liquid NH_3) [138].

[aaf]vs 2,3-dimethylpyridine in DMSO-d_6: $\delta^{15N}{}_{lig} = -64.4$ ppm (317.3 ppm in respect to liquid NH_3) [138].

[aag]vs 2,4-dimethylpyridine in $CDCl_3$: $\delta^{15N}{}_{lig} = -80.7$ ppm (301.0 ppm in respect to liquid NH_3) [138].

[aah]vs 2,6-dimethylpyridine in $CDCl_3$: $\delta^{15N}{}_{lig} = -70.5$ ppm (311.2 ppm in respect to liquid NH_3) [138].

[aai]^{15}N NMR data for 2,6-dimethylpyridine in CD_3CN are unavailable.

[aaj]vs 3,5-dimethylpyridine in $CDCl_3$: $\delta^{15N}{}_{lig} = -69.5$ ppm (312.2 ppm in respect to liquid NH_3) [138].

[aak]vs 4-dimethylaminopyridine in CD_3CN: $\delta^{15N}{}_{lig} = -105$ ppm [144].

[aal]Previous literature value for $[Au^{III}(4\text{-dimethylaminopyridine})_2]^+$ ($CF_3SO_3^-$ counterion), in $CDCl_3$: $\delta^{15N}{}_{compl} = -163$ ppm, $\Delta^{15N}{}_{coord} = -51$ ppm [144].

[aam]$\Delta^{15N}{}_{coord}$ cannot be determined because parent ligand 1-(pyrid-2-yl)-3-methyl-4-trifluoroacetylpyrazol-5-ol occurs in solid phase as zwitterion, having deprotonated –O^- group (with simultaneous conversion to ketone form and appearance of CF_3CO^- moiety, double-bonded to pyrazolic ring) and protonated pyridine nitrogen NH^+—as proved by its low $\delta^{15N}{}_{lig}$ value of −218.6 ppm (137.2 ppm in respect to solid $(NH_4)_2SO_4$) [139].

[aan]vs solid 2,4,6-trimethylpyridine: $\delta^{15N}{}_{lig} = -95.1$ ppm (260.7 ppm in respect solid $(NH_4)_2SO_4$) [139].
[aao]vs 2,2′-bipyridine in $CDCl_3$ (^{15}N NMR data in CD_3CN being unavailable): $\delta^{15N}{}_{lig} = -75.1$ ppm (own data).
[aap]vs 1,10-phenanthroline in DMSO-d_6: $\delta^{15N}{}_{lig} = -69.3$ ppm [68,69,98].
[aaq]vs N(1), N(1″) in 2,2′:6′,2″-terpyridine moiety in 4′-(pyrid-2-yl)-2,2′:6′,2″-terpyridine in DMSO-d_6: $\delta^{15N}{}_{lig} = -72.2$ ppm [141].
[aar]vs pyridine moiety in 4′-(pyrid-2-yl)-2,2′:6′,2″-terpyridine in DMSO-d_6: $\delta^{15N}{}_{lig} = -82.2$ ppm [141].
[aas]vs N(1), N(1″) in 2,2′:6′,2″-terpyridine moiety in 4′-((morpholin-4-yl)phenyl)-2,2′:6′,2″-terpyridine in DMSO-d_6: $\delta^{15N}{}_{lig} = -71.8$ ppm [141].
[aat]vs N(1′) in 2,2′:6′,2″-terpyridine moiety in 4′-((morpholin-4-yl)phenyl)-2,2′:6′,2″-terpyridine in DMSO-d_6: $\delta^{15N}{}_{lig} = -89.4$ ppm [141].
[aau]vs N(1), N(1″) in 2,2′:6′,2″-terpyridine moiety in 4′-((morpholin-4-yl)phenyl)-2,2′:6′,2″-terpyridine in CD_3OD: $\delta^{15N}{}_{lig} = -80.9$ ppm [141].
[aav]vs N(1′) in 2,2′:6′,2″-terpyridine moiety in 4′-((morpholin-4-yl)phenyl)-2,2′:6′,2″-terpyridine in CD_3OD: $\delta^{15N}{}_{lig} = -89.3$ ppm [141].
[aaw]vs pyridine in CD_2Cl_2: −67.0 ppm [142].
[A]X-ray structure of THF solvate: EHEBOK [135].
[B]X-ray structure of CH_2Cl_2 solvate: EHEBAW [135].
[C]X-ray structure: NEVREM [145].
[D]X-ray structure of CH_2ClCH_2Cl solvate: LULNIR [136].
[E]X-ray structure: NEVRAI [145].
[F]X-ray structure of CH_2ClCH_2Cl solvate: LULNOX [136].
[G]X-ray structure: NEVRIQ [145].
[H]X-ray structure of CH_2ClCH_2Cl solvate: LUKZAU [136].
[I]X-ray structure of CH_2ClCH_2Cl solvate: LUKZIC [136].
[J]X-ray structure: LUKZUO [136].
[K]X-ray structure: LULBEB [136].
[L]X-ray structure of H_2O solvate: AGPYNO (AGPYNO 01-02) [136,146,147].
[M]X-ray structure: QQQBUJ 01-02 [147,148].
[N]X-ray structure: UHAPIE [137].
[O]X-ray structure: GIJZAC [138].
[P]X-ray structure of analogous NO_3^- salt (H_2O solvate): YUCNIU [149].
[Q]X-ray structure: SOYXUB [139].
[R]X-ray structure: SOYXIP [139].
[S]X-ray structure of analogous ClO_4^- salt: MAYCIZ [150].
[T]X-ray structure of analogous $CF_3SO_3^-$ salt: MAYCUL [150].
[U]X-ray structure: UJOQAN [151].

Table 6 Square-planar Au(III) (d^8) complexes and organometallics with azines (**1–3**, **7** ring systems), ^{15}N NMR studied in 2012–20.

Compound	Parent ring system	^{15}N chemical shift (δ^{15N}_{compl}, ppm)	^{15}N coordination shift (Δ^{15N}_{coord}, ppm)	Nitrogen *trans* to
Au(III) compounds				
trans-[Au^{III}(**pyridine**)$_2Cl_2$]Cl^{CD2Cl2} [142][a]	**1**	N^{py}: −154.2	−87.2[aaa]	(N)[a]
trans-[Au^{III}(4-methyl**pyridine**)$_2Cl_2$]Cl [142][a]	**1**	N^{py}: −162.6	−91.0[aab]	(N)[a]
trans-[Au^{III}(4-methoxy**pyridine**)$_2Cl_2$]Cl^{CD2Cl2} [142][a]	**1**	N^{py}: −178.2	−92.2[aac]	(N)[a]
trans-[Au^{III}(4-trifluoromethyl**pyridine**)$_2Cl_2$]Cl^{CD2Cl2} [142][a]	**1**	N^{py}: −144.1	−86.1[aad]	(N)[a]
[Au^{III}(2-benzyl**pyridine***-κ^2-N,$C^{ph,C2}$)Cl_2]$^{DMSO\text{-}d6}$ [152][b,A]	**1**	N^{py}: −154.2[aa1,ab]	−93.3[aae,aaf]	Cl $^{X\text{-}ray}$
trans(N^{py},$CH_2^{NHC(CH3)OCH2CH2}$)-[Au^{III}(2-(4-methylphenyl)**pyridine***-κ^2-N,$C^{ph,C2}$)(NHC(CH_3)OCH_2CH_2)] CF_3COO^{CD2Cl2} [153][b,c,B]	**1**	N^{py}: −129	Unknown[aag]	$C^{CH2\ X\text{-}ray}$
[Au^{III}(2-phenylsulfanyl**pyridine**)Cl_3]^{CDCl3} [154][C]	**1**	N^{py}: −155.1	−79.1[aah]	Cl $^{X\text{-}ray}$
[Au^{III}(2-(4-methylphenyl)sulfanyl**pyridine**)Cl_3]^{CDCl3} [154]	**1**	N^{py}: −155.9	−80.2[aai]	(Cl)[aaaa]
[Au^{III}(2-(4-*tert*-butylphenyl)sulfanyl**pyridine**)Cl_3]^{CDCl3} [154]	**1**	N^{py}: −155.7	−80.0[aaj]	(Cl)[aaaa]
[Au^{III}(2-(4-bromophenyl)sulfanyl**pyridine**)Cl_3]^{CDCl3} [154][D]	**1**	N^{py}: −153.0	−78.9[aak]	Cl $^{X\text{-}ray}$
[Au^{III}(2-(4-nitrophenyl)sulfanyl**pyridine**)Cl_3]^{CDCl3} [154]	**1**	N^{py}: −146.1	−76.8[aal]	(Cl)[aaaa]
[Au^{III}(2-phenylsulfinyl**pyridine**)Cl_3]^{CDCl3} [154]	**1**	N^{py}: −151.5	−78.2[aam]	(Cl)[aaaa]
[Au^{III}(2-phenylselanyl**pyridine**)Cl_3]^{CDCl3} [154]	**1**	N^{py}: −147.3	−79.6[aan]	(Cl)[aaaa]
[Au^{III}(2-(1-(nitratomethyl)-2,2-dimethylpropylazanidocarbonyl)**pyridine**-κ^2-N^{py},N^N)Cl_2]^{CD3CN} [155][d,E]	**1**	N^{py}: −166.9	Unknown[aao1]	Cl $^{X\text{-}ray}$

[Au^{III}(2-(1-(trifluoroacetylomethyl)-2,2-dimethylpropylazanidocarbonyl)**pyridine**-κ^2-N^{py},N^N)Cl_2]^{CD3CN} [155][d]	**1**	N^{py}: −166.7	Unknown[aao2]	(Cl)[aaaa]
[Au^{III}(2-(1-(methoxymethyl)-2,2-dimethylpropylazanidocarbonyl)**pyridine**-κ^2-N^{py},N^N)Cl_2]^{CD3CN} [155][d]	**1**	N^{py}: −166.0	Unknown[aao3]	(Cl)[aaaa]
[Au^{III}(2-(1-(nitratomethyl)-2,2-dimethylpropylazanidocarbonyl)**pyridine**-κ^2-N^{py},N^N)Br_2]^{CD3CN} [155][d,F]	**1**	N^{py}: −167.0	Unknown[aao1]	Br [X-ray]
[Au^{III}(2-(1-(trifluoroacetylomethyl)-2,2-dimethylpropylazanidocarbonyl)**pyridine**-κ^2-N^{py},N^N)Br_2]^{CD3CN} [155][d,G]	**1**	N^{py}: −167.3	Unknown[aao2]	Br [X-ray]
[Au^{III}(2-(4-*tert*-butyl-4,5-dihydro-1,3-oxazol-2-yl)**pyridine**-κ^2-N^{py},N^{ox})Cl_2]NO_3^{CD2Cl2} [155]	**1**	N^{py}: −166.1	−98.4[aap]	(Cl)[aaaa]
[Au^{III}(2-(4-*tert*-butyl-4,5-dihydro-1,3-oxazol-2-yl)**pyridine**-κ^2-N^{py},N^{ox})Cl_2]NO_3^{CD3CN} [155]	**1**	N^{py}: −165.7	−100.0[aaq]	(Cl)[aaaa]
[Au^{III}(2-(4-*tert*-butyl-4,5-dihydro-1,3-oxazol-2-yl)**pyridine**-κ^2-N^{py},N^{ox})Cl_2]NO_3^{CD3OD} [155]	**1**	N^{py}: −166.1	−89.9[aar]	(Cl)[aaaa]
[Au^{III}(2-(4-*tert*-butyl-4,5-dihydro-1,3-oxazol-2-yl)**pyridine**-κ^2-N^{py},N^{ox})Cl_2]PF_6^{CD3CN} [155]	**1**	N^{py}: −153.3	−87.6[aaq]	(Cl)[aaaa]
[Au^{III}(2-(4-*tert*-butyl-4,5-dihydro-1,3-oxazol-2-yl)**pyridine**-κ^2-N^{py},N^{ox})Cl_2]PF_6^{CD3OD} [155]	**1**	N^{py}: −154.2	−78.0[aar]	(Cl)[aaaa]
[Au^{III}(2-(4-*tert*-butyl-4,5-dihydro-1,3-oxazol-2-yl)**pyridine**-κ^2-N^{py},N^{ox})Cl_2]BF_4^{CD3CN} [155][H]	**1**	N^{py}: −153.3	−87.6[aaq]	Cl [X-ray]

Continued

Table 6 Square-planar Au(III) (d^8) complexes and organometallics with azines (**1–3**, **7** ring systems), ^{15}N NMR studied in 2012–20.—cont'd

Compound	Parent ring system	^{15}N chemical shift (δ^{15N}_{compl}, ppm)	^{15}N coordination shift (Δ^{15N}_{coord}, ppm)	Nitrogen *trans* to
$[Au^{III}$(2-(4-*tert*-butyl-4,5-dihydro-1,3-oxazol-2-yl) **pyridine**-κ^2-N^{py},N^{ox})$Cl_2]BF_4$^{CD3OD} [155][H]	**1**	N^{py}: −152.9	−76.7[aar]	Cl [X-ray]
$[Au^{III}$(2-(4-*tert*-butyl-4,5-dihydro-1,3-oxazol-2-yl) **pyridine**-κ^2-N^{py},N^{ox})$Cl_2]CF_3COO$^{CD2Cl2} [155]	**1**	N^{py}: −165.3	−97.6[aap]	(Cl)[aaaa]
$[Au^{III}$(2-(4-*tert*-butyl-4,5-dihydro-1,3-oxazol-2-yl) **pyridine**-κ^2-N^{py},N^{ox})$Cl_2]CF_3COO$^{CD3CN} [155]	**1**	N^{py}: −165.8	−100.1[aaq]	(Cl)[aaaa]
$[Au^{III}$(2-(4-*tert*-butyl-4,5-dihydro-1,3-oxazol-2-yl) **pyridine**-κ^2-N^{py},N^{ox})$Cl_2]CF_3COO$^{CD3OD} [155]	**1**	N^{py}: −166.2	−90.0[aar]	(Cl)[aaaa]
$[Au^{III}$(2-(4-*tert*-butyl-4,5-dihydro-1,3-oxazol-2-yl) **pyridine**-κ^2-N^{py},N^{ox})$Cl_2]CF_3SO_3$^{CD2Cl2} [155]	**1**	N^{py}: −152.0	−84.3[aap]	(Cl)[aaaa]
$[Au^{III}$(2-(4-*tert*-butyl-4,5-dihydro-1,3-oxazol-2-yl) **pyridine**-κ^2-N^{py},N^{ox})$Cl_2]CF_3SO_3$^{CD3CN} [155]	**1**	N^{py}: −153.2	−87.5[aaq]	(Cl)[aaaa]
$[Au^{III}$(2-(4-*tert*-butyl-4,5-dihydro-1,3-oxazol-2-yl) **pyridine**-κ^2-N^{py},N^{ox})$Cl_2]CF_3SO_3$^{CD3OD} [155]	**1**	N^{py}: −153.0	−76.8[aar]	(Cl)[aaaa]
$[Au^{III}$(2-(4-*tert*-butyl-4,5-dihydro-1,3-oxazol-2-yl) **pyridine**-κ^2-N^{py},N^{ox})$Cl_2](CF_3SO_2)_2N$^{CD3CN} [155]	**1**	N^{py}: −153.1	−87.4[aaq]	(Cl)[aaaa]
$[Au^{III}$(2-(4-*tert*-butyl-4,5-dihydro-1,3-oxazol-2-yl) **pyridine**-κ^2-N^{py},N^{ox})$Cl_2](CF_3SO_2)_2N$^{CD3OD} [155]	**1**	N^{py}: −153.0	−76.8[aar]	(Cl)[aaaa]
$[Au^{III}$(2-(4-*tert*-butyl-4,5-dihydro-1,3-oxazol-2-yl) **pyridine**-κ^2-N^{py},N^{ox})$Cl_2]SbF_6$^{CD2Cl2} [155]	**1**	N^{py}: −152.7	−85.0[aap]	(Cl)[aaaa]
$[Au^{III}$(2-(4-*tert*-butyl-4,5-dihydro-1,3-oxazol-2-yl) **pyridine**-κ^2-N^{py},N^{ox})$Cl_2]SbF_6$^{CD3CN} [155]	**1**	N^{py}: −153.2	−87.5[aaq]	(Cl)[aaaa]

$[Au^{III}$(2-(4-*tert*-butyl-4,5-dihydro-1,3-oxazol-2-yl) **pyridine**-κ^2-N^{py},N^{ox})$Cl_2]SbF_6^{CD3OD}$ [155]	**1**	N^{py}: −153.0	−76.8[aar]	(Cl)[aaaa]
$[Au^{III}$(2-(4-*tert*-butyl-4,5-dihydro-1,3-oxazol-2-yl) **pyridine**-κ^2-N^{py},N^{ox})$Br_2]NO_3^{CD2Cl2}$ [155]	**1**	N^{py}: −165.5	−97.8[aap]	(Br)[aaaa]
$[Au^{III}$(2-(4-*tert*-butyl-4,5-dihydro-1,3-oxazol-2-yl) **pyridine**-κ^2-N^{py},N^{ox})$Br_2]NO_3^{CD3CN}$ [155]	**1**	N^{py}: −167.0	−101.3[aaq]	(Br)[aaaa]
$[Au^{III}$(2-(4-*tert*-butyl-4,5-dihydro-1,3-oxazol-2-yl) **pyridine**-κ^2-N^{py},N^{ox})$Br_2]PF_6^{CD2Cl2}$ [155]	**1**	N^{py}: −153.2	−85.5[aap]	(Br)[aaaa]
$[Au^{III}$(2-(4-*tert*-butyl-4,5-dihydro-1,3-oxazol-2-yl) **pyridine**-κ^2-N^{py},N^{ox})$Br_2]PF_6^{CD3CN}$ [155]	**1**	N^{py}: −152.9	−87.2[aaq]	(Br)[aaaa]
$[Au^{III}$(2-(4-*tert*-butyl-4,5-dihydro-1,3-oxazol-2-yl) **pyridine**-κ^2-N^{py},N^{ox})$Br_2]BF_4^{CD2Cl2}$ [155][I]	**1**	N^{py}: −152.5	−84.8[aap]	Br [X-ray]
$[Au^{III}$(2-(4-*tert*-butyl-4,5-dihydro-1,3-oxazol-2-yl) **pyridine**-κ^2-N^{py},N^{ox})$Br_2]BF_4^{CD3CN}$ [155][I]	**1**	N^{py}: −153.0	−87.3[aaq]	Br [X-ray]
$[Au^{III}$(2-(4-*tert*-butyl-4,5-dihydro-1,3-oxazol-2-yl) **pyridine**-κ^2-N^{py},N^{ox})$Br_2]BF_4^{CD3OD}$ [155][I]	**1**	N^{py}: −153.1	−76.9[aar]	Br [X-ray]
$[Au^{III}$(2-(4-*tert*-butyl-4,5-dihydro-1,3-oxazol-2-yl) **pyridine**-κ^2-N^{py},N^{ox})$Br_2]CF_3COO^{CD2Cl2}$ [155]	**1**	N^{py}: −165.4	−97.7[aap]	(Br)[aaaa]
$[Au^{III}$(2-(4-*tert*-butyl-4,5-dihydro-1,3-oxazol-2-yl) **pyridine**-κ^2-N^{py},N^{ox})$Br_2]CF_3COO^{CD3CN}$ [155]	**1**	N^{py}: −166.1	−100.4[aaq]	(Br)[aaaa]
$[Au^{III}$(2-(4-*tert*-butyl-4,5-dihydro-1,3-oxazol-2-yl) **pyridine**-κ^2-N^{py},N^{ox})$Br_2]CF_3COO^{CD3OD}$ [155]	**1**	N^{py}: −166.4	−90.2[aar]	(Br)[aaaa]

Continued

Table 6 Square-planar Au(III) (d^8) complexes and organometallics with azines (**1–3, 7** ring systems), ^{15}N NMR studied in 2012–20.—cont'd

Compound	Parent ring system	^{15}N chemical shift (δ^{15N}_{compl}, ppm)	^{15}N coordination shift (Δ^{15N}_{coord}, ppm)	Nitrogen *trans* to
$[Au^{III}$(2-(4-*tert*-butyl-4,5-dihydro-1,3-oxazol-2-yl) **pyridine**-κ²-N^{py},N^{ox})$Br_2]CF_3SO_3$^{CD2Cl2} [155]	**1**	N^{py}: −152.4	−84.7[aap]	(Br)[aaaa]
$[Au^{III}$(2-(4-*tert*-butyl-4,5-dihydro-1,3-oxazol-2-yl) **pyridine**-κ²-N^{py},N^{ox})$Br_2]CF_3SO_3$^{CD3CN} [155]	**1**	N^{py}: −152.8	−87.1[aaq]	(Br)[aaaa]
$[Au^{III}$(2-(4-*tert*-butyl-4,5-dihydro-1,3-oxazol-2-yl) **pyridine**-κ²-N^{py},N^{ox})$Br_2]CF_3SO_3$^{CD3OD} [155]	**1**	N^{py}: −151.0	−74.8[aar]	(Br)[aaaa]
$[Au^{III}$(2-(4-*tert*-butyl-4,5-dihydro-1,3-oxazol-2-yl) **pyridine**-κ²-N^{py},N^{ox})$Br_2](CF_3SO_2)_2N$^{CD2Cl2} [155]	**1**	N^{py}: −152.5	−84.8[aap]	(Br)[aaaa]
$[Au^{III}$(2-(4-*tert*-butyl-4,5-dihydro-1,3-oxazol-2-yl) **pyridine**κ²-N^{py},N^{ox})$Br_2](CF_3SO_2)_2N$^{CD3CN} [155]	**1**	N^{py}: −153.0	−87.3[aaq]	(Br)[aaaa]
$[Au^{III}$(2-(4-*tert*-butyl-4,5-dihydro-1,3-oxazol-2-yl) **pyridine**-κ²-N^{py},N^{ox})$Br_2](CF_3SO_2)_2N$^{CD3OD} [155]	**1**	N^{py}: −152.8	−76.6[aar]	(Br)[aaaa]
$[Au^{III}$(**2,2′-bipyridine**)$Cl_2]PF_6$$^{DMSO-d6}$ [152][J]	**2**	N^{bpy}: −151.4[aa2,ac]	−78.8[aas,aat]	Cl [X-ray]
$[Au^{III}$(4,4′-dimethyl-**2,2′-bipyridine**)$Cl_2]PF_6$$^{DMSO-d6}$ [152][K]	**2**	N^{bpy}: −157.4[aa3]	−75.8[aau]	Cl [X-ray]
$[Au^{III}$(**1,10-phenanthroline**)$Cl_2]PF_6$$^{DMSO-d6}$ [152][L,M]	**3**	N^{phen}: −137.4[aa4,ad]	−68.1[aav,aaw]	Cl [X-ray]
$[Au^{III}$(7,8-**benzoquinoline**)$Cl_3]$^{CD3CN} [156]	**7**	N^{bqn}: −136.1	−61.2[aax]	(Cl)[aaaa]
$[Au^{III}$(7,8-**benzoquinoline***-κ²-N,C^{10})$Cl_2]$$^{DMSO-d6}$ [156][e]	**7**	N^{bqn}: −143.3	−69.2[aay]	(Cl)[aaaa]

[a]These are probably *trans* complexes.

[b]2-Benzylpyridine* and 2-(4-methylphenyl)pyridine* are deprotonated (at the phenyl C(2) atom), anionic (charge −1) forms of 2-benzylpyridine and 2-(4-methylphenyl)pyridine, acting as bidentate (κ^2-N(1)^{py},C(2)^{ph}) ligands.

[c]NHC(CH_3)OCH_2CH_2 (i.e., NH=C(CH_3)–O–CH_2–CH_2) is anionic (charge −1) form of NH=C(CH_3)–O–CH_2–CH_3, deprotonated at methyl group, acting as bidentate (κ^2-N^{NH},C^{CH2}) ligand.

[d]Azanidocarbonylpyridines are anionic (charge −1) species, deprotonated at aliphatic nitrogen (–N–), acting as bidentate (κ^2-N(1)^{py},N^{N}) ligands.

[e]7,8-Benzoquinoline* is deprotonated (at C(10)), anionic (charge −1) form of 7,8-benzoquinoline, acting as bidentate (κ^2-N(1),C(10)) ligand.

[aa1-aa4]Original $\delta^{15N}{}_{compl}$ values (227.5 ppm ([aa1]), 230.3 ppm ([aa2]), 224.3 ppm ([aa3]), 244.3 ppm ([aa4])), primarily referenced to liquid NH_3 by de Paiva et al. [152] to liquid NH_3, were recalculated in respect to neat CH_3NO_2 by subtracting 381.7 ppm relative shift [37].

[ab]See footnote ([aab]).

[ac]See footnote ([aap]).

[ad]See footnote ([aas]).

[aaa]vs pyridine in CD_2Cl_2: −67.0 ppm [142].

[aab]vs 4-methylpyridine in CD_2Cl_2: −71.6 ppm [142].

[aac]vs 4-methoxypyridine in CD_2Cl_2: −86.0 ppm [142].

[aad]vs 4-trifluoromethylpyridine in CD_2Cl_2: −58.0 ppm [142].

[aae]vs 2-benzylpyridine in DMSO-d_6: $\delta^{15N}{}_{lig} = -60.9$ ppm [157].

[aaf]Previous literature value for [Au^{III}(2-benzylpyridine*)Cl_2], in DMSO-d_6: $\delta^{15N}{}_{compl} = -151.8$ ppm, $\Delta^{15N}{}_{coord} = -90.9$ ppm [157].

[aag]$\Delta^{15N}{}_{coord}$ cannot be determined due to the lack of $\delta^{15N}{}_{lig}$ for 2-(4-methylphenyl)pyridine.

[aah]vs 2-phenylsulfanylpyridine in $CDCl_3$: $\delta^{15N}{}_{lig} = -76.0$ ppm [154].

[aai]vs 2-(4-methylphenyl)sulfanylpyridine in $CDCl_3$: $\delta^{15N}{}_{lig} = -75.7$ ppm [154].

[aaj]vs 2-(4-*tert*-butylphenyl)sulfanylpyridine in $CDCl_3$: $\delta^{15N}{}_{lig} = -75.7$ ppm [154].

[aak]vs 2-(4-bromophenyl)sulfanylpyridine in $CDCl_3$: $\delta^{15N}{}_{lig} = -74.1$ ppm [154].

[aal]vs 2-(4-nitrophenyl)sulfanylpyridine in $CDCl_3$: $\delta^{15N}{}_{lig} = -69.3$ ppm [154].

[aam]vs 2-phenylsulfinylpyridine in $CDCl_3$: $\delta^{15N}{}_{lig} = -73.3$ ppm [154].

[aan]vs 2-phenylselanylpyridine in $CDCl_3$: $\delta^{15N}{}_{lig} = -67.7$ ppm [154].

[aao1-aao3]$\Delta^{15N}{}_{coord}$ cannot be determined due to the lack of $\delta^{15N}{}_{lig}$ for 2-(1-(nitratomethyl)-2,2-dimethylpropylaminocarbonyl)pyridine ([aao1]), 2-(1-(trifluoroacetylomethyl)-2,2-dimethylpropylaminocarbonyl)pyridine ([aao2]) and 2-(1-(methoxymethyl)-2,2-dimethylpropylaminocarbonyl)pyridine ([aao3]).

[aap]vs 2-(4-*tert*-butyl-4,5-dihydro-1,3-oxazol-2-yl)pyridine in CD_2Cl_2: $\delta^{15N}{}_{lig} = -67.7$ ppm [155].

[aaq]vs 2-(4-*tert*-butyl-4,5-dihydro-1,3-oxazol-2-yl)pyridine in CD_3CN: $\delta^{15N}{}_{lig} = -65.7$ ppm [155].

[aar]vs 2-(4-*tert*-butyl-4,5-dihydro-1,3-oxazol-2-yl)pyridine in CD_3OD: $\delta^{15N}{}_{lig} = -76.2$ ppm [155].

[aas]vs 2,2′-bipyridine in DMSO-d_6: $\delta^{15N}{}_{lig} = -72.6$ ppm [68,69,98].

[aat]Previous literature value for [Au^{III}(2,2′-bipyridine)Cl_2]$^+$ ($AuCl_4^-$ counterion), in DMSO-d_6: $\delta^{15N}{}_{compl} = -150.7$ ppm, $\Delta^{15N}{}_{coord} = -78.1$ ppm [69].

[aau]vs 4,4′-dimethyl-2,2′-bipyridine in DMSO-d_6: $\delta^{15N}{}_{lig} = -81.6$ ppm [130].

[aav]vs 1,10-phenanthroline in DMSO-d_6: $\delta^{15N}{}_{lig} = -69.3$ ppm [68,69,98].

Continued

Table 6 Square-planar Au(III) (d^8) complexes and organometallics with azines (**1–3**, **7** ring systems), ^{15}N NMR studied in 2012–20.—cont'd

[aaw]previous literature value for $[Au^{III}(1,10\text{-phenanthroline})Cl_2]^+$ ($AuCl_4^-$ counterion), in DMSO-d_6: $\delta^{15N}_{compl} = -151.1$ ppm, $\Delta^{15N}_{coord} = -81.8$ ppm [69].

[aax]vs 7,8-benzoquinoline in CD_3CN: $\delta^{15N}_{lig} = -74.9$ ppm [156].

[aay]vs 7,8-benzoquinoline in DMSO-d_6: $\delta^{15N}_{lig} = -74.1$ ppm [156].

[aaaa]The only stereomeric possibility, resulting from the molecular formula.

[A]X-ray structure: HOSHEE [158].

[B]X-ray structure: IPISEH [153].

[C]X-ray structure: DUCYEI [154].

[D]X-ray structure: DUCYIM [154].

[E]X-ray structure: LOYYIK [155].

[F]X-ray structure: LOYYAC [155].

[G]X-ray structure: LOYYOQ [155].

[H]X-ray structure of CH_3CN solvate: LOYXUV [155].

[I]X-ray structure of H_2O solvate: LOYYEG [155].

[J]X-ray structure: YIDLOP (YIDLOP01) [152,159].

[K]X-ray structure: KUMYEX [160].

[L]X-ray structure of analogous Cl^- salt: XIZDOC [159], also its H_2O solvate: QIRRAK [161].

[M]X-ray structure of analogous ClO_4^- salt: XIZFIY [159].

Table 7 Zn(II), Cd(II) and Hg(II) (d^{10}) complexes with azines (**1**, **3**, **5**, **11** ring systems), ^{15}N NMR studied in 2012–20.

Compound	Parent ring system	^{15}N chemical shift (δ^{15N}_{compl}, ppm)	^{15}N coordination shift (Δ^{15N}_{coord}, ppm)
Zn(II) compounds			
$[Zn^{II}$(2-(1-(**pyrid-2-yl**)-5-(4-methoxyphenyl)-*1H*-pyrazol-3-yl)**pyridine**-κ^3-N^{pyA},N^{pyB},$N^{pz,N2}$)$Cl_2]^{CDCl3}$ [11][a,b,A,#]	**1**	N^{pyA}: −106.3	−38.5[aaa]
		N^{pyB}: −105.8	−24.8[aab]
$[Zn^{II}$(2-(1-(**pyrid-2-yl**)-5-(4-methoxyphenyl)-*1H*-pyrazol-3-yl)**pyridine**-κ^3-N^{pyA},N^{pyB},$N^{pz,N2}$)$Cl_2]^{solid}$ [11][a,b,A,#]	**1**	N^{pyA}: −107.8	−40.0[aaa]
		N^{pyB}: −100.1	−19.1[aab]
$[Zn^{II}$(**1,10-phenanthroline**-5,6-dione)$_3](CF_3SO_3)_2^{CD3CN}$ [162][c]	**3**	N^{phen}: −131[aa]	−62[aac]
$[Zn^{II}$(8-(4-methylbenzenesulfonamidato-1,2-phenylenemethylimino)**quinoline**-κ^3-N^{quin},$N^{N=}$,N^{N-}) $(CH_3COO)]^{DMSO\text{-}d6}$ [163][d]	**5**	N^{quin}: −127.1[ab1]	−52.5[aad]
Cd(II) compounds			
$[Cd^{II}$(8-(4-methylbenzenesulfonamidato-1,2-phenylenemethylimino)**quinoline**-κ^4-N^{quin},$N^{N=}$,N^{N-},O^{SO2}) (8-(4-methylbenzenesulfonamidato-1,2-phenylenemethylimino)**quinoline**-κ^3-N^{quin},$N^{N=}$,N^{N-})$]^{DMSO\text{-}d6}$ [163][e,B]	**5**	N^{quin}: −108.7[ab2]	−34.1[aad]
trans$(NO_3,NO_3)(H_2O,H_2O)$-$[Cd^{II}$(2-amino-4-methyl**pyrimidine**-κ^1-$N^{pm,N1}$)$_2(H_2O)_2(NO_3)_2]^{solid}$ [164][c,C]	**11**	$N^{pm,N1}$: −194	−56[aae]
		$N^{pm,N3}$: −160[ac]	Not applicable
Hg(II) compounds	No interesting species		

[#] See also Table 13 for pyrazole-derived ligand(s).
[a] Coordination number 5 (trigonal-bipyramidal geometry).
[b] In this complex both pyridine rings, i.e., at pyrazolic C(3) atom (pyridine A) and at pyrazolic N(1) atom (pyridine B), are bound to Zn(II).
[c] Coordination number 6 (octahedral geometry; $N^{pm,N1}$ *trans* to N).
[d] Coordination number 4 (tetrahedral geometry).
[e] Coordination number 7 (pentagonal-bipyramidal geometry).
[aa-ab] Original δ^{15N}_{compl} values, primarily referenced to liquid NH_3 by the following authors, were recalculated in respect to neat CH_3NO_2 by subtracting 381.7 ppm (or 382 ppm) relative shift [37]:
[aa] 251 ppm by Wendlandt et al. [162].
[ab1-ab2] 254.6 ppm ([ab1]), 273.0 ppm ([ab2]) by Burlov et al. [163].
[ac] This δ^{15N}_{compl} concerns uncoordinated nitrogen.
[aaa] vs pyridine-A ring (at pyrazolic C(3) atom) in solid 2-(1-(pyrid-2-yl)-5-(4-methoxyphenyl)-1*H*-pyrazol-3-yl)pyridine: $\delta^{15N}_{lig} = -67.8$ ppm [11].
[aab] vs pyridine-B ring (at pyrazolic N(1) atom) in solid 2-(1-(pyrid-2-yl)-5-(4-methoxyphenyl)-1*H*-pyrazol-3-yl)pyridine: $\delta^{15N}_{lig} = -81.0$ ppm [11].
[aac] vs 1,10-phenanthroline-5,6-dione in CD_3CN: $\delta^{15N}_{lig} = -69$ ppm (313 ppm in respect to liquid NH_3) [162].
[aad] vs 8-(4-methylbenzenesulfonamido-1,2-phenylenemethylimino)quinoline in DMSO-d_6: $\delta^{15N}_{lig} = -74.6$ ppm (307.1 ppm in respect to liquid NH_3) [163].
[aae] vs N(1) in 2-amino-4-methylpyrimidine in DMSO-d_6 (^{15}N NMR for neat 2-amino-4-methylpyrimidine being unavailable): $\delta^{15N}_{lig} = -138.2$ ppm [77].
[A] X-ray structure: GOCROH [11].
[B] X-ray structure of $CH_2Cl_2{\cdot}H_2O$ solvate: BUJJUN [163].
[C] X-ray structure: UVOROO [164].

Table 8 Square-planar Ni(II), Pd(II) and Pt(II) (d^8), and octahedral Pt(IV) (d^6, low-spin) complexes and organometallics with azoles (**21–22**, **25**, **27–28** ring systems), ^{15}N NMR studied in 2012–20 *

Compound	Parent ring system	^{15}N chemical shift (δ^{15N}_{compl}, ppm)	^{15}N coordination shift (Δ^{15N}_{coord}, ppm)	Nitrogen *trans* to
Ni(II) compounds	No interesting species			
Pd(II) compounds				
$[Pd^{II}$(2-(1-(pyrid-2-yl)-5-(4-methoxyphenyl)-*1H*-**pyrazol**-3-yl)pyridine-κ^2-$N^{pyA},N^{pz,N2})Cl_2]^{solid}$ [11][A,#]	**21**	$N^{pz,N2}$: −159.7	−76.3[aaa]	Cl [X-ray]
trans-$[Pd^{II}$(2-(thiophen-2-yl)-*1H*-**imidazole**-κ^1-$N^3)_2Cl_2]^{DMF\text{-}d7}$ [29][B]	**22**	$N^{im,N3}$: −214.3[aa1]	−90.2[aab]	$N^{im,N3}$ [X-ray]
trans-$[Pd^{II}$(2-(thiophen-2-yl)-*1H*-**imidazole**-κ^1-$N^3)_2Cl_2]^{solid}$ [29][B]	**22**	$N^{im,N3}$: −220.2[aa2]	−80.5[aac]	$N^{im,N3}$ [X-ray]
$[Pd^{II}$(((1-(5-phenyl-(1,2-oxazol-3-yl)methyl)-*1H*-**1,2,3-triazol**-4-yl) methanol)-κ^1-$N^{tz,N3})(\mu\text{-}Cl)Cl]_2^{DMF\text{-}d7}$ [30]	**25**	$N^{tz,N3}$: −154[ab1]/−134[ab2]	Unknown[aad]	(Cl)[aaaa]
$[Pd^{II}$(((1-(5-phenyl-(1,2-oxazol-3-yl)methyl)-*1H*-**1,2,3-triazol**-4-yl)methyl 2,2-dimethylpropanoate)-κ^1-$N^{tz,N3})(\mu\text{-}Cl)Cl]_2^{DMF\text{-}d7}$ [30]	**25**	$N^{tz,N3}$: −147[ab3]/−129[ab4]	−120/−102[aae]	(Cl)[aaaa]
$[Pd^{II}$(2-(*1H*-**benzotriazol**-1-yl)pyridine-κ^2-$N^{py},N^{btz,N2})Cl_2]^{DMSO\text{-}d6}$ [10][#]	**27**	$N^{btz,N2}$: unknown[ad]	−26[aaf]	(Cl)[aaaa]
$[Pd^{II}$(1,3-bis-(2-(diphenylphosphino)ethyl)-**4,5,6,7-tetrahydrobenzotriazolium**-κ^3-N^2,P,P′)Cl]$(CF_3SO_3)_2^{CD3NO2}$ [31][a2,C]	**28**	$N^{thbtz,N2}$: −125.5	−80.5[aag]	Cl [X-ray]
$[Pd^{II}$(1,3-bis-(2-(diphenylphosphino)ethyl))-**4,5,6,7-tetrahydrobenzotriazolium**-κ^3-N^2,P,P′)$(CH_3)](CF_3SO_3)_2^{CD3NO2}$ [31][a2]	**28**	$N^{thbtz,N2}$: −95.3	−50.3[aag]	(CH_3)[aaaa]
$[Pd^{II}$(1,3-bis-(2-(diphenylphosphino)ethyl))-**4,5,6,7-tetrahydrobenzotriazolium**-κ^3-N^2,P,P′)$(P(C_6H_5)_3)](CF_3SO_3)_3^{CD3NO2}$ [31][a2]	**28**	$N^{thbtz,N2}$: −115.0	−70.0[aag]	(PPh_3)[aaaa]

Pt(II) compounds				
cis-[Pt^{II}(2-(thiophen-2-yl)-*1H*-**imidazole**-κ^1-N^3)$_2Cl_2$]^{solid} [29][b]	**22**	$N^{im,N3}$: −219.2[aa3]	−79.5[aac]	Cl[b]
[Pt^{II}(2-phenyl-6-(1-(12-hydroxo-*n*-dodecyl)-1*H*-**1,2,3-triazol**-4-yl)pyridine*-κ^3-N^{py},$C^{ph,C2}$,$N^{tz,N3}$)(pyridine)]$PF_6$$^{acetone-d6}$ [22][c,#]	**25**	$N^{tz,N3}$: −85[ac1]	Unknown[aah1]	($C^{ph,C2}$)[aaaa]
[Pt^{II}(2-phenyl-6-(1-(12-trimethylammonioethoxy-*n*-dodecyl)-1*H*-**1,2,3-triazol**-4-yl)pyridine*-κ^3-N^{py},$C^{ph,C2}$,$N^{tz,N3}$)(pyridine)]$(PF_6)_2$$^{acetone-d6}$ [22][d,#]	**25**	$N^{tz,N3}$: −86[ac2]	Unknown[aah2]	($C^{ph,C2}$)[aaaa]
[Pt^{II}(1-β-D-ribofuranosyl-4-(pyrid-2-yl)-*1H*-**1,2,3-triazole**-κ^2-N^{py},$N^{tz,N3}$)Cl_2]$^{DMF-d7}$ [23][#]	**25**	$N^{tz,N3}$: −129.6	−98.4[aai]	(Cl)[aaaa]
[Pt^{II}(1,3-bis-(2-(diphenylphosphino)ethyl)**benzotriazolium**-κ^3-N^2,P,P′)Cl]$(CF_3SO_3)_2$^{CD3NO2} [32][a1]	**27**	$N^{btz,N2}$: −134.7	−105.8[aaj]	(Cl)[aaaa]
[Pt^{II}(1,3-bis-(2-(diphenylphosphino)ethyl)**benzotriazolium**-κ^3-N^2,P,P′)(CH_3)]$(CF_3SO_3)_2$^{CD3NO2} [32][a1]	**27**	$N^{btz,N2}$: −92.7	−63.8[aaj]	(CH_3)[aaaa]
[Pt^{II}(1,3-bis-(2-(diphenylphosphino)ethyl)**benzotriazolium**-κ^3-N^2,P,P′)($P(C_6H_5)_3$)]$(CF_3SO_3)_3$^{CD3NO2} [32][a1]	**27**	$N^{btz,N2}$: −113.5	−84.6[aaj]	(PPh_3)[aaaa]
[Pt^{II}(1,3-bis-(2-(diphenylphosphino)ethyl))-**4,5,6,7-tetrahydrobenzotriazolium**-κ^3-N^2,P,P′)Cl]$(CF_3SO_3)_2$^{CD3NO2} [32][a2,D]	**28**	$N^{thbtz,N2}$: −152.9	−107.9[aag]	Cl [X-ray]
[Pt^{II}(1,3-bis-(2-(diphenylphosphino)ethyl))-**4,5,6,7-tetrahydrobenzotriazolium**-κ^3-N^2,P,P′)(CH_3)]$(CF_3SO_3)_2$^{CD3NO2} [32][a2,E]	**28**	$N^{thbtz,N2}$: −108.5	−63.5[aag]	CH_3 [X-ray]
[Pt^{II}(1,3-bis-(2-(diphenylphosphino)ethyl))-**4,5,6,7-tetrahydrobenzotriazolium**-κ^3-N^2,P,P′)($P(C_6H_5)_3$)]$(CF_3SO_3)_3$^{CD3NO2} [32][a2,F]	**28**	$N^{thbtz,N2}$: −129.9	−84.9[aag]	PPh_3 [X-ray]

Continued

Table 8 Square-planar Ni(II), Pd(II) and Pt(II) (d^8), and octahedral Pt(IV) (d^6, low-spin) complexes and organometallics with azoles (**21–22**, **25**, **27–28** ring systems), ^{15}N NMR studied in 2012–20 *—cont'd

Compound	Parent ring system	^{15}N chemical shift (δ^{15N}_{compl}, ppm)	^{15}N coordination shift (Δ^{15N}_{coord}, ppm)	Nitrogen *trans* to
Pt(IV) compounds				
$[Pt^{IV}$(1,3-bis-(2-(diphenylphosphino)ethyl)**benzotriazolium**-κ^3-N^2,P,P′)$Cl_3](CF_3SO_3)_2$^{CD3NO2} [31][a1]	**27**	$N^{btz,N2}$: −150.0	−121.1[aaj]	(Cl)[aaaa]
$[Pt^{IV}$(1,3-bis-(2-(diphenylphosphino)ethyl))-**4,5,6,7-tetrahydrobenzotriazolium**-κ^3-N^2,P,P′)$Cl_3](CF_3SO_3)_2$^{CD3NO2} [31][a2]	**28**	$N^{thbtz,N2}$: −167.8	−122.8[aag]	(Cl)[aaaa]

*Except for Pt(II) complexes with purines and Pt(II), Pt(IV) complexes with 1,2,4-triazolo-[1,5*a*]-pyrimidines which are listed in Tables 14 and 15, respectively.
#See also Table 1 or Table 2 for pyridine ligand(s).
[a1-a2]1,3-Bis-(2-(diphenylphosphino)ethyl)benzotriazolium ([a1]) and 1,3-bis-(2-(diphenylphosphino)ethyl)-4,5,6,7-tetrahydrobenzotriazolium ([a2]) are cationic ligands, due to the presence of a delocalized positive charge (+1) at all triazolic nitrogens.
[b]This is probably *cis* isomer although authors suggested *trans* geometry.
[c]2-Phenyl-6-(1-(12-hydroxo-*n*-dodecyl)-*1H*-1,2,3-triazol-4-yl)pyridine* is deprotonated (at the phenyl C(2) atom), anionic (charge −1) form of 2-phenyl-6-(1-(12-hydroxo-*n*-dodecyl)-*1H*-1,2,3-triazol-4-yl)pyridine, acting as tridentate (κ^3-$N(1)^{py}$,$C(2)^{ph}$,$N(3)^{tz}$) ligand.
[d]2-Phenyl-6-(1-(12-trimethylammonioethoxy-*n*-dodecyl)-*1H*-1,2,3-triazol-4-yl)pyridine* is zwitterion, having (+1) charge at quarternary nitrogen in the trimethylammonio-moiety and (−1) charge at the deprotonated phenyl C(2) atom, acting as tridentate (κ^3-$N(1)^{py}$,$C(2)^{ph}$,$N(3)^{tz}$) ligand.
[aa-ac]Original δ^{15N}_{compl} values, primarily referenced to liquid NH_3 by the following authors, were recalculated in respect to neat CH_3NO_2 by subtracting 381.7 ppm (or 382 ppm) relative shift [37]:
[aa1-aa3]167.4 ppm ([aa1]), 161.5 ppm ([aa2]), 162.5 ppm ([aa3]) by Zanvettor et al. [29].
[ab1-ab4]228 ppm ([ab1]), 248 ppm ([ab2]), 235 ppm ([ab3]), 253 ppm ([ab4]) by Bumagin et al. [30].
[ac1-ac2]297 ppm ([ad1]), 296 ppm ([ad2]) by Hebenbrock et al. [22].

[ad]Original $\delta^{15N}{}_{compl}$ value (22 ppm) [10] cannot be recalculated in respect to neat CH_3NO_2 due to the unknown ^{15}N NMR reference.
[aaa]vs N(2) in pyrazole moiety in solid 2-(1-(pyrid-2-yl)-5-(4-methoxyphenyl)-*1H*-pyrazol-3-yl)pyridine: $\delta^{15N}{}_{lig} = -83.4$ ppm [11].
[aab]vs N(3) in 2-(thiophen-2-yl)-*1H*-imidazole in DMF-d_7: $\delta^{15N}{}_{lig} = -124.1$ ppm (257.6 ppm in respect to liquid NH_3) [29].
[aac]vs N(3) in solid 2-(thiophen-2-yl)-*1H*-imidazole: $\delta^{15N}{}_{lig} = -139.7$ ppm (242.0 ppm in respect to liquid NH_3) [29].
[aad]$\Delta^{15N}{}_{coord}$ cannot be determined due to the lack of $\delta^{15N}{}_{lig}$ for (1-(5-phenyl-(1,2-oxazol-3-yl)methyl)-*1H*-1,2,3-triazol-4-yl)methanol (N(3) in 1,2,3-triazole moiety remained undetected) [30].
[aae]vs N(3) in 1,2,3-triazole moiety in (1-(5-phenyl-(1,2-oxazol-3-yl)methyl)-1*H*-1,2,3-triazol-4-yl)methyl 2,2-dimethylpropanoate in DMF-d_7: $\delta^{15N}{}_{lig} = -27$ ppm (355 ppm in respect to liquid NH_3) [30].
[aaf]$\Delta^{15N}{}_{coord}$ determined as difference of ca. −48 ppm (for $N^{btz,N3}$ in $[Pd^{II}$(2-(*1H*-benzotriazol-1-yl)pyridine)$Cl_2]$) and ca. −22 ppm (for $N^{btz,N2}$ in 2-(*1H*-benzotriazol-1-yl)pyridine), these values being published in respect to the unknown ^{15}N NMR reference standard [10].
[aag]vs N(2) in tetrahydrobenzotriazole moiety in (1,3-bis-(2-(diphenylphosphino)ethyl)-4,5,6,7-tetrahydrobenzotriazolium)$^+$ cation ($CF_3SO_3^-$ counterion) in $CDCl_3$ (^{15}N NMR data in CD_3NO_2 being unavailable): $\delta^{15N}{}_{lig} = -45.0$ ppm [32].
[aah1-aah2]$\Delta^{15N}{}_{coord}$ cannot be determined due to the lack of $\delta^{15N}{}_{lig}$ for 2-phenyl-6-(1-(12-hydroxo-*n*-dodecyl)-*1H*-1,2,3-triazol-4-yl)pyridine ([aah1]) and (2-phenyl-6-(1-(12-trimethylammonioethoxy-*n*-dodecyl)-*1H*-1,2,3-triazol-4-yl)pyridine)$^+$ cation ([aah2]).
[aai]vs N(3) in 1,2,3-triazole moiety in 1-β-D-ribofuranosyl-4-(pyrid-2-yl)-*1H*-1,2,3-triazole in DMF-d_7: $\delta^{15N}{}_{lig} = -31.2$ ppm [23].
[aaj]vs N(2) in benzotriazole moiety in (1,3-bis-(2-(diphenylphosphino)ethyl)benzotriazolium)$^+$ cation ($CF_3SO_3^-$ counterion) in $CDCl_3$ (^{15}N NMR data in CD_3NO_2 being unavailable): $\delta^{15N}{}_{lig} = -28.9$ ppm [32].
[aaaa]The only stereomeric possibility, resulting from the molecular formula.
[A]X-ray structure of CH_2Cl_2 solvate: GOCRUN [11].
[B]X-ray structure: KUYYAG [29].
[C]X-ray structure of CH_2Cl_2 solvate: UGUGEK [31].
[D]X-ray structure of CH_2Cl_2 solvate: VIXHER [32].
[E]X-ray structure of CH_2Cl_2 solvate: VIXGOA [32].
[F]X-ray structure of the mixed salt with CF_3SO_3K, $CH_3CH_2OCH_2CH_3$ solvate: VIXGUG [32].

Table 9 Square-planar Rh(I) and Ir(I) (d^8) complexes and organometallics with azoles (**21**, **25**, **27–28** ring systems), ^{15}N NMR studied in 2012–20.

Compound	Parent ring system	^{15}N chemical shift (δ^{15N}_{compl}, ppm)	^{15}N coordination shift (Δ^{15N}_{coord}, ppm)	Nitrogen *trans* to
Rh(I) compounds				
trans(Cl,C=C^{vinyl})-[Rh^I(1-vinyl-3,5-dimethyl-*1H*-**pyrazole**-κ^2-N^2,$C^{C=C}$)(1,3-bis(2,6-diisopropylphenyl)imidazol-2-ylidene-κ^1-C^2)Cl]^{C6D6} [56][A]	**21**	$N^{pz,N2}$: −146.1[aa]	−66.0[aaa]	$C^{ylidene,C2}$ [X-ray]
[Rh^I(1,3-bis-(2-(diphenylphosphino)ethyl)-*1H*-**1,2,3-triazolium**-κ^3-N^2,P,P′)(CO)]$(PF_6)_2$^{CD3NO2} [31][a1,B]	**25**	$N^{tz,N2}$: −99.2	−66.3[aab]	CO [X-ray]
[Rh^I(1,3-bis-(2-(diphenylphosphino)ethyl)**benzotriazolium**-κ^3-N^2,P,P′)Cl]PF_6^{CD3NO2} [32][a2]	**27**	$N^{btz,N2}$: −92.7	−63.8[aac]	(Cl)[aaaa]
[Rh^I(1,3-bis-(2-(diphenylphosphino)ethyl)**benzotriazolium**-κ^3-N^2,P,P′)I]PF_6^{CD3NO2} [32][a2]	**27**	$N^{btz,N2}$: −94.1	−65.2[aac]	(I)[aaaa]
[Rh^I(1,3-bis-(2-(diphenylphosphino)ethyl)**benzotriazolium**-κ^3-N^2,P,P′)(CO)]$(PF_6)_2$^{CD3NO2} [31][a2]	**27**	$N^{btz,N2}$: −92.1	−63.2[aac]	(CO)[aaaa]
[Rh^I(1,3-bis-(2-(diphenylphosphino)ethyl)**benzotriazolium**-κ^3-N^2,P,P′)(tetrahydrofurane)]$(PF_6)_2$$^{THF-d8}$ [32][a2]	**27**	$N^{btz,N2}$: −92.3	−63.4[aac]	(O)[aaaa]
[Rh^I(1,3-bis-(2-(diphenylphosphino)ethyl)**benzotriazolium**-κ^3-N^2,P,P′)(CH_2CH_2)]$(PF_6)_2$^{CD3NO2} [32][a2]	**27**	$N^{btz,N2}$: −93.5	−64.6[aac]	(C=C^{CH2CH2})[aaaa]
[Rh^I(1,3-bis-(2-(diphenylphosphino)ethyl)**benzotriazolium**-κ^3-N^2,P,P′)($P(C_6H_5)_3$)]$(PF_6)_2$^{CD3NO2} [32][a2,C]	**27**	$N^{btz,N2}$: −81.4	−52.5[aac]	PPh_3 [X-ray]

[Rh^{I}(1,3-bis-(2-(diphenylphosphino)ethyl)-**4,5,6,7-tetrahydrobenzotriazolium**-κ^{3}-N^{2},P,P′)Cl]PF_6^{CD3NO2} [32][a3]	**28**	N^{btz,N2}: −103.1	−58.3[aad]	(Cl)[aaaa]
[Rh^{I}(1,3-bis-(2-(diphenylphosphino)ethyl)-**4,5,6,7-tetrahydrobenzotriazolium**-κ^{3}-N^{2},P,P′)(CO)]$(PF_6)_2$^{CD3NO2} [31][a3,D]	**28**	N^{btz,N2}: −110.1	−65.3[aad]	CO $^{X\text{-}ray}$
[Rh^{I}(1,3-bis-(2-(diphenylphosphino)ethyl)-**4,5,6,7-tetrahydrobenzotriazolium**-κ^{3}-N^{2},P,P′)(tetrahydrofurane)]$(PF_6)_2$^{CD2Cl2} [32][a3]	**28**	N^{btz,N2}: −117.5	−72.7[aad]	(O)[aaaa]
[Rh^{I}(1,3-bis-(2-(diphenylphosphino)ethyl)-**4,5,6,7-tetrahydrobenzotriazolium**-κ^{3}-N^{2},P,P′)(CH_2CH_2)]$(PF_6)_2$^{CD3NO2} [32][a3,E]	**28**	N^{btz,N2}: −109.8	−65.0[aad]	C=C^{CH2CH2} X-ray
[Rh^{I}(1,3-bis-(2-(diphenylphosphino)ethyl)-**4,5,6,7-tetrahydrobenzotriazolium**-κ^{3}-N^{2},P,P′)(P$(C_6H_5)_3$)]$(PF_6)_2$^{CD3NO2} [32][a3]	**28**	N^{btz,N2}: −99.7	−54.9[aad]	(PPh_3)[aaaa]
Ir(I) compounds				
[Ir^{I}(1,3-bis-(2-(diphenylphosphino)ethyl)**benzotriazolium**-κ^{3}-N^{2},P,P′)Cl]PF_6^{CD3NO2} [169][a2]	**27**	N^{btz,N2}: −131.0	−102.1[aac]	(Cl)[aaaa]
[Ir^{I}(1,3-bis-(2-(diphenylphosphino)ethyl)**benzotriazolium**-κ^{3}-N^{2},P,P′)(CO)]$(PF_6)_2$^{CD3NO2} [169][a2,F]	**27**	N^{btz,N2}: −108.7	−79.8[aac]	CO $^{X\text{-}ray}$
[Ir^{I}(1,3-bis-(2-(diphenylphosphino)ethyl)**benzotriazolium**-κ^{3}-N^{2},P,P′)(CH_3CN)]$(PF_6)_2$^{CD3NO2} [169][a2]	**27**	N^{btz,N2}: −118.5	−89.6[aac]	(N^{CH3CN})[aaaa]

Continued

Table 9 Square-planar Rh(I) and Ir(I) (d^8) complexes and organometallics with azoles (**21**, **25**, **27–28** ring systems), ^{15}N NMR studied in 2012–20.—cont'd

[a1-a3]1,3-Bis-(2-(diphenylphosphino)ethyl)-*1H*-1,2,3-triazolium ([a1]), 1,3-bis-(2-(diphenylphosphino)ethyl)benzotriazolium ([a2]) and 1,3-bis-(2-(diphenylphosphino)ethyl)-4,5,6,7-tetrahydrobenzotriazolium ([a3]) are cationic ligands, due to the presence of a delocalized positive charge (+1) at all triazolic nitrogens.

[aa]Original $\delta^{15N}{}_{compl}$ value (235.6 ppm), primarily referenced to liquid NH_3 by Azpiroz et al. [56], was recalculated in respect to neat CH_3NO_2 by subtracting 381.7 ppm relative shift [37].

[aaa]vs N(2) in 1-vinyl-3,5-dimethyl-*1H*-pyrazole in $CDCl_3$ (^{15}N NMR data in C_6D_6 being unavailable): $\delta^{15N}{}_{lig} = -80.1$ ppm (301.6 ppm in respect to liquid NH_3) [170].

[aab]vs N(2) in 1,2,3-triazole moiety in (1,3-bis-(2-(diphenylphosphino)ethyl)-*1H*-1,2,3-triazolium)$^+$ cation (Br^- counterion) in $CDCl_3$ (^{15}N NMR data in case of PF_6^- counterions and/or in CD_3NO_2 being unavailable): $\delta^{15N}{}_{lig} = -32.9$ ppm [31].

[aac]vs N(2) in benzotriazole moiety in (1,3-bis-(2-(diphenylphosphino)ethyl)benzotriazolium)$^+$ cation ($CF_3SO_3^-$ counterions) in $CDCl_3$ (^{15}N NMR data in case of PF_6^- counterion and/or in CD_3NO_2 being unavailable): $\delta^{15N}{}_{lig} = -28.9$ ppm [32].

[aad]vs N(2) in tetrahydrobenzotriazole moiety in (1,3-bis-(2-(diphenylphosphino)ethyl)-4,5,6,7-tetrahydrobenzotriazolium)$^+$ cation (PF_6^- counterion) in $CDCl_3$ (^{15}N NMR data in CD_3NO_2 or CD_2Cl_2 being unavailable): $\delta^{15N}{}_{lig} = -44.8$ ppm [32].

[aaaa]The only stereomeric possibility, resulting from the molecular formula.

[A]X-ray structure: GUVQIZ [56].

[B]X-ray structure: UGUFAF [31].

[C]X-ray structure of THF solvate: VIXHIV [32].

[D]X-ray structure of $CHCl_3$ solvate: UGUFEJ [31].

[E]X-ray structure of THF·H_2O solvate: VIXHAN [32].

[F]X-ray structure of THF solvate: GOGNAU [169].

Table 10 Octahedral Co(III), Rh(III) and Ir(III) (d^6, low-spin) complexes and organometallics with azoles (**22**, **25**, **27** ring systems), ^{15}N NMR studied in 2012–20.

Compound	Parent ring system	^{15}N chemical shift (δ^{15N}_{compl}, ppm)	^{15}N coordination shift (Δ^{15N}_{coord}, ppm)	Nitrogen *trans* to
Co(III) compounds	No interesting species			
Rh(III) compounds				
$[Rh^{III}$(*1H,1′H*-2,2′-bi**imidazole**-κ^2-N^3,$N^{3'}$)(pentamethylcyclopentadienyl)Cl]$PF_6$$^{DMSO-d6}$ [125][a,A]	**22**	2 × $N^{im,N3,N3'}$: −173.8	Unknown[aaa]	C=C[dienyl X-ray]
$[Rh^{III}$(1-benzyl-4-phenyl-*1H*-**1,2,3-triazole***-κ^2-N^3,$C^{ph,C2}$)(pentamethycyclopentadienyl)Cl]^{CDCl3} [171][a,b,B]	**25**	$N^{tz,N3}$: −89.6[aa]	−52[aab]	C=C[dienyl X-ray]
$[Rh^{III}$(1,3-bis-(2-(diphenylphosphino)ethyl)**benzotriazolium-**κ^3-N^2,P,P′)Cl_3]PF_6^{CD3NO2} [31][c,C]	**27**	$N^{btz,N2}$: −113.1	−84.2[aac]	Cl [X-ray]
$[Rh^{III}$(1,3-bis-(2-(diphenylphosphino)ethyl)**benzotriazolium-**κ^3-N^2,P,P′)I_3]PF_6^{CD3NO2} [31][c]	**27**	$N^{btz,N2}$: −118.1	−89.2[aac]	(I)[aaaa]
trans-$[Rh^{III}$(1,3-bis-(2-(diphenylphosphino)ethyl) **benzotriazolium-**κ^3-N^2,P,P′)(CH_3)I_2]PF_6^{CD3NO2} [31][c,D]	**27**	$N^{btz,N2}$: −76.9	−48.0[aac]	CH_3 [X-ray]
cis-$[Rh^{III}$(1,3-bis-(2-(diphenylphosphino)ethyl) **benzotriazolium-**κ^3-N^2,P,P′)Cl$(H)_2$]PF_6^{CD3NO2} [169][c,d1]	**27**	$N^{btz,N2}$: −70.4	−41.5[aac]	(H)[d1]
Ir(III) compounds				
$[Ir^{III}$(*1H,1′H*-2,2′-bi**imidazole**-κ^2-N^3,$N^{3'}$)(pentamethylcyclopentadienyl)Cl]$PF_6$$^{DMSO-d6}$ [125][a,E]	**22**	2 × $N^{im,N3,N3'}$: −189.7	Unknown[aaa]	C=C[dienyl X-ray]
cis-$[Ir^{III}$(1,3-bis-(2-(diphenylphosphino)ethyl) **benzotriazolium-**κ^3-N^2,P,P′)Cl$(H)_2$]PF_6^{CD3NO2} [169][c,d1]	**27**	$N^{btz,N2}$: −98.3	−69.4[aac]	(H)[d1]

Continued

Table 10 Octahedral Co(III), Rh(III) and Ir(III) (d^6, low-spin) complexes and organometallics with azoles (**22, 25, 27** ring systems), ^{15}N NMR studied in 2012–20.—cont'd

Compound	Parent ring system	^{15}N chemical shift (δ^{15N}_{compl}, ppm)	^{15}N coordination shift (Δ^{15N}_{coord}, ppm)	Nitrogen *trans* to
cis-[Ir^{III}(1,3-bis-(2-(diphenylphosphino)ethyl) **benzotriazolium-κ**^{3}-N^2,P,P′)(CO)$(H)_2$]$(PF_6)_2$[CD3NO2] [169][c,d2,F]	**27**	$N^{btz,N2}$: −122.7	−93.8[aac]	H [X-ray]
cis-[Ir^{III}(1,3-bis-(2-(diphenylphosphino)ethyl) **benzotriazolium-κ**^{3}-N^2,P,P′)(CH_3CN)$(H)_2$]$(PF_6)_2$[CD3NO2] [169][c,d2,G]	**27**	$N^{btz,N2}$: −106.3	−77.4[aac]	H [X-ray]

[a]Pentamethylcyclopentadienyl ligand occupies three coordination sites, within octahedral geometry.
[b]1-Benzyl-4-phenyl-*1H*-1,2,3-triazole* is deprotonated (at the phenyl C(2) atom), anionic (charge −1) form of 1-benzyl-4-phenyl-*1H*-1,2,3-triazole, acting as bidentate (κ^2-N(3)^{tz},C(2)^{ph}) ligand.
[c]1,3-Bis-(2-(diphenylphosphino)ethyl)benzotriazolium is a cationic ligand, due to the presence of a delocalized positive charge (+1) at all triazolic nitrogens.
[d1-d2]The assumed *cis* isomers ([d1]) are analogous to the X-ray studied *cis*-[Ir^{III}(1,3-bis-(2-(diphenylphosphino)ethyl)benzotriazolium)(CO)$(H)_2]^{2+}$ and *cis*-[Ir^{III}(1,3-bis-(2-(diphenylphosphino)ethyl)benzotriazolium)(CH_3CN)$(H)_2]^{2+}$ and ([d2]) [169].
[aa]Original δ^{15N}_{compl} value (292.1 ppm), primarily referenced to liquid NH_3 by Rono et al. [171], was recalculated in respect to neat CH_3NO_2 by subtracting 381.7 ppm relative shift [37].
[aaa]Δ^{15N}_{coord} cannot be determined due to the lack of δ^{15N}_{lig} for *1H,1′H*-2,2′-biimidazole.
[aab]vs N(3) in 1-benzyl-4-phenyl-*1H*-1,2,3-triazole in $CDCl_3$: $\delta^{15N}_{lig} = -38$ ppm (344 ppm in respect to liquid NH_3) [171].
[aac]vs N(2) in benzotriazole moiety in (1,3-bis-(2-(diphenylphosphino)ethyl)benzotriazolium)$^+$ cation ($CF_3SO_3^-$ counterion) in $CDCl_3$ (^{15}N NMR data in case of PF_6^- counterions and/or in CD_3NO_2 being unavailable): $\delta^{15N}_{lig} = -28.9$ ppm [32].
[aaaa]The only stereomeric possibility, resulting from the molecular formula.
[A]X-ray structure: KULGEG [125].
[B]X-ray structure of $CHCl_3$ solvate: RONFOS [171].
[C]X-ray structure: UGUFIN [31].
[D]X-ray structure of analogous I^- salt, DMF solvate: UGUFOT [31].
[E]X-ray structure: KULGIK [125].
[F]X-ray structure of CH_2Cl_2 solvate: GOGNIC [169].
[G]X-ray structure of $CH_3CH_2OCH_2CH_3$ solvate: GOGNOI [169].

Table 11 Octahedral Fe(II), Ru(II) and Os(II) (d^6, low-spin) complexes and organometallics with azoles (**21–28** ring systems), ^{15}N NMR studied in 2012–20*

Compound	Parent ring system	^{15}N chemical shift (δ^{15N}_{compl}, ppm)	^{15}N coordination shift (Δ^{15N}_{coord}, ppm)	Nitrogen *trans* to
Fe(II) compounds	No interesting species			
Ru(II) compounds				
[Ru^{II}(1,3-bis(quinolin-8-yl)-*1H*-**pyrazole**-κ^3-N^{quin},$N^{quin'}$,$N^{pz,N2}$)$_2$](PF_6)$_2$^{CD3CN} [117][A,#]	**21**	2 × $N^{pz,N2}$: −138[aa]	Unknown[aaa]	2 × N^{quiin} [X-ray]
[Ru^{II}(1-phenyl-3-methyl-4-((*n*-hexyl)carbonyl)-*1H*-**pyrazol**-5-olate-μ^3-N^{N2},O^{C-O},$O^{C=O}$)(1-isopropyl-4-methylbenzene)]$_4$(CF_3SO_3)$_4$^{solid} [121][a]	**21**	4 × $N^{pz,N2}$: −143.0[ab1]	−8.4[aab]	($C{=}C^{bz}$)[aaaa]
[Ru^{II}(1-phenyl-3-methyl-4-*tert*-butylmethylcarbonyl-*1H*-**pyrazol**-5-olate-μ^3-$N^{,N2}$,O^{C-O},$O^{C=O}$)(1-isopropyl-4-methylbenzene)]$_4$(CF_3SO_3)$_4$^{solid} [121][a]	**21**	4 × $N^{pz,N2}$: −142.8[ab2]	−23.2[aac]	($C{=}C^{bz}$)[aaaa]
[Ru^{II}(tris(*1H*-**pyrazol**-1-yl)methanide-κ^3-$N^{pz1,N2}$,$N^{pz2,N2}$,$N^{pz3,N2}$)(cyclopentadienyl)]$^{THF\text{-}d8}$ [122][b1,B]	**21**	3 × $N^{pz,N2}$: −142[ac]	−63[aad]	$C{=}C^{dienyl}$ [X-ray]
[Ru^{II}(tris(*1H*-**pyrazol**-1-yl)methanenitrile-κ^3-$N^{pz1,N2}$,$N^{pz2,N2}$,$N^{pz3,N2}$)(cyclopentadienyl)]PF_6^{CD2Cl2} [123][b1]	**21**	3 × $N^{pz,N2}$: −145.6[ad1]	−68.9[aae]	($C{=}C^{dienyl}$)[aaaa]
[Ru^{II}(tris(*1H*-**pyrazol**-1-yl)methanenitrile-κ^3-$N^{pz1,N2}$,$N^{pz2,N2}$,$N^{pz3,N2}$)(cyclopentadienyl)]PF_6^{CD3CN} [123][b1]	**21**	3 × $N^{pz,N2}$: −147.1[ad2]	−70.4[aae]	($C{=}C^{dienyl}$)[aaaa]
trans(P,P),*trans*(P,N^{py})-[Ru^{II}(5-(pyrid-2-yl)**pyrazolate**-κ^2-N^{py},$N^{pz,N1}$)($P(CH_3)_3$)$_3$Cl]^{CDCl3} [106][c1,#]	**21**	$N^{pz,N1}$: −159.4	Unknown[aaf1]	(Cl)[c1]
trans(P,P),*trans*(P,N^{py})-[Ru^{II}(3-phenyl-5-(pyrid-2-yl)**pyrazolate**-κ^2-N^{py},$N^{pz,N1}$)($P(CH_3)_3$)$_3$Cl]^{CDCl3} [106][c2,C,#]	**21**	$N^{pz,N1}$: −145.1	Unknown[aaf2]	Cl [X-ray]
trans(P,P),*trans*(P,N^{py})-[Ru^{II}(4-bromo-5-(pyrid-2-yl)**pyrazolate**-κ^2-N^{py},$N^{pz,N1}$)($P(CH_3)_3$)$_3$Cl]^{CDCl3} [106][c2,D,#]	**21**	$N^{pz,N1}$: −156.5	Unknown[aaf3]	Cl [X-ray]

Continued

Table 11 Octahedral Fe(II), Ru(II) and Os(II) (d^6, low-spin) complexes and organometallics with azoles (**21–28** ring systems), ^{15}N NMR studied in 2012–20*—cont'd

Compound	Parent ring system	^{15}N chemical shift (δ^{15N}_{compl}, ppm)	^{15}N coordination shift (Δ^{15N}_{coord}, ppm)	Nitrogen *trans* to
trans(P,P),*trans*(P,N^{py})-[Ru^{II}(4-nitro-5-(pyrid-2-yl)**pyrazolate**-κ²-N^{py},$N^{pz,N1}$)($P(CH_3)_3)_3Cl$]^{CDCl3} [106][c2,E,#]	**21**	$N^{pz,N1}$: −149.1	Unknown[aaf4]	Cl [X-ray]
[Ru^{II}(1-methyl-2-diisopropylphosphino-*1H*-**imidazole**-κ²-N³,P)(pentamethylcyclopentadienyl)]$PF_6$$^{acetone-d6}$ [124][b2,d]	**22**	$N^{im,N3}$: −158.3	Unknown[aag1]	(C=C^{dienyl})[aaaa]
[Ru^{II}(1-methyl-2-diisopropylphosphino-*1H*-**imidazole**-κ²-N³,P)(tetramethyltrifluoromethylcyclopentadienyl)]$PF_6$$^{THF-d8}$ [124][b3,d]	**22**	$N^{im,N3}$: −168.8	Unknown[aag1]	(C=C^{dienyl})[aaaa]
[Ru^{II}(1-methyl-2-diisopropylphosphino-4-*tert*-butyl-*1H*-**imidazole**-κ²-N³,P)(pentamethylcyclopentadienyl)]$PF_6$$^{acetone-d6}$ [124][b2,d,F]	**22**	$N^{im,N3}$: −147.0	Unknown[aag2]	C=C^{dienyl} [X-ray]
[Ru^{II}(1-methyl-2-diisopropylphosphino-4-*tert*-butyl-*1H*-**imidazole**-κ²-N³,P)(tetramethyltrifluoromethylcyclopentadienyl)]$PF_6$$^{THF-D8}$ [124][b3,d,G]	**22**	$N^{im,N3}$: −155.7	Unknown[aag2]	C=C^{dienyl} [X-ray]
[Ru^{II}(*1H,1′H*-2,2′-bi**imidazole**-κ²-N³,N³′)(1-isopropyl-4-methylbenzene)Cl]$PF_6$$^{DMSO-d6}$ [125][a,H]	**22**	2 × $N^{im,N3,N3'}$: −181.7	Unknown[aah]	C=C^{bz} [X-ray]
[Ru^{II}(1,3-bis-(2-(diphenylphosphino)ethyl)-*1H*-**1,2,3-triazolium**-κ³-N²,P,P′)Cl(μ-Cl)]$_2$$(PF_6)_2$^{CD3NO2} [31][e1,I]	**25**	$N^{tz,N2}$: −94.0	−61.1[aai]	Cl [X-ray]
[Ru^{II}(1,3-bis-(2-(diphenylphosphino)ethyl)**benzotriazolium**-κ³-N²,P,P′)Cl(μ-Cl)]$_2$$(PF_6)_2$^{CD3NO2} [31][e2]	**27**	$N^{btz,N2}$: −75.7	−46.8[aaj]	(Cl)[aaaa]
[Ru^{II}(1,3-bis-(2-(diphenylphosphino)ethyl)-**4,5,6,7-tetrahydrobenzotriazolium**-κ³-N²,P,P′)Cl(μ-Cl)]$_2$$(PF_6)_2$^{CD3NO2} [31][e3]	**28**	$N^{thbtz,N2}$: −106.3	−61.5[aak]	(Cl)[aaaa]

trans(P,P),*trans*(P,N^{py})-[Ru^{II}(5-(pyrid-2-yl)-**1,2,4-triazolate**-κ^2-N^{py},$N^{tz,N1}$)($P(CH_3)_3)_3$Cl] [106]^{CDCl3} [J,#]	**26**	$N^{tz,N1}$: −158.1	Unknown[aaf5]	Cl [X-ray]
trans(P,P),*trans*(P,N^{py})-[Ru^{II}(3-methyl-5-(pyrid-2-yl)-**1,2,4-triazolate**-κ^2-N^{py},$N^{tz,N1}$)($P(CH_3)_3)_3$Cl]^{CDCl3} [106][K,#]	**26**	$N^{tz,N1}$: −160.2	Unknown[aaf6]	Cl [X-ray]
Os(II) compounds				
$N(CH_2CH_2CH_2CH_3)_4$(*cis*-[Os^{II}(*1H*-**pyrazole**-κ^1-N^2)(NO)Cl_4])$^{DMSO-d6}$ [126][f,L]	**21**	$N^{pz,N2}$: −202.0[ae1]	−111.5[aal]	Cl [X-ray]
$N(CH_2CH_2CH_2CH_3)_4$(*trans*-[Os^{II}(*1H*-**pyrazole**-κ^1-N^2)(NO)Cl_4])$^{DMSO-d6}$ [126][f,M]	**21**	$N^{pz,N2}$: −153.4[ae2]	−62.9[aal]	NO [X-ray]
$N(CH_2CH_2CH_2CH_3)_4$(*cis*-[Os^{II}(*1H*-**imidazole**-κ^1-N^3)(NO)Cl_4])$^{DMSO-d6}$ [126][f,g2,N]	**22**	$N^{im,N3}$: −229.9[ae3]	−92[aam]	Cl [X-ray]
$N(CH_2CH_2CH_2CH_3)_4$(*cis*-[Os^{II}(*1H*-**indazole**-κ^1-N^2)(NO)Cl_4])$^{DMSO-d6}$ [127][f,O]	**23**	$N^{ind,N2}$: −187.5[ae4]	−110.0[aan]	Cl [X-ray]
$N(CH_2CH_2CH_2CH_3)_4$(*trans*-[Os^{II}(*1H*-**indazole**-κ^1-N^2)(NO)Cl_4)$^{DMSO-d6}$ [127][f,P]	**23**	$N^{ind,N2}$: −143.0[ae5]	−65.5[aan]	NO [X-ray]
$N(CH_2CH_2CH_2CH_3)_4$(*cis*-[Os^{II}(*1H*-**benzoimidazole**-κ^1-N^3)(NO)Cl_4])$^{DMSO-d6}$ [126][f,g1]	**24**	$N^{bim,N3}$: −243.9[ae6]	−90.4[aao]	(Cl)[g1]

[*]Except for Ru(II) complexes with 1,2,4-triazolo-[1,5*a*]-pyrimidines which are listed in Table 15.
[#]See also Table 4 for pyridine and quinoline ligand(s).
[a]1-Isopropyl-4-methylbenzene ligand occupies three coordination sites, within octahedral geometry.
[b1-b3]Cyclopentadienyl ([b1]), pentamethylcyclopentadienyl ([b2]) and tetramethyltrifluoromethylcyclopentadienyl ([b3]) ligands occupy three coordination sites, within octahedral geometry.
[c1-c2]The assumed *trans*(P,P),*trans*(P,N^{py}) isomer ([c1]) is analogous to the X-ray studied *trans*(P,P),*trans*(P,N^{py})-[Ru^{II}(3-phenyl-5-(pyrid-2-yl)pyrazolate)($P(CH_3)_3)_3$Cl], *trans*(P,P),*trans*(P,N^{py})-[Ru^{II}(4-bromo-5-(pyrid-2-yl)pyrazolate)($P(CH_3)_3)_3$Cl], *trans*(P,P),*trans*(P,N^{py})-[Ru^{II}(4-nitro-5-(pyrid-2-yl)pyrazolate)($P(CH_3)_3)_3$Cl] ([c2]) [106].
[d]Coordination number 5.
[e1-e3]1,3-Bis-(2-(diphenylphosphino)ethyl)-*1H*-1,2,3-triazolium ([e1]), 1,3-bis-(2-(diphenylphosphino)ethyl)benzotriazolium ([e2]) and 1,3-bis-(2-(diphenylphosphino)ethyl)-4,5,6,7-tetrahydrobenzotriazolium ([e3]) are cationic ligands, due to the presence of a delocalized positive charge (+1) at all triazolic nitrogens.

Continued

Table 11 Octahedral Fe(II), Ru(II) and Os(II) (d^6, low-spin) complexes and organometallics with azoles (**21–28** ring systems), ^{15}N NMR studied in 2012–20*—cont'd

[f]In these osmium complexes the NO ligands are cationic (NO^+ nitrosonium ions), due to shift of the unpaired electron from the originally neutral, paramagnetic nitrous oxide (NO) molecule towards the previously +3 charged (Os^{III}) central ion; in consequence, the formed coordination compounds are diamagnetic and can be treated as anionic (charge −1) Os^{II} species, having the following partial electric charges: +1 at NO, +2 at Os, 4 × (−1) at Cl.

[g1-g2]The assumed *cis* isomer ([g1]) is analogous to the X-ray studied *cis*-[Os^{II}(*1H*-imidazole)(NO)Cl_4]$^-$ ([g2]) [126].

[aa-ae]Original δ^{15N}_{compl} values, primarily referenced to liquid NH_3 by the following authors, were recalculated in respect to neat CH_3NO_2 by subtracting 381.7 ppm (or 382 ppm) relative shift [37]:

[aa]244 ppm by Jarenmark et al. [117].

[ab1-ab2]238.7 ppm ([ab1]), 238.9 ppm ([ab2]) by Pettinari et al. [121].

[ac]240 ppm by Gonzalez-Gallardo et al. [122].

[ad1-ad2]236.1 ppm ([ad1]), 234.6 ppm ([ad2]) by Trapp et al. [123].

[ae1-ae6]179.7 ppm ([ae1]), 228.3 ppm ([ae2]), 151.8 ppm ([ae3]), 194.2 ppm ([ae4]), 238.7 ppm ([ae5]), 137.8 ppm ([ae6]) by Buchel and Gavriluta et al. [126,127].

[aaa]Δ^{15N}_{coord} cannot be determined due to the lack of δ^{15N}_{lig} for 1,3-bis(quinolin-8-yl)-*1H*-pyrazole.

[aab]vs N(2) in solid 1-phenyl-3-methyl-4-((*n*-hexyl)carbonyl)-*1H*-pyrazol-5-ol: $\delta^{15N}_{lig} = -134.6$ ppm (247.1 ppm in respect to liquid NH_3) [121].

[aac]vs N(2) in solid 1-phenyl-3-methyl-4-*tert*-butylmethylcarbonyl-*1H*-pyrazol-5-ol: $\delta^{15N}_{lig} = -119.6$ ppm (262.1 ppm in respect to liquid NH_3) [121].

[aad]vs N(2) in tris(*1H*-pyrazol-1-yl)methane in $CDCl_3$ (^{15}N NMR data in THF being unavailable): $\delta^{15N}_{lig} = -79.0$ ppm [123].

[aae]vs N(2) in tris(*1H*-pyrazol-1-yl)methanenitrile in DMSO-d_6 (^{15}N NMR data in CD_2Cl_2 or CD_3CN being unavailable): $\delta^{15N}_{lig} = -76.7$ ppm (305.0 ppm in respect to liquid NH_3) [123].

[aaf1-aaf6]Δ^{15N}_{coord} cannot be determined due to the lack of δ^{15N}_{lig} for (5-(pyrid-2-yl)pyrazolate)$^-$ ([aaf1]), (3-phenyl-5-(pyrid-2-yl)pyrazolate)$^-$ ([aaf2]), (4-bromo-5-(pyrid-2-yl)pyrazolate)$^-$ ([aaf3]), (4-nitro-5-(pyrid-2-yl)pyrazolate)$^-$ ([aaf4]), (5-(pyrid-2-yl)-1,2,4-triazolate)$^-$ ([aaf5]) and (3-methyl-5-(pyrid-2-yl)-1,2,4-triazolate)$^-$ ([aaf6]) anions, whereas their determination in respect to δ^{15N}_{lig} for neutral molecules of these *1H*-pyrazole- or *1H*-1,2,4-triazole-derivatives has no sense because of the presence of hydrogen at N(1) atoms.

[aag1-aag2]Δ^{15N}_{coord} cannot be determined due to the lack of δ^{15N}_{lig} for 1-methyl-2-diisopropylphosphino-*1H*-imidazole ([aag1]) and 1-methyl-2-diisopropylphosphino-4-*tert*-butyl-*1H*-imidazole ([aag2]).

[aah]Δ^{15N}_{coord} cannot be determined due to the lack of δ^{15N}_{lig} for *1H,1'H*-2,2'-biimidazole.

[aai]vs N(2) in 1,2,3-triazole moiety in 1,3-bis-(2-(diphenylphosphino)ethyl)-*1H*-1,2,3-triazolium cations (Br^- counterions) in $CDCl_3$ (^{15}N NMR data in case of PF_6^- counterions and/or in CD_3NO_2 being unavailable): $\delta^{15N}_{lig} = -32.9$ ppm [31].

[aaj]vs N(2) in benzotriazole moiety in 1,3-bis-(2-(diphenylphosphino)ethyl)benzotriazolium cations ($CF_3SO_3^-$ counterions) in $CDCl_3$ (^{15}N NMR data in case of PF_6^- counterions and/or in CD_3NO_2 being unavailable): $\delta^{15N}_{lig} = -28.9$ ppm [32].

[aak]vs N(2) in tetrahydrobenzotriazole moiety in 1,3-bis-(2-(diphenylphosphino)ethyl)-4,5,6,7-tetrahydrobenzotriazolium cations (PF_6^- counterions) in $CDCl_3$ (^{15}N NMR data in CD_3NO_2 being unavailable): $\delta^{15N}{}_{lig} = -44.8$ ppm [32].
[aal]vs N(2) in solid *1H*-pyrazole: $\delta^{15N}{}_{lig} = -90.5$ ppm [128].
[aam]vs N(3) in solid *1H*-imidazole: $\delta^{15N}{}_{lig} = -138$ ppm [128].
[aan]vs N(2) in solid *1H*-indazole: $\delta^{15N}{}_{lig} = -77.5$ ppm [128].
[aao]vs N(3) in *1H*-benzoimidazole in CD_3OD at 178 K: $\delta^{15N}{}_{lig} = -153.5$ ppm [128].
[aaaa]The only stereomeric possibility, resulting from the molecular formula.
[A]X-ray structure: UJOSOD [117].
[B]X-ray structure: MITYEV [122].
[C]X-ray structure: TILLIL [106].
[D]X-ray structure: VIQVUO [106].
[E]X-ray structure: VIQVOI [106].
[F]X-ray structure: HOTTUI [124].
[G]X-ray structure: HOTVIY [124].
[H]X-ray structure of H_2O solvate: KULPAL [125].
[I]X-ray structure of analogous $(Br^-)_2$ salt, CH_3NO_2 solvate: UGUFUZ [31].
[J]X-ray structure: VIQVIC [106].
[K]X-ray structure: SARQIO [106].
[L]X-ray structure: LINTEJ [126].
[M]X-ray structure: LINSUY [126].
[N]X-ray structure: LINTAF [126].
[O]X-ray structure: LINSEI [127].
[P]X-ray structure: LINSIM [127].

Table 12 Cu(I), Ag(I), Au(I) (d^{10}) and Au(III) (d^8) complexes with azoles (**21–22**, **24**, **26** ring systems), ^{15}N NMR studied in 2012–20.

Compound	Parent ring system	^{15}N chemical shift (δ^{15N}_{compl}, ppm)	^{15}N coordination shift (Δ^{15N}_{coord}, ppm)
Cu(I) compounds	No interesting species		
Ag(I) compounds			
[Ag^I(1-phenyl-3-methyl-4-benzoyl**pyrazolate**-5-one-κ^1-N^2)(*1H*-**imidazole**-κ^1-N^3)]^{solid} [139]^{a,b}	**21, 22**	$N^{pz,N2}$: −148.2^{aa1}	−45.1^{aaa}
		$N^{im,N3}$: −178.6^{aa2}	−43.3^{aab}
[Ag^I(1-phenyl-3-methyl-4-benzoyl**pyrazolate**-5-one-κ^1-N^2)(1-methyl-*1H*-**imidazole**-κ^1-N^3)$_2$]^{solid} [139]^{a,c,A}	**21, 22**	$N^{pz,N2}$: −138.6^{aa3}	−35.5^{aaa}
		$N^{im1,N3}$: −153.9^{aa4}	−13.1^{aac}
		$N^{im2,N3}$: −157.2^{aa5}	−16.4^{aac}
[Ag^I(1-phenyl-3-methyl-4-benzoyl**pyrazolate**-5-one-κ^1-N^2)(1-benzyl-*1H*-**imidazole**-κ^1-N^3)]^{solid} [139]^{a,b}	**21, 22**	$N^{pz,N2}$: −151.0^{aa6}	−47.9^{aaa}
		$N^{im,N3}$: −167.8^{aa7}	−46.3^{aad}
[Ag^I(1-phenyl-3-methyl-4-benzoyl-*1H*-**pyrazol**-5-olate-μ-N^2, $O^{C5\text{-}O}$)]$_n^{solid}$ [139]^{a,b}	**21**	$N^{pz,N2}$: −157.5^{aa8}	−54.4^{aaa}
[Ag^I(1-(pyrid-2-yl)-3-methyl-4-trifluoroacetyl**pyrazolate**-5-one-κ^2-N^{py},$N^{pz,N2}$)(2,4,6-trimethylpyridine)]^{solid} [139]$^{c,d,B,\#}$	**21**	$N^{pz,N2}$: −137.0^{aa9}	−20.9^{aae}
[Ag^I(1-(pyrid-2-yl)-3-methyl-4-trifluoroacetyl**pyrazolate**-5-one-κ^2-N^{py},$N^{pz,N2}$)(2-ethyl-*1H*-**imidazole**-κ^1-N^3)]^{solid} [139]$^{c,d,C,\#}$	**21, 22**	$N^{pz,N2}$: −128.2^{aa10}	−12.1^{aae}
		$N^{im,N3}$: −154.7^{aa11}	−17.2^{aaf}
[Ag^I(1-(pyrid-2-yl)-3-methyl-4-trifluoroacetyl**pyrazolate**-5-one-κ^2-N^{py},$N^{pz,N2}$)(1-benzyl-*1H*-**imidazole**-κ^1-N^3)]^{solid} [139]$^{c,d,\#}$	**21, 22**	$N^{pz,N2}$: −126.1^{aa12}	−10.0^{aae}
		$N^{im,N3}$: −154.0^{aa13}	−32.5^{aad}

$[Ag^{I}$(1-(pyrid-2-yl)-3-methyl-4-trifluoroacetyl**pyrazolate**-5-one-κ^2-N^{py},$N^{pz,N2}$(1-benzyl-2-methyl-*1H*-**imidazole**-κ^1-N^3))$]^{solid}$ [139][c,d,#]	**21, 22**	$N^{pz,N2}$: −137.2[aa14]	−21.1[aae]
		$N^{im,N3}$: −155.0[aa15]	−34.1[aag]
$[Ag^{I}$(1-(pyrid-2-yl)-3-methyl-4-trifluoroacetyl**pyrazolate**-5-one-κ^2-N^{py},$N^{pz,N2}$)(2-trifluoromethyl-*1H*-**benzoimidazole**-κ^1-N^3)$]^{solid}$ [139][c,d,#]	**21, 24**	$N^{pz,N2}$: −140.6[aa16]	−24.5[aae]
		$N^{bim,N3}$: −154.9[aa17]	−9.4[aah]
$[Ag^{I}_2$(1-(pyrid-2-yl)-3-methyl-4-trifluoroacetyl**pyrazolate**-5-one-κ^2-N^{py},$N^{pz,N2}$)$_2$(1,1′-carbonyl-*1H,1H′*-di**imidazole**-μ^2-N^3,$N^{3'}$)$]^{solid}$ [139][c,d,#]	**21, 22**	$N^{pz,N2}$: −123.7[aa18]	−7.6[aae]
		$N^{im,N3,N3'}$: −157.5[aa19]	−22.3[aai]
$[Ag^{I}$(1-(pyrid-2-yl)-3-methyl-4-trifluoroacetyl-*1H*-**pyrazol**-5-olate-μ^3-N^{py},$N^{pz,N2}$,$O^{pz,C5-O}]_n^{solid}$ [139][c,d,#]	**21**	$N^{pz1,N2}$: −119.5[aa20]	−3.4[aae]
		$N^{pz2,N2}$: −123.8[aa21]	−7.7[aae]
$[Ag^{I}$(*1H*-**imidazole**-κ^1-N^3)$_2$](1-phenyl-3-methyl-4-benzoylpyrazolate-5-one)^{solid} [139][D]	**22**	$N^{im,N3}$: −173.9[aa22]	−38.6[aab]
Au(I) compounds	No interesting species		
Au(III) compounds			
$[Au^{III}$((*R,S*)-1-(4-chlorophenyl)-3-((*1H*-**1,2,4-triazol**-1-yl)methyl)-4,4-dimethylpentan-3-ol-κ^1-N^4)$Cl_3]^{CDCl3}$ [143][e]	**26**	$N^{tz,N4}$: −198.9[ab]	−67.0[aaj]

[#]See also Table 5 for pyridine ligand(s).
[a]1-Phenyl-3-methyl-4-benzoylpyrazolate-5-one and 1-phenyl-3-methyl-4-benzoyl-*1H*-pyrazol-5-olate are ketone and enol forms of the same anion (charge −1) formed by deprotonation of 1-phenyl-3-methyl-4-benzoyl-*1H*-pyrazol-5-ol.
[b]Coordination number 2 (linear geometry).
[c]Coordination number 3 (trigonal geometry).
[d]1-(Pyrid-2-yl)-3-methyl-4-trifluoroacetylpyrazolate-5-one and 1-(pyrid-2-yl)-3-methyl-4-trifluoroacetyl-*1H*-pyrazol-5-olate are ketone and enol forms of the same anion (charge −1) formed by deprotonation of 1-(pyrid-2-yl)-3-methyl-4-trifluoroacetyl-*1H*-pyrazol-5-ol.

Continued

Table 12 Cu(I), Ag(I), Au(I) (d^{10}) and Au(III) (d^{8}) complexes with azoles (**21–22**, **24**, **26** ring systems), ^{15}N NMR studied in 2012–20.—cont'd

[e]Coordination number 4 (square-planar geometry, $N^{tz,N4}$ *trans* to Cl).

[aa1-aa22]Original $\delta^{15N}{}_{compl}$ values (207.6 ppm ([aa1]), 177.2 ppm ([aa2]), 217.2 ppm ([aa3]), 201.9 ppm ([aa4]), 198.6 ppm ([aa5]), 204.8 ppm ([aa6]), 188.0 ppm ([aa7]), 198.3 ppm ([aa8]), 218.8 ppm ([aa9]), 227.6 ppm ([aa10]), 201.1 ppm ([aa11]), 229.7 ppm ([aa12]), 201.8 ppm ([aa13]), 218.6 ppm ([aa14]), 200.8 ppm ([aa15]), 215.2 ppm ([aa16]), 200.9 ppm ([ab17]), 232.1 ppm ([aa18]), 198.3 ppm ([aa19]), 236.3 ppm ([aa20]), 232.0 ppm ([aa21]), 181.9 ppm ([aa22])), primarily referenced to solid $(NH_4)_2SO_4$ by Marchetti et al. [139], were recalculated in respect to neat CH_3NO_2 by subtracting 355.8 ppm relative shift [139].

[ab]Original $\delta^{15N}{}_{compl}$ value (182.8 ppm), primarily referenced to liquid NH_3 by Golubyatnikova et al. [143], was recalculated in respect to neat CH_3NO_2 by subtracting 381.7 ppm relative shift [37].

[aaa]vs N(2) in pyrazole moiety in solid 1-phenyl-3-methyl-4-benzoyl-*1H*-pyrazol-5-ol: $\delta^{15N}{}_{lig} = -103.1$ ppm (average from -102.3 ppm and -103.9 ppm; 253.5 ppm and 251.9 ppm in respect to solid $(NH_4)_2SO_4$) [139].

[aab]vs N(3) in solid *1H*-imidazole: $\delta^{15N}{}_{lig} = -135.3$ ppm (220.5 ppm in respect to solid $(NH_4)_2SO_4$) [139].

[aac]vs N(3) in solid 1-methyl-*1H*-imidazole: $\delta^{15N}{}_{lig} = -140.8$ ppm (215.0 ppm in respect to solid $(NH_4)_2SO_4$) [139].

[aad]vs N(3) in solid 1-benzyl-*1H*-imidazole: $\delta^{15N}{}_{lig} = -121.5$ ppm (234.3 ppm in respect to solid $(NH_4)_2SO_4$) [139].

[aae]vs N(2) in pyrazole moiety in solid 1-(pyrid-2-yl)-3-methyl-4-trifluoroacetyl-*1H*-pyrazol-5-ol: $\delta^{15N}{}_{lig} = -116.1$ ppm (239.7 ppm in respect to solid $(NH_4)_2SO_4$) [139].

[aaf]vs N(3) in solid 2-ethyl-*1H*-imidazole: $\delta^{15N}{}_{lig} = -137.5$ ppm (218.3 ppm in respect to solid $(NH_4)_2SO_4$) [139].

[aag]vs N(3) in solid 1-benzyl-2-methyl-*1H*-imidazole: $\delta^{15N}{}_{lig} = -120.9$ ppm (234.9 ppm in respect to solid $(NH_4)_2SO_4$) [139].

[aah]vs N(3) in solid 2-trifluoromethyl-*1H*-benzoimidazole: $\delta^{15N}{}_{lig} = -145.5$ ppm (210.3 ppm in respect to solid $(NH_4)_2SO_4$) [139].

[aai]vs N(3),N(3′) in solid 1,1′-carbonyl-*1H,1H′*-diimidazole: $\delta^{15N}{}_{lig} = -135.2$ ppm (220.6 ppm in respect to solid $(NH_4)_2SO_4$) [139].

[aaj]vs N(4) in 1,2,4-triazole moiety in (*R,S*)-1-(4-chlorophenyl)-3-((*1H*-1,2,4-triazol-1-yl)methyl)-4,4-dimethylpentan-3-ol in $CDCl_3$: $\delta^{15N}{}_{lig} = -131.9$ ppm (249.8 ppm in respect to liquid NH_3) [143].

[A]X-ray structure: SOYXEL [139].

[B]X-ray structure: SOYXUB [139].

[C]X-ray structure: SOYXIP [139].

[D]X-ray structure: SOYXAH [139].

Table 13 Zn(II), Cd(II) and Hg(II) (d^{10}) complexes with azoles (**21–22** ring systems), ^{15}N NMR studied in 2012–20.

Compound	Parent ring system	^{15}N chemical shift (δ^{15N}_{compl}, ppm)	^{15}N coordination shift (Δ^{15N}_{coord}, ppm)
Zn(II) compounds			
$[Zn^{II}$(2-(1-(pyrid-2-yl)-5-(4-methoxyphenyl)-*1H*-**pyrazol**-3-yl)pyridine-κ^3-N^{pyA},N^{pyB},$N^{pz,N2}$)$Cl_2]^{CDCl3}$ [11][a,b,A,#]	**21**	$N^{pz,N2}$: −132.7	−49.3[aaa]
$[Zn^{II}$(2-(1-(pyrid-2-yl)-5-(4-methoxyphenyl)-*1H*-**pyrazol**-3-yl)pyridine-κ^3-N^{pyA},N^{pyB},$N^{pz,N2}$)$Cl_2]^{solid}$ [11][a,b,A,#]	**21**	$N^{pz,N2}$: −132.0	−48.6[aaa]
$[Zn^{II}$(1-methyl-*1H*-**imidazole**-κ^1-$N^3)_6]((CF_3SO_2)_2N)_2^{neat,343K}$ [165][c,B]	**22**	$N^{im,N3}$: −165.9[aa1]	−40.4[aab]
$[Zn^{II}$(1-ethyl-*1H*-**imidazole**-κ^1-$N^3)_6]((CF_3SO_2)_2N)_2^{neat,313K}$ [165][c,C]	**22**	$N^{im,N3}$: −166.6[aa2]	−46.2[aac]
$[Zn^{II}$(1-*n*-butyl-*1H*-**imidazole**-κ^1-$N^3)_6]((CF_3SO_2)_2N)_2^{neat,313K}$ [165][c]	**22**	$N^{im,N3}$: −177.7[aa3]	−57.2[aad]
Cd(II) compounds	No interesting species		
Hg(II) compounds	No interesting species		

[#] See also Table 7 for pyridine ligand(s).

[a] Coordination number 5 (trigonal-bipyramidal geometry).

[b] In this complex both pyridine rings, i.e., at pyrazolic C(3) atom (pyridine A) and at pyrazolic N(1) atom (pyridine B), are bound to Zn(II).

[c] Coordination number 6 (octahedral geometry; $N^{im,N3}$ *trans* to N).

[aa1-aa3] Original δ^{15N}_{compl} values (215.8 ppm ([aa1]), 215.1 ppm ([aa2]), 204.0 ppm ([aa3])), primarily referenced to liquid NH_3 by Steichen et al. [165], were recalculated in respect to neat CH_3NO_2 by subtracting 381.7 ppm relative shift [37].

[aaa] vs N(2) in pyrazole moiety in solid 2-(1-(pyrid-2-yl)-5-(4-methoxyphenyl)-*1H*-pyrazol-3-yl)pyridine: $\delta^{15N}_{lig} = -83.4$ ppm [11].

[aab] vs N(3) in neat 1-methyl-*1H*-imidazole: $\delta^{15N}_{lig} = -125.5$ ppm [166].

[aac] vs N(3) in 1-ethyl-*1H*-imidazole in DMSO-d_6 (^{15}N NMR data for neat liquid being unavailable): $\delta^{15N}_{lig} = -120.4$ ppm [167].

[aad] vs N(3) in 1-*n*-butyl-*1H*-imidazole in CD_2Cl_2 or CD_3CN (^{15}N NMR data in the solid phase being unavailable): $\delta^{15N}_{lig} = -120.5$ ppm (average from −121 ppm and −120 ppm, respectively) [168].

[A] X-ray structure: GOCROH [11].

[B] X-ray structure: KOLXAM [165].

[C] X-ray structure: KOLXEQ [165].

Table 14 Square-planar Pt(II) (d^8) complexes with purines (**29** ring systems), ^{15}N NMR studied in 2012–20.

Complex	^{15}N chemical shift (δ^{15N}_{compl}, ppm)	^{15}N coordination shift (Δ^{15N}_{coord}, ppm)	Nitrogen *trans* to
Pt(II) complexes			
$[Pt^{II}$(2-chloro-6-(furan-2-yl)methylamino-9-isopropyl-*9H*-**purine**-κ^1-$N^7)_2(C_2O_4)]^{DMF\text{-}d7}$ [33] [a]	$N^{pu,N7}$: −254.5[aa1]	−114.3[aaa1]	(O)[aaaa]
$[Pt^{II}$(2-chloro-6-(5-methylfuran-2-yl)methylamino-9-isopropyl-*9H*-**purine**-κ^1-$N^7)_2(C_2O_4)]^{DMF\text{-}d7}$ [33][a]	$N^{pu,N7}$: −252.9[aa2]	−111.3[aaa2]	(O)[aaaa]
trans-$[Pt^{II}$(6-(2-chlorophenyl)methylamino-9-β-D-ribofuranosyl-*9H*-**purine**-κ^1-$N^7)_2Cl_2]^{DMF\text{-}d7}$ [34][b,c]	$N^{pu,N7}$: −246.6[aa3]	−106.0[aab]	(N^{pu})[c]
trans-$[Pt^{II}$(6-(2-hydroxyphenyl)methylamino-9-β-D-ribofuranosyl-*9H*-**purine**-κ^1-$N^7)_2Cl_2]^{DMF\text{-}d7}$ [34][b,c]	$N^{pu,N7}$: −246.8[aa4]	−105.7[aac]	(N^{pu})[c]
trans-$[Pt^{II}$(6-(2-methoxyphenyl)methylamino-9-β-D-ribofuranosyl-*9H*-**purine**-κ^1-$N^7)_2Cl_2]^{DMF\text{-}d7}$ [34][b,c]	$N^{pu,N7}$: −245.8[aa5]	−105.0[aad]	(N^{pu})[c]
trans-$[Pt^{II}$(6-(3-hydroxyphenyl)methylamino-9-β-D-ribofuranosyl-*9H*-**purine**-κ^1-$N^7)_2Cl_2]^{DMF\text{-}d7}$ [34][b,c]	$N^{pu,N7}$: −247.0[aa6]	−105.8[aae]	(N^{pu})[c]
trans-$[Pt^{II}$(6-(4-methylphenyl)methylamino-9-β-D-ribofuranosyl-*9H*-**purine**-κ^1-$N^7)_2Cl_2]^{DMF\text{-}d7}$ [34][b,c]	$N^{pu,N7}$: −246.7[aa7]	−106.1[aaf]	(N^{pu})[c]
trans-$[Pt^{II}$(6-(4-fluorophenyl)methylamino-9-β-D-ribofuranosyl-*9H*-**purine**-κ^1-$N^7)_2Cl_2]^{DMF\text{-}d7}$ [34][b,c]	$N^{pu,N7}$: −246.1[aa8]	−105.4[aag]	(N^{pu})[c]
trans-$[Pt^{II}$(6-(4-chlorophenyl)methylamino-9-β-D-ribofuranosyl-*9H*-**purine**-κ^1-$N^7)_2Cl_2]^{DMF\text{-}d7}$ [34][b,c]	$N^{pu,N7}$: −247.3[aa9]	−106.7[aah]	(N^{pu})[c]

trans-[Pt^{II}(6-(4-methoxyphenyl)methylamino-9-β-D-ribofuranosyl-*9H*-**purine**-κ^{1}-N^{7})$_2$Cl$_2$]$^{DMF-d7}$ [34][b,c]	N^{pu,N7}: −246.1[aa10]	−105.0[aai]	(N^{pu})[c]
trans-[Pt^{II}(6-(2-hydroxy-3-methoxyphenyl)methylamino-9-β-D-ribofuranosyl-*9H*-**purine**-κ^{1}-N^{7})$_2$Cl$_2$]$^{DMF-d7}$ [34][b,c]	N^{pu,N7}: −246.7[aa11]	−106.2[aaj]	(N^{pu})[c]
trans-[Pt^{II}(2-chloro-6-(3-hydroxyphenyl)methylamino-9-β-D-ribofuranosyl-*9H*-**purine**-κ^{1}-N^{7})$_2$Cl$_2$]$^{DMF-d7}$ [34][b,c]	N^{pu,N7}: −247.0[aa12]	−106.5[aak]	(N^{pu})[c]
trans-[Pt^{II}(2-chloro-6-(4-hydroxyphenyl)methylamino-9-β-D-ribofuranosyl-*9H*-**purine**-κ^{1}-N^{7})$_2$Cl$_2$]$^{DMF-d7}$ [34][b,c]	N^{pu,N7}: −246.3[aa13]	−106.2[aal]	(N^{pu})[c]
trans-[Pt^{II}(2-chloro-6-(2-hydroxy-5-methylphenyl)methylamino-9-β-D-ribofuranosyl-*9H*-**purine**-κ^{1}-N^{7})$_2$Cl$_2$]$^{DMF-d7}$ [34][b,c]	N^{pu,N7}: −246.0[aa14]	−105.8[aam]	(N^{pu})[c]
trans-[Pt^{II}(2-chloro-6-(2-hydroxy-3-methoxyphenyl)methylamino-9-β-D-ribofuranosyl-*9H*-**purine**-κ^{1}-N^{7})$_2$Cl$_2$]$^{DMF-d7}$ [34][b,c]	N^{pu,N7}: −245.6[aa15]	−105.7[aan]	(N^{pu})[c]
[Pt^{II}(6-(2-chlorophenyl)methylamino-9-β-D-ribofuranosyl-*9H*-**purine**-κ^{1}-N^{7})$_2$(C$_2$O$_4$)]$^{DMF-d7}$ [35][b]	N^{pu,N7}: −254.1[aa16]	−113.5[aab]	(O)[c]
[Pt^{II}(6-(2-methoxyphenyl)methylamino-9-β-D-ribofuranosyl-*9H*-**purine**-κ^{1}-N^{7})$_2$(C$_2$O$_4$)]$^{DMF-d7}$ [35][b]	N^{pu,N7}: −254.1[aa17]	−113.3[aad]	(O)[c]
[Pt^{II}(6-(4-methylphenyl)methylamino-9-β-D-ribofuranosyl-*9H*-**purine**-κ^{1}-N^{7})$_2$(C$_2$O$_4$)]$^{DMF-d7}$ [35][b]	N^{pu,N7}: −253.7[aa18]	−113.1[aaf]	(O)[c]
[Pt^{II}(6-(4-fluorophenyl)methylamino-9-β-D-ribofuranosyl-*9H*-**purine**-κ^{1}-N^{7})$_2$(C$_2$O$_4$)]$^{DMF-d7}$ [35][b]	N^{pu,N7}: −254.2[aa19]	−113.5[aag]	(O)[c]
[Pt^{II}(6-(4-methoxyphenyl)methylamino-9-β-D-ribofuranosyl-*9H*-**purine**-κ^{1}-N^{7})$_2$(C$_2$O$_4$)]$^{DMF-d7}$ [35][b]	N^{pu,N7}: −253.9[aa20]	−112.8[aai]	(O)[c]

Continued

Table 14 Square-planar Pt(II) (d^8) complexes with purines (**29** ring systems), ^{15}N NMR studied in 2012–20.—cont'd

Complex	^{15}N chemical shift (δ^{15N}_{compl}, ppm)	^{15}N coordination shift (Δ^{15N}_{coord}, ppm)	Nitrogen *trans* to
cis-$[Pt^{II}$(6-(furan-2-yl)methylamino-9-β-D-ribofuranosyl-*9H*-purine-κ^1-N^7)$(CH_3SOCH_3)Cl_2]^{DMF-d7}$ [23][d,e]	$N^{pu,N7}$: −224.0	−84.1[aao]	(Cl)[e]
$[Pt^{II}$(6-methylamino-9-methylpurinate-κ^2-N^7,N^{NCH3})(1,2-bis(diphenylphosphino)ethane-κ^2-P,P′)$]^{CD2Cl2}$ [36][f]	$N^{pu,N7}$: −196.7	Unknown[aap]	(P)[aaaa]

[a](Furan-2-yl)methyl group and 6-amino-*9H*-purine ring system are commonly known as furfuryl group and adenine ring system; thus, 2-chloro-6-(furan-2-yl)methylamino-9-isopropyl-*9H*-purine and 2-chloro-6-(5-methylfuran-2-yl)methylamino-9-isopropyl-*9H*-purine are 2-chloro-(N^{NH}-furfuryl)-9-isopropyladenine and 2-chloro-(N^{NH}-(5-methylfurfuryl))-9-isopropyladenine, respectively.

[b]6-Amino-9-β-D-ribofuranosyl-*9H*-purine ring system is commonly known as adenosine (AD) ring system; thus, 6-phenylmethylamino-9-β-D-ribofuranosyl-*9H*-purine derivatives are N^{NH}-benzyladenosine (BzAD) derivatives.

[c]These are probably *trans* isomers.

[d]6-(Furan-2-yl)methylamino-9-β-D-ribofuranosyl-*9H*-purine, i.e., N^{NH}-furfuryladenosine, is commonly known as kinetin riboside.

[e]This is probably *cis* isomer.

[f]6-Methylamino-9-methylpurinate is anionic (charge −1), bidentate (κ^2-N^7,N^{NCH3}) ligand, formed by deprotonation of 6-methylamino-9-methyl-*9H*-purine.

[aa1–aa20]Original δ^{15N}_{compl} values (127.2 ppm ([aa1]), 128.8 ppm ([aa2]), 135.1 ppm ([aa3]), 134.9 ppm ([aa4]), 135.9 ppm ([aa5]), 134.7 ppm ([aa6]), 135.0 ppm ([aa7]), 135.6 ppm ([aa8]), 134.4 ppm ([aa9]), 135.6 ppm ([aa10]), 135.0 ppm ([aa11]), 134.7 ppm ([aa12]), 135.4 ppm ([aa13]), 135.7 ppm ([aa14]), 136.1 ppm ([aa15]), 127.6 ppm ([aa16]), 127.6 ppm ([aa17]), 128.0 ppm ([aa18]), 127.5 ppm ([aa19]), 127.8 ppm ([aa20])), primarily referenced to liquid NH_3 by Krikavova et al. and Starha et al. [33–35], were recalculated in respect to neat CH_3NO_2 by subtracting 381.7 ppm relative shift [37].

[aaa1–aaa2]Δ^{15N}_{coord} were published directly (with no δ^{15N}_{lig}, which can be determined as $\delta^{15N}_{compl} - \Delta^{15N}_{coord}$ differences: N(7) in purine moiety in 2-chloro-6-(furan-2-yl)methylamino-9-isopropyl-*9H*-purine −140.2 ppm ([aaa1]), N(7) in purine moiety in 2-chloro-6-(5-methylfuran-2-yl)methylamino-9-isopropyl-*9H*-purine −141.6 ([aaa2]), both in DMF-d_7) [33].

[aab]vs N(7) in purine moiety in 6-(2-chlorophenyl)methylamino-9-β-D-ribofuranosyl-*9H*-purine in DMF-d_7: $\delta^{15N}_{lig} = -140.6$ ppm (241.1 ppm in respect to liquid NH_3) [34].

[aac]vs N(7) in purine moiety in 6-(2-hydroxyphenyl)methylamino-9-β-D-ribofuranosyl-*9H*-purine in DMF-d_7: $\delta^{15N}_{lig} = -141.1$ ppm (240.6 ppm in respect to liquid NH_3) [34].

[aad]vs N(7) in purine moiety in 6-(2-methoxyphenyl)methylamino-9-β-D-ribofuranosyl-*9H*-purine in DMF-d_7: $\delta^{15N}_{lig} = -140.8$ ppm (240.9 ppm in respect to liquid NH_3) [34].

[aae]vs N(7) in purine moiety in 6-(3-hydroxyphenyl)methylamino-9-β-D-ribofuranosyl-*9H*-purine in DMF-d_7: $\delta^{15N}_{lig} = -141.2$ ppm (240.5 ppm in respect to liquid NH_3) [34].

[aaf]vs N(7) in purine moiety in 6-(4-methylphenyl)methylamino-9-β-D-ribofuranosyl-*9H*-purine in DMF-d_7: $\delta^{15N}{}_{lig} = -140.6$ ppm (241.1 ppm in respect to liquid NH_3) [34].
[aag]vs N(7) in purine moiety in 6-(4-fluorophenyl)methylamino-9-β-D-ribofuranosyl-*9H*-purine in DMF-d_7: $\delta^{15N}{}_{lig} = -140.7$ ppm (241.0 ppm in respect to liquid NH_3) [34].
[aah]vs N(7) in purine moiety in 6-(4-chlorophenyl)methylamino-9-β-D-ribofuranosyl-*9H*-purine in DMF-d_7: $\delta^{15N}{}_{lig} = -140.6$ ppm (241.1 ppm in respect to liquid NH_3) [34].
[aai]vs N(7) in purine moiety in 6-(4-methoxyphenyl)methylamino-9-β-D-ribofuranosyl-*9H*-purine in DMF-d_7: $\delta^{15N}{}_{lig} = -141.1$ ppm (240.6 ppm in respect to liquid NH_3) [34].
[aaj]vs N(7) in purine moiety in 6-(2-hydroxy-3-methoxyphenyl)methylamino-9-β-D-ribofuranosyl-*9H*-purine in DMF-d_7: $\delta^{15N}{}_{lig} = -140.5$ ppm (241.2 ppm in respect to liquid NH_3) [34].
[aak]vs N(7) in purine moiety in 2-chloro-6-(3-hydroxyphenyl)methylamino-9-β-D-ribofuranosyl-*9H*-purine in DMF-d_7: $\delta^{15N}{}_{lig} = -140.5$ ppm (241.2 ppm in respect to liquid NH_3) [34].
[aal]vs N(7) in purine moiety in 2-chloro-6-(4-hydroxyphenyl)methylamino-9-β-D-ribofuranosyl-*9H*-purine in DMF-d_7: $\delta^{15N}{}_{lig} = -140.1$ ppm (241.6 ppm in respect to liquid NH_3) [34].
[aam]vs N(7) in purine moiety in 2-chloro-6-(2-hydroxy-5-methylphenyl)methylamino-9-β-D-ribofuranosyl-*9H*-purine in DMF-d_7: $\delta^{15N}{}_{lig} = -140.2$ ppm (241.5 ppm in respect to liquid NH_3) [34].
[aan]vs N(7) in purine moiety in 2-chloro-6-(2-hydroxy-3-methoxyphenyl)methylamino -9-β-D-ribofuranosyl-*9H*-purine in DMF-d_7: $\delta^{15N}{}_{lig} = -139.9$ ppm (241.8 ppm in respect to liquid NH_3) [34].
[aao]vs N(7) in purine moiety in 6-(furan-2-yl)methylamino-9-β-D-ribofuranosyl-*9H*-purine in DMF-d_7 (at 343 K): $\delta^{15N}{}_{lig} = -139.9$ ppm [23].
[aaaa]The only stereomeric possibility, resulting from the molecular formula.

Table 15 Square-planar Pt(II) (d^8), and octahedral Pt(IV) and Ru(II) (d^6, low-spin) complexes with 1,2,4-triazolo-[1,5*a*]-pyrimidines (**30** ring systems), ^{15}N NMR studied in 2012–20.

Complex	^{15}N chemical shift (δ^{15N}_{compl}, ppm)	^{15}N coordination shift (Δ^{15N}_{coord}, ppm)	Nitrogen *trans* to
Pt(II) complexes			
cis-[Pt^{II}(5-methyl-7-isobutyl-1,2,4-triazolo-[1,5*a*]-pyrimidine-κ^1-N^3)$_2Cl_2$]$^{DMF\text{-}d7}$ [38][A]	$N^{TP,N3}$: −245.3	−93.3[aaa]	Cl [X-ray]
cis-[Pt^{II}(5,7-diethyl-1,2,4-triazolo-[1,5*a*]-pyrimidine-κ^1-N^3)$_2$ Cl_2]$^{DMF\text{-}d7}$ [39][a]	$N^{TP,N3}$: −244.8	−93.0[aab]	(Cl)[a]
cis-[Pt^{II}(5-methyl-7-isobutyl-1,2,4-triazolo-[1,5*a*]-pyrimidine-κ^1-N^3)(NH_3)Cl_2]$^{DMF\text{-}d7}$ [38][B]	$N^{TP,N3}$: −246.4	−94.4[aaa]	Cl [X-ray]
cis-[Pt^{II}(5,7-diethyl-1,2,4-triazolo-[1,5*a*]-pyrimidine-κ^1-N^3) (NH_3)Cl_2]$^{DMF\text{-}d7}$ [39][a]	$N^{TP,N3}$: −246.3	−94.5[aab]	(Cl)[a]
cis-[Pt^{II}(5-methyl-7-isobutyl-1,2,4-triazolo-[1,5*a*]-pyrimidine-κ^1-N^3)(CH_3SOCH_3)Cl_2]$^{DMF\text{-}d7}$ [38][b2,C]	$N^{TP,N3}$: −238.4	−86.4[aaa]	Cl [X-ray]
trans-[Pt^{II}(5-methyl-7-isobutyl-1,2,4-triazolo-[1,5*a*]-pyrimidine-κ^1-N^3)(CH_3SOCH_3)Cl_2]$^{DMF\text{-}d7}$ [38][b1]	$N^{TP,N3}$: −228.8	−76.8[aaa]	(S)[b1]
cis-[Pt^{II}(5,7-diethyl-1,2,4-triazolo-[1,5*a*]-pyrimidine-κ^1-N^3) (CH_3SOCH_3)Cl_2]$^{DMF\text{-}d7}$ [39][D]	$N^{TP,N3}$: −237.2	−85.4[aab]	Cl [X-ray]
cis-[Pt^{II}(5-methyl-7-isobutyl-1,2,4-triazolo-[1,5*a*]-pyrimidine-κ^1-N^3)$_2I_2$]$^{DMF\text{-}d7}$ [40][a]	$N^{TP,N3}$: −233.4	−81.4[aaa]	(I)[a]
cis-[Pt^{II}(5,7-diethyl-1,2,4-triazolo-[1,5*a*]-pyrimidine-κ^1-N^3)$_2I_2$]$^{DMF\text{-}d7}$ [40][a]	$N^{TP,N3}$: −233.1	−81.3[aab]	(I)[a]
cis-[Pt^{II}(5,6,7-trimethyl-1,2,4-triazolo-[1,5*a*]-pyrimidine-κ^1-N^3)$_2I_2$]$^{DMF\text{-}d7}$ [40][a]	$N^{TP,N3}$: −234.7	−82.5[aac]	(I)[a]

cis-[Pt^{II}(5,7-di-*tert*-butyl-1,2,4-triazolo-[1,5*a*]-pyrimidine-κ^1-N^3)$_2$ (CH_3COO)$_2$]^{CDCl3} [41][a,c]	$N^{TP,N3}$: −258.0/−253.9[c]	−102.9/−98.8[aad]	(O)[a]
cis-[Pt^{II}(5,7-dimethyl-1,2,4-triazolo-[1,5*a*]-pyrimidine-κ^1-N^3)$_2$ ($CH_2(COO)_2$)]$^{DMF\text{-}d7}$ [42][E]	$N^{TP,N3}$: −258.7	−104.8[aae]	O [X-ray]
cis-[Pt^{II}(5-methyl-7-isobutyl-1,2,4-triazolo-[1,5*a*]-pyrimidine-κ^1-N^3)$_2$($CH_2(COO)_2$)]$^{acetone\text{-}d6}$ [42]	$N^{TP,N3}$: −256.1	−104.1[aaa,aaf]	(O)[aaaa]
cis-[Pt^{II}(5,7-di-*tert*-butyl-1,2,4-triazolo-[1,5*a*]-pyrimidine-κ^1-N^3)$_2$($CH_2(COO)_2$)]^{CDCl3} [42][F]	$N^{TP,N3}$: −254.7	−99.6[aad]	O [X-ray]
[Pt^{II}(5,7-dimethyl-1,2,4-triazolo-[1,5*a*]-pyrimidine-κ^1-N^3)$_2$ ($OOCCH_2CH_2CH_2COO$))]^{D2O} [43]	$N^{TP,N3}$: −257.0	−103.1[aae,aag]	(O)[aaaa]
[Pt^{II}(5-methyl-7-isobutyl-1,2,4-triazolo-[1,5*a*]-pyrimidine-κ^1-N^3)$_2$($OOCCH_2CH_2CH_2COO$))]$^{acetone\text{-}d6}$ [43]	$N^{TP,N3}$: −254.1	−102.1[aaa,aaf]	(O)[aaaa]
[Pt^{II}(5,7-di-*tert*-butyl-1,2,4-triazolo-[1,5*a*]-pyrimidine-κ^1-N^3)$_2$ ($OOCCH_2CH_2CH_2COO$))]^{CDCl3} [43]	$N^{TP,N3}$: −255.0	−99.9[aad]	(O)[aaaa]
[Pt^{II}(5,7-dimethyl-1,2,4-triazolo-[1,5*a*]-pyrimidine-κ^1-N^3)$_2$ (cyclobutane-1,1-dicarboxylate)]^{D2O} [43]	$N^{TP,N3}$: −257.0	−103.1[aae,aag]	(O)[aaaa]
[Pt^{II}(5-methyl-7-isobutyl-1,2,4-triazolo-[1,5*a*]-pyrimidine-κ^1-N^3)$_2$(cyclobutane-1,1-dicarboxylate)]$^{acetone\text{-}d6}$ [43]	$N^{TP,N3}$: −255.5	−103.5[aaa,aaf]	(O)[aaaa]
[Pt^{II}(5,7-di-*tert*-butyl-1,2,4-triazolo-[1,5*a*]-pyrimidine-κ^1-N^3)$_2$ (cyclobutane-1,1-dicarboxylate)]^{CDCl3} [43][G]	$N^{TP,N3}$: −253.9	−98.8[aad]	O [X-ray]
[Pt^{II}(5,7-dimethyl-1,2,4-triazolo-[1,5*a*]-pyrimidine-κ^1-N^3) (CH_3SOCH_3)($CH_2(COO)_2$)]$^{DMF\text{-}d7}$ [44]	$N^{TP,N3}$: −252.2	−98.3[aae]	(O)[aaaa]

Continued

Table 15 Square-planar Pt(II) (d^8), and octahedral Pt(IV) and Ru(II) (d^6, low-spin) complexes with 1,2,4-triazolo-[1,5*a*]-pyrimidines (**30** ring systems), ^{15}N NMR studied in 2012–20.—cont'd

Complex	^{15}N chemical shift (δ^{15N}_{compl}, ppm)	^{15}N coordination shift (Δ^{15N}_{coord}, ppm)	Nitrogen *trans* to
$[Pt^{II}$(5,7-dimethyl-1,2,4-triazolo-[1,5*a*]-pyrimidine-κ^1-N^3)(CH_3SOCH_3)(cyclobutane-1,1-dicarboxylate)$]^{DMF\text{-}d7}$ [44]	$N^{TP,N3}$: −251.6	-100.9^{aah}	$(O)^{aaaa}$
$[Pt^{II}$(5,7-diphenyl-1,2,4-triazolo-[1,5*a*]-pyrimidine-κ^1-N^3)(CH_3SOCH_3)($CH_2(COO)_2$)$]^{DMF\text{-}d7}$ [44]	$N^{TP,N3}$: −251.7	-97.8^{aae}	$(O)^{aaaa}$
$[Pt^{II}$(5,7-diphenyl-1,2,4-triazolo-[1,5*a*]-pyrimidine-κ^1-N^3)(CH_3SOCH_3)(cyclobutane-1,1-dicarboxylate)$]^{DMF\text{-}d7}$ [44]	$N^{TP,N3}$: −251.6	-100.9^{aah}	$(O)^{aaaa}$
cis-$[Pt^{II}$(4,7-dihydro-5-methyl-1,2,4-triazolo-[1,5*a*]-pyrimidin-7-one-κ^1-$N^3)_2I_2]^{DMSO\text{-}d6}$ [40][a]	$N^{TP,N3}$: −256.3	-90.6^{aai}	$(I)^{a}$
Pt(IV) complexes			
cis-$[Pt^{IV}$(5-methyl-7-isobutyl-1,2,4-triazolo-[1,5*a*]-pyrimidine-κ^1-$N^3)_2Cl_4]^{DMF\text{-}d7}$ [45][H]	$N^{TP,N3}$: −247.7	-95.7^{aaa}	(Cl) $^{X\text{-}ray}$
Ru(II) complexes			
$[Ru^{II}$(1,2,4-triazolo-[1,5*a*]-pyrimidine-κ^1-N^3)(1-isopropyl-4-methylbenzene)$Cl_2]^{CDCl3}$ [46][d]	$N^{TP,N3}$: −213.4	-63.5^{aaj1}	$(C{=}C^{bz})^{aaaa}$
$[Ru^{II}$(5,7-dimethyl-1,2,4-triazolo-[1,5*a*]-pyrimidine-κ^1-N^3)(1-isopropyl-4-methylbenzene)$Cl_2]^{CDCl3}$ [46][d,I]	$N^{TP,N3}$: −217.1	-63.1^{aaj2}	$C{=}C^{bz\ X\text{-}ray}$
$[Ru^{II}$(5-methyl-7-isobutyl-1,2,4-triazolo-[1,5*a*]-pyrimidine-κ^1-N^3)(1-isopropyl-4-methylbenzene)$Cl_2]^{CDCl3}$ [46][d,J]	$N^{TP,N3}$: −217.4	-63.4^{aaj3}	$C{=}C^{bz\ X\text{-}ray}$
$[Ru^{II}$(5,7-di-*tert*-butyl-1,2,4-triazolo-[1,5*a*]-pyrimidine-κ^1-N^3)(1-isopropyl-4-methylbenzene)$Cl_2]^{CDCl3}$ [46][d]	$N^{TP,N3}$: −218.0	-60.9^{aaj5}	$(C{=}C^{bz})^{aaaa}$

$[Ru^{II}$(5,7-diphenyl-1,2,4-triazolo-[1,5*a*]-pyrimidine-κ^1-N^3)(1-isopropyl-4-methylbenzene)$Cl_2]^{CDCl3}$ [46][d,K]	$N^{TP,N3}$: −215.4	−54.5[aaj6]	C=C[bz X-ray]
trans(Cl,S),(Cl,S)-$[Ru^{II}$(1,2,4-triazolo-[1,5*a*]-pyrimidine-κ^1-N^3)$(CH_3SOCH_3)_3Cl_2]^{CDCl3}$ [47][e1]	$N^{TP,N3}$: −210.0	−60.1[aaj1]	(S)[e1]
trans(Cl,S),(Cl,S)-$[Ru^{II}$(5,7-dimethyl-1,2,4-triazolo-[1,5*a*]-pyrimidine-κ^1-N^3)$(CH_3SOCH_3)_3Cl_2]^{CDCl3}$ [47][e1]	$N^{TP,N3}$: −214.6	−63.6[aaj2]	(S)[e1]
trans(Cl,S),(Cl,S)-$[Ru^{II}$(5-methyl-7-isobutyl-1,2,4-triazolo-[1,5*a*]-pyrimidine-κ^1-N^3)$(CH_3SOCH_3)_3Cl_2]^{CDCl3}$ [47][e2,L]	$N^{TP,N3}$: −215.2	−61.2[aaj3]	S [X-ray]
trans(Cl,S),(Cl,S)-$[Ru^{II}$(5,6,7-trimethyl-1,2,4-triazolo-[1,5*a*]-pyrimidine-κ^1-N^3)$(CH_3SOCH_3)_3Cl_2]^{CDCl3}$ [48][e2,M]	$N^{TP,N3}$: −216.2	−64.0[aak]	S [X-ray]
trans(Cl,S),(Cl,N)-$[Ru^{II}$(5,7-diethyl-1,2,4-triazolo-[1,5*a*]-pyrimidine-κ^1-$N^3)_2(CH_3SOCH_3)_2Cl_2]^{CDCl3}$ [47][f2,N]	$N^{TP-A,N3}$, $N^{TP-B,N3}$: both −205.5	−51.8[aaj4]	Cl [X-ray] S [X-ray]
trans(Cl,S),(Cl,N)-$[Ru^{II}$(5,7-di-*tert*-butyl-1,2,4-triazolo-[1,5*a*]-pyrimidine-κ^1-$N^3)_2(CH_3SOCH_3)_2Cl_2]^{CDCl3}$ [47][f2,O]	$N^{TP-A,N3}$, $N^{TP-B,N3}$: both −206.8	−49.7[aaj5]	Cl [X-ray] S [X-ray]
trans(Cl,S),(Cl,N)-$[Ru^{II}$(5,7-diphenyl-1,2,4-triazolo-[1,5*a*]-pyrimidine-κ^1-$N^3)_2(CH_3SOCH_3)_2Cl_2]^{CDCl3}$ [47][f1]	$N^{TP-A,N3}$, $N^{TP-B,N3}$: both −205.0	−44.1[aaj6]	(Cl)[f1] (S)[f1]
trans(Cl,S),(Cl,S)-$[Ru^{II}$(4,7-dihydro-5-methyl-1,2,4-triazolo-[1,5*a*]-pyrimidin-7-one-κ^1-N^3)$(CH_3SOCH_3)_3Cl_2]^{DMSO\text{-}d6}$ [49][e1]	$N^{TP,N3}$: −232.2	−66.5[aai]	(S)[e1]

[a]These are probably *cis* isomers.
[b1-b2]The assumed *trans* form ([b1]) is isomeric to the X-ray studied *cis* species ([b2]) [38].
[c]Two rotamers are present in $CDCl_3$.
[d]1-Isopropyl-4-methylbenzene ligand occupies three coordination sites, within octahedral geometry.
[e1-e2]The assumed *trans*(Cl,S),(Cl,S) isomers ([e1]) are analogous to the X-ray studied *trans*(Cl,S),(Cl,S)-$[Ru^{II}$(5-methyl-7-isobutyl-1,2,4-triazolo-[1,5*a*]-pyrimidine--κ^1-N^3)$(CH_3SOCH_3)_3Cl_2]$ and *trans*(Cl,S),(Cl,S)-$[Ru^{II}$(5,6,7-trimethyl-1,2,4-triazolo-[1,5*a*]-pyrimidine-κ^1-N^3)$(CH_3SOCH_3)_3Cl_2]$ ([e2]) [47,48].
[f1-f2]The assumed *trans*(Cl,S),(Cl,N) isomer ([f1]) is analogous to the X-ray studied *trans*(Cl,S),(Cl,N)-$[Ru^{II}$(5,7-diethyl-1,2,4-triazolo-[1,5*a*]-pyrimidine-κ^1-$N^3)_2(CH_3SOCH_3)_2Cl_2]$ and *trans*(Cl,S),(Cl,N)-$[Ru^{II}$(5,7-di-*tert*-butyl-1,2,4-triazolo-[1,5*a*]-pyrimidine-κ^1-$N^3)_2(CH_3SOCH_3)_2Cl_2]$ ([f2]) [47].

Continued

Table 15 Square-planar Pt(II) (d^8), and octahedral Pt(IV) and Ru(II) (d^6, low-spin) complexes with 1,2,4-triazolo-[1,5*a*]-pyrimidines (**30** ring systems), ^{15}N NMR studied in 2012–20.—cont'd

[aaa]vs N(3) in 5-methyl-7-isobutyl-1,2,4-triazolo-[1,5*a*]-pyrimidine in DMF-d_7: $\delta^{15N}{}_{lig} = -152.0$ [38,45].

[aab]vs N(3) in 5,7-diethyl-1,2,4-triazolo-[1,5*a*]-pyrimidine in DMF-d_7: $\delta^{15N}{}_{lig} = -151.8$ ppm [39].

[aac]$\Delta^{15N}{}_{coord}$ was published directly (with no $\delta^{15N}{}_{lig}$, which can be determined as $\delta^{15N}{}_{compl} - \Delta^{15N}{}_{coord}$ difference: N(3) in 5,6,7-trimethyl-1,2,4-triazolo-[1,5*a*]-pyrimidine −152.2 ppm, in DMF-d_7) [40].

[aad]vs N(3) in 5,7-di-*tert*-butyl-1,2,4-triazolo-[1,5*a*]-pyrimidine in $CDCl_3$: $\delta^{15N}{}_{lig} = -155.1$ ppm [42,50].

[aae]vs N(3) in 5,7-dimethyl-1,2,4-triazolo-[1,5*a*]-pyrimidine in DMF-d_7: $\delta^{15N}{}_{lig} = -153.9$ ppm [51].

[aaf]^{15}N NMR data for 5-methyl-7-isobutyl-1,2,4-triazolo-[1,5*a*]-pyrimidine in acetone-d_6 are unavailable.

[aag]^{15}N NMR data for 5,7-dimethyl-1,2,4-triazolo-[1,5*a*]-pyrimidine in D_2O are unavailable.

[aah]vs N(3) in 5,7-diphenyl-1,2,4-triazolo-[1,5*a*]-pyrimidine in DMF-d_7: $\delta^{15N}{}_{lig} = -150.7$ ppm [52].

[aai]vs N(3) in 4,7-dihydro-5-methyl-1,2,4-triazolo-[1,5*a*]-pyrimidin-7-one in DMSO-d_6: $\delta^{15N}{}_{lig} = -165.7$ ppm [53,54].

[aaj1-aaj6]$\Delta^{15N}{}_{coord}$ were published directly (with no $\delta^{15N}{}_{lig}$, which can be determined as $\delta^{15N}{}_{compl} - \Delta^{15N}{}_{coord}$ differences: N(3) in 1,2,4-triazolo-[1,5*a*]-pyrimidine −149.9 ppm ([aaj1]), N(3) in 5,7-dimethyl-1,2,4-triazolo-[1,5*a*]-pyrimidine −154.0 ppm or −154.0 ppm ([aaj2]), N(3) in 5-methyl-7-isobutyl-1,2,4-triazolo-[1,5*a*]-pyrimidine −154.0 ppm ([aaj3]), N(3) in 5,7-diethyl-1,2,4-triazolo-[1,5*a*]-pyrimidine −153.7 ppm ([aaj4]), N(3) in 5,7-di-*tert*-butyl-1,2,4-triazolo-[1,5*a*]-pyrimidine −157.1 ppm ([aaj5]), N(3) in 5,7-diphenyl-1,2,4-triazolo-[1,5*a*]-pyrimidine −160.9 ([aaj6]), all in $CDCl_3$) [47,49].

[aak]vs N(3) in 5,6,7-trimethyl-1,2,4-triazolo-[1,5*a*]-pyrimidine in DMF-d_7 (^{15}N NMR data in $CDCl_3$ being unavailable): $\delta^{15N}{}_{lig} = -152.2$ ppm [40].

[aaaa]The only stereomeric possibility, resulting from the molecular formula.

[A]X-ray structure: ZEDQUX [38].

[B]X-ray structure: ZEDREI [38].

[C]X-ray structure: ZEDRAE [38].

[D]X-ray structure: IVUTIE [39].

[E]X-ray structure of H_2O solvate: ZUJYUA [42].

[F]X-ray structure of $CHCl_3$ solvate: ZUJZAH [42].

[G]X-ray structure of $CHCl_3$ solvate: SEBRUP [43].

[H]X-ray structure: ZISNEX [45].

[I]X-ray structure: TUZRIR [49].

[J]X-ray structure: TUZROX [49].

[K]X-ray structure: TUZRUD [49].

[L]X-ray structure of CH_3OH solvate: JUDBOC [47].

[M]X-ray structure: MEJTUT [48].

[N]X-ray structure: JUCQEG [47].

[O]X-ray structure of CH_3OH solvate: JUCQIK [47].

In the first column the molecular formulae are listed, together with the respective counterions for ionic species. In superscript the solvent is given, and the temperature—however, except the common range 293–303 K (if it is within this range, there is no need to specify it, especially that in many papers it was not reported at all or described simply as "room temperature," "ambient conditions," etc.). The coordination mode is also shown, except for cases when it is obvious and typical for a given ligand. If the X-ray structure is available, it is indicated by the respective footnote (A, B, C, ...), together with its reference code from the Cambridge Crystallographic Database (June 2020).

In the second column, the parent azine or azole ring system of a given (usually variously substituted) heterocycle is indicated by the appropriate number (**1–28**), except for Tables 14 and 15 where this is not necessary as they simply contain derivatives of **29** or **30**.

In the third (the second for Tables 14 and 15) column the ^{15}N NMR chemical shifts are listed, all referenced to neat nitromethane; if the original δ^{15N} values were published in respect to any other NMR references, they have been recalculated to CH_3NO_2 basing on the known relative shifts [37].

In the fourth column (the third for Tables 14 and 15) the ^{15}N NMR coordination shifts, calculated by subtracting the ^{15}N NMR chemical shifts for the reviewed complexes or organometallics and the respective ligands (preferably in the same or at least similar solvents) are listed.

In the fifth (the fourth for Tables 14 and 15), optional column the type of a donor atom in *trans* position in respect to the concerned nitrogen (upon square-planar or octahedral geometry) is indicated.

For nearly all reviewed complexes and organometallics the ^{15}N shielding effects occur, thus the $\Delta^{15N}{}_{coord}$ parameters are negative, and it is more convenient to use their absolute values, $|\Delta^{15N}{}_{coord}|$; hence, they have been discussed in the main text in this form.

The reason for ^{15}N shielding effect upon transition metal complexation of azines and azoles is known. Generally, this phenomenon is analogous to that observed upon protonation of pyridine-like nitrogens [172] and caused by engagement of lone-pair electrons of such N: atoms in the σ covalent bonding (with H^+ or M^{n+}). It results in the replacement of the most pronounced $n_N \rightarrow \pi^*$ electronic transition by a weaker but much more energetic $\pi \rightarrow \pi^*$ one, followed by the increase of mean energy of electron excitation (ΔE) and the decrease of absolute value of paramagnetic contribution ($|\sigma_{para,}{}^{15N}|$) in the ^{15}N shielding constant (σ^{15N})—as $|\sigma_{para}|$ is proportional to $1/\Delta E$, in accordance with the Karplus-Pople equation (its alternative form is called the Ramsey equation) [173–175]:

$$|\sigma_{para}| = \frac{e^2}{m^2}\frac{h^2}{c^2}\Delta E^{-1} r_{2p}^{-3}\left[Q_{AA} + \sum Q_{AX}\right]$$

where r_{2p} is the distance between the 2p electron and the ^{15}N nucleus, while $[Q_{AA} + \Sigma Q_{AX}]$ reflects charge density and the presence of multiple bonds around the ^{15}N nucleus.

The detailed discussion of all physical quantities in the above equations are beyond the scope of this review, nevertheless, the decrease of $|\sigma_{para,}{}^{15N}|$ contribution is evident. Since in the whole expression for nuclear shielding the paramagnetic term appears with a negative sign, the total ^{15}N shielding constant (σ^{15N}) increases; thus the ^{15}N chemical shift (δ^{15N}) decreases and the respective ^{15}N coordination shift ($\Delta^{15N}{}_{coord}$) is negative.

Moreover, for the most heavy metal complexes (from the last transition series, i.e., Pt, Ir, Os, Au, Hg) at least partly the observed ^{15}N shielding effects are probably the result of scalar and spin-orbit relativistic effects.

1. Transition metal complexes and organometallics with azines

In Review 2013, 273 ^{15}N NMR-studied transition metal complexes or organometallics with azines (Ni(II)—0, Pd(II)—75, Pt(II)—44, Pd(IV)—1, Pt(IV)—8, Rh(I)—16, Ir(I)—0, Co(III)—12, Rh(III)—39, Ir(III)—14, Fe(II)—7, Ru(II)—11, Os(II)—3, Cu(I)—2, Ag(I)—12, Au(I)—0, Au(III)—20, Zn(II)—3, Cd(II)—5, Hg(II)—1) were discussed in respect to their ^{15}N chemical and coordination shifts (paragraph 1, including tables 2.1–2.14). In the present paper 277 ^{15}N NMR-studied species are described (Ni(II)—0, Pd(II)—67, Pt(II)—37, Pt(IV)—0, Rh(I)—11, Ir(I)—4, Co(III)—3, Rh(III)—2, Ir(III)—31, Fe(II)—6, Ru(II)—62, Os(II)—3, Cu(I)—2, Ag(I)—18, Au(I)—1, Au(III)—25, Zn(II)—3, Cd(II)—2, Hg(II)—0).

There are no Ni(II) species, whereas those of Pd(II) are very numerous (67 examples; Table 1). Ten *trans*(Cl,$C^{ph,C2}$)-[Pd^{II}(2-phenylpyridine*-κ^2-N,$C^{ph,C2}$)LCl] organometallics (L = NH_3, pyridine, 2-, 3-, 4-methylpyridine, 2,3-, 2,4-, 3,5-, 2,6-dimethylpyridine, 2,4,6-trimethylpyridine), [Pd^{II}(2-phenylpyridine*-κ^2-N,$C^{ph,C2}$)(μ-Cl)]$_2$ dimer and [Pd^{II}(2-phenylpyridine*-κ^2-N,$C^{ph,C2}$)(DMSO)$_2$]$^+$ [Pd^{II}(2-phenylpyridine*-κ^2-N,$C^{ph,C2}$)Cl_2]$^-$ ionic pair {2-phenylpyridine* (2ppy*): deprotonated, at the phenyl C(2) atom, anionic (−1), $N(1)^{py}$,$C(2)^{ph}$-chelating form of 2-phenylpyridine} were studied by Niedzielska et al. [2] and Pawlak et al. [3]. Two *trans*-[Pd^{II}(4-methylpyridine)(ylidene)I_2] complexes with imidazol-2-ylidenes were reported by Martinez-Olid et al. [4,5], while seven *trans*-[Pd^{II}(4-*tert*-butylpyridine)$_2X_2$] and *trans*-[Pd^{II}(4-*tert*-butylpyridine)$_2$XY] ones (X = N-methoxy-2,6-difluorobenzamidate; Y = CH_3COO, C_6H_5COO) by Tereniak et al. [6]. *Trans*-[Pd^{II}(3-chloropyridine)XCl_2] (X = derivative of 1,2,3-triazolium-5-yl zwitterion) was described by Virant et al. [7]. Six [Pd^{II}(LL^1-κ^2-N^{py},$N^{NH2/NH/N}$)Cl_2] complexes with 2-arylaminopyridines (LL^1) were characterized by Requet et al. [8], while five [Pd^{II}(LL^2-κ^2-N^{py}, $N^{N=}$)(CH_3)Cl] complexes with 2-aryliminopyridines (LL^2) by Rosar et al. [9]. [Pd^{II}(2-(*1H*-benzotriazol-1-yl)pyridine-κ^2-N^{py},$N^{btz,N2}$)Cl_2] was investigated by Dikmen et al. [10]. Two [Pd^{II}(2-(1-(pyrid-2-yl)-5-(4-methoxyphenyl)-*1H*-pyrazol-3-yl)pyridine-κ^2-$N^{pyA/pyB}$,$N^{pz,N2}$)Cl_2] isomers were researched by Soria et al. [11].

[Pd^{II}(2,2′-bipyridine)(mefenamate)$_2$] {mefenemate = 2-(2,3-dimethylphenylamino)benzoate} was studied by Carvalho et al. [12]. Five [Pd^{II}(2,2′-bipyridine)(thiourea#)$_2$]$^{2+}$ and five [Pd^{II}(1,10-phenanthroline)(thiourea#)$_2$]$^{2+}$ cations {thiourea#: S-bonded thiourea, its N-alkyl or N,N′-dialkyl derivative} were reported by Rotondo et al. [13]. [Pd^{II}(bis(*tert*-butyl) 2,2′-bipyridine-3,3′-dicarboxylate)$_2$]$^{2+}$ was described by Campos-Carrasco et al. [14]. Three Pd(II) organometallics with 8-methylquinoline*-κ^2-N,C^{CH2} {8-methylquinoline*: deprotonated, at the methyl side group, anionic (−1), $N(1)^{quin}$,C^{CH2}-chelating form of 8-methylquinoline} and [Pd^{II}_2(quinoline-8-carboxylate-κ^2-N,O)$_2$(μ-CH_3COO)$_2$] were characterized by Wenzel et al. [15], while another Pd(II) complex with 8-substituted quinoline derivative by Filipovic et al. [16]. Two *trans*-[$Pd^{II}L_2(CH_3COO)_2$] complexes (L = 4-(2-bromophenoxy)quinoline, 4-(2-bromophenoxy)-7-bromoquinoline) were investigated by Shanahan et al. [17]. Twelve Pd(II) carboxylate complexes with 9,9-dimethyl-4,5-diaza-*9H*-fluorene-κ^2-N^4,N^5 or 4,5-diaza-*9H*-fluoren-9-one-κ^1-N^4/-κ^2-N^4,N^5/-μ-N^4,N^5 were researched by White et al. [18] and Jaworski et al. [19].

Pt(II) species are very numerous as well (37 examples; Table 1). [Pt^{II}(2-phenylpyridine*-κ^2-N,$C^{ph,C2}$)(μ-Cl)]$_2$ dimer was studied by Pawlak et al. [3]. [Pt^{II}(2-iminomethylpyridine-κ^2-N^{py},$N^{NH=}$)Cl_2] was reported by Pinter et al. [20]. Three [Pt^{II}(L-κ^1-N^{py})$_2Cl_2$] complexes with *n*-dodecylureaylpyridines (L) were described by Dawn et al. [21]. Two Pt(II) organometallics with 2-phenylpyridine* derivatives and unsubstituted pyridine were characterized by Hebenbrock et al. [22], while [Pt^{II}(1-β-

D-ribofuranosyl-4-(pyrid-2-yl)-*1H*-1,2,3-triazole-κ^{2}-N^{py},N^{tz,N3})Cl$_2$] by Jakubowski et al. [23].

Five [Pt^{II}(2,2′-bipyridine)(thiourea#)$_2$]$^{2+}$ and five [Pt^{II}(1,10-phenanthroline)(thiourea#)$_2$]$^{2+}$ cations were studied by Rotondo et al. [13]. *Cis*-[Pt^{II}(4-azaindole-κ^{1}-N^{4})$_2$I$_2$] and *cis*-[Pt^{II}(4-azaindole-κ^{1}-N^{4})(NH$_3$)I$_2$], as well as eight *cis*-[Pt^{II}(L-κ^{1}-N^{7})$_2$Cl$_2$], three *trans*-[Pt^{II}(L-κ^{1}-N^{7})$_2$Cl$_2$] and six [Pt^{II}(L-κ^{1}-N^{7})$_2$(C$_2$O$_4$)] complexes with 7-azaindole derivatives (L) were reported by Starha et al. [24–28].

In contrast, there are no Pt(IV) species.

Rh(I) species are moderately numerous (11 examples; Table 2). *Trans* (Cl,C=C^{cyclooct})-[Rh^{I}(pyridine)(ylidene)(cyclooctene)Cl] was studied by Rubio-Perez et al. [55], while two *trans*(Cl,C=C^{vinyl})-[Rh^{I}(pyridine)(ylidene)(vinylazole)Cl] complexes (vinylazole = 1-vinyl-*1H*-pyrazole, 1-vinyl-3,5-dimethyl-*1H*-pyrazole) and *trans*(Cl,C=C^{vinyl})-[Rh^{I}(2-vinylpyridine-κ^{2}-N,C$^{C=C}$)(ylidene)Cl] organometallic by Azpiroz et al. [56,57] (in all: ylidene = 1,3-bis(2,6-diisopropylphenyl)imidazol-2-ylidene). *Trans* (Cl,C=C^{cyclooct})-[Rh^{I}(pyrid-2-ylacetonitrile-κ^{1}-N^{py})(ylidene)(cyclooctene)Cl], *trans*(CO,O)-[Rh^{I}(pyrid-2-ylmethanolate-κ^{2}-N,O)(ylidene)(CO)] and *trans*(CO,S)-[Rh^{I}(pyrid-2-ylmethanol-κ^{1}-N)(ylidene)(CO)(C$_6$H$_5$CH$_2$S)] (ylidene = 1,3-bis(2,6-diisopropylphenyl)imidazol-2-ylidene) were reported by Palacios et al. [58,59], who described also three [Rh^{I}(quinolin-8-olate-κ^{2}-N,O)(1,3-bis(2,6-diisopropylphenyl)imidazol-2-ylidene)(alkene)] complexes (alkene = ethylene, cyclooctene) [60]. Rh(I) complex with 1,5-naphthyridine derivative was characterized by Sieb et al. [61].

Ir(I) species are noticeably less abundant than those of Rh(I), being very rare (4 examples; Table 2). Two [Ir^{I}(LL-κ^{2}-N,O)(diene)] complexes (LL = quinolin-8-olate, 2-methylquinolin-8-olate) were studied by Rampersadh et al. [62]. [Ir^{I}(pyridazine-κ^{1}-N^{1})(ylidene)(diene)]$^{+}$ and [Ir^{I}(phthalazine-κ^{1}-N^{2})(ylidene)(diene)]$^{+}$ (ylidene = 1,3-bis(2,4,6-trimethylphenyl)imidazol-2-ylidene; diene = 1,5-cyclooctadiene) were reported by Appleby et al. [63].

Co(III) and Rh(III) species are noticeably less abundant than those of Rh(I), being very rare (3 and 2 examples, respectively; Table 3). *Mer*-[Co^{III}(2-(2-(trifluoromethyl)vinyl-2-olate)pyridine-κ^{2}-N,O)$_3$] was studied by Buyukyazi et al. [84]. [Co^{III}(2,5-bis(pyrid-2-yl)-3,4-dimethylpyrrolate-κ^{3}-N^{py},N$^{py'}$,N^{pr})$_2$]$^{2+}$ was reported by McPherson et al. [85]. *Trans*(N^{bpy,N1},NH^{X}), *trans*(N$^{bpy,N1'}$,NO^{Y})-[Co^{III}(2,2′-bipyridine)XY] (X = NHCOC(*CN*)NO, Y = NHC(OCH$_2$CH$_3$)C(*CN*)NO) was described by Vitushkina et al. [86]. *Trans*(H,N^{py}),*trans*(Cl,CCSi)-[Rh^{III}(pyridine)$_2$(1,3-bis(2,6-diisopropylphenyl) imidazol-2-ylidene)((CH_3)$_3$SiCC)(H)Cl] was characterized by Rubio-Perez et al. [55]. [Rh^{III}(5,5′-bis(pyren-1-yl)-2,2′-bipyridine)(pentamethyl-cyclopentadienyl)Cl]$^+$ was investigated by Wisser et al. [87].

Ir(III) species are significantly more abundant than those of Co(III), Rh(III) and Ir(I), being very numerous (31 examples; Table 3). Eight *fac*-[Ir^{III}(pyridine)$_3$(ylidene)(H)$_2$]$^+$ cations with imidazol-2-ylidenes or dihydroimidazol-2-ylidenes were studied by Lloyd et al. [88] and Wong et al. [89]. *Trans*(H,N^{py}),*trans*(H,N^{CH3CN})-[Ir^{III}(pyridine)(1,3-bis(2,4,6-trimethylphenyl)imidazol-2-ylidene)(P(C_6H_5)$_3$)(CH_3CN)(H)$_2$]$^+$ was reported by Fekete et al. [90]. *Trans*(H,N^{py}),*trans*(H,Cl)-[Ir^{III}(pyridine)$_2$(cyclooctene)(H)$_2$Cl] was described by Iali et al. [91]. Four *trans*(H,N^{py})- or *trans*(H,Cl)-[Ir^{III}(pyridine)(diphosphinephenyl)(H)Cl] complexes (diphosphinephenyl = 2,6-bis(di-*tert*-butylphosphinomethyl)phenyl, 2,6-bis(di-*tert*-butylphosphinite)phenyl, 2,6-bis(di-*tert*-butylphosphinite)-4-(ethoxycarbonyl)phenyl) were characterized by Titova et al. [92] and Osipova et al. [93]. Ir(III) organometallic with 2-((3-methyl-1,3-dihydro-2*H*-imidazol-2-ylidene)-1-yl)-4,6-di-*tert*-butylpyridine was investigated by Specht et al. [94].

Two *trans*(*H*,N^{LL,N1}),*trans*(H,N$^{LL,N1'}$)-[Ir^{III}(LL)(L)(1,3-bis(2,4,6-trimethylphenyl)imidazol-2-ylidene)(H)$_2$]$^+$ cations (LL = 2,2′-bipyridine, 1,10-phenanthroline; L = pyridine, pyridine-3-carbonamide) were studied by Mewis et al. [95]. Two *trans*(H,O)-[Ir^{III}(2-methylquinolin-8-olate-κ^2-N,O)(1,5-cyclooctadiene)(M(C_6H_5)$_3$)(H)] complexes (M = Si, Sn) were reported by Rampersadh et al. [62]. Two [Ir^{III}(LL^1-κ^2-N^{py},N^{ind,N7})(dienyl)Cl]$^+$ complexes and four [Ir^{III}(LL^{2*}-κ^2-N^{ind,N7},C$^{ph,C2/th,C3}$)(dienyl)Cl] organometallics (LL^1 = 1-(pyrid-2-yl)-7-azaindole, LL^2 = 1-phenyl-7-azaindole, 1-(thiophen-2-yl)-7-azaindole; dienyl = pentamethylcyclopentadienyl, tetramethylphenylcyclopentadienyl) {1-phenyl-7-azaindole* and 1-(thiophen-2-yl)-7-azaindole*: deprotonated, at the phenyl C(2) or thiophenyl C(3) atom, anionic (−1), N(7)^{azaind},C(2)^{ph}- or N(7)^{azaind},C(3)^{thph}-chelating forms of 1-phenyl-7-azaindole and 1-(thiophen-2-yl)-7-azaindole, respectively} were described by Starha et al. [96].

Trans(H,N^{pd})-[Ir^{III}(pyridazine-κ^1-N^1)(ylidene)(diene)(H)$_2$]$^+$ and *trans*(H,N^{phth})-[Ir^{III}(phthalazine-κ^1-N^2)(ylidene)(diene)(H)$_2$]$^+$ (diene = 1,5-cyclooctadiene), as well as *fac*-[Ir^{III}(pyridazine-κ^1-N^1)$_3$(ylidene)(H)$_2$]$^+$ and *fac*-[Ir^{III}(phthalazine-κ^1-N^2)$_3$(ylidene)(H)$_2$]$^+$ cations were studied by Appleby [63], while *trans*(H,Cl),*trans*(H,N^{qnz})-[Ir^{III}(quinazoline-κ^1-N^3)$_2$(ylidene)(H)$_2$Cl] and *fac*-[Ir^{III}(quinazoline-κ^1-N^3)$_3$(ylidene)(H)$_2$]$^+$ by Richards et al. [97] (in all: ylidene = 1,3-bis(2,4,6-trimethylphenyl)imidazol-2-ylidene).

Fe(II) species are moderately rare (6 examples; Table 4). Three $[Fe^{II}(LLL^1\text{-}\kappa^3\text{-}N^{pyA},N^{pyB1},N^{pyB2})(LLL^2\text{-}\kappa^3\text{-}N^{pyC},N^{pyD1},N^{pyD2})]^{2+}$ cations (LLL^1, LLL^2 = 2,6-bis(pyrid-2-ylcarbonyl)pyridine or 2,6-bis(pyrid-2-ylamino(methyl))pyridine) were studied by Mengel et al. [102]. $[Fe^{II}$(2-((pyrid-2-yl)methyl(hydroxo)iminemethyl)pyridine-κ^3-$N^{pyA},N^{N=},N^{pyB})_2]^{2+}$ was reported by Markiewicz et al. [103]. Two *trans*(CO,X),*trans*(CO,N^{py})-$[Fe^{II}$(2-(1-(2,4,6-trimethylphenylimino)ethyl)-6-(di-*tert*-butylphosphinomethyl)pyridine-κ^3-$N^{py},N^{N=}$,P)$(CO)_2$X] complexes (X = H, F) were described by Butschke et al. [104].

Ru(II) species are noticeably more abundant than those of Fe(II), being very numerous (62 examples; Table 4). Two *trans*(Cl,Cl)-$[Ru^{II}$(2-(pyrid-2-yl)pyrimidine-4-carboxylic acid-κ^2-$N^{py},N^{pm,N1/N3})(CO)_2Cl_2]$ and four $[Ru^{II}$(2-(pyrid-2-yl)-4-methylpyrimidine-κ^2-$N^{py},N^{pm,N1})(DMSO)_2Cl_2]$ complexes of various geometry were studied by Battistin et al. [105]. Six *trans*(P,P),*trans*(P,N^{py})-$[Ru^{II}$(LL-κ^2-$N^{py},N^{pz,N1/tz,N1})(P(CH_3)_3)_3Cl]$ complexes with 5-(pyrid-2-yl)pyrazolate, 5-(pyrid-2-yl)-1,2,4-triazolate or their derivatives (LL) were reported by Muller et al. [106]. Twenty $[Ru^{II}$(LL*-κ^2-$N^{py/pm,N1},C^{tz,C4/C5}$)(1-isopropyl-4-methylbenzene)Cl$]^+$ organometallics {LL*: variously substituted 2-(*1H*-1,2,3-triazolium-1-yl)pyridine*, 2-(*1H*-1,2,3-triazolium-5-yl)pyridine*, 2-(*1H*-1,2,3-triazolium-1-yl)methylpyridine* or 2-(*1H*-1,2,3-triazolium-1-yl)pyrimidine* zwitterions, having both charges within the 1,2,3-triazolic ring and acting as N(1)^{py},C(5)^{tz}-, N(1)^{py},C(4)^{tz}-, N(1)^{py},C(5)^{tz}- or N(1)^{pm},C(5)^{tz}-chelating ligands, respectively} were described by Bolje et al. [107,120].

$[Ru^{II}$(2,2′-bipyridine)$_2$(4-cyano-4′-methyl-2,2′-bipyridine)$]^{2+}$ was studied by Ortiz et al. [108]. Six $[Ru^{II}$(LL)(2,2′:6′,2″-terpyridine)Cl$]^+$ and five $[Ru^{II}$(LL)(2,2′:6′,2″-terpyridine)$(H_2O)]^{2+}$ cations (LL = 2,2′-bipyridine, 4,4′-dihydroxo-2,2′-bipyridine, 6,6′-dihydroxo-2,2′-bipyridine, 4,4′-dimethoxy-2,2′-bipyridine, 6,6′-dimethoxy-2,2′-bipyridine, 1,10-phenanthroline) were reported by Marelius et al. [109] and Chrzanowska et al. [110,113]. A mixture of *mer*- and *fac*-$[Ru^{II}$(4-methyl-4′-((pyrid-4-yl)vinyl)-2,2′-bipyridine)$_3]^{2+}$ was described by Keisar et al. [111], while $[Ru^{II}$(4,4′-azobis(2,2′-bipyridine))$(CN)_4]^{2-}$ by Abate et al. [112]. Four $[Ru^{II}(LL^1)_x(LL^2)_y]^{2x-4}$ species (LL^1 = 4,7-diphenyl-1,10-phenanthroline, LL^2 = 4,7-bis(3-sulfonatophenyl)-1,10-phenanthroline; x = 3, 2, 1, 0, and y = 3 − x) were characterized by Babak et al. [114]. Two $[Ru^{II}$(2,2′:6′,2″-terpyridine)(XX)Cl$]^+$ cations (XX = $NH_2CH_2CH_2NH_2$, $(CH_3)_2NCH_2CH_2N(CH_3)_2$) were investigated by Chrzanowska et al. [110,113], while six $[Ru^{II}(LLL^1)(LLL^2\text{-}\kappa^3\text{-}N^{pyA},N^{pyB1},N^{pyB2})]^{2+}$ ones (LLL^1 = 2,2′:6′,2″-terpyridine-4′-carboxylic acid, 2,2′:6′,2″-terpyridine-4,4′,4″-tricarboxylic acid and their methyl or ethyl esters; LLL^2 = 2,6-bis(pyrid-2-ylamino(-methyl))pyridine, 2,6-bis(pyrid-2-ylamino(methyl))-4-aminopyridine) by Breivogel et al. [115,116]. $[Ru^{II}$(1,3-bis(quinolin-8-yl)-*1H*-pyrazole-κ^3-N^{quinA},N^{quinB},$N^{pz,N2})_2]^{2+}$ was researched by Jarenmark et al. [117], while two other Ru(II) complexes with quinoline, isoquinoline and/or 1,1′-biisoqunoline derivatives by Gobec et al. [118] and Scherrer et al. [119].

Os(II) species are significantly less abundant than the Ru(II) ones and slightly less abundant than those of Fe(II), being very rare (3 examples; Table 4). Two $[Os^{II}$(4-methyl-4′-((pyrid-4-yl)vinyl)-2,2′-bipyridine$)_3]^{2+}$ cations were studied by Keisar et al. [111]. $[Os^{II}$(LL-κ^2-$N^{pm,N1}$,$C^{tz,C5}$*) (1-isopropyl-4-methylbenzene)Cl$]^+$ {LL*: derivative of 2-(*1H*-1,2,3-triazolium-1-yl)pyrimidine* zwitterion having both charges within the 1,2,3-triazolic ring and acting as $N(1)^{pm}$,$C(5)^{tz}$-chelating ligands} was reported by Bolje et al. [120].

Cu(I) species are very rare (2 examples; Table 5). Two organometallics with 2-(2,3,4,5-tetrafluorophenyl)pyridine*-κ^2-N,$C^{ph,C6}$ and 1,3-bis(2,6-diisopropylphenyl)imidazole-2-ylidene or bis(2-(diphenylphosphino)-1-phenyl) ether were studied by Molteni et al. [135].

Ag(I) species are noticeably more abundant than those of Cu(I), being moderately numerous (18 examples; Table 5). Six $[Ag^IL_2]^+$ cations (L = pyridine, 2,3-dimethylpyridine, 2,4-dimethylpyridine, 2,6-dimethylpyridine, 3,5-dimethylpyridine, 4-dimethylaminopyridine; the first known—ref. [168] in Review 2013) and $[Ag^I$(2,2′-bipyridine$)_2]^+$ were studied by Bedin et al. [136], Zopes et al. [137], and Sakharov et al. [138]. Seven Ag(I) complexes with 1-(pyrid-2-yl)-3-methyl-4-trifluoroacetylpyrazolate-5-one-κ^2-N^{py},$N^{pz,N2}$ (i.e., 1-(pyrid-2-yl)-3-methyl-4-trifluoroacetyl-*1H*-pyrazol-5-olate-κ^2-N^{py},$N^{pz,N2}$) anions and pyridine or imidazole derivatives were reported by Marchetti et al. [139]. Ag(I) complex with 1,10-phenanthroline was described by Silva et al. [140]. Three Ag(I) complexes with 2,2′:6′,2″-terpyridine derivatives were described by Mansour et al. [141].

Au(I) species are noticeably less abundant than those of Ag(I), being very rare (1 example; Table 5). $[Au^I$(pyridine$)_2]^+$ was studied by Reiersolmoen et al. [142].

Au(III) species are noticeably more abundant than those of Au(I), being moderately numerous (25 examples; Table 6). Four *trans*-$[Au^{III}L_2Cl_2]^+$ cations (L = pyridine, 4-methylpyridine, 4-methoxypyridine, 4-trifluoromethylpyridine) were studied by Reiersolmoen et al. [142]. The known $[Au^{III}$(2-benzylpyridine*-κ^2-N,$C^{ph,C2}$)$Cl_2]$ organometallic (ref. [5] in Review 2013) was reported by de Paiva et al. [152], while another

Au(III) organometallic with 2-(4-methylphenyl)pyridine*-κ^2-N,C^{ph,C2} by Holmsen et al. [153] {2-benzylpyridine* and 2-(4-methylphenyl)pyridine*: deprotonated, at the phenyl C(2) atom, anionic (−1), N(1)^{py}, C(2)^{ph}-chelating forms of 2-benzylpyridine and 2-(4-methylphenyl) pyridine}. Seven [Au^{III}LCl$_3$] complexes (L=2-phenylsulfanylpyridine and its 4′-derivatives, 2-phenylsulfinylpyridine, 2-phenylselanylpyridine) were described by Mruk et al. [154]. Three Au(III) chloride and two Au(III) bromide complexes with azanidocarbonylpyridine anions, as well as two [Au^{III}(2-(4-*tert*-butyl-4,5-dihydro-1,3-oxazol-2-yl)pyridine-κ^2-N^{py}, N^{ox})X$_2$]$^+$ cations (X=Cl, Br) were characterized by Reiersolmoen et al. [155]. Three known [Au^{III}(LL)Cl$_2$]$^+$ cations (LL=2,2′-bipyridine, 4,4′-dimethyl-2,2′-bipyridine, 1,10-phenanthroline; ref. [87] in Review 2013) were investigated by de Paiva et al. [152]. [Au^{III}(7,8-benzoquinoline)Cl$_3$] and [Au^{III}(7,8-benzoquinoline*-κ^2-N,C^{10})Cl$_2$] {7,8-benzoquinoline*: deprotonated, at the C(10) atom, anionic (−1), N(1),C(10)-chelating form of 7,8-benzoquinoline} were researched by Niedzielska et al. [156].

Zn(II) and Cd(II) species are very rare (3 and 2 examples, respectively; Table 7), being only complexes (no organometallics). [Zn^{II}(2-(1-(pyrid-2-yl)-5-(4-methoxyphenyl)-*1H*-pyrazol-3-yl)pyridine-κ^3-N^{pyA},N^{pyB},N^{pz,N2}) Cl$_2$] was studied by Soria et al. [11]. [Zn^{II}(1,10-phenanthroline-5, 6-dione)$_3$]$^{2+}$ was reported by Wendlandt et al. [162]. One Zn(II) and one Cd(II) complex with quinoline derivatives were described by Burlov et al. [163]. *Trans*(NO$_3$,NO$_3$)(H$_2$O,H$_2$O)-[Cd^{II}(2-amino-4-methylpyrimidine-κ^1-N^{pm,N1})$_2$(H$_2$O)$_2$(NO$_3$)$_2$] was characterized by Klai et al. [164].

In contrast, there are no Hg(II) species.

1.1 ^{15}N NMR coordination shifts in square-planar (d^8) Pd(II)- and Pt(II)-azine complexes or organometallics

The |Δ^{15N}_{coord}| parameters are in the following ranges—Pd(II) ca. 50–100 ppm, Pt(II) ca. 75–140 ppm (Table 1) [2–28], in consistency with the previously reviewed data (Pd(II) ca. 35–110 ppm, Pt(II) ca. 65–115 ppm) [1].

For *trans*(Cl,C^{ph,C2})-[Pd^{II}(2-phenylpyridine*-κ^2-N,C^{ph,C2})LCl] organometallics (L=NH$_3$, pyridine, 2-, 3-, 4-methylpyridine, 2,3-, 2,4-, 3,5-, 2,6-dimethylpyridine, 2,4,6-trimethylpyridine) |Δ^{15N}_{coord}| are ca. 53–65 ppm for L (except for NH$_3$) and ca. 68–70 ppm for 2-phenylpyridine*, their similarity reflecting the same N^{py} *trans* to N^{2ppy*} geometry [2]. For [Pd^{II} (2-phenylpyridine*-κ^2-N,C^{ph,C2})(DMSO-*d*$_6$)$_2$]$^+$ (N^{2ppy*} *trans* to S)

$|\Delta^{15N}{}_{coord}|$ is ca. 69 ppm, whereas for [Pd^{II}(2-phenylpyridine*-κ^2-N,C^{ph,C2})Cl_2]$^-$ and [Pd^{II}(2-phenylpyridine*-κ^2-N,C^{ph,C2})(μ-Cl)]$_2$ (in both: N^{2ppy*} *trans* to Cl)—ca. 73–74 ppm [3].

For *trans*-[Pd^{II}(4-methylpyridine)(ylidene)I_2] and *trans*-[Pd^{II}(3-chloropyridine)(1-(4-methylphenyl)-3-methyl-4-phenyl-1,2,3-triazolium-5-yl)Cl_2] $|\Delta^{15N}{}_{coord}|$ are comparable due to the same N^{py} *trans* to C geometry, being ca. 59–65 ppm [4,5] and ca. 69 ppm [7], respectively. In contrast, for *trans*-[Pd^{II}(4-*tert*-butylpyridine)$_2X_2$] and *trans*-[Pd^{II}(4-*tert*-butylpyridine)$_2$XY] (X = N-methoxy-2,6-difluorobenzamidate, Y = CH_3COO, C_6H_5COO) $|\Delta^{15N}{}_{coord}|$ are as much as ca. 95–97 ppm, reflecting N^{py} *trans* to N geometry [6].

For [Pd^{II}(LL-κ^2-N^{py},N$^{NH2/NH/N}$)Cl_2] complexes (LL = 2-arylaminopyridines; N^{py} *trans* to Cl) $|\Delta^{15N}{}_{coord}|$ are ca. 87–93 ppm [8]. In case of two [Pd^{II}(2-phenyliminomethylpyridine-κ^2-N^{py},N$^{N=}$)(CH_3)Cl] isomers, $|\Delta^{15N}{}_{coord}|$ for the X-ray studied *trans*(Cl,N^{py}) species is larger (by ca. 33 ppm) than for the assumed *trans*(CH_3,N^{py}) one: 81.8 ppm vs 48.4 ppm [9]; it follows the known relation for isomeric Pd(II) species containing nitrogen *trans* to Cl or *trans* to CH_3 [1,176]. For some other *trans*(Cl,N$^{N=}$)-[Pd^{II}(LL-κ^2-N^{py},N$^{N=}$)(CH_3)Cl] complexes (LL = 2-aryliminopyridines; N^{py} *trans* to CH_3) $|\Delta^{15N}{}_{coord}|$ cannot be determined due to the lack of $\delta^{15N}{}_{lig}$ for LL, $\delta^{15N}{}_{compl}$ ranging from ca. −113 ppm to ca. −111 ppm [9]. For [Pd^{II}(2-(*1H*-benzotriazol-1-yl)pyridine-κ^2-N^{py},N^{btz,N2})Cl_2] (N^{py} *trans* to Cl) $|\Delta^{15N}{}_{coord}|$ is ca. 54 ppm [10].

For two [Pd^{II}(2-(1-(pyrid-2-yl)-5-(4-methoxyphenyl)-*1H*-pyrazol-3-yl)pyridine-κ^2-N$^{pyA/pyB}$,N^{pz,N2})Cl_2] isomers differing in the coordinated pyridine ring (py-A vs py-B; in both N^{py} *trans* to Cl) $|\Delta^{15N}{}_{coord}|$ for the Pd(II)-bonded pyridine moieties are ca. 75–94 ppm, whereas analogous $\delta^{15N}{}_{lig} \rightarrow \delta^{15N}{}_{compl}$ changes for the uncoordinated pyridine rings are much smaller and of opposite sign, i.e., $\delta^{15N}{}_{compl}$ decreases (e.g., −81.0 ppm → −65.0/−62.0 ppm) [11].

For [Pd^{II}(2,2′-bipyridine)(mefenamate)$_2$] (N^{bpy} *trans* to O) $|\Delta^{15N}{}_{coord}|$ is ca. 90 ppm [12]. For [Pd^{II}(2,2′-bipyridine)(thiourea#)$_2$]$^{2+}$ and [Pd^{II}(1,10-phenanthroline)(thiourea#)$_2$]$^{2+}$ (N$^{bpy/phen}$ *trans* to S) $|\Delta^{15N}{}_{coord}|$ are ca. 65–67 ppm, with nearly no difference between both ligands [13] (in contrast to the known [Pd^{II}(2,2′-bipyridine)Cl_2] and [Pd^{II}(1,10-phenanthroline)Cl_2] complexes, as well as their analogs [1,68,130,177]). For [Pd^{II}(bis(*tert*-butyl) 2,2′-bipyridine-3,3′-dicarboxylate)$_2$]$^{2+}$ (N^{bpy} *trans* to N) $|\Delta^{15N}{}_{coord}|$ is ca. 84 ppm [14].

For Pd(II) organometallics with 8-methylquinoline*-κ^2-N,C^{CH2} (N^{quin} *trans* to P, NO_2 or O) $|\Delta^{15N}{}_{coord}|$ are ca. 51–73 ppm, while for $[Pd^{II}_2$(quinoline-8-carboxylate-κ^2-N,O$)_2(\mu$-$CH_3COO)_2]$ (N^{quin} *trans* to O)—ca. 95 ppm [15].

For $[Pd^{II}$(8-(ethoxycarbonylmethylaminoiminomethyl)quinoline-κ^2-N^{quin},$N^{N=})Cl_2]$ (N^{quin} *trans* to Cl) $|\Delta^{15N}{}_{coord}|$ cannot be determined due to the lack of $\delta^{15N}{}_{lig}$ for 8-(ethoxycarbonylmethylaminoiminomethyl)quinoline; however, comparison of $\delta^{15N}{}_{compl}$ (−146.2 ppm, in DMSO-d_6) [16] to $\delta^{15N}{}_{lig}$ for relatively similar 8-methylquinoline (−76 ppm, in $CDCl_3$) [15] allows to estimate it as ca. 70 ppm.

For both *trans*-$[Pd^{II}L_2(CH_3COO)_2]$ complexes (L = 4-(2-bromophenoxy) quinoline, 4-(2-bromophenoxy)-7-bromoquinoline; N^{quin} *trans* to N) $|\Delta^{15N}{}_{coord}|$ are ca. 95–97 ppm [17].

For Pd(II) carboxylate complexes with LL = 9,9-dimethyl-4,5-diaza-*9H*-fluorene-κ^2-N^4,N^5 or 4,5-diaza-*9H*-fluoren-9-one-κ^1-N^4/-κ^2-N^4,N^5/ -μ-N^4,N^5 ($N^{dafl,N4}$, $N^{dafl,N5}$ *trans* to O or *trans* to N) $|\Delta^{15N}{}_{coord}|$ cannot be determined due to the lack of $\delta^{15N}{}_{lig}$ for LL (and also for unsubstituted 4,5-diaza-*9H*-fluorene). However, in some 4,5-diaza-*9H*-fluoren-9-one compounds only N(4) is Pd(II)-bonded, while N(5) remains uncoordinated; thus, owing to equivalency of both nitrogens in the free ligand ($\delta^{N(4)}{}_{lig} = \delta^{N(5)}{}_{lig}$), the $\delta^{N(4)}{}_{compl} - \delta^{N(5)}{}_{compl}$ differences can be regarded as ^{15}N NMR *quasi*-coordination shifts: *trans/anti*-$[Pd^{II}$(4,5-diaza-*9H*-fluoren-9-one-κ^1-$N^4)_2(CH_3COO)_2]$ −99.6 ppm (−174.6 ppm vs −75.0 ppm), *trans/syn*-$[Pd^{II}$(4,5-diaza-*9H*-fluoren-9-one-κ^1-$N^4)_2(CH_3COO)_2]$ −103.0 ppm (−175.4 ppm vs −72.4 ppm), *cis/anti*-$[Pd^{II}$(4,5-diaza-*9H*-fluoren-9-one-κ^1-$N^4)_2(CH_3COO)_2]$ −100.6 ppm (−178.0 ppm vs −77.4 ppm), $[Pd^{II}{}_2$(4,5-diaza-*9H*-fluoren-9-one-κ^1-$N^4)_2(\mu$-$CH_3COO)_2(CH_3COO)_2]$ −115.3 ppm (−192.4 ppm vs −77.1 ppm), Pd^{II}(4,5-diaza-*9H*-fluoren-9-one-κ^1-N^4) $(CH_3COO)_3]$ −106.0 ppm (−181.8 ppm vs −75.8 ppm) [18,19].

For $[Pt^{II}$(2-phenylpyridine*-κ^2-N,$C^{ph,C2})(\mu$-Cl$)]_2$ (N2ppy* *trans* to Cl) $|\Delta^{15N}{}_{coord}|$ is ca. 92 ppm [3]. For $[Pt^{II}$(2-iminomethylpyridine-κ^2-N^{py}, $N^{NH=})Cl_2]$ (N^{py} *trans* to Cl) $|\Delta^{15N}{}_{coord}|$ cannot be determined due to the lack of $\delta^{15N}{}_{lig}$ for 2-iminomethylpyridine; however, comparison of $\delta^{15N}{}_{compl}$ (−171.9 ppm, in DMF-d_7) [20] to $\delta^{15N}{}_{lig}$ for relatively similar 2-aminomethylpyridine (−76.4 ppm, in H_2O; originally 305.3 ppm in respect to liquid NH_3) [178] allows to estimate it as ca. 95 ppm. For *cis*-$[Pt^{II}$(L-κ^1-$N^{py})_2Cl_2]$ complexes (L = *n*-dodecylureaylpyridines; N^{py} *trans* to Cl) $|\Delta^{15N}{}_{coord}|$ are surprisingly as much as ca. 132–139 ppm [21].

In case of [Pt^{II}(LLL*-κ^3-N^{py},$C^{ph,C2}$,$N^{tz,N3}$)(pyridine)]$^{+/2+}$ organometallics (LLL = 2-phenyl-6-(1-(12-hydroxo-*n*-dodecyl)-*1H*-1,2,3-triazol-4-yl) pyridine, 2-phenyl-6-(1-(12-(trimethylammonioethoxy)-*n*-dodecyl)-*1H*-1,2,3-triazol-4-yl)pyridine$^+$; N^{py} *trans* to Npy in 2p6tpy*) $|\Delta^{15N}{}_{coord}|$ for pyridine ligands are ca. 106 ppm, whereas for LLL* they cannot be determined due to the lack of $\delta^{15N}{}_{lig}$; nevertheless, comparison of $\delta^{15N}{}_{compl}$ for LLL* (both ca. −167 ppm, in acetone-d_6) to $\delta^{15N}{}_{lig}$ for relatively similar 2-phenyl-6-(1-(12-(4-methylphenylsulfonato)-*n*-dodecyl)-*1H*-1,2,3-triazol-4-yl)pyridine and 2-phenyl-6-(1-(12-(dimethylaminoethoxy)-*n*-dodecyl)-*1H*-1,2,3-triazol-4-yl)pyridine (−88 and −86 ppm, respectively, both in $CDCl_3$; originally 294 and 296 ppm in respect to liquid NH_3) allows to estimate them as ca. 80 ppm [22].

For [Pt^{II}(1-β-D-ribofuranosyl-4-(pyrid-2-yl)-*1H*-1,2,3-triazole-κ^2-N^{py}, $N^{tz,N3}$)Cl_2] (N^{py} *trans* to Cl) $|\Delta^{15N}{}_{coord}|$ for pyridine ring is ca. 102 ppm [23].

For [Pt^{II}(2,2′-bipyridine)(thiourea#)$_2$]$^{2+}$ and [Pt^{II}(1,10-phenanthroline)(thiourea#)$_2$]$^{2+}$ ($N^{bpy/phen}$ *trans* to S) $|\Delta^{15N}{}_{coord}|$ are ca. 77–79 ppm, with nearly no difference between both ligands [13] (in contrast to the known [Pt^{II}(2,2′-bipyridine)Cl_2] and [Pt^{II}(1,10-phenanthroline)Cl_2] complexes [1,68,130,177]).

For *cis*-[Pt^{II}(4-azaindole-κ^1-N^4)$_2I_2$] and *cis*-[Pt^{II}(4-azaindole-κ^1-N^4)(NH_3)I_2] ($N^{azaind,N4}$ *trans* to I) $|\Delta^{15N}{}_{coord}|$ are ca. 91–93 ppm [24], while for *cis*-[Pt^{II}(L-κ^1-N^7)$_2Cl_2$] complexes (L = 7-azaindole derivatives; $N^{azaind,N7}$ *trans* to Cl)—predominantly ca. 101–103 ppm [25,27]. For *trans*-[Pt^{II}(L-κ^1-N^7)$_2Cl_2$] isomers ($N^{azaind,N7}$ *trans* to N) $|\Delta^{15N}{}_{coord}|$ are ca. 104–110 ppm [26], being larger (by ca. 2–7 ppm) than for the relevant *cis* forms: 106.7 ppm vs 101.5 ppm for L = 3-chloro-7-azaindole, 109.5 ppm vs 102.7 ppm for L = 5-bromo-7-azaindole and 104.1 ppm vs 102.4 ppm for 3-iodo-7-azaindole [25,26]; this ^{15}N shielding weakening upon "N *trans* to N" → "N *trans* to Cl" geometry change follows the known tendency [1,68,71–73,179,180]. For [Pt^{II}(L-κ^1-N^7)$_2$(C_2O_4)] complexes ($N^{azaind,N7}$ *trans* to O) $|\Delta^{15N}{}_{coord}|$ are even larger, being ca. 114–117 ppm [26,28].

For the corresponding M(II) compounds, Pd(II) → Pt(II) transition results in ^{15}N shielding enhancement (by ca. 11–18 ppm), following the known relationship [1,68,71–73,179,180]. As examples may serve both [M^{II}(2-phenylpyridine*-κ^2-N,$C^{ph,C2}$)(μ-Cl)]$_2$ dimers ($|\Delta^{15N}{}_{coord}|$: ca. 74 ppm → ca. 92 ppm) [2], as well as [M^{II}(2,2′-bipyridine)(thiourea#)$_2$]$^{2+}$ ($|\Delta^{15N}{}_{coord}|$: ca. 65–67 ppm → ca. 77–78 ppm) and [M^{II}(1,10-phenanthroline)(thiourea#)$_2$]$^{2+}$ ($|\Delta^{15N}{}_{coord}|$: ca. 65–66 ppm → ca. 78–79 ppm) cations [13].

1.2 ^{15}N NMR coordination shifts in square-planar (d^8) Rh(I)- and Ir(I)-azine complexes or organometallics

The $|\Delta^{15N}{}_{coord}|$ parameters are in the following ranges—Rh(I) ca. 30–70 ppm, Ir(I) ca. 35–55 ppm (Table 2) [55–63], in the former case in consistency with the previously reviewed data (Rh(I) ca. 40–90 ppm) [1].

For *trans*(Cl,C=C^{cyclooct})-[Rh^I(pyridine)(ylidene)(cyclooctene)Cl], *trans*(Cl,C=C^{vinyl})-[Rh^I(pyridine)(ylidene)(vinylazole)Cl] (vinylazole = 1-vinyl-*1H*-pyrazole, 1-vinyl-3,5-dimethyl-*1H*-pyrazole) and *trans*(Cl,C=C^{vinyl})-[Rh^I(2-vinylpyridine-κ^2-N,C$^{C=C}$)(ylidene)Cl) (ylidene = 1,3-bis (2,6-diisopropylphenyl)imidazol-2-ylidene) $|\Delta^{15N}{}_{coord}|$ are ca. 45–63 ppm, reflecting the same N^{py} *trans* to C geometry [55–57].

For *trans*(Cl,C=C^{cyclooct})-[Rh^I(pyrid-2-ylacetonitrile-κ^1-N^{py})(ylidene) (cyclooctene)Cl], *trans*(CO,O)-[Rh^I(pyrid-2-ylmethanolate-κ^2-N,O)(ylidene) (CO)] and *trans*(CO,S)-[Rh^I(pyrid-2-ylmethanol-κ^1-N)(ylidene)(CO) ($C_6H_5CH_2S$)] (ylidene = 1,3-bis(2,6-diisopropylphenyl)imidazol-2-ylidene; N^{py} *trans* to C) $|\Delta^{15N}{}_{coord}|$ cannot be determined due to the lack of $\delta^{15N}{}_{lig}$ for pyrid-2-ylacetonitrile and pyrid-2-ylmethanol, $\delta^{15N}{}_{compl}$ ranging from ca. −128 ppm to ca. −111 ppm [58,59]. For *trans*(O,C=C^{alkene})- and *trans*(N^{quin}, C=C^{alkene})-[Rh^I(quinolin-8-olate-κ^2-N,O)(1,3-bis(2,6-diisopropylphenyl)-imidazol-2-ylidene)(alkene)] complexes (alkene = ethylene, cyclooctene) $|\Delta^{15N}{}_{coord}|$ are ca. 32–38 ppm, reflecting N^{quin} *trans* to C or *trans* to C=C geometry [60].

For *trans*(Cl,C=C^{ind})-[Rh^I(4-(*1H*-inden-3-yl)-1,5-naphthyridine-κ^2-N^5,C$^{ind,C2=C3}$)(CO)Cl] (N^{naphth,N5} *trans* to CO) $|\Delta^{N(5)}{}_{coord}|$ is 67.7 ppm [61]. In this organometallic only N(5) is Rh(I)-bonded, while N(1) remains uncoordinated, the $\delta^{N(5)}{}_{compl} - \delta^{N(1)}{}_{compl}$ difference being ca. −64 ppm (−136.8 ppm vs −72.8 ppm); for comparison, in case of 4-(*1H*-inden-3-yl)-1,5-naphthyridine the analogous $\delta^{N(5)}{}_{lig} - \delta^{N(1)}{}_{lig}$ difference is much more slight and of opposite sign, being ca. +2 ppm (−69.1 ppm vs −71.2 ppm) [61].

For [Ir^I(quinolin-8-olate-κ^2-N,O)(1,5-cyclooctadiene)] $|\Delta^{15N}{}_{coord}|$ is 52.5 ppm and a comparable value probably concerns [Ir^I(2-methylquinolin-8-olate-κ^2-N,O)(1,5-cyclooctadiene)], due to their $\delta^{15N}{}_{compl}$ similarity (−148.2 ppm vs −145.6 ppm); both reflect the same N^{quin} *trans* to C=C geometry [62].

For N(1) in [Ir^I(pyridazine-κ^1-N^1)(ylidene)(1,5-cyclooctadiene)]$^+$ and N (2) in [Ir^I(phthalazine-κ^1-N^2)(ylidene)(1,5-cyclooctadiene)]$^+$ (ylidene = 1,3-bis(2,4,6-trimethylphenyl)imidazol-2-ylidene) $|\Delta^{15N}{}_{coord}|$ are ca. 37–38 ppm, reflecting $N^{pd/phth}$ *trans* to C=C geometry [63]. In both cations only one nitrogen (N(1) in pyridazine, N(2) in phthalazine) is Ir(I)-bonded, while another (N(2) in pyridazine, N(3) in phthalazine) remains uncoordinated (whereas in free ligands they are equivalent). The $\delta^{N(1)}{}_{compl} - \delta^{N(2)}{}_{compl}$ (for pyridazine) and $\delta^{N(2)}{}_{compl} - \delta^{N(3)}{}_{compl}$ (for phthalazine) differences are as follows—[Ir^{I}(pyridazine-κ^1-N^1)(ylidene)(1,5-cyclooctadiene)]$^+$: −74.3 ppm (−69.9 ppm vs +4.4 ppm) and [Ir^{I}(phthalazine-κ^1-N^2)(ylidene)(1,5-cyclooctadiene)]$^+$: −74.5 ppm (−70.2 ppm vs +4.3 ppm), their absolute magnitudes being much larger than the $|\Delta^{15N}{}_{coord}|$ parameters [63].

1.3 ^{15}N NMR coordination shifts in octahedral (d^6) Co(III)-, Rh(III)-, and Ir(III)-azine complexes or organometallics

The $|\Delta^{15N}{}_{coord}|$ parameters are in the following ranges—Co(III) ca. 65–110 ppm, Rh(III) ca. 40–60 ppm, Ir(III) ca. 20–115 ppm (Table 3) [55,62,63,84–97], in consistency with the previously reviewed data (all M(III) ca. 40–125 ppm) [1].

For *mer*-[Co^{III}(2-(2-(trifluoromethyl)vinyl-2-olate)pyridine-κ^2-N,O)$_3$] $|\Delta^{15N}{}_{coord}|$ cannot be determined due to the lack of $\delta^{15N}{}_{lig}$ for 2-(2-(trifluoromethyl)vinyl-2-ol)pyridine; however, it must be larger (by ca. 8–15 ppm) for N^{py} *trans* to O than for N^{py} *trans* to N, as the respective $\delta^{15N}{}_{compl}$ values are: −202 ppm < −187 ppm or −194 ppm [84]. For [Co^{III}(2,5-bis(pyrid-2-yl)-3,4-dimethylpyrrolate-κ^3-N^{py},$N^{py'}$,N^{pr})$_2$]$^{2+}$ (N^{py} *trans* to N) $|\Delta^{15N}{}_{coord}|$ is 109.0 ppm [85]. For *trans*($N^{bpy,N1}$,NH^X), *trans*($N^{bpy,N1'}$,NO^Y)-[Co^{III}(2,2′-bipyridine)XY] (X=NHCOC(*CN*)NO, Y= NHC(OCH$_2$CH$_3$)C(*CN*)NO) $|\Delta^{15N}{}_{coord}|$ for $N^{bpy,N1}$ *trans* to NH is 81.9 ppm, while for $N^{bpy,N1'}$ *trans* to NO is 67.6 ppm, differing by ca. 14 ppm [86].

For *trans*(H,N^{py}),*trans*(Cl,CCSi)-[Rh^{III}(pyridine)$_2$(1,3-bis(2,6-diisopropylphenyl)imidazol-2-ylidene)((CH$_3$)$_3$SiCC)(H)Cl] $|\Delta^{15N}{}_{coord}|$ for N^{py} *trans* to H is smaller (by ca. 22 ppm) than for N^{py} *trans* to C: 39.8 ppm vs 61.8 ppm [55]. For [Rh^{III}(5,5′-bis(pyren-1-yl)-2,2′-bipyridine)(pentamethylcyclopentadienyl)Cl]$^+$ $|\Delta^{15N}{}_{coord}|$ cannot be determined due to the lack of $\delta^{15N}{}_{lig}$ for 5,5′-bis(pyren-1-yl)-2,2′-bipyridine, $\delta^{15N}{}_{compl}$ being ca. −134 ppm [87].

For *fac*-[Ir^{III}(pyridine)$_3$(ylidene)(H)$_2$]$^+$ cations (ylidene = 1,3-dimethylimidazol-2-ylidene, 1,3-diisopropyl-4,5-dimethylimidazol-2-ylidene, 1,3-bis(cyclohexyl)imidazol-2-ylidene, 1,3-bis(2,6-diisopropylphenyl)imidazol-2-ylidene, 1,3-bis(2,4,6-trimethylphenyl)imidazol-2-ylidene, 1-methyl-3-

phenylimidazol-2-ylidene, 1,3-bis(2,6-diisopropylphenyl)-4,5-dihydroimidazol-2-ylidene, 1,3-bis(2,4,6-trimethylphenyl)-4,5-dihydroimidazol-2-ylidene) $|\Delta^{15N}{}_{coord}|$ for N^{py} *trans* to H are again smaller (by ca. 11–18 ppm) than for N^{py} *trans* to C: ca. 47–60 ppm vs ca. 60–77 ppm [88,89].

For *trans*(H,N^{py}),*trans*(H,N^{CH3CN})-[Ir^{III}(pyridine)(1,3-bis(2,4,6-trimethylphenyl)imidazol-2-ylidene)(P(C_6H_5)$_3$)(CH_3CN)(H)$_2$]$^+$ (N^{py} *trans* to H) $|\Delta^{15N}{}_{coord}|$ is 74.9 ppm [90]. For *trans*(H,N^{py}),*trans*(H,Cl)-[Ir^{III} (pyridine)$_2$(cyclooctene)(H)$_2$Cl] $|\Delta^{15N}{}_{coord}|$ for N^{py} *trans* to H is smaller (by ca. 30 ppm) than for N^{py} *trans* to C=C: 58.3 ppm vs 88.3 ppm [91].

For *trans*(H,N^{py})-[Ir^{III}(pyridine)(diphosphinephenyl)(H)Cl] complexes (N^{py} *trans* to H) $|\Delta^{15N}{}_{coord}|$ are smaller (by ca. 11–14 ppm) than for the only *trans*(H,Cl)-[Ir^{III}(pyridine)(diphosphinephenyl)(H)Cl] one (N^{py} *trans* to C): ca. 70–74 ppm vs ca. 84 ppm (diphosphinephenyl = 2,6-bis(di-*tert*-butylphosphinomethyl)phenyl, 2,6-bis(di-*tert*-butylphosphinite)phenyl, 2,6-bis(di-*tert*-butylphosphinite)-4-(ethoxycarbonyl)phenyl), this relation being well exemplified for both isomers of [Ir^{III}(pyridine)(2,6-bis(di-*tert*-butylphosphinomethyl)phenyl)(H)Cl]: 73.4 ppm vs 84.2 ppm [92,93].

For [Ir^{III}(2-((3-methyl-1,3-dihydro-*2H*-imidazol-2-ylidene)-1-yl)-4,6-di-*tert*-butylpyridine-κ^2-N^{py},$C^{im,2}$)(pentamethylcyclopentadienyl)Cl] organometallic (N^{py} *trans* to C=C) $|\Delta^{15N}{}_{coord}|$ is 76.9 ppm [94].

In case of *trans*(H,$N^{bpy,N1}$),*trans*(H,$N^{bpy,N1'}$)-[Ir^{III}(2,2′-bipyridine)(pyridine-3-carbonamide-N^{py})(ylidene)(H)$_2$]$^+$ and *trans*(H,$N^{phen,N1}$),*trans* (H,$N^{phen,N1'}$)-[Ir^{III}(1,10-phenanthroline)(pyridine)(ylidene)(H)$_2$]$^+$ cations (ylidene = 1,3-bis(2,4,6-trimethylphenyl)imidazol-2-ylidene) $|\Delta^{15N}{}_{coord}|$ for 2,2′-bipyridine or 1,10-phenanthroline ($N^{bpy/phen}$ *trans* to H) are smaller (by ca. 28 ppm) than for pyridine or pyridine-3-carbonamide (N^{py} *trans* to C): ca. 39 ppm vs ca. 67 ppm and ca. 36 ppm vs ca. 64 ppm, respectively [95].

For *trans*(H,O)-[Ir^{III}(2-methylquinolin-8-olate-κ^2-N,O)(1,5-cyclooctadiene)(M(C_6H_5)$_3$)(H)] complexes (M = Si, Sn) $|\Delta^{15N}{}_{coord}|$ cannot be determined due to the lack of $\delta^{15N}{}_{lig}$ for 2-methylquinolin-8-ol; however, comparison of the respective $\delta^{15N}{}_{compl}$ values (−153.2 ppm, in C_6D_6; −163.0 ppm, in $CDCl_3$) [62] to $\delta^{15N}{}_{lig}$ for relatively similar quinolin-8-ol (−95.7 ppm, in $CDCl_3$) [66] allows to estimate them as ca. 55–70 ppm, reflecting N^{quin} *trans* to C=C geometry.

For [Ir^{III}(1-(pyrid-2-yl)-7-azaindole-κ^2-N^{py},$N^{ind,N7}$)(dienyl)Cl]$^+$ complexes as well as [Ir^{III}(1-phenyl-7-azaindole*-κ^2-$N^{ind,N7}$,$C^{ph,C2}$)(dienyl)Cl] and [Ir^{III}(1-(thiophen-2-yl)-7-azaindole*-κ^2-$N^{ind,N7}$,$C^{th,C3}$)(dienyl)Cl] organometallics (dienyl = pentamethylcyclopentadienyl, tetramethylphenylcyclopentadienyl) $|\Delta^{15N}{}_{coord}|$ for N(7) in 7-azaindole moiety are ca.

94–113 ppm, while those for N(1) in pyridine side groups are ca. 94–97 ppm, their comparable magnitudes reflecting analogous $N^{azaind,N7}$ or N^{py} *trans* to C=C geometries [96].

For N(1) in *trans*(H,N^{pd})-[Ir^{III}(pyridazine-κ^1-N^1)(ylidene)(diene)$(H)_2]^+$ and N(2) in *trans*(H,N^{phth})-[Ir^{III}(phthalazine-κ^1-N^2)(ylidene)(diene)$(H)_2]^+$ cations (ylidene = 1,3-bis(2,4,6-trimethylphenyl)imidazol-2-ylidene, diene = 1,5-cyclooctadiene; $N^{pd/phth}$ *trans* to H) $|\Delta^{15N}{}_{coord}|$ are 55.0 and 83.0 ppm, respectively, differing by ca. 28 ppm [63]. Similar differences (by ca. 27 ppm) are observed when to compare $|\Delta^{15N}{}_{coord}|$ for N(1) in *fac*-[Ir^{III}(pyridazine-κ^1-N^1)$_3$(ylidene)$(H)_2]^+$ and N(2) in *fac*-[Ir^{III}(phthalazine-κ^1-N^2)$_3$(ylidene)$(H)_2]^+$ (23.1 ppm vs 49.9 ppm for N *trans* to H; 43.5 ppm vs 70.8 ppm for N *trans* to C); moreover, a general relation that $|\Delta^{15N}{}_{coord}|$ for nitrogen *trans* to H is smaller (by ca. 20–21 ppm) than for nitrogen *trans* to C is observed (*fac*-[Ir^{III}(pyridazine-κ^1-N^1)$_3$(ylidene)$(H)_2]^+$: 23.1 ppm vs 43.5 ppm; *fac*-[Ir^{III}(phthalazine-κ^1-N^2)$_3$(ylidene)$(H)_2]^+$: 49.9 ppm vs 70.8 ppm) [63].

In these four complexes only one nitrogen (N(1) in pyridazine, N(2) in phthalazine) is Ir(III)-bonded, while another (N(2) in pyridazine, N(3) in phthalazine) remains uncoordinated (whereas in free ligands they are equivalent). The $\delta^{N(1)}{}_{compl} - \delta^{N(2)}{}_{compl}$ (for pyridazine) and $\delta^{N(2)}{}_{compl} - \delta^{N(3)}{}_{compl}$ (for phthalazine) differences are as follows—*trans*(H,N^{pd})-[Ir^{III}(pyridazine-κ^1-N^1)(ylidene)(diene)$(H)_2]^+$: −105.8 ppm (−87.4 ppm vs +18.4 ppm), *trans*(H,N^{pd})-[Ir^{III}(phthalazine-κ^1-N^2)(ylidene)(diene)$(H)_2]^+$: −102.3 ppm (−115.1 ppm vs −12.8 ppm), *fac*-[Ir^{III}(pyridazine-κ^1-N^1)$_3$(ylidene)$(H)_2]^+$: −63.8 ppm (−55.5 ppm vs +8.3 ppm) and −86.4 ppm (−75.9 ppm vs +10.5 ppm), *fac*-[Ir^{III}(phthalazine-κ^1-N^2)$_3$(ylidene)$(H)_2]^+$: −61.2 ppm (−82.0 ppm vs −20.8 ppm) and −84.0 ppm (−102.9 ppm vs −18.9 ppm), their absolute magnitudes being much larger than the $|\Delta^{15N}{}_{coord}|$ parameters [63].

Also for *trans*(H,Cl),*trans*(H,N^{qnz})-[Ir^{III}(quinazoline-κ^1-N^3)$_2$(ylidene)$(H)_2$Cl] and *fac*-[Ir^{III}(quinazoline-κ^1-N^3)$_3$(ylidene)$(H)_2]^+$ (ylidene = 1,3-bis(2,4,6-trimethylphenyl)imidazol-2-ylidene) $|\Delta^{15N}{}_{coord}|$ for $N^{qnz,N3}$ *trans* to H are smaller (by ca. 12–20 ppm) than for $N^{qnz,N3}$ *trans* to C (44.8 ppm vs 65.1 ppm; 53.5 ppm vs 66.0 ppm) [97]. In both complexes only N(3) is Ir(III)-bonded, while N(1) remains uncoordinated, the $\delta^{N(3)}{}_{compl} - \delta^{N(1)}{}_{compl}$ differences being as follows—*trans*(H,Cl),*trans*(H,N^{qnz})-[Ir^{III}(quinazoline-κ^1-N^3)$_2$(ylidene)$(H)_2$Cl]: −37.1 ppm (−135.2 ppm vs −98.1 ppm) and −56.7 ppm (−155.5 ppm vs −98.8 ppm), *fac*-[Ir^{III}(quinazoline-κ^1-N^3)$_3$(ylidene)$(H)_2]^+$: −44.7 ppm (−143.9 ppm vs −99.2 ppm) and −59.8 ppm (−156.4 ppm vs −96.6 ppm) [97]; in contrast, for free quinazoline the analogous $\delta^{N(3)}{}_{lig} - \delta^{N(1)}{}_{lig}$ difference is of opposite sign: +10.5 ppm (−90.4 ppm vs −100.9 ppm) [75].

1.4 ^{15}N NMR coordination shifts in octahedral (d^6) Fe(II)-, Ru(II)-, and Os(II)-azine complexes or organometallics

The $|\Delta^{15N}{}_{coord}|$ parameters are in the following ranges—Fe(II) ca. 50–55 ppm, Ru(II) ca. 0–80 ppm, Os(II) ca. 70–75 ppm (Table 4) [102–120], in consistency with the previously reviewed data (Fe(II) ca. 10–65 ppm, Ru(II) ca. 20–75 ppm, Os(II) ca. 35–90 ppm) [1].

For $[Fe^{II}(LLL^1\text{-}\kappa^3\text{-}N^{pyA},N^{pyB1},N^{pyB2})(LLL^2\text{-}\kappa^3\text{-}N^{pyC},N^{pyD1},N^{pyD2})]^{2+}$ cations (LLL^1, LLL^2 = 2,6-bis(pyrid-2-ylcarbonyl)pyridine or 2,6-bis (pyrid-2-ylamino(methyl))pyridine; all N^{py} *trans* to N) $|\Delta^{15N}{}_{coord}|$ cannot be determined due to the lack of $\delta^{15N}{}_{lig}$ for LLL^1 and LLL^2, $\delta^{15N}{}_{compl}$ ranging from ca. −170 ppm to ca. −110 ppm [102]. Owing to the same reason, $|\Delta^{15N}{}_{coord}|$ cannot be found for $[Fe^{II}$(2-((pyrid-2-yl)methyl(hydroxo) iminemethyl)pyridine-κ^3-$N^{pyA},N^{N=},N^{pyB})_2]^{2+}$ (both N^{py} *trans* to N), while $\delta^{15N}{}_{compl}$ are ca. −145 ppm and ca. −134 ppm (unassigned) [103]. For *trans*(CO,X),*trans*(CO,N^{py})-$[Fe^{II}$(2-(1-(2,4,6-trimethylphenylimino)-ethyl)-6-(di-*tert*-butylphosphinomethyl)pyridine-κ^3-$N^{py},N^{N=}$,P)$(CO)_2X]$ complexes (X = H, F; N^{py} *trans* to CO) $|\Delta^{15N}{}_{coord}|$ are ca. 53–54 ppm [104].

For both *trans*(Cl,Cl)-$[Ru^{II}$(2-(pyrid-2-yl)pyrimidine-4-carboxylic acid-κ^2-$N^{py},N^{pm,N1/N3})(CO)_2Cl_2]$ linkage isomers (differing in the coordinated pyrimidinic nitrogen: $N(1)^{pm}$ vs $N(3)^{pm}$) $|\Delta^{15N}{}_{coord}|$ cannot be determined due to the lack of $\delta^{15N}{}_{lig}$ for 2-(pyrid-2-yl)pyrimidine-4-carboxylic acid, $\delta^{15N}{}_{compl}$ for N^{py}, $N^{pm,N(1)}$ and $N^{pm,N(3)}$ ranging from ca. −144 ppm to ca. −86 ppm. For each isomer $\delta^{15N}{}_{compl}$ for the Ru(II)-bonded pyrimidinic nitrogen (N^{pm} *trans* to CO) is lower (by ca. 21 ppm or ca. 57 ppm) than for the uncoordinated one: Ru-$N(1)^{pm}$ −130.2 ppm vs N (3)^{pm} −108.9 ppm or Ru-$N(3)^{pm}$ −143.6 ppm vs $N(1)^{pm}$ −86.2 ppm. Moreover, in both complexes $\delta^{15N}{}_{compl}$ for the coordinated pyrimidinic nitrogen is lower (by ca. 46–48 ppm) than $\delta^{15N}{}_{lig}$ for the same N atom in 2-(pyrid-2-yl)pyrimidine-4-carboxylic acid·HNO_3 adduct (a nitrate salt containing protonated pyridinic nitrogen and unprotonated pyrimidinic nitrogens): Ru-$N(1)^{pm}$ −130.2 ppm vs $N(1)^{pm}$ −84.1 ppm or Ru-$N(3)^{pm}$ −143.6 ppm vs $N(3)^{pm}$ −95.5 ppm [105].

For various isomers of $[Ru^{II}$(2-(pyrid-2-yl)-4-methylpyrimidine-κ^2-N^{py}, $N^{pm,N1})(DMSO)_2Cl_2]$ $|\Delta^{15N}{}_{coord}|$ are ca. 42–74 ppm (ca. 53–74 ppm for pyridinic nitrogens and ca. 42–59 ppm for the pyrimidinic ones); generally, they are larger for nitrogens *trans* to Cl than *trans* to S (ca. 52–74 ppm vs ca. 42–60 ppm), which is observed also separately for pyridine ring (ca. 64–74 ppm vs 53–60 ppm) and N(1) in pyrimidine ring (ca. 52–59 ppm vs 42–47 ppm) [105]. Moreover, for each complex $\delta^{15N}{}_{compl}$ for the Ru(II)-bonded pyrimidinic N(1) atom is by as much as ca. 54–70 ppm lower than for the uncoordinated N(3) one: −160.4 ppm vs −90.2 ppm, −153.4 ppm vs −87.7 ppm, −147.6 ppm vs −88.7 ppm, −143.3 ppm vs −89.5 ppm, whereas for N(1) and N(3) in 2-(pyrid-2-yl)-4-methylpyrimidine the analogous $\delta^{15N}{}_{lig}$ difference is only ca. 9 ppm (−101.2 ppm vs −92.5 ppm) [105].

For *trans*(P,P),*trans*(P,N^{py})-[Ru^{II}(LL-κ^2-N^{py},N$^{pz,N1/tz,N1}$)(P(CH$_3$)$_3$)$_3$Cl] complexes (LL = 5-(pyrid-2-yl)pyrazolate, 5-(pyrid-2-yl)-1,2,4-triazolate or their derivatives; N^{py} *trans* to P) $\delta^{15N}{}_{compl}$ for pyridine moieties (N^{py} *trans* to N) range from ca. −173 ppm to ca. −137 ppm, while $|\Delta^{15N}{}_{coord}|$ can be determined only in case of LL = 5-(pyrid-2-yl)pyrazolate, 4-bromo-5-(pyrid-2-yl)pyrazolate and 3-methyl-5-(pyrid-2-yl)-1,2,4-triazolate, being ca. 47–50 ppm [106].

For [Ru^{II}(LL*)(1-isopropyl-4-methylbenzene)Cl]$^+$ organometallics (LL = variously substituted 2-(*1H*-1,2,3-triazolium-1-yl)pyridine*-κ^2-N^{py}, C^{tz,C5}, 2-(*1H*-1,2,3-triazolium-5-yl)pyridine*-κ^2-N^{py},C^{tz,C4}, 2-(*1H*-1,2,3-triazolium-1-yl)methylpyridine*-κ^2-N^{py},C^{tz,C5} or 2-(*1H*-1,2,3-triazolium-1-yl)pyrimidine*-κ^2-N^{pm,N1},C^{tz,C5} zwitterions) $|\Delta^{15N}{}_{coord}|$ for pyridinic or pyrimidinic N(1) atoms are ca. 60–62 ppm, ca. 66–67 ppm, ca. 76–77 ppm and ca. 52 ppm, respectively, reflecting N^{py} or N^{pm,N1} *trans* to C=C geometries [107,120]. Moreover, in case of [Ru^{II}(2-(*1H*-1,2,3-triazolium-1-yl)pyrimidine*-κ^2-N^{pm,N1},C^{tz,C5})(1-isopropyl-4-methylbenzene)Cl]$^+$ $\delta^{15N}{}_{compl}$ for the Ru(II)-bonded pyrimidinic N(1) atom is by ca. 50 ppm lower than for the uncoordinated N(3) one: ca. −164 ppm vs ca. −114 ppm, whereas for N(1) and N(3) in the parent cationic ligand 2-(3-methyl-4-phenyl-*1H*-1,2,3-triazolium-1-yl)pyrimidine$^+$ the $\delta^{15N}{}_{lig}$ values are identical (ca. −112 ppm) [120].

In case of [Ru^{II}(2,2′-bipyridine)$_2$(4-cyano-4′-methyl-2,2′-bipyridine)]$^{2+}$ (all N^{bpy} *trans* to N) $|\Delta^{15N}{}_{coord}|$ for all nitrogens in both 2,2′-bipyridine ligands are the same (57.0 ppm), whereas for 4-cyano-4′-methyl-2,2′-bipyridine ligand they cannot be determined due to the lack of $\delta^{15N}{}_{lig}$ ($\delta^{15N}{}_{compl}$ are ca. −141 ppm and ca. −123 ppm; N^{bpy} *trans* to N) [108].

For [Ru^{II}(2,2′-bipyridine)(2,2′:6′,2″-terpyridine)Cl]$^+$ and [Ru^{II}(2,2′-bipyridine)(2,2′:6′,2″-terpyridine)(H$_2$O)]$^{2+}$ cations $|\Delta^{15N}{}_{coord}|$ are highly dependent on the type of nitrogen. In case of 2,2′-bipyridine $|\Delta^{N(1)}{}_{coord}|$ (N^{bpy} *trans* to N) is ca. 44–49 ppm, while $|\Delta^{N(1')}{}_{coord}|$ (N^{bpy} *trans* to Cl or *trans* to O)—ca. 55–58 ppm or ca. 65 ppm, being surprisingly larger (by ca. 10–11 ppm and ca. 16 ppm, respectively). In contrast, for 2,2′:6′,2″-terpyridine $|\Delta^{N(1),N(1'')}{}_{coord}|$ is as much as ca. 56–66 ppm, whereas $|\Delta^{N(1')}{}_{coord}|$—only ca. 9–19 ppm, independently on the fact that all N^{tpy} atoms are *trans* to N [109,110]; such a relation between $|\Delta^{15N}{}_{coord}|$ for the outer N(1), N(1″) and the inner N(1′) atoms was already reported for the previously reviewed Fe(II), Ru(II) and Os(II) complexes with 2,2′:6′,2″-terpyridine [1,129].

Similarly, for analogous $[Ru^{II}(LL)(2,2′:6′,2″\text{-terpyridine})Cl]^+$ and $[Ru^{II}(LL)(2,2′:6′,2″\text{-terpyridine})(H_2O)]^{2+}$ cations (LL = 4,4′-dihydroxo-2,2′-bipyridine, 6,6′-dihydroxo-2,2′-bipyridine, 4,4′-dimethoxy-2,2′-bipyridine, 6,6′-dimethoxy-2,2′-bipyridine) $|\Delta^{15N}{}_{coord}|$ concerning 2,2′:6′,2″-terpyridine are as much as ca. 55–66 ppm for N(1), N(1″) and only ca. 6–15 ppm for N(1′) [109]. In contrast, $|\Delta^{15N}{}_{coord}|$ for LL ligands cannot be determined due to the lack of $\delta^{15N}{}_{lig}$; nevertheless, for each $[Ru^{II}(LL)(2,2′:6′,2″\text{-terpyridine})Cl]^+$ or $[Ru^{II}(LL)(2,2′:6′,2″\text{-terpyridine})(H_2O)]^{2+}$ species $\delta^{15N}{}_{compl}$ decreases (thus, ^{15}N shielding is enhanced) in order "$N^{bpy/phen}$ *trans* to N" → "$N^{bpy/phen}$ *trans* to Cl" (−147.9 ppm → −163.8 ppm, −161.4 ppm → −178.3 ppm, −143.6 ppm → −158.4 ppm, −149.8 ppm → −160.9 ppm; thus, by ca. 11–17 ppm) or "N^{LL} *trans* to N" → "N^{LL} *trans* to O" (−150.1 ppm → −172.5 ppm, −161.3 ppm → −181.3 ppm, −145.8 ppm → −166.7 ppm, −154.4 ppm → −173.5 ppm; thus, by ca. 19–22 ppm) [109], following the same relations for above mentioned $[Ru^{II}(2,2′\text{-bipyridine})(2,2′:6′,2″\text{-terpyridine})Cl]^+$ (−119.8 ppm → −130.7 ppm or −117.2 ppm → −127.6 ppm) and $[Ru^{II}(2,2′\text{-bipyridine})(2,2′:6′,2″\text{-terpyridine})(H_2O)]^{2+}$ (−121.7 ppm → −137.9 ppm) cations [109,110].

In case of $[Ru^{II}(1,10\text{-phenanthroline})(2,2′:6′,2″\text{-terpyridine})Cl]^+$ the patterns of $|\Delta^{15N}{}_{coord}|$ changes are similar: for 1,10-phenanthroline $|\Delta^{N(1)}{}_{coord}|$ (N^{phen} *trans* to N) is ca. 51 ppm and $|\Delta^{N(1')}{}_{coord}|$ (N^{phen} *trans* to Cl)—ca. 62 ppm, while for 2,2′:6′,2″-terpyridine $|\Delta^{N(1),N(1'')}{}_{coord}|$ is ca. 57 ppm and $|\Delta^{N(1')}{}_{coord}|$—ca. 10 ppm (all N^{tpy} *trans* to N) [113].

For mixture of *mer*- and *fac*-$[Ru^{II}(4\text{-methyl-}4′\text{-((pyrid-4-yl)vinyl)-}2,2′\text{-bipyridine})_3]^{2+}$ (all N^{bpy} *trans* to N) $|\Delta^{15N}{}_{coord}|$ cannot be determined due to the lack of $\delta^{15N}{}_{lig}$ for 4-methyl-4′-((pyrid-4-yl)vinyl)-2,2′-bipyridine;

$\delta^{15N}{}_{compl}$ are −140 and −133 ppm but their assignment to the respective nitrogens and/or isomers remains unknown [111].

For $[Ru^{II}$(4,4′-azobis(2,2′-bipyridine))$(CN)_4]^{2-}$ (both N^{bpy} *trans* to CN) $|\Delta^{15N}{}_{coord}|$ cannot be determined due to the lack of $\delta^{15N}{}_{lig}$ for 4,4′-azobis(2,2′-bipyridine); $\delta^{15N}{}_{compl}$ for the Ru(II)-bonded 2,2′-bipyridine moiety (N^{bpyA}) are −116.7 and −129.1 ppm (unattributed), being lower (by ca. 30–50 ppm) than for the uncoordinated 2,2′-bipyridine moiety (N^{bpyB}: −79.6 and −88.6 ppm, also unassigned) [112].

For $[Ru^{II}$(4,7-diphenyl-1,10-phenanthroline)$_3]^{2+}$ (N^{phen} *trans* to N) $|\Delta^{15N}{}_{coord}|$ is ca. 67 ppm, whereas for $[Ru^{II}$(4,7-diphenyl-1,10-phenanthroline)$_2$(4,7-bis(3-sulfonatophenyl)-1,10-phenanthroline)], $[Ru^{II}$ (4,7-diphenyl-1,10-phenanthroline)(4,7-bis(3-sulfonatophenyl)-1,10-phenanthroline)$_2]^{2-}$ and $[Ru^{II}$(4,7-bis(3-sulfonatophenyl)-1,10-phenanthroline)$_3]^{4-}$ $|\Delta^{15N}{}_{coord}|$ cannot be determined due to the absence of clear ^{15}N assignments to both ligands and the lack of $\delta^{15N}{}_{lig}$ for 4,7-bis(3-sulfonatophenyl)-1,10-phenanthroline. Nevertheless, $\delta^{15N}{}_{compl}$ for all four $[Ru^{II}(LL^1)_x(LL^2)_y]^{2x-4}$ species are nearly the same (ca. −140 ppm to ca. −138 ppm) [114].

For $[Ru^{II}$(2,2′:6′,2″-terpyridine)(XX)Cl$]^+$ cations (XX = $NH_2CH_2CH_2NH_2$, $(CH_3)_2NCH_2CH_2N(CH_3)_2$) the pattern of $|\Delta^{15N}{}_{coord}|$ changes is again the same like for the other 2,2′:6′,2″-terpyridine complexes: $|\Delta^{15N}{}_{coord}|$ for N(1), N(1″) is ca. 55–64 ppm, whereas for N(1′) it is less than 7 ppm (all N^{tpy} *trans* to N); however, this comparison is misleading as $\Delta^{N(1')}{}_{coord}$ is either positive (+4.1 ppm) or negative (−7.1 ppm) [110,113].

For $[Ru^{II}(LLL^1)(LLL^2\text{-}\kappa^3\text{-}N^{pyA},N^{pyB1},N^{pyB2})]^{2+}$ cations (LLL^1 = 2,2′:6′,2″-terpyridine-4′-carboxylic acid, 2,2′:6′,2″-terpyridine-4,4′,4″-tricarboxylic acid and their methyl or ethyl esters; LLL^2 = 2,6-bis(pyrid-2-ylamino(methyl))pyridine, 2,6-bis(pyrid-2-ylamino(methyl))-4-aminopyridine; all N^{tpy} or N^{py} *trans* to N) $|\Delta^{15N}{}_{coord}|$ cannot be determined due to the lack of $\delta^{15N}{}_{lig}$ for LLL^1 and LLL^2, $\delta^{15N}{}_{compl}$ ranging from ca. −144 ppm to ca. −132 ppm for N(1), N(1″) and from ca. −88 ppm to ca. −84 ppm for N(1′) in LLL^1, as well as from ca. −200 ppm to ca. −165 ppm for all N^{py} in LLL^2 [115,116].

For $[Ru^{II}$(1,3-bis(quinolin-8-yl)-*1H*-pyrazole-κ^3-N^{quinA},N^{quinB},$N^{pz,N2}$)$_2]^{2+}$ (N^{quin} *trans* to N) $|\Delta^{15N}{}_{coord}|$ cannot be determined due to the lack of $\delta^{15N}{}_{lig}$ for 1,3-bis(quinolin-8-yl)-*1H*-pyrazole, $\delta^{15N}{}_{compl}$ for N^{quin} ranging from ca. −152 ppm to ca. −145 ppm (unassigned) [117].

Similarly, for [Ru^{II}(5-chloro-7-iodoquinolin-8-olate-κ^2-N,O)(1-isopropyl-4-methylbenzene)Cl] (N^{quin} *trans* to C=C) and *trans*-[Ru^{II}(6-bromoisoquinoline)$_2$(1,1′-biisoquinoline-3,3′-dicarboxylate-κ^4-N,N′,O,O′)] (N^{isoquin} *trans* to N, N^{biisoquin} *trans* to O) $|\Delta^{15N}{}_{coord}|$ cannot be determined due to the lack of $\delta^{15N}{}_{lig}$ for 5-chloro-7-iodoquinolin-8-ol, 6-bromoisoquinoline and 1,1′-biisoquinoline-3,3′-dicarboxylic acid, $\delta^{15N}{}_{compl}$ ranging from ca. −144 ppm to ca. −117 ppm [118,119].

For mixture of *mer*- and *fac*-[Os^{II}(4-methyl-4′-((pyrid-4-yl)vinyl)-2,2′-bipyridine)$_3$]$^{2+}$ (all N^{bpy} *trans* to N) $|\Delta^{15N}{}_{coord}|$ cannot be determined due to the lack of $\delta^{15N}{}_{lig}$ for 4-methyl-4′-((pyrid-4-yl)vinyl)-2,2′-bipyridine; $\delta^{15N}{}_{compl}$ are −161 and −154 ppm but their assignment to the respective nitrogens and/or isomers remains unknown [111].

For [Os^{II}(2-(*1H*-1,2,3-triazolium-1-yl)pyrimidine*-κ^2-N^{pm,N1},C^{tz,C5})(1-isopropyl-4-methylbenzene)Cl]$^+$ $|\Delta^{15N}{}_{coord}|$ for pyrimidinic N(1) atom is ca. 74 ppm, reflecting N^{pm,N1} *trans* to C=C geometry; unfortunately, it is not possible to compare $\delta^{15N}{}_{compl}$ for the Os(II)-bonded pyrimidinic N(1) atom to the uncoordinated N(3) one as the latter nitrogen remained undetected [120].

For the corresponding M(II) compounds, Ru(II)→Os(II) transition results in ^{15}N shielding enhancement (by ca. 21–22 ppm), following the known relationship [1,98,129]. As examples may serve [M^{II}(2-(*1H*-1,2,3-triazolium-1-yl)pyrimidine*-κ^2-N^{pm,N1},C^{tz,C5})(1-isopropyl-4-methylbenzene)Cl]$^+$ (increase of $|\Delta^{15N}{}_{coord}|$ for pyrimidinic N(1) atom: 52 ppm→74 ppm) [120] and *mer-/fac*-[M^{II}(4-methyl-4′-((pyrid-4-yl)vinyl)-2,2′-bipyridine)$_3$]$^{2+}$ (decrease of $\delta^{15N}{}_{compl}$: −140 and −133 ppm→ −161 and −154 ppm) [111].

1.5 ^{15}N NMR coordination shifts in (d^{10}) Cu(I)- and Ag(I)-azine complexes or organometallics

The $|\Delta^{15N}{}_{coord}|$ parameters are in the following ranges—Cu(I) ca. 15–35 ppm, Ag(I) ca. 20–65 ppm, Au(I) ca. 90 ppm (Table 5) [135–142], in consistency with the previously reviewed data (Cu(I) ca. 35 ppm, Ag(I) ca. 10–55 ppm) [1].

For [Cu^I(2-(2,3,4,5-tetrafluorophenyl)pyridine*-κ^2-N,C^{ph,C6})(1,3-bis(2,6-diisopropylphenyl)imidazole-2-ylidene)] and [Cu^I(2-(2,3,4,5-tetrafluorophenyl)pyridine*-κ^2-N,C^{ph,C6})(bis(2-(diphenylphosphino)-1-phenyl)ether)] $|\Delta^{15N}{}_{coord}|$ are ca. 18–34 ppm [135].

For $[Ag^{I}L_2]^+$ cations (L = pyridine, 2,3-dimethylpyridine, 2,4-dimethylpyridine, 2,6-dimethylpyridine, 3,5-dimethylpyridine, 4-dimethylaminopyridine) $|\Delta^{15N}{}_{coord}|$ are ca. 24–64 ppm, being in agreement with the previously reviewed literature data (for L = pyridine and 4-dimethylaminopyridine [1,144]), while for $[Ag^{I}(2,2'\text{-bipyridine})_2]^+$—ca. 23–65 ppm [136–138].

For Ag(I) complexes with 1-(pyrid-2-yl)-3-methyl-4-trifluoroacetylpyrazolate-5-one-κ^2-N^{py},$N^{pz,N2}$ (i.e., 1-(pyrid-2-yl)-3-methyl-4-trifluoroacetyl-*1H*-pyrazol-5-olate-κ^2-N^{py},$N^{pz,N2}$) anions and pyridine or imidazole derivatives $|\Delta^{15N}{}_{coord}|$ cannot be determined due to the lack of $\delta^{15N}{}_{lig}$ for LL, $\delta^{15N}{}_{compl}$ ranging from ca. −182 ppm to ca. −147 ppm; additionally, in case of $[Ag^{I}$(1-(pyrid-2-yl)-3-methyl-4-trifluoroacetylpyrazolate-5-one)(2,4,6-trimethylpyridine)] $|\Delta^{15N}{}_{coord}|$ for 2,4,6-trimethylpyridine is 23.8 ppm [139].

In case of $[Ag^{I}$(1,10-phenanthroline)(2-(NH_2CSNHNCH)pyridine-κ^2-$N^{N=}$,S)$]^+$ $|\Delta^{15N}{}_{coord}|$ for 1,10-phenanthroline is ca. 43 ppm, being the same for both nitrogens; their equivalency indicates rather tetrahedral than square-planar geometry [140].

For the studied in situ $[Ag^{I}$(4′-(pyrid-2-yl)-2,2′:6′,2″-terpyridine)(DMSO-d_6)$]^+$, $[Ag^{I}$(4′-((morpholin-4-yl)phenyl)-2,2′:6′,2″-terpyridine)(DMSO-d_6)$]^+$ and $[Ag^{I}$(4′-((morpholin-4-yl)phenyl)-2,2′:6′,2″-terpyridine)(CD_3OD)$]^+$ cations $|\Delta^{15N}{}_{coord}|$ for all N^{tpy} and N^{py} atoms are ca. 23–31 ppm [141].

For $[Au^{I}(\text{pyridine})_2]^+$ $|\Delta^{15N}{}_{coord}|$ is ca. 89 ppm [142]. Comparison to $[Ag^{I}(\text{pyridine})_2]^+$ (upon the same BF_4^- counterion and CD_2Cl_2 solvent) exhibits that Ag(I) → Au(I) transition results in ^{15}N shielding enhancement ($|\Delta^{15N}{}_{coord}|$ increase by ca. 27–32 ppm): 61.2 ppm or 57 ppm → 88.6 ppm [136,137,142].

1.6 ^{15}N NMR coordination shifts in square-planar (d^8) Au(III)-azine complexes or organometallics

The $|\Delta^{15N}{}_{coord}|$ parameters are in the range ca. 75–105 ppm (Table 6) [142,152–156], in consistency with the previously reviewed data (Au(III) ca. 75–105 ppm) [1].

For *trans*-$[Au^{III}L_2Cl_2]^+$ cations (L = pyridine, 4-methylpyridine, 4-methoxypyridine, 4-trifluoromethylpyridine) $|\Delta^{15N}{}_{coord}|$ are ca. 86–92 ppm, reflecting N^{py} *trans* to N geometry [142].

For $[Au^{III}$(2-benzylpyridine*-κ^2-N,$C^{ph,C2}$)Cl_2] organometallic and $[Au^{III}(LL)Cl_2]^+$ complexes (LL = 2,2′-bipyridine, 4,4′-dimethyl-2,2′-bipyridine,

1,10-phenanthroline) $|\Delta^{15N}{}_{coord}|$ are ca. 93 ppm and ca. 68–79 ppm, being in agreement with the previously reviewed literature data [1,69,157], respectively, and reflecting N^{py} or $N^{bpy/phen}$ *trans* to Cl geometry [152].

For *trans*(N^{py},CH_2^{XX})-[Au^{III}(2-(4-methylphenyl)pyridine*-κ^2-N, $C^{ph,C2}$)(XX)]$^+$ (XX=NHC(CH_3)OCH_2CH_2) organometallic (N^{py} *trans* to C) $|\Delta^{15N}{}_{coord}|$ cannot be determined due to the lack of $\delta^{15N}{}_{lig}$ for 2-(4-methylphenyl)pyridine, $\delta^{15N}{}_{compl}$ being −129 ppm [153].

For [Au^{III}LCl_3] complexes (L=2-phenylsulfanylpyridine and its derivatives, 2-phenylsulfinylpyridine, 2-phenylselanylpyridine; N^{py} *trans* to Cl) $|\Delta^{15N}{}_{coord}|$ are ca. 77–80 ppm [154].

For [Au^{III}(LL-κ^2-N^{py},N^N)X_2] complexes (LL=azanidocarbonylpyridine anions, X=Cl, Br; N^{py} *trans* to X) $|\Delta^{15N}{}_{coord}|$ cannot be determined due to the lack of $\delta^{15N}{}_{lig}$ for parent aminocarbonylpyridines (LLH), $\delta^{15N}{}_{compl}$ ranging from ca. −168 ppm to ca. −166 ppm, with no significant distinction between chloride and bromide species [155]. Similarly, for [Au^{III}(2-(4-*tert*-butyl-4,5-dihydro-1,3-oxazol-2-yl)pyridine-κ^2-N^{py},N^{ox})X_2]$^+$ cations (X=Cl, Br; N^{py} *trans* to X) $|\Delta^{15N}{}_{coord}|$ are comparable for both halides: ca. 76–100 ppm for X=Cl and ca. 75–102 ppm for X=Br (more expressed variability is noted upon counterions or solvents change) [155].

In case of [Au^{III}(7,8-benzoquinoline*-κ^2-N,C^{10})Cl_2] and [Au^{III}(7,8-benzoquinoline)Cl_3] (in both: N^{bqn} *trans* to Cl) the ^{15}N shielding effect is more enhanced for the former organometallic than for the latter complex ($|\Delta^{15N}{}_{coord}|$: 69.2 ppm vs 61.2 ppm) [156], following the known relation for many [Au^{III}(LL*)Cl_2] and [Au^{III}(LL)Cl_3] species [1,74,157,181].

Comparison of *trans*-[Au^{III}(pyridine)$_2Cl_2$]$^+$ to [Au^I(pyridine)$_2$]$^+$ cations reveals nearly no impact on ^{15}N shielding effect upon Au(I)→Au(III) transition, as exhibited by negligible $|\Delta^{15N}{}_{coord}|$ change: 88.6 ppm→87.2 ppm [142].

1.7 ^{15}N NMR coordination shifts in (d^{10}) Zn(II)- and Cd(II)-azine complexes

The $|\Delta^{15N}{}_{coord}|$ parameters are in the following ranges—Zn(II) ca. 20–60 ppm, Cd(II) ca. 35–55 ppm (Table 7) [11,162–164], in consistency with the previously reviewed data (Zn(II) ca. 35–50 ppm, Cd(II) ca. 35–75 ppm) [1].

For [Zn^{II}(2-(1-(pyrid-2-yl)-5-(4-methoxyphenyl)-*1H*-pyrazol-3-yl)pyridine-κ^3-N^{pyA},N^{pyB},$N^{pz,N2}$)Cl_2] $|\Delta^{15N}{}_{coord}|$ are ca. 19–40 ppm [11]. For

$[Zn^{II}(1,10\text{-phenanthroline-5,6-dione})_3]^{2+}$ $|\Delta^{15N}{}_{coord}|$ is ca. 62 ppm [162]. For $[Zn^{II}$(8-(4-methylbenzenesulfonamidato-1,2-phenylenemethylimino)quinoline-κ^3-N^{quin},$N^{N=}$,N^{N-})$(CH_3COO)]$ and $[Cd^{II}$(8-(4-methylbenzenesulfonamidato-1,2-phenylenemethylimino)quinoline-κ^4-N^{quin},$N^{N=}$,N^{N-},O^{SO2})(8-(4-methylbenzenesulfonamidato-1,2-phenylenemethylimino)quinoline-κ^3-N^{quin},$N^{N=}$,N^{N-})] $|\Delta^{15N}{}_{coord}|$ for N^{quin} are ca. 34–53 ppm [163]. For *trans*$(NO_3,NO_3)(H_2O,H_2O)$-$[Cd^{II}$(2-amino-4-methylpyrimidine-κ^1-$N^{pm,N1}$)$_2(H_2O)_2(NO_3)_2]$ ($N^{pm,N1}$ *trans* to N) $|\Delta^{15N}{}_{coord}|$ for $N^{pm,N1}$ is ca. 56 ppm; moreover, $\delta^{15N}{}_{compl}$ for the Ru(II)-bonded pyrimidinic N(1) atom is lower (by ca. 34 ppm) than for the uncoordinated N(3) one: −194 ppm vs −160 ppm, whereas for N(1) and N(3) in 2-amino-4-methylpyrimidine the analogous $\delta^{15N}{}_{lig}$ difference is only ca. 8 ppm (−138.2 ppm vs −130.3 ppm) [164].

2. Transition metal complexes and organometallics with azoles

In Review 2013, 38 ^{15}N NMR-studied transition metal complexes or organometallics with azoles (Ni(II)—0, Pd(II)—15, Pt(II)—1, Co(III)—0, Rh(I)—3, Ir(I)—1, Rh(III)—0, Ir(III)—1, Fe(II)—0, Ru(II)—2, Os(II)—0, Cu(I)—2, Ag(I)—5, Au(I)—5, Au(III)—0, Zn(II)—2, Cd(II)—1, Hg(II)—0) were discussed in respect to their ^{15}N chemical and coordination shifts (paragraph 2, including tables 2.15–2.18). In the present paper 89 analogous ^{15}N NMR-studied species are described (Ni(II)—0, Pd(II)—9, Pt(II)—10, Pt(IV)—2, Rh(I)—13, Ir(I)—3, Co(III)—0, Rh(III)—6, Ir(III)—4, Fe(II)—0, Ru(II)—19, Os(II)—6, Cu(I)—0, Ag(I)—12, Au(I)—0, Au(III)—1, Zn(II)—4, Cd(II)—0, Hg(II)—0); it must be noted, however, that 20 of them are simultaneously transition metal-azine complexes or organometallics, which were discussed in the earlier part of this review.

There are no Ni(II) species, whereas those of Pd(II) are moderately rare (9 examples; Table 8). $[Pd^{II}$(2-(1-(pyrid-2-yl)-5-(4-methoxyphenyl)-*1H*-pyrazol-3-yl)pyridine-κ^2-N^{pyA},$N^{pz,N2}$)$Cl_2]$ was studied by Soria et al. [11]. *Trans*-$[Pd^{II}$(2-(thiophen-2-yl)-*1H*-imidazole-κ^1-N^3)$_2Cl_2]$ was reported by Zanvettor et al. [29]. Two $[Pd^{II}$(LL-κ^1-$N^{tz,N3}$)(μ-Cl)Cl$]_2$ dimers with 1,2,3-triazole derivatives (LL) were described by Bumagin et al. [30]. $[Pd^{II}$(2-(*1H*-benzotriazol-1-yl)pyridine-κ^2-N^{py},$N^{btz,N2}$)$Cl_2]$ was characterized by Dikmen et al. [10]. Three $[Pd^{II}$(1,3-bis-(2-(diphenylphosphino)ethyl)-4,5,6,7-tetrahydrobenzotriazolium-κ^3-N^2,P,P′)

$X]^{2+/3+}$ cations (X=Cl, CH_3, $P(C_6H_5)_3$) were investigated by Tulchinsky et al. [31].

Pt(II) species are moderately rare as well (10 examples; Table 8). [Pt^{II}(2-(thiophen-2-yl)-*1H*-imidazole-κ^1-N^3)$_2$Cl$_2$] was studied by Zanvettor et al. [29]. Two Pt(II) organometallics with 1,2,3-triazole derivatives were reported by Hebenbrock et al. [22]. [Pt^{II}(1-β-D-ribofuranosyl-4-(pyrid-2-yl)-*1H*-1,2,3-triazole-κ^2-N^{py},N^{tz,N3})Cl$_2$] was described by Jakubowski et al. [23]. Three [Pt^{II}(1,3-bis-(2-(diphenylphosphino)ethyl)benzotriazolium-κ^3-N^2,P,P′)X]$^{2+/3+}$ and three [Pt^{II}(1,3-bis-(2-(diphenylphosphino)ethyl)-4,5,6,7-tetrahydrobenzotriazolium-κ^3-N^2,P,P′)X]$^{2+/3+}$ cations (X=Cl, CH_3, $P(C_6H_5)_3$) were characterized by Tulchinsky et al. [32].

Pt(IV) species are slightly less abundant than those of Pt(II), being very rare (2 examples; Table 8). [Pt^{IV}(1,3-bis-(2-(diphenylphosphino)ethyl) benzotriazolium-κ^3-N^2,P,P′)Cl$_3$]$^{2+}$ and [Pt^{IV}(1,3-bis-(2-(diphenylphosphino)ethyl)-4,5,6,7-tetrahydrobenzotriazolium-κ^3-N^2,P,P′)Cl$_3$]$^{2+}$ were studied by Tulchinsky et al. [31].

Rh(I) species are moderately numerous (13 examples, Table 9). *Trans* (Cl,C=C^{vinyl})-[Rh^I(1-vinyl-3,5-dimethyl-*1H*-pyrazole-κ^2-N^2,C$^{C=C}$) (1,3-bis(2,6-diisopropylphenyl)imidazol-2-ylidene)Cl] was studied by Azpiroz et al. [56]. [Rh^I(1,3-bis-(2-(diphenylphosphino)ethyl)-*1H*-1,2,3-triazolium-κ^3-N^2,P,P′)(CO)]$^{2+}$, six [Rh^I(1,3-bis-(2-(diphenylphosphino) ethyl)benzotriazolium-κ^3-N^2,P,P′)X]$^{+/2+}$ (X=Cl, I, CO, THF, CH_2CH_2, $P(C_6H_5)_3$) and five [Rh^I(1,3-bis-(2-(diphenylphosphino)ethyl)-4,5,6,7-tetrahydrobenzotriazolium-κ^3-N^2,P,P′)X]$^{+/2+}$ (X=Cl, CO, THF, CH_2 CH_2, $P(C_6H_5)_3$) were reported by Tulchinsky et al. [31,32].

Ir(I) species are noticeably less abundant than those of Rh(I), being very rare (3 examples; Table 9). Three [Ir^I(1,3-bis-(2-(diphenylphosphino)ethyl) benzotriazolium-κ^3-N^2,P,P′)X]$^{+/2+}$ cations (X=Cl, CO, CH_3CN) were studied by Vahav et al. [169].

There are no Co(III) species, whereas those of Rh(III) are slightly less abundant than those of Rh(III), being moderately rare (6 examples; Table 10). [Rh^{III}(*1H,1′H*-2,2′-biimidazole-κ^2-N^3,N$^{3'}$)(pentamethylcyclopentadienyl)Cl]$^+$ was studied by Gilewska et al. [125]. [Rh^{III}(1-benzyl-4-phenyl-*1H*-1,2,3-triazole*-κ^2-N^3,C^{ph,C2})(pentamethycyclopentadienyl)Cl] {1-benzyl-4-phenyl-*1H*-1,2,3-triazole*: deprotonated, at the phenyl C(2) atom, anionic (−1), N(3)^{tz},C(2)^{ph}-chelating form of 1-benzyl-4-phenyl-*1H*-1,2,3-triazole} was reported by Rono et al. [171]. Four Rh(III) complexes with 1,3-bis-(2-(diphenylphosphino)ethyl)benzotriazolium-κ^3-N^2,P,P′$^+$ cations were described by Tulchinsky et al. [31] and Vahav et al. [169].

Ir(III) species are slightly less abundant than those of Rh(III), being very rare (4 examples; Table 10), similarly to the Ir(I) ones. $[Ir^{III}(1H,1'H\text{-}2,2'\text{-biimidazole-}\kappa^2\text{-}N^3,N^{3'})(\text{pentamethylcyclopentadienyl})Cl]^+$ was studied by Gilewska et al. [125]. Three *cis*-$[Ir^{III}(1,3\text{-bis-}(2\text{-}(\text{diphenylphosphino})\text{ethyl})\text{benzotriazolium-}\kappa^3\text{-}N^2,P,P')X(H)_2]^{+/2+}$ (X = Cl, CO, CH_3CN) were reported by Vahav et al. [169].

There are no Fe(II) species, whereas those of Ru(II) are moderately numerous (19 examples; Table 11). $[Ru^{II}(1,3\text{-bis(quinolin-8-yl)-}1H\text{-pyrazole-}\kappa^3\text{-}N^{quin},N^{quin'},N^{pz,N2})_2]^{2+}$ was studied by Jarenmark et al. [117]. Two $[Ru^{II}(LL\text{-}\mu^3\text{-}N^{N2},O^{C\text{-}O},O^{C=O})(1\text{-isopropyl-4-methylbenzene})]_4^{4+}$ cations with pyrazol-5-olate derivatives (LL) were reported by Pettinari et al. [121]. $[Ru^{II}(\text{tris}(1H\text{-pyrazol-1-yl})\text{methanide-}\kappa^3\text{-}N^{pz1,N2},N^{pz2,N2},N^{pz3,N2})(\text{cyclopentadienyl})]$ and $[Ru^{II}(\text{tris}(1H\text{-pyrazol-1-yl})\text{methanenitrile-}\kappa^3\text{-}N^{pz1,N2},N^{pz2,N2},N^{pz3,N2})(\text{cyclopentadienyl})]^+$ were described by Gonzalez-Gallardo et al. [122] and Trapp et al. [123]. Six $[Ru^{II}(LL\text{-}\kappa^2\text{-}N^{py},N^{pz,N1/tz,N1})(P(CH_3)_3)_3Cl]$ complexes with 5-(pyrid-2-yl) pyrazolate, 5-(pyrid-2-yl)-1,2,4-triazolate or their derivatives (LL) were characterized by Muller et al. [106]. Four $[Ru^{II}(LL\text{-}\kappa^2\text{-}N^3,P)(\text{dienyl})]^+$ cations (LL = 1-methyl-2-diisopropylphosphino-*1H*-imidazole, 1-methyl-2-diisopropylphosphino-4-*tert*-butyl-*1H*-imidazole; dienyl = pentamethylcyclopentadienyl, tetramethyltrifluoromethylcyclopentadienyl) were investigated by Paulson et al. [124]. $[Ru^{II}(1H,1'H\text{-}2,2'\text{-biimidazole})(1\text{-isopropyl-4-methylbenzene})Cl]^+$ was researched by Gilewska et al. [125]. Three $[Ru^{II}(LLL\text{-}\kappa^3\text{-}N^2,P,P')Cl(\mu\text{-}Cl)]_2^{2+}$ cations (LLL = 1,3-bis-(2-(diphenylphosphino)ethyl)-*1H*-1,2,3-triazolium, 1,3-bis-(2-(diphenylphosphino) ethyl)benzotriazolium, 1,3-bis-(2-(diphenylphosphino)ethyl)-4,5,6,7-tetrahydrobenzotriazolium) were analyzed by Tulchinsky et al. [31].

Os(II) species are slightly less abundant than those of Ru(II), being moderately rare (6 examples; Table 11). Six $[Os^{II}L(NO)Cl_4]^-$ anions (L = $1H\text{-pyrazole-}\kappa^1\text{-}N^2$, $1H\text{-imidazole-}\kappa^1\text{-}N^3$, $1H\text{-indazole-}\kappa^1\text{-}N^2$, $1H\text{-benzoimidazole-}\kappa^1\text{-}N^3$) were studied by Buchel et al. [126] and Gavriluta et al. [127].

There are no Cu(I) species, whereas those of Ag(I) are moderately numerous (12 examples; Table 12), being only complexes (no organometallics). Eleven Ag(I) complexes with $1\text{-phenyl-3-methyl-4-benzoylpyrazolate-5-one-}\kappa^1\text{-}N^2$ (i.e., $1\text{-phenyl-3-methyl-4-benzoyl-}1H\text{-pyrazol-5-olate-}\kappa^1\text{-}N^2$) or $1\text{-(pyrid-2-yl)-3-methyl-4-trifluoroacetylpyrazolate-5-one-}\kappa^2\text{-}N^{py},N^{pz,N2}$ (i.e., $1\text{-(pyrid-2-yl)-3-methyl-4-trifluoroacetyl-}1H\text{-pyrazol-5-olate-}\kappa^2\text{-}N^{py},N^{pz,N2}$) anions and $1H\text{-imidazole-}\kappa^1\text{-}N^3$ derivatives, as well as $[Ag^I(1H\text{-imidazole-}\kappa^1\text{-}N^3)_2]^+$ were studied by Marchetti et al. [139].

In contrast, there are no Au(I) species, while the only Au(III) one is trichloride complex with 1,2,4-triazole derivative, studied by Golubyatnikova et al. [143].

Zn(II) species are very rare (4 examples; Table 13), being only complexes (no organometallics). [Zn^{II}(2-(1-(pyrid-2-yl)-5-(4-methoxyphenyl)-*1H*-pyrazol-3-yl)pyridine-κ^3-N^{pyA},N^{pyB},$N^{pz,N2}$)Cl_2] was studied by Soria et al. [11], while three [Zn^{II}(L-κ^1-N^3)$_6$]$^{2+}$ cations (L = 1-methyl-*1H*-imidazole, 1-ethyl-*1H*-imidazole, 1-*n*-butyl-*1H*-imidazole) by Steichen et al. [165].

In contrast, there are no Cd(II) or Hg(II) species.

2.1 ^{15}N NMR coordination shifts in square-planar (d^8) Pd(II)- and Pt(II)-azole complexes or organometallics

The $|\Delta^{15N}{}_{coord}|$ parameters are in the following ranges—Pd(II) ca. 25–120 ppm, Pt(II) ca. 60–110 ppm (Table 8, top) [10,11,29–32], in consistency with the previously reviewed data (Pd(II) ca. 45–110 ppm, Pt(II) ca. 120 ppm) [1].

For [Pd^{II}(2-(1-(pyrid-2-yl)-5-(4-methoxyphenyl)-*1H*-pyrazol-3-yl) pyridine-κ^2-N^{pyA},$N^{pz,N2}$)Cl_2] $|\Delta^{15N}{}_{coord}|$ for pyrazolic N(2) atom is 76.3 ppm, reflecting $N^{pz,N2}$ *trans* to Cl geometry [11]. For *trans*-[Pd^{II}(2-(thiophen-2-yl)-*1H*-imidazole-κ^1-N^3)$_2Cl_2$] $|\Delta^{15N}{}_{coord}|$ for imidazolic N(3) atom is ca. 80–90 ppm, corresponding to $N^{im,N3}$ *trans* to N geometry [29].

In case of [Pd^{II}(LL-κ^1-$N^{tz,N3}$)(μ-Cl)Cl]$_2$ dimers (LL = 1-(5-phenyl-(1,2-oxazol-3-yl)methyl)-*1H*-1,2,3-triazol-4-ylmethanol, 1-(5-phenyl-(1,2-oxazol-3-yl)methyl)-*1H*-1,2,3-triazol-4-ylmethyl 2,2-dimethylpropanoate) $\delta^{15N}{}_{compl}$ for triazolic N(3) atom ranges from ca. −154 ppm to ca. −129 ppm; $|\Delta^{15N}{}_{coord}|$ can be determined only for the latter complex (ca. 102–120 ppm; $N^{tz,N3}$ *trans* to Cl) [30]. For [Pd^{II}(2-(*1H*-benzotriazol-1-yl) pyridine-κ^2-N^{py},$N^{btz,N2}$)Cl_2] ($N^{btz,N2}$ *trans* to Cl) $|\Delta^{15N}{}_{coord}|$ for benzotriazolic N(2) atom is surprisingly only 26 ppm [10].

For [Pd^{II}(LLL-κ^3-N^2,P,P′)X]$^{2+/3+}$ cations (LLL = 1,3-bis-(2-(diphenylphosphino)ethyl)-4,5,6,7-tetrahydrobenzotriazolium; X = Cl, CH_3, $P(C_6H_5)_3$) $|\Delta^{15N}{}_{coord}|$ for tetrahydrobenzotriazolic N(2) atoms are ca. 50–81 ppm, increasing in order "$N^{thbtz,N2}$ *trans* to CH_3" → "$N^{thbtz,N2}$ *trans* to P" → "$N^{thbtz,N2}$ *trans* to Cl" (50.3 ppm → 70.0 ppm → 80.5 ppm) [31]; a more expressed ^{15}N shielding effect for nitrogens *trans* to Cl comparing to those *trans* to CH_3 was already reported for many previously rewieved Pd(II) complexes with pyrazole derivatives [1].

In contrast, for three other Pd(II)-LLL complexes, where N(2) remained uncoordinated (κ^2-P,P′ chelates; no Pd(II)-N bonds), $\delta^{thbtz,N(2)}{}_{compl}$ were comparable to $\delta^{thbtz,N(2)}{}_{lig}$: $[Pd^{II}(LLL\text{-}\kappa^2\text{-}P,P')Cl_2]^+$ −46.1 ppm, $[Pd^{II}(LLL\text{-}\kappa^2\text{-}P,P')(CH_3)Cl]^+$ −49.0 ppm, $[Pd^{II}(LLL\text{-}\kappa^2\text{-}P,P')(P(C_6H_5)_3)Cl]^{2+}$ −50.1 ppm (all $CF_3SO_3^-$ salts in CD_3NO_2) [31] vs $(LLL)^+$ −45.0 ppm ($CF_3SO_3^-$ salt in $CDCl_3$) [32].

For $[Pt^{II}(2\text{-(thiophen-2-yl)-}1H\text{-imidazole-}\kappa^1\text{-}N^3)_2Cl_2]$ $|\Delta^{15N}{}_{coord}|$ for imidazolic N(3) atom is 79.5 ppm, being nearly the same like for *trans*-$[Pd^{II}(2\text{-(thiophen-2-yl)-}1H\text{-imidazole-}\kappa^1\text{-}N^3)_2Cl_2]$ (80.5 ppm) [29]. The authors treated this similarity as a proof for analogous *trans* geometry (exhibited for the Pd(II) species by X-ray studies) but they were probably wrong, as replacement of Pd(II) by Pt(II) usually results in increase of $|\Delta^{15N}{}_{coord}|$. Most likely, this compound is *cis*-$[Pt^{II}(2\text{-(thiophen-2-yl)-}1H\text{-imidazole-}\kappa^1\text{-}N^3)_2Cl_2]$ because simultaneous Pd(II) → Pt(II) and *trans* → *cis* transitions have opposite impacts (the former: ^{15}N shielding enhancement; the latter: ^{15}N shielding weakening due to "$N^{im,N3}$ *trans* to N" → "$N^{im,N3}$ *trans* to Cl" geometry change), resulting in nearly zero net effect. A similar phenomenon was already reported for *trans*-$[Pd^{II}L_2Cl_2]$ and *cis*-$[Pd^{II}L_2Cl_2]$ complexes with pyridine and its methyl or phenyl derivatives [1,68,71–73,179,180].

For $[Pt^{II}(LLL^*\text{-}\kappa^3\text{-}N^{py},C^{ph,C2},N^{tz,N3})(pyridine)]^{+/2+}$ organometallics (LLL = 2-phenyl-6-(1-(12-hydroxo-*n*-dodecyl)-*1H*-1,2,3-triazol-4-yl)pyridine, 2-phenyl-6-(1-(12-(trimethylammonioethoxy)-*n*-dodecyl)-*1H*-1,2,3-triazol-4-yl)pyridine$^+$; $N^{tz,N3}$ *trans* to $C^{ph,C2}$) $|\Delta^{15N}{}_{coord}|$ cannot be determined due to the lack of $\delta^{15N}{}_{lig}$ for both LLL; nevertheless, comparison of $\delta^{15N}{}_{compl}$ for triazolic N(3) atoms (−85 and −86 ppm, respectively, both in acetone-d_6) to $\delta^{15N}{}_{lig}$ for relatively similar 2-phenyl-6-(1-(12-(4-methylphenylsulfonato)-*n*-dodecyl)-*1H*-1,2,3-triazol-4-yl)pyridine and 2-phenyl-6-(1-(12-(dimethylaminoethoxy)-*n*-dodecyl)-*1H*-1,2,3-triazol-4-yl)pyridine (both −37 ppm, in $CDCl_3$; originally 345 ppm in respect to liquid NH_3) allows to estimate them as ca. 50 ppm [22].

For $[Pt^{II}(1\text{-}\beta\text{-D-ribofuranosyl-4-(pyrid-2-yl)-}1H\text{-1,2,3-triazole-}\kappa^2\text{-}N^{py},N^{tz,N3})Cl_2]$ ($N^{tz,N3}$ *trans* to Cl) $|\Delta^{15N}{}_{coord}|$ for triazolic N(3) atom is ca. 98 ppm [23].

For $[Pt^{II}(LLL\text{-}\kappa^3\text{-}N^2,P,P')X]^{2+/3+}$ cations (LLL^1 = 1,3-bis-(2-(diphenylphosphino)ethyl)benzotriazolium, LLL^2 = 1,3-bis-(2-(diphenylphosphino)ethyl)-4,5,6,7-tetrahydrobenzotriazolium; X = Cl, CH_3, $P(C_6H_5)_3$) $|\Delta^{15N}{}_{coord}|$ for benzotriazolic and tetrahydrobenzotriazolic N(2) atoms are ca. 63–108 ppm, increasing in order "$N^{btz/thbtz,N2}$ *trans* to CH_3" → "$N^{btz/thbtz,N2}$ *trans* to P" → "$N^{btz/thbtz,N2}$ *trans* to Cl" (63.8 ppm → 84.6 ppm → 105.8 ppm and 63.5 ppm → 84.9 ppm → 107.9 ppm); $|\Delta^{15N}{}_{coord}|$ differences between the corresponding $[Pt^{II}(LLL^1\text{-}\kappa^3\text{-}N^2,P,P')X]^{2+/3+}$ and $[Pt^{II}(LLL^2\text{-}\kappa^3\text{-}N^2,P,P')X]^{2+/3+}$ species (with the same X) are below 2 ppm [32].

In contrast, for two other Pt(II)-LLL complexes, where N(2) remained uncoordinated (κ^2-P,P′ chelates; no Pt(II)-N bonds), $\delta^{btz/thbtz,N(2)}{}_{compl}$ were comparable to $\delta^{btz/thbtz,N(2)}{}_{lig}$: $[Pt^{II}(LLL^1\text{-}\kappa^2\text{-}P,P')(CH_3)_2]^+$ −27.9 ppm and $[Pt^{II}(LLL^2\text{-}\kappa^2\text{-}P,P')(CH_3)_2]^+$ −41.9 ppm (both $CF_3SO_3^-$ salts in CD_3NO_2) [32] vs $(LLL^1)^+$ −28.9 ppm and $(LLL^2)^+$ −45.0 ppm (both $CF_3SO_3^-$ salts in $CDCl_3$) [32].

For the corresponding $[M^{II}(LLL\text{-}\kappa^3\text{-}N^2,P,P')X]^{2+/3+}$ cations (LLL = 1,3-bis-(2-(diphenylphosphino)ethyl)-4,5,6,7-tetrahydrobenzotriazolium), Pd(II) → Pt(II) transition results in ^{15}N shielding enhancement ($|\Delta^{15N}{}_{coord}|$ increase by ca. 13–28 ppm): 80.5 ppm → 107.9 ppm for X = Cl, 50.3 ppm → 63.5 ppm for X = CH_3 and 70.0 ppm → 84.9 ppm for X = $P(C_6H_5)_3$ [31,32].

2.2 ^{15}N NMR coordination shifts in octahedral (d^6) Pt(IV)-azole complexes

The $|\Delta^{15N}{}_{coord}|$ parameters are in the range ca. 120–125 ppm (Table 8, bottom) [31].

For $[Pt^{IV}(LLL\text{-}\kappa^3\text{-}N^2,P,P')Cl_3]^{2+}$ cations (LLL = 1,3-bis-(2-(diphenylphosphino)ethyl)benzotriazolium, 1,3-bis-(2-(diphenylphosphino)ethyl)-4,5,6,7-tetrahydrobenzotriazolium) $|\Delta^{15N}{}_{coord}|$ for benzotriazolic and tetrahydrobenzotriazolic N(2) atoms are ca. 121–123 ppm, reflecting $N^{btz/thbtz,N2}$ *trans* to Cl geometry; $|\Delta^{15N}{}_{coord}|$ difference between both species is below 2 ppm [31].

Comparison of the corresponding $[Pt^{IV}(LLL\text{-}\kappa^3\text{-}N^2,P,P')Cl_3]^{2+}$ and $[Pt^{II}(LLL\text{-}\kappa^3\text{-}N^2,P,P')Cl]^{2+}$ cations (LLL = 1,3-bis-(2-(diphenylphosphino)ethyl)benzotriazolium, 1,3-bis-(2-(diphenylphosphino)ethyl)-4,5,6,7-tetrahydrobenzotriazolium) reveals ^{15}N shielding enhancement ($|\Delta^{15N}{}_{coord}|$ increase by ca. 15 ppm) upon Pt(II) → Pt(IV) transition: 105.8 ppm → 121.1 ppm and 107.9 ppm → 122.8 ppm [31,32]; the reason may be either higher oxidation state of platinum or presence of two additional chlorides (with maintaining $N^{btz,N2}$ *trans* to Cl geometry).

2.3 ^{15}N NMR coordination shifts in square-planar (d^8) Rh(I)- and Ir(I)-azole complexes or organometallics

The $|\Delta^{15N}{}_{coord}|$ parameters are in the following ranges—Rh(I) ca. 50–75 ppm, Ir(I) ca. 80–105 ppm (Table 9) [31,32,56,169], in the former case in consistency with the previously reviewed data (Rh(I) ca. 65–80 ppm) [1].

For *trans*(Cl,C=C^{vinyl})-[Rh^I(1-vinyl-3,5-dimethyl-*1H*-pyrazole-κ^2-N^2, C$^{C=C}$)(1,3-bis(2,6-diisopropylphenyl)imidazol-2-ylidene)Cl] $|\Delta^{15N}{}_{coord}|$ for pyrazolic N(2) atom is 66.0 ppm, reflecting N^{pz,N2} *trans* to C geometry [56].

For [Rh^I(LLL^1-κ^3-N^2,P,P′)(CO)]$^{2+}$, [Rh^I(LLL^2-κ^3-N^2,P,P′)X]$^{+/2+}$ (X=Cl, I, CO, THF, CH_2CH_2, $P(C_6H_5)_3$) and [Rh^I(LLL^3-κ^3-N^2, P,P′)X]$^{+/2+}$ (X=Cl, CO, THF, CH_2CH_2, $P(C_6H_5)_3$) cations (LLL^1= 1,3-bis-(2-(diphenylphosphino)ethyl)-*1H*-1,2,3-triazolium, LLL^2=1, 3-bis-(2-(diphenylphosphino)ethyl)benzotriazolium, LLL^3=1,3-bis-(2-(diphenylphosphino)ethyl)-4,5,6,7-tetrahydrobenzotriazolium) $|\Delta^{15N}{}_{coord}|$ for triazolic, benzotriazolic and tetrahydrobenzotriazolic N(2) atoms are ca. 52–67 ppm, unlike for Pd(II) and Pt(II) with no significant dependence on X ligands (and their donor atoms) in *trans* position to N$^{btz/thbtz,N2}$; $|\Delta^{15N}{}_{coord}|$ differences between the corresponding [Rh^I(LLL^2-κ^3-N^2,P,P′)X]$^{+/2+}$ and [Rh^I(LLL^3-κ^3-N^2,P,P′)X]$^{+/2+}$ species (with the same X) are below 10 ppm [31,32].

In contrast, for three other Rh(I)-LLL complexes, where N(2) remained uncoordinated (κ^2-P,P′ chelates; no Rh(I)-N bonds), $\delta^{tz/btz/thbtz,N(2)}{}_{compl}$ were comparable to $\delta^{tz/btz/thbtz,N(2)}{}_{lig}$: [Rh^I(LLL^1-κ^2-P,P′)(CO)Cl]$^+$ −39.8 ppm, [Rh^I(LLL^2-κ^2-P,P′)(CO)Cl]$^+$ −33.0 ppm and [Rh^I(LLL^3-κ^2-P,P′)(CO)Cl]$^+$ −50.9 ppm (all PF_6^- salts in CD_3NO_2) [31] vs (LLL^1)$^+$ −32.9 ppm, (LLL^2)$^+$ −28.9 ppm and (LLL^3)$^+$ −45.0 ppm or −44.8 ppm (various salts in $CDCl_3$) [31,32].

For [Ir^I(LLL-κ^3-N^2,P,P′)X]$^{+/2+}$ cations (LLL=1,3-bis-(2-(diphenylphosphino)ethyl)benzotriazolium; X=Cl, CO, CH_3CN) $|\Delta^{15N}{}_{coord}|$ for benzotriazolic N(2) atoms are ca. 80–102 ppm, increasing in order "N^{btz,N2} *trans* to CO"→"N^{btz,N2} *trans* to N"→"N^{btz,N2} *trans* to Cl" (79.8 ppm→89.6 ppm→102.1 ppm) [169].

In contrast, for another Ir(I)-LLL complex, where N(2) remained uncoordinated (κ^2-P,P′ chelate; no Ir(I)-N bond), $\delta^{btz,N(2)}{}_{compl}$ was comparable to $\delta^{N(2)}{}_{lig}$: [Ir^I(LLL-κ^2-P,P′)(CO)Cl]$^+$ −31.5 ppm (PF_6^- salt in CD_3NO_2) [169] vs (LLL)$^+$ −28.9 ppm ($CF_3SO_3^-$ salt in $CDCl_3$) [32].

For the corresponding $[M^{I}(LLL\text{-}\kappa^3\text{-}N^2,P,P')X]^{+/2+}$ cations (LLL = 1,3-bis-(2-(diphenylphosphino)ethyl)benzotriazolium), Rh(I) → Ir(I) transition results in ^{15}N shielding enhancement ($|\Delta^{15N}{}_{coord}|$ increase by ca. 16–38 ppm): 63.8 ppm → 102.1 ppm for X = Cl and 63.2 ppm → 79.8 ppm for X = CO [31,32,169].

2.4 ^{15}N NMR coordination shifts in octahedral (d^6) Rh(III)- and Ir(III)-azole complexes or organometallics

The $|\Delta^{15N}{}_{coord}|$ parameters are in the following ranges—Rh(III) ca. 40–90 ppm, Ir(III) ca. 70–95 ppm (Table 10) [31,125,169,171], in the latter case in consistency with the previously reviewed data (Ir(III) ca. 80–95 ppm) [1].

In case of both $[M^{III}$(*1H,1′H*-2,2′-biimidazole-κ^2-$N^3,N^{3'}$) (pentamethylcyclopentadienyl)Cl$]^+$ cations (M = Rh, Ir; $N^{im,N3,N3'}$ *trans* to C=C) $|\Delta^{15N}{}_{coord}|$ for imidazolic N(3), N(3′) atoms cannot be determined due to the lack of $\delta^{15N}{}_{lig}$ for *1H,1′H*-2,2′-biimidazole; nevertheless, comparison of $\delta^{15N}{}_{compl}$ exhibits ^{15}N shielding enhancement (by ca. 16 ppm) upon Rh(III) → Ir(III) transition: −173.8 ppm → − 189.7 ppm [125].

For $[Rh^{III}$(1-benzyl-4-phenyl-*1H*-1,2,3-triazole*-κ^2-$N^3,C^{ph,C2}$)(pentamethycyclopentadienyl)Cl] $|\Delta^{15N}{}_{coord}|$ for triazolic N(3) atom is ca. 52 ppm, reflecting $N^{tz,N3}$ *trans* to C=C geometry [171].

For $[Rh^{III}(LLL\text{-}\kappa^3\text{-}N^2,P,P')X_3]^+$ (X = Cl, I), *trans*-$[Rh^{III}(LLL\text{-}\kappa^3\text{-}N^2,P,P')(CH_3)I_2]^+$ and *cis*-$[Rh^{III}(LLL\text{-}\kappa^3\text{-}N^2,P,P')Cl(H)_2]^+$ cations (LLL = 1, 3-bis-(2-(diphenylphosphino)ethyl)benzotriazolium) $|\Delta^{15N}{}_{coord}|$ for benzotriazolic N(2) atoms are ca. 41–89 ppm, being larger (by ca. 35–50 ppm) for $N^{btz,N2}$ *trans* to Cl or I (ca. 84–89 ppm) than $N^{btz,N2}$ *trans* to CH_3 or H (ca. 41–48 ppm) [31,169].

For *cis*-$[Ir^{III}$(1,3-bis-(2-(diphenylphosphino)ethyl)benzotriazolium-κ^3-$N^2,P,P')X(H)_2]^{+/2+}$ cations (LLL = 1,3-bis-(2-(diphenylphosphino) ethyl)benzotriazolium; X = Cl, CO, CH_3CN) $|\Delta^{15N}{}_{coord}|$ for benzotriazolic N(2) atoms are in much more narrow range ca. 69–94 ppm, reflecting the same $N^{btz,N2}$ *trans* to H geometry [169].

For the corresponding *cis*-$[M^{III}(LLL\text{-}\kappa^3\text{-}N^2,P,P')Cl(H)_2]^+$ cations (LLL = 1,3-bis-(2-(diphenylphosphino)ethyl)benzotriazolium), Rh(III) → Ir(III) transition results in ^{15}N shielding enhancement ($|\Delta^{15N}{}_{coord}|$ increase by ca. 28 ppm): 41.5 ppm → 69.4 ppm [169].

Comparison of the corresponding $[Rh^{III}(LLL\text{-}\kappa^3\text{-}N^2,P,P')X_3]^+$ and $[Rh^{I}(LLL\text{-}\kappa^3\text{-}N^2,P,P')X]^+$ cations (LLL = 1,3-bis-(2-(diphenylphosphino) ethyl)benzotriazolium; the same X) reveals ^{15}N shielding enhancement ($|\Delta^{15N}{}_{coord}|$ increase by ca. 20–24 ppm) upon Rh(I) → Rh(III) transition: 63.8 ppm → 84.2 ppm for X = Cl and 65.2 ppm → 89.2 ppm for X = I [31,32]; the reason may be either higher oxidation state of rhodium or presence of two additional halides (with maintaining $N^{btz,N2}$ *trans* to X geometry).

In contrast, comparison of the corresponding *cis*-$[Ir^{III}(LLL\text{-}\kappa^3\text{-}N^2,P,P')X(H)_2]^{+/2+}$ and $[Ir^{I}(LLL\text{-}\kappa^3\text{-}N^2,P,P')X]^{+/2+}$ cations (LLL = 1,3-bis-(2-(diphenylphosphino)ethyl)benzotriazolium; the same X) reveals either $|\Delta^{15N}{}_{coord}|$ decrease or increase upon Ir(I) → Ir(III) transition: 102.1 ppm → 69.4 ppm for X = Cl, 79.8 ppm → 93.8 ppm for X = CO and 89.6 ppm → 77.4 ppm for X = CH_3CN [169]; the reason for this variability may be combination of higher oxidation state of iridium, presence of two additional hydrides and different consequences of geometry changes ("$N^{btz,N2}$ *trans* to Cl" → "$N^{btz,N2}$ *trans* to H", "$N^{btz,N2}$ *trans* to CO" → "$N^{btz,N2}$ *trans* to H" and "$N^{btz,N2}$ *trans* to N" → "$N^{btz,N2}$ *trans* to H").

2.5 ^{15}N NMR coordination shifts in octahedral (d^6) Ru(II)- and Os(II)-azole complexes or organometallics

The $|\Delta^{15N}{}_{coord}|$ parameters are in the following ranges—Ru(II) ca. 10–70 ppm, Os(II) ca. 60–110 ppm (Table 11) [31,106,117,121–127], in the former case in consistency with the previously reviewed data (Ru(II) ca. 50–75 ppm) [1].

For $[Ru^{II}(1,3\text{-bis(quinolin-8-yl)-}1H\text{-pyrazole-}\kappa^3\text{-}N^{quin},N^{quin'},N^{pz,N2})_2]^{2+}$ ($N^{pz,N2}$ *trans* to N) $|\Delta^{15N}{}_{coord}|$ cannot be determined due to the lack of $\delta^{15N}{}_{lig}$ for 1,3-bis(quinolin-8-yl)-*1H*-pyrazole, $\delta^{15N}{}_{compl}$ for $N^{pz,N2}$ being ca. −138 ppm [117].

For $[Ru^{II}(LLL\text{-}\mu^3\text{-}N^{N2},O^{C-O},O^{C=O})(1\text{-isopropyl-4-methylbenzene})]_4{}^{4+}$ cations (LLL = 1-phenyl-3-methyl-4-((*n*-hexyl)carbonyl)-*1H*-pyrazol-5-olate, 1-phenyl-3-methyl-4-*tert*-butylmethylcarbonyl-*1H*-pyrazol-5-olate) $|\Delta^{15N}{}_{coord}|$ for pyrazolic N(2) atom are ca. 8–23 ppm, reflecting $N^{pz,N2}$ *trans* to C=C geometry; moreover, the respective $\delta^{pz,N2}{}_{compl}$ values (−143.0 ppm and −142.8 ppm) are by ca. 14–18 ppm lower than for both $[Ru^{II}(LL\text{-}\kappa^2\text{-}O^{C-O},O^{C=O})(1\text{-isopropyl-4-methylbenzene})Cl]$ complexes, where N(2) remained uncoordinated (−125.2 and −128.4 ppm; originally 256.5 and 253.3 ppm in respect to liquid NH_3) [121].

For [Ru^{II}(tris(*1H*-pyrazol-1-yl)methanide-κ^3-$N^{pz1,N2}$,$N^{pz2,N2}$,$N^{pz3,N2}$)(cyclopentadienyl)] and [Ru^{II}(tris(*1H*-pyrazol-1-yl)methanenitrile-κ^3-$N^{pz1,N2}$,$N^{pz2,N2}$,$N^{pz3,N2}$)(cyclopentadienyl)]$^+$ $|\Delta^{15N}{}_{coord}|$ for pyrazolic N(2) atoms are ca. 63–70 ppm, reflecting $N^{pz,N2}$ *trans* to C=C geometry [122,123].

For *trans*(P,P),*trans*(P,N^{py})-[Ru^{II}(LL-κ^2-N^{py},$N^{pz,N1/tz,N1}$)($P(CH_3)_3)_3$Cl] complexes (LL = 5-(pyrid-2-yl)pyrazolate, 5-(pyrid-2-yl)-1,2,4-triazolate or their derivatives; $N^{pz,N1/tz,N1}$ *trans* to Cl) $|\Delta^{15N}{}_{coord}|$ cannot be determined due to the lack of $\delta^{15N}{}_{lig}$ for LL anions, whereas their determination in respect to $\delta^{15N}{}_{lig}$ for neutral LLH molecules has no sense because of the presence of hydrogen at pyrazolic or triazolic N(1) atoms; nevertheless, $\delta^{pz,N1/tz,N1}{}_{compl}$ for Ru(II)-bonded pyrazolic or triazolic N(1) atoms (ca. −160 ppm to ca. −145 ppm and ca. −160 ppm to ca. −158 ppm, respectively) are in case of each compound by ca. 67–90 ppm lower than $\delta^{pz,N2/tz,N2}{}_{compl}$ for the uncoordinated N(2) ones (ca. −78 ppm to ca. −69 ppm and ca. −85 ppm to ca. −83 ppm, respectively) [106].

For [Ru^{II}(LL-κ^2-N^3,P)(dienyl)]$^+$ cations (LL = 1-methyl-2-diisopropylphosphino-*1H*-imidazole, 1-methyl-2-diisopropylphosphino-4-*tert*-butyl-*1H*-imidazole; dienyl = pentamethylcyclopentadienyl, tetramethyltrifluoromethylcyclopentadienyl; coordination number 5) $|\Delta^{15N}{}_{coord}|$ cannot be determined due to the lack of $\delta^{15N}{}_{lig}$ for LL, $\delta^{15N}{}_{compl}$ for imidazolic N(3) atoms ranging from ca. −169 ppm to ca. −147 ppm [124].

For [Ru^{II}(*1H,1'H*-2,2'-biimidazole-κ^2-N^3,$N^{3'}$)(1-isopropyl-4-methylbenzene)Cl]$^+$ ($N^{im,N3,N3'}$ *trans* to C=C) $|\Delta^{15N}{}_{coord}|$ cannot be determined due to the lack of $\delta^{15N}{}_{lig}$ for *1H,1'H*-2,2'-biimidazole, $\delta^{15N}{}_{compl}$ for imidazolic N(3), N(3') atoms being −181.7 ppm [125].

For [Ru^{II}(LLL-κ^3-N^2,P,P')Cl(μ-Cl)]$_2^{2+}$ cations (LLL = 1,3-bis-(2-(diphenylphosphino)ethyl)-*1H*-1,2,3-triazolium, 1,3-bis-(2-(diphenylphosphino)ethyl)benzotriazolium, 1,3-bis-(2-(diphenylphosphino)ethyl)-4,5,6,7-tetrahydrobenzotriazolium) $|\Delta^{15N}{}_{coord}|$ for benzotriazolic N(2) atoms are ca. 47–62 ppm, reflecting $N^{btz,N2}$ *trans* to Cl geometry [31].

In case of [Os^{II}L(NO)Cl_4]$^-$ anions (L = *1H*-pyrazole-κ^1-N^2, *1H*-imidazole-κ^1-N^3, *1H*-indazole-κ^1-N^2, *1H*-benzoimidazole-κ^1-N^3) $|\Delta^{15N}{}_{coord}|$ for pyrazolic N(2), imidazolic N(3), indazolic N(2) and benzoimidazolic N(3) atoms are ca. 90–112 ppm for *cis* and ca. 63–66 ppm for *trans* species, reflecting the respective N^L *trans* to Cl and N^L *trans* to NO geometry [126,127]. For both isomers of [Os^{II}(*1H*-pyrazole-κ^1-N^2)(NO)Cl_4]$^-$ or [Os^{II}(*1H*-indazole-κ^1-N^2)(NO)Cl_4]$^-$ ^{15}N shielding weakening (by ca. 44–49 ppm) is observed upon "N *trans* to Cl" → "N *trans* to NO" geometry change (111.5 ppm → 62.9 ppm and 110.0 ppm → 65.5 ppm, respectively) [126,127].

2.6 ^{15}N NMR coordination shifts in (d^{10}) Ag(I)- and square-planar (d^8) Au(III)-azole complexes

The $|\Delta^{15N}_{coord}|$ parameters are in the following ranges—Ag(I) ca. 0–55 ppm, Au(III) ca. 65–70 ppm (Table 12) [139,143], in the former case in consistency with the previously reviewed data (Ag(I) ca. 10–55 ppm) [1].

For Ag(I) complexes with 1-phenyl-3-methyl-4-benzoylpyrazolate-5-one-κ^1-N^2 anions and *1H*-imidazole-κ^1-N^3, 1-methyl-*1H*-imidazole-κ^1-N^3 or 1-benzyl-*1H*-imidazole-κ^1-N^3 $|\Delta^{15N}_{coord}|$ for pyrazolic N(2) and imidazolic N(3) atoms are ca. 35–55 ppm and ca. 13–46 ppm, respectively [139]. For Ag(I) complexes with 1-(pyrid-2-yl)-3-methyl-4-trifluoroacetylpyrazolate-5-one-κ^2-N^{py},$N^{pz,N2}$ anions and 2-ethyl-*1H*-imidazole-κ^1-N^3, 1-benzyl-*1H*-imidazole-κ^1-N^3, 1-benzyl-2-methyl-*1H*-imidazole-κ^1-N^3, 2-trifluoromethyl-*1H*-benzoimidazole-κ^1-N^3 or 1,1′-carbonyl-*1H,1H′*-diimidazole-μ^2-N^3,$N^{3'}$ $|\Delta^{15N}_{coord}|$ for pyrazolic N(2) and imidazolic N(3) atoms are ca. 3–25 ppm and ca. 9–34 ppm, respectively [139]. For $[Ag^I(\mathit{1H}\text{-imidazole-}\kappa^1\text{-}N^3)_2]^+$ $|\Delta^{15N}_{coord}|$ for imidazolic N(3) atom is 38.6 ppm [139].

For $[Au^{III}$((*R*,*S*)-1-(4-chlorophenyl)-3-((*1H*-1,2,4-triazol-1-yl)methyl)-4,4-dimethyl-pentan-3-ol-κ^1-$N^4)Cl_3]$ $|\Delta^{15N}_{coord}|$ for triazolic N(4) atom is 67.0 ppm, reflecting $N^{tz,N4}$ *trans* to Cl geometry [143].

2.7 ^{15}N NMR coordination shifts in (d^{10}) Zn(II)-azole complexes

The $|\Delta^{15N}_{coord}|$ parameters are in the range ca. 40–60 ppm [11,165], in consistency with the previously reviewed data (Zn(II) ca. 15–40 ppm) [1].

For $[Zn^{II}$(2-(1-(pyrid-2-yl)-5-(4-methoxyphenyl)-*1H*-pyrazol-3-yl)pyridine-κ^3-N^{pyA},N^{pyB},$N^{pz,N2})Cl_2]$ $|\Delta^{15N}_{coord}|$ for pyrazolic N(2) atom is ca. 48–50 ppm [11]. For $[Zn^{II}(L\text{-}\kappa^1\text{-}N^3)_6]^{2+}$ cations (L = 1-methyl-*1H*-imidazole, 1-ethyl-*1H*-imidazole, 1-*n*-butyl-*1H*-imidazole; $N^{im,N3}$ *trans* to N) $|\Delta^{15N}_{coord}|$ for imidazolic N(3) atoms are ca. 40–57 ppm [165].

3. Transition metal complexes with purines

In Review 2013, 52 ^{15}N NMR-studied transition metal complexes with purines (Pd(II)—15, Pt(II)—31, Pt(IV)—4, Au(I)—2) were discussed in respect to their ^{15}N chemical and coordination shifts (paragraph 3, including table 2.19). In the present paper 22 analogous ^{15}N NMR-studied species (22 Pt(II) complexes; no organometallics) are described (Table 14).

Two $[Pt^{II}(L^1\text{-}\kappa^1\text{-}N^7)_2(C_2O_4)]$ complexes with $(N^{NH}$-furfuryl)-9-isopropyladenine derivatives (L^1) were studied by Krikavova et al. [33]. Thirteen *trans*-$[Pt^{II}(L^2\text{-}\kappa^1\text{-}N^7)_2Cl_2]$ and five $[Pt^{II}(L^2\text{-}\kappa^1\text{-}N^7)_2(C_2O_4)]$ complexes with $(N^{NH}$-benzyl)adenosine derivatives (L^2) were reported by Starha et al. [34,35]. *cis*-$[Pt^{II}(L^3\text{-}\kappa^1\text{-}N^7)(DMSO)Cl_2]$ with $(N^{NH}$-furfuryl) adenosine (L^3) was described by Jakubowski et al. [23]. $[Pt^{II}$(6-methylamino-9-methylpurinate-κ^2-N^7,N^{NCH3})(1,2-bis(diphenylphosphino) ethane)] was characterized by Mueller et al. [36]. In all these molecules Pt(II) coordination occurred via N(7) [23,33–36].

Generally, in case of Pt(II) species $|\Delta^{15N}{}_{coord}|$ for N(7) atoms are predominantly ca. 105–115 ppm, being generally larger (by ca. 5–10 ppm) for $[Pt^{II}(L\text{-}\kappa^1\text{-}N^7)_2(C_2O_4)]$ (ca. 111–115 ppm; $N^{pu,N7}$ *trans* to O) than *trans*-$[Pt^{II}(L\text{-}\kappa^1\text{-}N^7)_2Cl_2]$ (ca. 105–107 ppm; $N^{pu,N7}$ *trans* to N) compounds [33–35]. This relation is well observed for the corresponding Pt(II) oxalate and *trans*-dichloride complexes with the same $(N^{NH}$-benzyl)adenosine derivative: 113.5 ppm vs 106.0 ppm for L = 2-chloro(N^{NH}-benzyl)adenosine, 113.3 ppm vs 105.0 ppm for L = 2-methoxy(N^{NH}-benzyl)adenosine, 113.1 ppm vs 106.1 ppm for L = 4-methyl(N^{NH}-benzyl)adenosine, 113.5 ppm vs 105.4 ppm for L = 4-fluoro(N^{NH}-benzyl)adenosine and 112.8 ppm vs 105.0 ppm for L = 4-methoxy(N^{NH}-benzyl)adenosine [34,35].

In contrast, for *cis*-$[Pt^{II}(N^{NH}$-furfuryl)adenosine-κ^1-N^7)(DMSO)$Cl_2]$ ($N^{pu,N7}$ *trans* to Cl) $|\Delta^{15N}{}_{coord}|$ for N(7) atom is only ca. 84 ppm [23].

For $[Pt^{II}$(6-methylamino-9-methylpurinate-κ^2-N^7,N^{NCH3})(1,2-bis (diphenylphosphino)ethane)] $|\Delta^{15N}{}_{coord}|$ for N(7) atom cannot be determined due to the lack of $\delta^{15N}{}_{lig}$ for 6-methylamino-9-methylpurine, $\delta^{15N}{}_{compl}$ being ca. −197 ppm [36].

4. Transition metal complexes with 1,2,4-triazolo-[1,5*a*]-pyrimidines

In Review 2013, 32 ^{15}N NMR-studied transition metal complexes with 1,2,4-triazolo-[1,5*a*]-pyrimidine (TP) and its derivatives (Pd(II)—5, Pt(II)—20, Pt(IV)—4, Au(III)—3) were discussed in respect to their ^{15}N chemical and coordination shifts (paragraph 4, including table 2.20). In the present paper 39 analogous ^{15}N NMR-studied species (25 Pt(II), 1 Pt(IV) and 13 Ru(II) complexes; no organometallics) are described (Table 15).

Two *cis*-[Pt^{II}(L-κ^1-N^3)$_2$Cl$_2$], two *cis*-[Pt^{II}(L-κ^1-N^3)(NH$_3$)Cl$_2$], two *cis*-[Pt^{II}(L-κ^1-N^3)(DMSO)Cl$_2$], one *trans*-[Pt^{II}(L-κ^1-N^3)(DMSO)Cl$_2$], three *cis*-[Pt^{II}(L-κ^1-N^3)$_2$I$_2$], one *cis*-[Pt^{II}(L-κ^1-N^3)$_2$(CH$_3$COO)$_2$], nine [Pt^{II}(L-κ^1-N^3)$_2$(dicarboxylate)] (dicarboxylate = malonate, glutarate, cyclobutane-1, 1-dicarboxylate) and four [Pt^{II}(L-κ^1-N^3)(DMSO)(dicarboxylate)] (dicarboxylate = malonate, cyclobutane-1,1-dicarboxylate) complexes (L = 5,7-dimethyl-, 5-methyl-7-isobutyl, 5,7-diethyl-, 5,6,7-trimethyl-, 5,7-di-*tert*-butyl-, 5,7-diphenyl-TP), as well as *cis*-[Pt^{II}(4,7-dihydro-5-methyl-1,2,4-triazolo-[1,5*a*]-pyrimidin-7-one-κ^1-N^3)$_2$I$_2$] were studied by Lakomska et al., Hoffmann et al., and Jakubowski et al. [38–44], while the only *cis*-[Pt^{IV}(L-κ^1-N^3)$_2$Cl$_4$] complex (L = 5-methyl-7-isobutyl-TP) by Lakomska et al. [45].

Five [Ru^{II}(L-κ^1-N^3)(1-isopropyl-4-methylbenzene)Cl$_2$], four *trans*(Cl, S),(Cl,S)-[Ru^{II}(L-κ^1-N^3)(DMSO)$_3$Cl$_2$] and three *trans*(Cl,S),(Cl,N)-[Ru^{II}(L-κ^1-N^3)$_2$(DMSO)$_2$Cl$_2$] complexes (L = TP and 5,7-dimethyl-, 5-methyl-7-isobutyl-, 5,7-diethyl-, 5,7-di-*tert*-butyl-, 5,6,7-trimethyl-, 5,7-diphenyl-TP), as well as *trans*(Cl,S),(Cl,S)-[Ru^{II}(4,7-dihydro-5-methyl-1,2,4-triazolo-[1,5*a*]-pyrimidin-7-one-κ^1-N^3)(DMSO)$_3$Cl$_2$] were reported by Fandzloch et al. and Lakomska et al. [46–49]. In all these molecules Pt(II), Pt(IV) or Ru(II) coordination occurred via N(3) [38–49].

In case of Pt(II) species $|\Delta^{15N}{}_{coord}|$ for N(3) atoms are ca. 77–105 ppm, being as follows: ca. 85–95 ppm for *cis*-[Pt^{II}(L-κ^1-N^3)$_2$Cl$_2$], *cis*-[Pt^{II}(L-κ^1-N^3)(NH$_3$)Cl$_2$], *cis*-[Pt^{II}(L-κ^1-N^3)(DMSO)Cl$_2$] (N^{TP,N3} *trans* to Cl) [38,39], ca. 77 ppm for *trans*-[Pt^{II}(L-κ^1-N^3)(DMSO)Cl$_2$] (N^{TP,N3} *trans* to S) [38], ca. 81–83 ppm for *cis*-[Pt^{II}(L-κ^1-N^3)$_2$I$_2$] (N^{TP,N3} *trans* to I) [40] and ca. 98–105 ppm for *cis*-[Pt^{II}(L-κ^1-N^3)$_2$(CH$_3$COO)$_2$], [Pt^{II}(L-κ^1-N^3)$_2$(CH$_2$(COO)$_2$)], [Pt^{II}(L-κ^1-N^3)$_2$(OOCCH$_2$CH$_2$CH$_2$COO)], [Pt^{II}(L-κ^1-N^3)$_2$(cyclobutane-1,1-dicarboxylate)], [Pt^{II}(L-κ^1-N^3)(DMSO)(CH$_2$(COO)$_2$)], [Pt^{II}(L-κ^1-N^3)(DMSO)(cyclobutane-1,1-dicarboxylate)] (N^{TP,N3} *trans* to O) [41–44]. $|\Delta^{15N}{}_{coord}|$ for N(3) in *cis*-[Pt^{II}(4,7-dihydro-5-methyl-1,2,4-triazolo-[1,5*a*]-pyrimidin-7-one-κ^1-N^3)$_2$I$_2$] is ca. 91 ppm [40], being larger (by ca. 8–9 ppm) than for the other *cis*-[Pt^{II}(L-κ^1-N^3)$_2$I$_2$] complexes, despite the same N^{TP,N3} *trans* to I geometry.

Comparison of the corresponding *cis*-[Pt^{II}(L-κ^1-N^3)$_2$X$_2$] halide complexes (with the same L) reveals ^{15}N shielding weakening ($|\Delta^{15N}{}_{coord}|$ decrease by ca. 12 ppm) upon Cl → I replacement: 93.3 ppm → 81.4 ppm for L = 5-methyl-7-isobutyl-TP and 93.0 ppm → 81.3 ppm for L = 5,7-diethyl-TP [38–40], reflecting “N^{TP,N3} *trans* to Cl” → “N^{TP,N3} *trans* to I” geometry change and following the same relation for the previously rewieved *cis*-[Pt^{II}(L-κ^1-N^3)$_2X_2$] and *cis*-[Pt^{II}(L-κ^1-N^3)(NH_3)X_2] complexes (X = Cl, I) with L = 5,7-di-*tert*-butyl-TP or 5,7-diphenyl-TP [1,52,182,183]. An analogous dependency, although much less expressed ($|\Delta^{15N}{}_{coord}|$ decrease by 1.5 ppm) is observed when to compare already reviewed *cis*-[Pt^{II}(4,7-dihydro-5-methyl-1,2,4-triazolo-[1,5*a*]-pyrimidin-7-one-κ^1-N^3)$_2Cl_2$] and the presently discussed *cis*-[Pt^{II}(4,7-dihydro-5-methyl-1,2,4-triazolo-[1,5*a*]-pyrimidin-7-one-κ^1-N^3)$_2I_2$]: 92.1 ppm → 90.6 ppm [40,53].

Comparison of the corresponding *cis*- and *trans*-[Pt^{II}(L-κ^1-N^3)(DMSO)Cl_2] isomers (L = 5-methyl-7-isobutyl-TP) exhibits ^{15}N shielding weakening ($|\Delta^{15N}{}_{coord}|$ decrease by ca. 10 ppm): 86.4 ppm → 76.8 ppm [38], reflecting "$N^{TP,N3}$ *trans* to Cl" → "$N^{TP,N3}$ *trans* to S" geometry change and following the same relation for the previously reviewed pair of isomeric [Pt^{II}(L-κ^1-N^3)(DMSO)Cl_2] complexes with L = 5,7-diphenyl-TP [1,184].

In case of the only Pt(IV) complex (*cis*-[Pt^{IV}(5-methyl-7-isobutyl-TP-κ^1-N^3)$_2Cl_4$]) $|\Delta^{15N}{}_{coord}|$ for N(3) atom is 95.7 ppm, being larger (by ca. 2–3 ppm) than for its Pt(II) analogue, i.e., *cis*-[Pt^{II}(5-methyl-7-isobutyl-TP-κ^1-N^3)$_2Cl_2$] (93.3 ppm) [38,45]; the reason may be either higher oxidation state of platinum or presence of two additional chlorides (with maintaining $N^{TP,N3}$ *trans* to Cl geometry). The same relation was observed for the previously rewieved *cis*-[Pt^{IV}(L-κ^1-N^3)$_2Cl_4$] and *cis*-[Pt^{II}(L-κ^1-N^3)$_2Cl_2$] complexes (L = TP and 5,7-dimethyl-, 5,7-di-*tert*-butyl-TP) [1,51,52,54,180].

In case of Ru(II) species $|\Delta^{15N}{}_{coord}|$ for N(3) atoms are ca. 44–67 ppm, being comparable for [Ru^{II}(L-κ^1-N^3)(1-isopropyl-4-methylbenzene)Cl_2] (ca. 54–64 ppm) [46], *trans*(Cl,S),(Cl,S)-[Ru^{II}(L-κ^1-N^3)$(DMSO)_3Cl_2$] (ca. 60–64 ppm) [47,48], *trans*(Cl,S),(Cl,N)-[Ru^{II}(L-κ^1-N^3)$_2(DMSO)_2Cl_2$] (ca. 44–52 ppm) [47] and *trans*(Cl,S),(Cl,S)-[Ru^{II}(4,7-dihydro-5-methyl-1,2,4-triazolo-[1,5*a*]-pyrimidin-7-one-κ^1-N^3)$(DMSO)_3Cl_2$] (ca. 66–67 ppm) [49], despite variable geometries ($N^{TP,N3}$ *trans* to C=C, *trans* to S, *trans* to Cl),]; they are generally smaller than for Pt(II) complexes.

Particularly, comparison of *trans*(Cl,S),(Cl,S)-[Ru^{II}(L-κ^1-N^3)$(DMSO)_3Cl_2$] to the previously reviewed *trans*-[Pt^{II}(L-κ^1-N^3)(DMSO)Cl_2] compounds (in all: $N^{TP,N3}$ *trans* to S) exhibits ca. 10–25 ppm differences (60.1 ppm vs 84.9 ppm for L = TP; 63.6 ppm vs 74.7 ppm for L = 5,7-dimethyl-TP) [47,184].

Surprisingly, for *trans*(Cl,S),(Cl,N)-[Ru^{II}(L-κ^1-N^3)$_2$(DMSO)$_2$$Cl_2$] (L=5,7-diethyl-, 5,7-di-*tert*-butyl-, 5,7-diphenyl-TP) only one set of ^{15}N signals was recorded [47]; thus, it is not clear whether the published $|\Delta^{15N}{}_{coord}|$ parameters correspond to N(3) atoms *trans* to S or *trans* to Cl.

5. Conclusions

^{15}N NMR coordination shifts ($\Delta^{15N}{}_{coord}$) in diamagnetic transition metal complexes and organometallics with nitrogen-containing heterocycles are important parameters related to the molecular structure.

For azines and azoles, metal binding results in significant ^{15}N shielding of the respective N atom (highly negative$\Delta^{15N}{}_{coord}$ values), which allows to identify coordination site(s) in ligands containing more than one heterocyclic nitrogen and to distinguish possible linkage isomers). The absolute magnitude of ^{15}N shielding effect, reflected by the $|\Delta^{15N}{}_{coord}|$ parameters, is highly dependent on two principal factors: the type of metal and the type of donor ligand in *trans* position in respect to the N atom.

Generally, for analogous complexes or organometallics the Pd(II)$\rightarrow$Pt(II), Rh(I)$\rightarrow$Ir(I), Rh(III)$\rightarrow$Ir(III), Ru(II)$\rightarrow$Os(II) and Ag(I)$\rightarrow$Au(I) transitions result in ^{15}N shielding enhancement ($|\Delta^{15N}{}_{coord}|$ become larger). The increase of oxidation state (Pt(II)$\rightarrow$Pt(IV), Rh(I)$\rightarrow$Rh(III), Ir(I)$\rightarrow$Ir(III)), upon presence of the same auxiliary ligands, predominantly enhances the ^{15}N shielding effect as well, although some exceptions are noted.

On the other hand, ^{15}N shielding phenomenon is usually weakened in order: O$\rightarrow$N$\rightarrow$Cl$\rightarrow$I, S, P$\rightarrow$C, C=C, CO$\rightarrow$H ($|\Delta^{15N}{}_{coord}|$ become smaller), however, with some variations; this relationship allows to distinguish geometric isomers for square-planar and octahedral species. An important exclusion is 2,2′:6′,2″-terpyridine ring system, in this case $|\Delta^{15N}{}_{coord}|$ for the outer N(1), N(1″) atoms being always much larger than for the inner N (1′) one.

References

[1] L. Pazderski, ^{15}N and ^{31}P NMR coordination shifts in transition metal complexes with nitrogen- and phosphorus-containing heterocycles, Annu. Rep. NMR Spectrosc. 80 (2013) 33–180.

[2] D. Niedzielska, T. Pawlak, A. Wojtczak, L. Pazderski, E. Szłyk, Structural and ^{1}H, ^{13}C, ^{15}N NMR spectroscopic studies of Pd(II) chloride organometallics with 2-phenylpyridine and ammonia, pyridine or its methyl derivatives, Polyhedron 92 (2015) 41–51.

[3] T. Pawlak, D. Niedzielska, J. Vícha, R. Marek, L. Pazderski, Dimeric Pd(II) and Pt(II) chloride organometallics with 2-phenylpyridine and their solvolysis in dimethylsulfoxide, J. Organomet. Chem. 759 (2014) 58–66.
[4] F. Martinez-Olid, R. Andres, J.C. Flores, P. Gomez-Sal, Trimetallic PEPPSI-type palladium N-heterocyclic carbene complexes—improved catalyst lifetime in the Mizoroki–Heck coupling reaction, Eur. J. Inorg. Chem. (2015) 4076–4087.
[5] F. Martinez-Olid, R. Andres, E. de Jesus, J.C. Flores, P. Gomez-Sal, K. Heuze, L. Vellutini, Magnetically recoverable catalysts based on mono- or bis-(NHC) complexes of palladium for the Suzuki–Miyaura reaction in aqueous media: two NHC–Pd linkages are better than one, Dalton Trans. 45 (2016) 11633–11638.
[6] S.J. Tereniak, S.S. Stahl, Mechanistic basis for efficient, site-selective, aerobic catalytic turnover in Pd-catalyzed C−H imidoylation of heterocycle-containing molecules, J. Am. Chem. Soc. 139 (2017) 14533–14541.
[7] M. Virant, M. Mihelac, M. Gazvoda, A.E. Cotman, A. Frantar, B. Pinter, J. Kossmrlj, Pyridine wingtip in $[Pd(Py\text{-}tzNHC)_2]^{2+}$ complex is a proton shuttle in the catalytic hydroamination of alkynes, Org. Lett. 22 (2020) 2157–2161.
[8] A. Requet, O. Colin, F. Bourdreux, S.M. Salim, S. Marque, C. Thomassigny, C. Greck, J. Farjon, D. Prim, Pyridylalkylamine ligands and their palladium complexes: structure and reactivity revisited by NMR, Magn. Reson. Chem. 52 (2014) 273–278.
[9] V. Rosar, D. Dedeic, T. Nobile, F. Fini, G. Balducci, E. Alessio, C. Carfagna, B. Milani, Palladium complexes with simple iminopyridines as catalysts for polyketone synthesis, Dalton Trans. 45 (2016) 14609–14619.
[10] G. Dikmen, D. Hur, Palladium (II) complex: Synthesis, spectroscopic studies and DFT calculations, Chem. Phys. Lett. 716 (2019) 49–60.
[11] L. Soria, P. Ovejero, M. Cano, J.V. Heras, M.R. Torres, R. Claramunt, P. Cornago, Coordination behaviour of new dipyridylpyrazole ligands towards $ZnCl_2$ and $PdCl_2$ fragments. Crystalline structural characterization and multinuclear NMR studies as evidence of linkage and conformational isomers, RSC Adv. 4 (2014) 9383–9394.
[12] M.A. Carvalho, E.G.R. Arruda, D.M. Profirio, A.F. Gomes, F.C. Gozzo, A.L.B. Formiga, P.P. Corbi, Chemical and spectroscopic characterizations, ESI-QTOF mass spectrometric measurements and DFT studies of new complexes of palladium(II) with tryptamine and mefenamic acid, J. Mol. Struct. 1100 (2015) 6–13.
[13] A. Rotondo, S. Barresi, M. Cusumano, E. Rotondo, P. Donato, L. Mondello, NMR characterisation and dynamic behaviour of $[Pt(bipy)(R\text{-}Thiourea)_2]Cl_2$ and $[Pt(phen)(R\text{-}Thiourea)_2]Cl_2$ complexes, Inorg. Chim. Acta 410 (2014) 1–10.
[14] A. Campos-Carrasco, M. Bruce, M. Reguero, A.M. Masdeu-Bulto, Pd(II) complexes with 3,3′-substituted bipyridine ligands: synthesis, crystal structure, DFT calculations and catalytic studies in compressed carbon dioxide, Inorg. Chim. Acta 409 (2014) 285–295.
[15] M.N. Wenzel, P.K. Owens, J.T.W. Bray, J.M. Lynam, P.M. Aguiar, C. Reed, J.D. Lee, J.F. Hamilton, A.C. Whitwood, I.J.S. Fairlamb, Redox couple involving NO_x in aerobic Pd-catalyzed oxidation of p^3-C−H bonds: direct evidence for Pd−NO_3^-/NO_2^- interactions involved in oxidation and reductive elimination, J. Am. Chem. Soc. 139 (2017) 1177–1190.
[16] N. Filipovic, S. Grubisic, M. Jovanovic, M. Dulovic, I. Markovic, O. Klisuric, A. Marinkovic, D. Mitic, K. Andelkovic, T. Todorovic, Palladium(II) complexes with N-heteroaromatic bidentate hydrazone ligands: the effect of the chelate ring size and lipophilicity on in vitro cytotoxic activity, Chem. Biol. Drug Des. 84 (2014) 333–341.
[17] R.M. Shanahan, A. Hickey, L.M. Bateman, M.E. Light, G.P. McGlacken, One-pot cross-coupling/C−H functionalization reactions: quinoline as a substrate and ligand through N−Pd interaction, J. Org. Chem. 85 (2020) 2585–2596.

[18] P.B. White, J.N. Jaworski, C.G. Fry, B.S. Dolinar, I.A. Guzei, S.S. Stahl, Structurally diverse diazafluorene-ligated palladium(II) complexes and their implications for aerobic oxidation reactions, J. Am. Chem. Soc. 138 (2016) 4869–4880.
[19] J.N. Jaworski, C.V. Kozack, S.J. Tereniak, S.M.M. Knapp, C.R. Landis, J.T. Miller, S.S. Stahl, Operando spectroscopic and kinetic characterization of aerobic allylic C−H acetoxylation catalyzed by Pd(OAc)2/4,5-diazafluoren-9-one, J. Am. Chem. Soc. 141 (2019) 10462–10474.
[20] B. Pinter, D. Urankar, A. Pevec, F. De Proft, J. Kosmrlj, Platinum-mediated dinitrogen liberation from 2-picolyl azide through a putative Pt=N double bond containing intermediate, Inorg. Chem. 52 (2013) 4528–4533.
[21] A. Dawn, K.S. Andrew, D.S. Yufit, Y. Hong, J.P. Reddy, C.D. Jones, J.A. Aguilar, J.W. Steed, Supramolecular gel control of cisplatin crystallization: identification of a new solvate form using a cisplatin-mimetic gelator, Cryst. Growth Des. 15 (2015) 4591–4599.
[22] M. Hebenbrock, D. Gonzalez-Abradelo, C.A. Strassert, J. Muller, DNA Groove-binding ability of luminescent platinum(II) complexes based on a family of tridentate N∧N∧C ligands bearing differently substituted alkyl tethers, Z. Anorg. Allg. Chem. 644 (2018) 671–682.
[23] M. Jakubowski, I. Lakomska, J. Sitkowski, M. Pokrywczynska, P. Dabrowski, G. Framski, T. Ostrowski, Multinuclear magnetic resonance characterization and antiproliferative studies of novel dichlorido platinum(II) complexes containing kinetin riboside and 1-β-D-ribofuranosyl-4-(2-pyridyl)-*1H*-1,2,3-triazole, Polyhedron 180 (2020) 114428.
[24] P. Starha, Z. Travnícek, J. Vanco, Z. Dvorak, In vitro anticancer active cis-Pt(II)-diiodido complexes containing 4-azaindoles, J. Biol. Inorg. Chem. 24 (2019) 257–269.
[25] P. Starha, Z. Travnícek, A. Popa, I. Popa, T. Muchova, V. Brabec, How to modify 7-azaindole to form cytotoxic Pt(II) complexes: Highly in vitro anticancer effective cisplatin derivatives involving halogeno-substituted 7-azaindole, J. Inorg. Biochem. 115 (2012) 57–63.
[26] P. Starha, J. Hosek, J. Vanco, Z. Dvorak, P. Suchy, I. Popa, G. Prazanova, Z. Travnicek, Pharmacological and molecular effects of platinum(II) complexes involving 7-azaindole derivatives, PLoS One 9 (2014) e90341.
[27] P. Starha, Z. Dvorak, Z. Travnícek, Highly and broad-spectrum in vitro antitumor active *cis*-dichloridoplatinum(II) complexes with 7-azaindoles, PLoS One 10 (2015) e0136338.
[28] P. Starha, Z. Travnícek, I. Popa, Z. Dvorak, Synthesis, characterization and in vitro antitumor activity of platinum(II) oxalato complexes involving 7-azaindole derivatives as coligands, Molecules 19 (2014) 10832–10844.
[29] N.T. Zanvettor, D.H. Nakahata, R.E.F. de Paiva, M.A. Ribeiro, A. Cuin, P.P. Corbi, A.L.B. Formiga, Copper(II), palladium(II) and platinum(II) complexes with 2,2-thiophen-yl-imidazole: synthesis, spectroscopic characterization, X-ray crystallographic studies and interactions with calf-thymus DNA, Inorg. Chim. Acta 443 (2016) 304–315.
[30] N.A. Bumagin, A.V. Kletskov, S.K. Petkevich, I.A. Kolesnik, A.S. Lyakhov, L.S. Ivashkevich, A.V. Baranovsky, P.V. Kurman, V.I. Potkin, Substituted 1-(isoxazol-3-yl)methyl-1H-1,2,3-triazoles: synthesis, palladium(II) complexes, and high-turnover catalysis in aqueous media, Tetrahedron 74 (2018) 3578–3588.
[31] Y. Tulchinsky, S. Kozuch, P. Saha, A. Mauda, G. Nisnevich, M. Botoshansky, L.J.W. Shimon, M. Gandelman, Coordination chemistry of N-heterocyclic nitrenium-based ligands, Chem. A Eur. J. 21 (2015) 7099–7110.
[32] Y. Tulchinsky, S. Kozuch, P. Saha, M. Botoshansky, L.J.W. Shimon, M. Gandelman, Cation-cation bonding in nitrenium metal complexes, Chem. Sci. 5 (2014) 1305–1311.

[33] R. Krikavova, L. Hanouskova, Z. Dvorak, Z. Travnícek, Dichlorido-platinum(II) complexes with kinetin derivatives as promising cytotoxic agents avoiding resistance of cancer cells: contrasting results between cisplatin and oxaliplatin analogues, Polyhedron 90 (2015) 7–17.
[34] P. Starha, I. Popa, Z. Travnícek, J. Vanco, N6-benzyladenosine derivatives as novel N-donor ligands of Platinum(II) dichlorido complexes, Molecules 18 (2013) 6990–7003.
[35] P. Starha, I. Popa, Z. Travnícek, Platinum(II) oxalato complexes involving adenosine-based N-donor ligands: synthesis, characterization and cytotoxicity evaluation, Molecules 19 (2014) 3832–3847.
[36] K. Mueller, C. Schuetz, T. Ruffer, M. Bette, G.N. Kaluderovic, D. Steinborn, H. Schmidt, Synthesis, characterization, structures and in vitro antitumor activity of platinum(II) complexes bearing adeninato or methylated adeninato ligands, Inorg. Chim. Acta 507 (2020) 119539.
[37] R. Marek, A. Lycka, ^{15}N NMR spectroscopy in structural analysis, Curr. Org. Chem. 6 (2002) 35–66.
[38] I. Lakomska, M. Babinska, A. Wojtczak, A. Kozakiewicz, J. Sitkowski, A.A. Jarzecki, Experimental and theoretical investigation of the complexation of 5-methyl-7-isobutyl-1,2,4-triazolo[1,5-a]pyrimidine with platinum(II) ions, New J. Chem. 41 (2017) 7775–7782.
[39] I. Lakomska, M. Babinska, A. Wojtczak, J. Sitkowski, Synthesis, characterization and in vitro cytotoxicity of three types of platinum(II) complexes containing 5,7-diethyl-1,2,4-triazolo[1,5-a]pyrimidine, Inorg. Chim. Acta 453 (2016) 516–521.
[40] I. Lakomska, K. Hoffmann, T. Muziol, J. Sitkowski, Multinuclear magnetic resonance and X-ray characterization of platinum(II) complexes with substituted-1,2,4-triazolo [1,5-a]pyrimidines, J. Mol. Struct. 1056–1057 (2014) 146–151.
[41] K. Hoffmann, I. Lakomska, J. Wisniewska, A. Kaczmarek-Kedziera, J. Wietrzyk, Acetate platinum(II) compound with 5,7-ditertbutyl-1,2,4-triazolo[1,5-a]pyrimidine that overcomes cisplatin resistance: structural characterization, in vitro cytotoxicity, and kinetic studies, J. Coord. Chem. 68 (2015) 3193–3208.
[42] I. Lakomska, K. Hoffmann, A. Wojtczak, J. Sitkowski, E. Maj, J. Wietrzyk, Cytotoxic malonate platinum(II) complexes with 1,2,4-triazolo[1,5-a] pyrimidine derivatives: structural characterization and mechanism of the suppression of tumor cell growth, J. Inorg. Biochem. 141 (2014) 188–197.
[43] K. Hoffmann, J. Wiśniewska, A. Wojtczak, J. Sitkowski, A. Denslow, J. Wietrzyk, M. Jakubowski, I. Lakomska, Rational design of dicarboxylato platinum(II) complexes with purinemimetic ligands as novel anticancer agents, J. Inorg. Biochem. 172 (2017) 34–45.
[44] M. Jakubowski, I. Lakomska, J. Sitkowski, J. Wisniewska, Dicarboxylato platinum(II) complexes containing dimethyl sulfoxide and triazolopyrimidine as potential anticancer agents: synthesis, structural and biological studies in solution, New J. Chem. 42 (2018) 8113–8122.
[45] I. Lakomska, M. Jakubowski, M. Barwiolek, T. Muzioł, Different bonding of triazolopyrimidine to platinum(IV). Structural and in vitro cytotoxicity studies, Polyhedron 160 (2019) 123–129.
[46] I. Lakomska, K. Stefanczak, M. Fandzloch, J. Sitkowski, B. Filip-Psurska, A. Wojtczak, Use of 1,2,4-triazolo[1,5-a]pyrimidines to design new "piano-stool" ruthenium(II) compounds, Polyhedron 109 (2016) 33–39.
[47] M. Fandzloch, A. Jaromin, M. Zaremba-Czogalla, A. Wojtczak, A. Lewinska, J. Sitkowski, J. Wisniewska, I. Lakomska, J. Gubernator, Nanoencapsulation of a ruthenium(II) complex with triazolopyrimidine in liposomes as a tool for improving its anticancer activity against melanoma cell lines, Dalton Trans. 49 (2020) 1207–1219.

[48] M. Fandzloch, J.M.M. Arriaga, M. Sanchez-Moreno, A. Wojtczak, J. Jezierska, J. Sitkowski, J. Wisniewska, J.M. Salas, I. Lakomska, Strategies for overcoming tropical disease by ruthenium complexes with purine analog: application against *Leishmania spp.* and *Trypanosoma cruzi*, J. Inorg. Biochem. 176 (2017) 144–155.
[49] M. Fandzloch, A. Wojtczak, J. Sitkowski, I. Lakomska, Interaction of ruthenium(II) and ruthenium(III) ions with 5-methyl-1,2,4-triazolo[1,5-a]pyrimidin-7(4H)-one, Polyhedron 67 (2014) 410–415.
[50] E. Szlyk, A. Grodzicki, L. Pazderski, A. Wojtczak, J. Chatlas, G. Wrzeszcz, J. Sitkowski, B. Kamienski, Palladium(II) chloride complexes with 1,2,4-triazolo [1,5-*a*]-pyrimidines: X-ray, ^{15}N–^{1}H NMR and ^{15}N CP MAS studies, J. Chem. Soc. Dalton Trans. (2000) 867–872.
[51] E. Szlyk, L. Pazderski, I. Lakomska, A. Surdykowski, T. Glowiak, J. Sitkowski, L. Kozerski, The X-ray structure of bis(5,7-dimethyl-1,2,4-triazolo-[1,5α]-pyrimidinium)hexachloroplatinate(IV) and spectroscopic properties of Pt(II) and Pt(IV) chloride complexes with 1,2,4-triazolo-[1,5α]-pyrimidines, Polyhedron 21 (2002) 343–348.
[52] I. Lakomska, E. Szlyk, J. Sitkowski, L. Kozerski, J. Wietrzyk, M. Pelczynska, A. Nasulewicz, A. Opolski, Multinuclear NMR spectroscopy and antitumor activity of novel platinum(II) complexes with 5,7-disubstituted-1,2,4-triazolo[1,5-α]pyrimidines, J. Inorg. Biochem. 98 (2004) 167–172.
[53] I. Lakomska, A. Wojtczak, J. Sitkowski, L. Kozerski, E. Szlyk, X-ray structure and multinuclear NMR studies of platinum(II) complexes with 5-methyl-1,2,4-triazolo [1,5-*a*]pyrimidin-7(4H)-one, Polyhedron 26 (2007) 803–810.
[54] I. Lakomska, A. Wojtczak, J. Sitkowski, L. Kozerski, E. Szlyk, Platinum(IV) complexes with purine analogs. Studies of molecular structure and antiproliferative activity *in vitro*, Polyhedron 27 (2008) 2765–2770.
[55] L. Rubio-Perez, R. Azpiroz, A. Di Giuseppe, V. Polo, R. Castarlenas, J.J. Perez-Torrente, L.A. Oro, Pyridine-enhanced head-to-tail dimerization of terminal alkynes by a rhodium–N-heterocyclic-carbene catalyst, Chem. A Eur. J. 19 (2013) 15304–15314.
[56] R. Azpiroz, L. Rubio-Perez, A. Di Giuseppe, V. Passarelli, F.J. Lahoz, R. Castarlenas, J.J. Perez-Torrente, L.A. Oro, Rhodium(I)-N-heterocyclic carbene catalyst for selective coupling of N-vinylpyrazoles with alkynes via C–H activation, ACS Catal. 4 (2014) 4244–4253.
[57] R. Azpiroz, A. Di Giuseppe, V. Passarelli, J.J. Perez-Torrente, L.A. Oro, R. Castarlenas, Rhodium – N-heterocyclic carbene catalyzed hydroalkenylation reactions with 2-vinylpyridine and 2-vinylpyrazine: preparation of nitrogen-bridgehead heterocycles, Organometallics 37 (2018) 1695–1707.
[58] L. Palacios, A. Di Giuseppe, R. Castarlenas, F.J. Lahoz, J.J. Perez-Torrente, L.A. Oro, Pyridine versus acetonitrile coordination in rhodium–N-heterocyclic carbene square-planar complexes, Dalton Trans. 44 (2015) 5777–5789.
[59] L. Palacios, Y. Meheut, M. Galiana-Cameo, M.J. Artigas, A. Di Giuseppe, F.J. Lahoz, V. Polo, R. Castarlenas, J.J. Perez-Torrente, L.A. Oro, Design of highly selective alkyne hydrothiolation RhI-NHC catalysts: carbonyl-triggered nonoxidative mechanism, Organometallics 36 (2017) 2198–2207.
[60] L. Palacios, A. Di Giuseppe, M.J. Artigas, V. Polo, F.J. Lahoz, R. Castarlenas, J.J. Perez-Torrente, L.A. Oro, Mechanistic insight into the pyridine enhanced α-selectivity in alkyne hydrothiolation catalysed by quinolinolate – rhodium(I) – N-heterocyclic carbene complexes, Cat. Sci. Technol. 6 (2016) 8548–8561.
[61] D. Sieb, K. Schuhen, M. Morgen, H. Herrmann, H. Wadepohl, N.T. Lucas, R.W. Baker, M. Enders, Synthesis and complexation behavior of indenyl and cyclopentadienyl ligands functionalized with a naphthyridine unit, Organometallics 31 (2012) 356–364.

[62] T. Rampersadh, M.A. Fernandes, L. Carlton, Iridium complexes of aromatic N,O and N,N chelating ligands. Structures of [Ir(8OQ)(cod)], [Ir(8OQd)(cod)], $[Ir_2(7AI)_2(cod)_2]$ and $[Ir(8OQd)(H)(EPh_3)(cod)]$ (8OQ = 8-oxyquinolinate, 8OQd = 2-Methyl-8-oxyquinolinate, 7AI = 7-Azaindolate; E = Si, Sn; cod = 1,5-cyclooctadiene), J. Chem. Crystallogr. 44 (2014) 151–160.

[63] K.M. Appleby, R.E. Mewis, A.M. Olaru, G.G.R. Green, I.J.S. Fairlamb, S.B. Duckett, Investigating pyridazine and phthalazine exchange in a series of iridium complexes in order to define their role in the catalytic transfer of magnetisation from para-hydrogen, Chem. Sci. 6 (2015) 3981–3993.

[64] R. Zhou, J.A. Aguilar, A. Charlton, S.B. Duckett, P.I.P. Elliott, R. Kandiah, Parahydrogen derived illumination of pyridine based coordination products obtained from reactions involving rhodium phosphine complexes, Dalton Trans. (2005) 3773–3779.

[65] B. Wrackmeyer, G. Kehr, Geminal coupling constants $^2J(^{15}N,^{13}C)$ in 2-substituted pyridines, measured by hahn-echo extended (HEED) pulse sequences, Z. Naturforsch. B 48 (1993) 1433–1436.

[66] V.K. Jain, J. Mason, R.C. Mehrotha, Studies of chelation of the 8-quinolinato ligand in triorganoantimony(V) complexes in solution by multinuclear (^1H, ^{15}N, ^{13}C) NMR and electronic spectroscopy, J. Organomet. Chem. 309 (1986) 45–54.

[67] K. Dysz, K. Szmigiel-Bakalarz, M. Nentwig, O. Oeckler, M. Malik-Gajewska, E. Turlej, D. Michalska, B. Morzyk-Ociepa, A comparative study on cisplatin analogs containing 7-azaindole (7AIH) and its seven halogeno-derivatives: vibrational spectra, DFT calculations and in vitro antiproliferative activity. Crystal and molecular structure of *cis*-$[PtCl_2(4Br7AIH)_2]$•DMF, Polyhedron 173 (2019) 114136.

[68] L. Pazderski, E. Szlyk, J. Sitkowski, B. Kamienski, L. Kozerski, J. Tousek, R. Marek, Experimental and quantum-chemical studies of ^{15}N NMR coordination shifts in palladium and platinum chloride complexes with pyridine, 2,2'-bipyridine and 1,10-phenanthroline, Magn. Reson. Chem. 44 (2006) 163–170.

[69] L. Pazderski, J. Tousek, J. Sitkowski, L. Kozerski, R. Marek, E. Szlyk, ^1H, ^{13}C and ^{15}N NMR coordination shifts in gold(III), cobalt(III), rhodium(III) chloride complexes with pyridine, 2,2'-bipyridine and 1,10-phenanthroline, Magn. Reson. Chem. 45 (2007) 24–36.

[70] L. Pazderski, A. Surdykowski, M. Pazderska-Szablowicz, J. Sitkowski, L. Kozerski, B. Kamienski, E. Szlyk, ^1H, ^{13}C, ^{15}N NMR and ^{13}C, ^{15}N CPMAS studies of cobalt(III)-chloride-pyridine complexes, spontaneous py → Cl substitution in *trans*-$[Co(py)_4Cl_2]Cl$, and a new synthesis of *mer*-$[Co(py)_3Cl_3]$, Cent. Eur. J. Chem. 6 (2008) 55–64.

[71] L. Pazderski, J. Tousek, J. Sitkowski, L. Kozerski, E. Szlyk, ^1H, ^{13}C and ^{15}N NMR coordination shifts in Au(III), Pd(II) and Pt(II) chloride complexes with phenylpyridines, Magn. Reson. Chem. 47 (2009) 658–665.

[72] L. Pazderski, J. Tousek, J. Sitkowski, K. Malinakova, L. Kozerski, E. Szlyk, Experimental and quantum-chemical studies of ^1H, ^{13}C and ^{15}N NMR coordination shifts in Au(III), Pd(II) and Pt(II) chloride complexes with picolines, Magn. Reson. Chem. 47 (2009) 228–238.

[73] L. Pazderski, T. Pawlak, J. Sitkowski, L. Kozerski, E. Szlyk, Structural correlations for ^1H, ^{13}C and ^{15}N NMR coordination shifts in Au(III), Pd(II) and Pt(II) chloride complexes with lutidines and collidine, Magn. Reson. Chem. 48 (2010) 417–426.

[74] L. Pazderski, T. Pawlak, J. Sitkowski, L. Kozerski, E. Szlyk, ^1H, ^{13}C, ^{15}N and ^{195}Pt NMR studies of Au(III) and Pt(II) chloride organometallics with 2-phenylpyridine, Magn. Reson. Chem. 47 (2009) 932–941.

[75] A. Dokalik, H. Kalchhauser, W. Mikenda, G. Schweng, NMR spectra of nitrogen-containing compounds. Correlations between experimental and GIAO calculated data, Magn. Reson. Chem. 37 (1999) 895–902.

[76] M. Kline, S. Cheatham, A robust method for determining 1H–15N long-range correlations: 15N optimized CIGAR-HMBC experiments, Magn. Reson. Chem. 41 (2003) 307–314.
[77] W. Stadeli, W. von Philipsborn, ^{15}N NMR substituent effects in pyridines and pyrimidines, Org. Magn. Reson. 15 (1981) 106–109.
[78] H.J. Jakobsen, P.I. Yang, W.S. Brey, Natural abundance proton-coupled ^{15}N NMR spectra of pyridines observed from proton polarization transfer experiments, Org. Magn. Reson. 17 (1981) 290–295.
[79] E.C. Constable, A.M.W.C. Thompson, T.A. Leese, D.G.F. Reese, D.A. Tocher, Cyclometallation reactions of 2-phenylpyridine; crystal and molecular structure of (2-{2-pyridyl} phenyl)palladium(II) tetramer and (2-{2-pyridyl} phenyl)mercury(II) tetramer, Inorg. Chim. Acta 182 (1991) 93–100.
[80] U. Klement, Crystal structure of di-μ -chloro-bis(2-phenylpyridinato-C2,N-palladium(II)), $((C_{11}H_8N)PdCl)_2$, Z. Kristallogr. 208 (1993) 299–301.
[81] J.E. Bercaw, A.C. Durrell, H.B. Gray, J.C. Green, N. Hazari, J.A. Labinger, J.R. Winkler, Electronic structures of Pd^{II} dimers, Inorg. Chem. 49 (2010) 1801–1810.
[82] K. Ha, Crystal structure of di(μ_2-chlorido)bis[2-(2-pyridyl)phenyl-κ^2-N,C^1]dipalladium(II), $C_{22}H_{16}Cl_2N_2Pd_2$, Z. Kristallogr. New Cryst. Struct. 231 (2016) 331–332.
[83] A.V. Eremin, D.D. Vaulina, M.A. Stepanova, A.N. Belyaev, Synthesis and crystal structure of Palladium(II) mononuclear complex with thiourea, Russ. J. Gen. Chem. 81 (2011) 959.
[84] M. Buyukyazi, C. Hegemann, T. Lehnen, W. Tyrra, S. Mathur, Molecular Co(II) and Co(III) heteroarylalkenolates as efficient precursors for chemical vapor deposition of Co_3O_4 nanowires, Inorg. Chem. 53 (2014) 10928–10936.
[85] J.N. McPherson, R.W. Hogue, F.S. Akogun, L. Bondì, E.T. Luis, J.R. Price, A.L. Garden, S. Brooker, S.B. Colbran, Predictable substituent control of CoIII/II redox potential and spin crossover in bis(dipyridylpyrrolide)cobalt complexes, Inorg. Chem. 58 (2019) 2218–2228.
[86] S. Vitushkina, M. Teslenko, L. Vahovska, L. Findorakova, M. Vilkova, I. Potocnak, Low-dimensional compounds containing cyanido groups. Part XXXI. First simultaneous nucleophilic addition of water and ethanol to dicyanonitrosomethanide anions in the presence of Co(II), Inorg. Chim. Acta 456 (2017) 49–54.
[87] F.M. Wisser, M. Duguet, Q. Perrinet, A.C. Ghosh, M. Alves-Favaro, Y. Mohr, C. Lorentz, E.A. Quadrelli, R. Palkovits, D. Farrusseng, C. Mellot-Draznieks, V. de Waele, J. Canivet, Molecular porous photosystems tailored for long-term photocatalytic CO_2 reduction, Angew. Chem. Int. Ed. 59 (2020) 5116–5122.
[88] L.S. Lloyd, A. Asghar, M.J. Burns, A. Charlton, S. Coombes, M.J. Cowley, G.J. Dear, S.B. Duckett, G.R. Genov, G.G.R. Green, L.A.R. Highton, A.J.J. Hooper, M. Khan, I.G. Khazal, R.J. Lewis, R.E. Mewis, A.D. Roberts, A.J. Ruddlesden, Hyperpolarisation through reversible interactions with *para*hydrogen, Cat. Sci. Technol. 4 (2014) 3544–3554.
[89] C.M. Wong, M. Fekete, R. Nelson-Forde, M.R.D. Gatus, P.J. Rayner, A.C. Whitwood, S.B. Duckett, B.A. Messerle, Harnessing asymmetric N-heterocyclic carbene ligands to optimise SABRE hyperpolarization, Cat. Sci. Technol. 8 (2018) 4925–4933.
[90] M. Fekete, O. Bayfield, S.B. Duckett, S. Hart, R.E. Mewis, N. Pridmore, P.J. Rayner, A. Whitwood, Iridium(III) hydrido N-heterocyclic carbene–phosphine complexes as catalysts in magnetization transfer reactions, Inorg. Chem. 52 (2013) 13453–13461.
[91] W. Iali, G.G.R. Green, S.J. Hart, A.C. Whitwood, S.B. Duckett, Iridium cyclooctene complex that forms a hyperpolarization transfer catalyst before converting to a binuclear C – H bond activation product responsible for hydrogen isotope exchange, Inorg. Chem. 55 (2016) 11639–11643.

[92] E.M. Titova, G.A. Silantyev, O.A. Filippov, E.S. Gulyaeva, E.I. Gutsul, F.M. Dolgushin, N.V. Belkova, PCP pincer iridium chemistry—coordination of pyridines to [(tBuPCP)IrH(Cl)], Eur. J. Inorg. Chem. (2016) 56–63.
[93] E.S. Osipova, E.S. Gulyaeva, O.A. Filippov, S.V. Safronov, A.A. Pavlov, A.V. Polukeev, V.A. Kirkina, E.M. Titova, E.S. Shubina, N.V. Belkova, Effect of ligands on the lewis acidity of the metal and the binding of N-bases to iridium pincer complexes, Eur. J. Inorg. Chem. (2019) 1389–1397.
[94] Z.G. Specht, D.B. Grotjahn, C.E. Moore, A.L. Rheingold, Effects of hindrance in N-pyridyl imidazolylidenes coordinated to iridium on structure and catalysis, Organometallics 32 (2013) 6400–6409.
[95] R.E. Mewis, M. Fekete, G.G.R. Green, A.C. Whitwood, S.B. Duckett, Deactivation of signal amplification by reversible exchange catalysis, progress towards in vivo application, Chem. Commun. 51 (2015) 9857–9859.
[96] P. Starha, Z. Travnicek, H. Crlíkova, J. Vanco, J. Kasparkova, Z. Dvorak, Half-sandwich Ir(III) complex of N^1-Pyridyl-7-azaindole exceeds cytotoxicity of cisplatin at various human cancer cells and 3D multicellular tumor spheroids, Organometallics 37 (2018) 2749–2759.
[97] J.E. Richards, A.J.J. Hooper, O.W. Bayfield, M.C.R. Cockett, G.J. Dear, A.J. Holmes, R.O. John, R.E. Mewis, N. Pridmore, A.D. Roberts, A.C. Whitwood, S.B. Duckett, Using hyperpolarised NMR and DFT to rationalize the unexpected hydrogenation of quinazoline to 3,4-dihydroquinazoline, Chem. Commun. 54 (2018) 10375–10378.
[98] L. Pazderski, T. Pawlak, J. Sitkowski, L. Kozerski, E. Szlyk, ^{1}H NMR assignment corrections and ^{1}H, ^{13}C, ^{15}N NMR coordination shifts structural correlations in Fe(II), Ru(II) and Os(II) cationic complexes with 2,2'-bipyridine and 1,10-phenanthroline, Magn. Reson. Chem. 48 (2010) 450–457.
[99] E. Haslinger, M. Schlederer, W. Robien, P. Wolschann, Hydrogen bonding and tautomeric equilibria of complexes of pyridine with various proton donors. A nitrogen-15 nuclear magnetic resonance study, Monatsh. Chem. 115 (1984) 1345–1351.
[100] N.S. Rao, G.B. Rao, B.N. Murthy, M.M. Das, T. Prabhakar, M. Lalitha, Natural abundance nitrogen-15 nuclear magnetic resonance spectral studies on selected donors, Spectrochim. Acta A 58 (2002) 2737–2757.
[101] V.A. Sharnin, V.V. Aleksandriysky, S.V. Dushina, G.A. Gamova, Stability of H-complexes of nicotinamide nitrogen heteroatom with water and ethanol in mixed solvents by ^{13}C NMR probing, Magn. Reson. Chem. 51 (2013) 193–198.
[102] A.K.C. Mengel, C. Forster, A. Breivogel, K. Mack, J.R. Ochsmann, F. Laquai, V. Ksenofontov, K. Heinze, A heteroleptic push–pull substituted iron(II) bis(tridentate) complex with low-energy charge-transfer states, Chem. A Eur. J. 21 (2015) 704–714.
[103] G. Markiewicz, M. Piechocki, A. Walczak, E.A. Połomska, J. Harrowfield, A.R. Stefankiewicz, Generation and transformation of a hemi-iminal-based metalorganic Fe(II) structure obtained via subcomponent self-assembly in water, Dalton Trans. 46 (2017) 14826–14830.
[104] B. Butschke, K.L. Fillman, T. Bendikov, L.J.W. Shimon, Y. Diskin-Posner, G. Leitus, S.I. Gorelsky, M.L. Neidig, D. Milstein, How innocent are potentially redox non-innocent ligands? Electronic structure and metal oxidation states in iron-PNN complexes as a representative case study, Inorg. Chem. 54 (2015) 4909–4926.
[105] F. Battistin, G. Balducci, N. Demitri, E. Iengo, B. Milani, E. Alessio, ^{15}N NMR spectroscopy unambiguously establishes the coordination mode of the diimine linker 2-(2'-pyridyl)pyrimidine-4-carboxylic acid (cppH) in Ru(II) complexes, Dalton Trans. 44 (2015) 15671–15682.
[106] K. Muller, Y. Sun, A. Heimermann, F. Menges, G. Niedner-Schatteburg, C. van Wüllen, W.R. Thiel, Structure–reactivity relationships in the hydrogenation of carbon

dioxide with ruthenium complexes bearing pyridinylazolato ligands, Chem. A Eur. J. 19 (2013) 7825–7834.
[107] A. Bolje, S. Hohloch, D. Urankar, A. Pevec, M. Gazvoda, B. Sarkar, J. Kossmrlj, Exploring the scope of pyridyl- and picolyl-functionalized 1,2,3-triazol-5-ylidenes in bidentate coordination to ruthenium(II) cymene chloride complexes, Organometallics 33 (2014) 2588–2598.
[108] J.H.M. Ortiz, N. Vega, D. Comedi, M. Tirado, I. Romero, X. Fontrodona, T. Parella, F.E.M. Vieyra, C.D. Borsarelli, N.E. Katz, Improving the photosensitizing properties of ruthenium polypyridyl complexes using 4-methyl-2,2′-bipyridine-4′-carbonitrile as an auxiliary ligand, Inorg. Chem. 52 (2013) 4950–4962.
[109] D.C. Marelius, S. Bhagan, D.J. Charboneau, K.M. Schroeder, J.M. Kamdar, A.R. McGettigan, B.J. Freeman, C.E. Moore, A.L. Rheingold, A.L. Cooksy, D.K. Smith, J.J. Paul, E.T. Papish, D.B. Grotjahn, How do proximal hydroxy or methoxy groups on the bidentate ligand affect [(2,2′;6′,2″-Terpyridine)Ru-(N,N)X] water-oxidation catalysts? Synthesis, characterization, and reactivity at acidic and near-neutral pH, Eur. J. Inorg. Chem. (2014) 676–689.
[110] M. Chrzanowska, A. Katafias, O. Impert, A. Kozakiewicz, A. Surdykowski, P. Brzozowska, A. Franke, A. Zahl, R. Puchta, R. van Eldik, Structure and reactivity of [Ru^{II}(terpy)(N^N)Cl]Cl complexes: consequences for biological applications, Dalton Trans. 46 (2017) 10264–10280.
[111] H. Keisar, G. de Ruiter, A.H. Velders, P. Milko, A. Gulino, G. Evmenenko, L.J.W. Shimon, Y. Diskin-Posner, M. Lahav, M.E. van der Boom, Sorting of molecular building blocks from solution to surface, J. Am. Chem. Soc. 140 (2018) 8162–8171.
[112] P.O. Abate, G. Pourrieux, F.E.M. Vieyra, C.D. Borsarelli, T. Parella, M.M. Vergara, N.E. Katz, A novel mononuclear complex of 4,4″-azobis-(2,2′-bipyridine) coordinated to tetracyanoruthenium(II) can behave as a "molecular switch", Polyhedron 174 (2019) 114149.
[113] M. Chrzanowska, A. Katafias, A. Kozakiewicz, R. van Eldik, Steric and electronic tuning of the reactivity of [Ru^{II}(terpy)(N^N)Cl]Cl complexes, Inorg. Chim. Acta 504 (2020) 119449.
[114] M.V. Babak, P. Le Faouder, X. Trivelli, G. Venkatesan, S.I. Bezzubov, M. Kajjout, A.L. Gushchin, M. Hanif, O. Poizat, H. Vezin, C. Rolando, Heteroleptic ruthenium(II) complexes with bathophenanthroline and bathophenanthroline disulfonate disodium salt as fluorescent dyes for in-gel protein staining, Inorg. Chem. 59 (2020) 4527–4535.
[115] A. Breivogel, S. Wooh, J. Dietrich, T.Y. Kim, Y.S. Kang, K. Char, K. Heinze, Anchor-functionalized push-pull-substituted bis(tridentate) Ruthenium(II) polypyridine chromophores: photostability and evaluation as photosensitizers, Eur. J. Inorg. Chem. (2014) 2720–2734.
[116] A. Breivogel, M. Meister, C. Förster, F. Laquai, K. Heinze, Excited state tuning of bis(tridentate) ruthenium(II) polypyridine chromophores by push–pull effects and bite angle optimization: a comprehensive experimental and theoretical study, Chem. A Eur. J. 19 (2013) 13745–13760.
[117] M. Jarenmark, G. Carlstrom, L.A. Fredin, J.H. Wallenstein, I. Doverbratt, M. Abrahamsson, P. Persson, Diastereomerization dynamics of a bistridentate Ru^{II} complex, Inorg. Chem. 55 (2016) 3015–3022.
[118] M. Gobec, J. Kljun, I. Sosic, I. Mlinaric-Rascan, M. Ursic, S. Gobec, I. Turel, Structural characterization and biological evaluation of a clioquinol–ruthenium complex with copper-independent antileukaemic activity, Dalton Trans. 43 (2014) 9045–9051.
[119] D. Scherrer, M. Schilling, S. Luber, T. Fox, B. Spingler, R. Alberto, C.J. Richmond, Ruthenium water oxidation catalysts containing the non-planar tetradentate ligand, biisoquinoline dicarboxylic acid ($biqaH_2$), Dalton Trans. 45 (2016) 19361–19367.

[120] A. Bolje, S. Hohloch, J. Kosmrlj, B. Sarkar, Ru^{II}, Ir^{III} and Os^{II} mesoionic carbene complexes: efficient catalysts for transfer hydrogenation of selected functionalities, Dalton Trans. 45 (2016) 15983–15993.
[121] R. Pettinari, F. Marchetti, C. Pettinari, F. Condello, B.W. Skelton, A.H. White, M.R. Chierottie, R. Gobetto, Self-assembly of arene ruthenium acylpyrazolone fragments to tetranuclear metallacycles. Molecular structures and solid-state ^{15}N CPMAS NMR correlations, Dalton Trans. 45 (2016) 3974–3982.
[122] S. Gonzalez-Gallardo, I. Kuzu, P. Ona-Burgos, T. Wolfer, C. Wang, K.W. Klinkhammer, W. Klopper, S. Brase, F. Breher, Coinage metal complexes of tris(pyrazolyl)methanide-based redox-active metalloligands, Organometallics 33 (2014) 941–951.
[123] I. Trapp, S. Gonzalez-Gallardo, S. Hohnstein, D. Garnier, P. Ona-Burgos, F. Breher, A metallacyclic alkyl-amido carbene complex (MCAAC), Dalton Trans. 43 (2014) 4313–4319.
[124] E.R. Paulson, C.E. Moore, A.L. Rheingold, D.P. Pullman, R.W. Sindewald, A.L. Cooksy, D.B. Grotjahn, Dynamic π-bonding of imidazolyl substituent in a formally 16 electron $Cp^{*}Ru(\kappa^{2}\text{-}P,N)^{+}$ catalyst allows dramatic rate increases in (E)-selective monoisomerization of alkenes, ACS Catal. 9 (2019) 7217–7231.
[125] A. Gilewska, B. Barszcz, J. Masternak, K. Kazimierczuk, J. Sitkowski, J. Wietrzyk, E. Turlej, Similarities and differences in d^6 low-spin ruthenium, rhodium and iridium half-sandwich complexes: synthesis, structure, cytotoxicity and interaction with biological targets, J. Biol. Inorg. Chem. 24 (2019) 591–606.
[126] G.E. Buchel, A. Gavriluta, M. Novak, S.M. Meier, M.A. Jakupec, O. Cuzan, C. Turta, J.B. Tommasino, E. Jeanneau, G. Novitchi, D. Luneau, V.B. Arion, Striking difference in antiproliferative activity of ruthenium- and osmium-nitrosyl complexes with azole heterocycles, Inorg. Chem. 52 (2013) 6273–6285.
[127] A. Gavriluta, G.E. Buchel, L. Freitag, G. Novitchi, J.B. Tommasino, E. Jeanneau, P.S. Kuhn, L. Gonzalez, V.B. Arion, D. Luneau, Mechanism elucidation of the cis–trans isomerization of an azole ruthenium–nitrosyl complex and its osmium counterpart, Inorg. Chem. 52 (2013) 6260–6272.
[128] R.M. Claramunt, C. Lopez, D. Sanz, I. Alkorta, J. Elguero, A multinuclear NMR spectral study of parent azoles and benzazoles: experimental results and GIAO *ab initio* calculations, Heterocycles 55 (2001) 2109–2121.
[129] L. Pazderski, T. Pawlak, J. Sitkowski, L. Kozerski, E. Szlyk, ^{1}H, ^{13}C, ^{15}N NMR coordination shifts in Fe(II), Ru(II) and Os(II) cationic complexes with 2,2':6',2"-terpyridine, Magn. Reson. Chem. 49 (2011) 237–241.
[130] L. Pazderski, J. Tousek, J. Sitkowski, L. Kozerski, E. Szlyk, Experimental and quantum-chemical studies of ^{1}H, ^{13}C and ^{15}N NMR coordination shifts in Pd(II) and Pt(II) chloride complexes with methyl and phenyl derivatives of 2,2'-bipyridine and 1,10-phenanthroline, Magn. Reson. Chem. 45 (2007) 1045–1058.
[131] N. Nickita, G. Gasser, P. Pearson, M.J. Belousoff, L.Y. Goh, A.M. Bond, G.B. Deacon, L. Spiccia, Ruthenium(II) complexes incorporating 2-(2′-Pyridyl)pyrimidine-4-carboxylic acid, Inorg. Chem. 48 (2009) 68–81.
[132] X.J. Yang, F. Drepper, B. Wu, W.H. Sun, W. Haehnel, C. Janiak, From model compounds to protein binding: syntheses, characterizations and fluorescence studies of [RuII(bipy)(terpy)L]2+ complexes (bipy = 2,2'-bipyridine; terpy = 2,2':6',2"-terpyridine; L = imidazole, pyrazole and derivatives, cytochrome), Dalton Trans. (2005) 256–267.
[133] L. Duan, G.F. Manbeck, M. Kowalczyk, D.J. Szalda, J.T. Muckerman, Y. Himeda, E. Fujita, Noninnocent proton-responsive ligand facilitates reductive deprotonation and hinders CO2 reduction catalysis in $[Ru(tpy)(6DHBP)(NCCH_3)]^{2+}$ (6DHBP = 6,6′-$(OH)_2$bpy), Inorg. Chem. 55 (2016) 4582–4594.

[134] T. Meng, Q.P. Qin, Z.L. Chen, H.H. Zou, K. Wang, F.P. Liang, Discovery of a high in vitro and in vivo antitumor activities of organometallic ruthenium(II)-arene complexes with 5,7-dihalogenated-2-methyl-8-quinolinol, Dalton Trans. 48 (2019) 5352–5360.

[135] R. Molteni, K. Edkins, M. Haehnel, A. Steffen, C – H activation of fluoroarenes: synthesis, structure, and luminescence properties of copper(I) and gold(I) complexes bearing 2-phenylpyridine ligands, Organometallics 35 (2016) 629–640.

[136] M. Bedin, A. Karim, M. Reitti, A.C.C. Carlsson, F. Topic, M. Cetina, F. Pan, V. Havel, F. Al-Ameri, V. Sindelar, K. Rissanen, J. Gräfenstein, M. Erdelyi, Counterion influence on the N–I–N halogen bond, Chem. Sci. 6 (2015) 3746–3756.

[137] D. Zopes, C. Hegemann, J. Schläfer, W. Tyrra, S. Mathur, Single-source precursors for alloyed gold – silver nanocrystals—a molecular metallurgy approach, Inorg. Chem. 54 (2015) 3781–3787.

[138] S.G. Sakharov, V.V. Kovalev, Y.E. Gorbunova, Y.V. Kokunov, Structures of the silver complexes with lutidines according to the NMR data. Crystal structure of $[AgNO_3(3,5\text{-}Lut)_2]$, Russ. J. Coord. Chem. 39 (2013) 187–193.

[139] F. Marchetti, J. Palmucci, C. Pettinari, R. Pettinari, F. Condello, S. Ferraro, M. Marangoni, A. Crispini, S. Scuri, I. Grappasonni, M. Cocchioni, M. Nabissi, M.R. Chierotti, R. Gobetto, Novel composite plastics containing silver(I) acylpyrazolonato additives display potent antimicrobial activity by contact, Chem. A Eur. J. 21 (2015) 836–850.

[140] D.E.S. Silva, A.B. Becceneri, M.C. Solcia, J.V.B. Santiago, M.B. Moreira, J.A.G. Neto, F.R. Pavan, M.R. Cominetti, J.C.M. Pereira, A.V.G. Netto, Cytotoxic and apoptotic effects of ternary silver(I) complexes bearing 2-formylpyridine thiosemicarbazones and 1,10-phenanthroline, Dalton Trans. 49 (2020) 5264–5275.

[141] A.M. Mansour, K. Radacki, Structural studies, antimicrobial activity and protein interaction of photostable terpyridine silver(I) complexes, Eur. J. Inorg. Chem. (2019) 4020–4030.

[142] A.C. Reiersolmoen, D. Csokas, S. Oien-Odegaard, A. Vanderkooy, A.K. Gupta, A.C.C. Carlsson, A. Orthaber, A. Fiksdahl, I. Papai, M. Erdelyi, Catalytic activity of trans-bis(pyridine)gold complexes, J. Am. Chem. Soc. 142 (2020) 6439–6446.

[143] L.G. Golubyatnikova, R.A. Khisamutdinov, A.N. Lobov, Y.I. Murinov, Extraction of Gold(III) with (RS)-1-(4-Chlorophenyl)-4,4-dimethyl-3-(1H-1,2,4-triazol-1-yl-methyl)-pentan-3-ol from Hydrochloric acid solutions, Russ. J. Inorg. Chem. 58 (2013) 491–498.

[144] R. Kleinmaier, S. Arenz, A. Karim, A.C.C. Carlsson, M. Erdelyi, Solvent effects on ^{15}N NMR coordination shifts, Magn. Reson. Chem. 51 (2013) 46–53.

[145] C.Y. Chen, J.Y. Zeng, H.M. Lee, Argentophilic interaction and anionic control of supramolecular structures in simple silver pyridine complexes, Inorg. Chim. Acta 360 (2007) 21–30.

[146] S. Menchetti, G. Rossi, V. Tazzoli, La struttura cristallina del $Ag(Py)_2NO_3H_2O$, Rendiconti Istituto Lombardo, Accademia di Scienze e Lettere. Series A, Scienze Matematiche, Fisiche, Chimiche e Geologiche 104 (1970) 309–316.

[147] G.A. Bowmaker, K.C. Effendy, B.W. Lim, D. Skelton, A.H.W. Sukarianingsih, Syntheses, structures and vibrational spectroscopy of some 1:2 and 1:3 adducts of silver(I) oxyanion salts with pyridine and piperidine bases containing non-coordinating 2(,6)-substituents, Inorg. Chim. Acta 358 (2005) 4342–4370.

[148] L.M. Engelhardt, C. Pakawatchai, A.H. White, P.C. Healy, Lewis-base adducts of group I B Metal(i) compounds. Part 12. Structural studies of some Bis(methyl-substituted pyridine)-copper(1) and -silver(l) nitrates and perchlorates, J. Chem. Soc. Dalton Trans. (1985) 117–123.

[149] A.A. Massoud, V. Langerb, M.A.M. Abu-Youssef, Bis[4-(dimethylamino)pyridine-κ N^1]-silver(I) nitrate dehydrate, Acta Crystallogr. C 65 (2009) m352–m354.
[150] G.A. Bowmaker, S. Effendy, B.W. Marfuah, A.H.W. Skelton, Syntheses, structures and vibrational spectroscopy of some 1:1 and 1:2 adducts of silver(I) oxyanion salts with 2,2'-bis(pyridine) chelates, Inorg. Chim. Acta 358 (2005) 4371–4388.
[151] R. Corbo, G.F. Ryan, M.A. Haghighatbin, C.F. Hogan, D.J.D. Wilson, M.D. Hulett, P.J. Barnard, J.L. Dutton, Access to the parent tetrakis(pyridine)gold(III) trication, facile formation of rare Au(III) terminal hydroxides, and preliminary studies of biological properties, Inorg. Chem. 55 (2016) 2830–2839.
[152] R.E.F. de Paiva, Z. Du, D.H. Nakahata, F.A. Lima, P.P. Corbi, N.P. Farrell, Gold-catalyzed C–S aryl-group transfer in zinc finger proteins, Angew. Chem. Int. Ed. 57 (2018) 9305–9309.
[153] M.S.M. Holmsen, A. Nova, D. Balcells, E. Langseth, S.O. Odegaard, E.A. Traseth, R.H. Heync, M. Tilset, Small-molecule activation at Au(III): metallacycle construction from ethylene, water, and acetonitrile, Dalton Trans. 45 (2016) 14719–14724.
[154] J. Mruk, L. Pazderski, J. Scianowski, A. Wojtczak, Structural and NMR spectroscopic studies of 2-phenylsulfanylpyridine and its analogues or derivatives, and their Au(III) chloride complexes, Inorg. Chim. Acta 500 (2020) 119182.
[155] A.C. Reiersolmoen, D. Csokas, I. Papai, A. Fiksdahl, M. Erdelyi, Mechanism of Au(III)-mediated alkoxycyclization of a 1,6-enyne, J. Am. Chem. Soc. 141 (2019) 18221–18229.
[156] D. Niedzielska, L. Pazderski, A. Wojtczak, M. Kurzawa, J. Scianowski, E. Szlyk, Structural and spectroscopic studies of Au(III) chloride compounds with 7,8-benzoquinoline, Polyhedron 139 (2018) 155–171.
[157] D. Niedzielska, T. Pawlak, M. Bozejewicz, A. Wojtczak, L. Pazderski, E. Szlyk, Structural and spectroscopic studies of Au(III) and Pd(II) chloride complexes and organometallics with 2-benzylpyridine, J. Mol. Struct. 1032 (2013) 195–202.
[158] H. von Wachenfeldt, A.V. Polukeev, N. Loganathan, F. Paulsen, P. Roese, M. Garreau, O.F. Wendt, D. Stand, Cyclometallated gold(III) aryl-pyridine complexes as efficient catalysts for three-component synthesis of substituted oxazoles, Dalton Trans. 44 (2015) 5347–5353.
[159] R.T. Mertens, J.H. Kim, W.C. Jennings, S. Parkin, S.G. Awuah, Revisiting the reactivity of tetrachloroauric acid with N,N-bidentate ligands: structural and spectroscopic insights, Dalton Trans. 48 (2019) 2093–2099.
[160] A. Casini, M.C. Diawara, R. Scopelliti, S.M. Zakeeruddin, M. Grätzel, P.J. Dyson, Synthesis, characterisation and biological properties of gold(III) compounds with modified bipyridine and bipyridylamine ligands, Dalton Trans. 39 (2010) 2239–2245.
[161] F. Abbate, P. Orioli, B. Bruni, G. Marcon, L. Messori, Crystal structure and solution chemistry of the cytotoxic complex 1,2-dichloro(*o*-phenanthroline)gold(III) chloride, Inorg. Chim. Acta 311 (2000) 1–5.
[162] A.E. Wendlandt, S.S. Stahl, Bioinspired aerobic oxidation of secondary amines and nitrogen heterocycles with a bifunctional quinone catalyst, J. Am. Chem. Soc. 136 (2014) 506–512.
[163] A.S. Burlov, V.G. Vlasenko, N.I. Makarova, K.A. Lyssenko, V.V. Chesnokov, G.S. Borodkin, I.S. Vasilchenko, A.I. Uraev, D.A. Garnovskii, A.V. Metelitsa, E.L. Mukhanov, T.V. Lifintseva, I.V. Pankov, Chemical and electrochemical synthesis, molecular structures, DFT calculations and optical properties of metal-chelates of 8-(2-tosylaminobenzilideneimino)quinoline, Polyhedron 107 (2016) 153–162.
[164] K. Klai, K. Kaabi, W. Kaminsky, C. Jelsch, F. Lefebvre, C.B. Nasr, A Hirshfeld surface analysis, crystal structure and physicochemical studies of a new Cd(II) complex with the 2-amino-4-methylpyrimidine ligand, J. Mol. Struct. 1128 (2017) 378–384.

[165] M. Steichen, N.R. Brooks, L. Van Meervelt, J. Fransaer, K. Binnemans, Homoleptic and heteroleptic N-alkylimidazole zinc(II)-containing ionic liquids for high current density electrodeposition, Dalton Trans. 43 (2014) 12329–12341.
[166] O. Vogl, A. Rehman, P. Zarras, Polycationic salts V [1]. ^{15}N NMR spectra of amines, ammonium salt monomers, and polymers of styrene based trialkylammonium salts, Monatsh. Chem. 131 (2000) 437–449.
[167] V.K. Voronov, M.A. Borovik, I.A. Ushakov, N.I. Ivanova, L.V. Malakhova, NMR spectra of thiophosphoryl- and phosphorylazoles, Russ. Chem. Bull., Int. Ed. 58 (2009) 1516–1520.
[168] M. Hog, M. Schneider, G. Studer, M. Bauerle, S.A. Fohrenbacher, H. Scherer, I. Krossing, An investigation of the symmetric and asymmetric cleavage products in the system aluminum trihalide—1-butylimidazole, Chem. A Eur. J. 23 (2017) 11054–11066.
[169] H.L. Vahav, A. Pogoreltsev, Y. Tulchinsky, N. Fridman, A. Borner, M. Gandelman, Synthesis and characteristics of iridium complexes bearing N-heterocyclic nitrenium cationic ligands, Organometallics 38 (2019) 2494–2501.
[170] A.V. Afonin, V.K. Voronov, E.S. Domnina, L.A. Eskova, L.V. Baikalova, B.V. Trzhtsinskaya, E.I. Enikeeva, A.V. Vashchenko, ^{15}N NMR spectra and specific intramolecular interactions in N-vinylazoles, Chem. Heterocycl. Compd. 27 (1991) 845–848.
[171] C.K. Rono, W.K. Chu, J. Darkwa, D. Meyer, B.C.E. Makhubela, Triazolyl Ru^{II}, Rh^{III}, Os^{II}, and Ir^{III} complexes as potential anticancer agents: synthesis, structure elucidation, cytotoxicity, and DNA model interaction studies, Organometallics 38 (2019) 3197–3211.
[172] M. Witanowski, L. Stefaniak, G.A. Webb, Nitrogen NMR spectroscopy, Annu. Rep. NMR Spectrosc. 25 (1993) 1–480.
[173] M. Karplus, J.A. Pople, Theory of carbon NMR chemical shifts in conjugated molecules, J. Chem. Phys. 38 (1963) 2803–2807.
[174] J.A. Pople, The theory of carbon chemical shifts in NMR, Mol. Phys. 7 (1964) 301–306.
[175] J. Mason (Ed.), Chapter 12 (Nitrogen), in: Multinuclear NMR, first ed., Plenum Press, New York, 1987.
[176] A. Satake, H. Koshino, T. Nakata, Synthesis and structural analysis of palladium(II) pyridinylpyrazole complexes by ^{1}H-, ^{13}C-, ^{15}N-NMR and X-ray diffraction. Comparison of binuclear methylpalladium, chloromethylpalladium, and dichloropalladium complexes by ^{15}N-NMR, J. Organomet. Chem. 595 (2000) 208–214.
[177] T. Pawlak, L. Pazderski, J. Sitkowski, L. Kozerski, E. Szlyk, ^{1}H, ^{13}C, ^{195}Pt and ^{15}N NMR structural correlations in Pd(II) and Pt(II) chloride complexes with various alkyl and aryl derivatives of 2,2'-bipyridine and 1,10-phenanthroline, Magn. Reson. Chem. 49 (2011) 59–64.
[178] G. Anderegg, K. Popov, P.S. Pregosin, Nitrogen-NMR studies on the protonation of 2-aminomethylpyridine and Tris[(2-pyridyl)methyl]amine, Helv. Chim. Acta 69 (1986) 329–332.
[179] L. Pazderski, ^{15}N NMR Coordination Shifts in Pd(II), Pt(II), Au(III), Co(III), Rh(III), Ir(III), Pd(IV), Pt(IV) complexes with pyridine, 2,2'-bipyridine, 1,10-phenanthroline, quinoline, isoquinoline, 2,2'-biquinoline, 2,2':6',2''-terpyridine and their alkyl or aryl derivatives, Magn. Reson. Chem. 46 (2008) S3–S15.
[180] L. Pazderski, Application of ^{15}N NMR spectroscopy to determine coordination sphere geometry in Pd(II), Pt(II), Pt(IV), Co(III) and Rh(III) complexes with azines, Polish J. Chem. 83 (2009) 1241–1253.
[181] D. Niedzielska, T. Pawlak, T. Czubachowski, L. Pazderski, ^{1}H, ^{13}C and ^{15}N NMR studies of Au(III) and Pd(II) chloride complexes and organometallics with 2-acetylpyridine and 2-benzoylpyridine, J. Spectrosc. 982832 (2013).

[182] I. Lakomska, M. Fandzloch, B. Poplawska, J. Sitkowski, Platinum(II) complexes with 5,7-disubstituted-1,2,4-triazolo[1,5-*a*]pyrimidines: Spectroscopical characterization and cytotoxic activity *in vitro*, Spectrochim. Acta A 91 (2012) 126–129.
[183] I. Lakomska, M. Fandzloch, T. Muzioł, J. Sitkowski, J. Wietrzyk, Structure-cytotoxicity relationship for different types of mononuclear platinum(II) complexes with 5,7-ditertbutyl-1,2,4-triazolo[1,5-*a*]pyrimidine, J. Inorg. Biochem. 115 (2012) 100–105.
[184] E. Szlyk, I. Lakomska, A. Surdykowski, T. Glowiak, L. Pazderski, J. Sitkowski, L. Kozerski, The X-ray structure and spectroscopy of platinum(II) complexes with 1,2,4-triazolo[1,5-*a*]pyrimidines and dimethylsulfoxide, Inorg. Chim. Acta 333 (2002) 93–99.

CHAPTER FOUR

Update on ^{27}Al NMR studies of aluminosilicate glasses

Mattias Edén*
Physical Chemistry Division, Department of Materials and Environmental Chemistry, Arrhenius Laboratory, Stockholm University, Stockholm, Sweden
*Corresponding author: e-mail address: mattias.eden@mmk.su.se

Contents

Annual Reports on NMR Spectroscopy, Volume 101
ISSN 0066-4103
https://doi.org/10.1016/bs.arnmr.2020.07.002

Abstract

Aluminosilicate glasses and melts are of great geological and technological importance. Significant efforts have been spent for enhancing the insight into their structures. Here magic-angle-spinning (MAS) NMR that exploits the spin-5/2 ^{27}Al as probe nucleus constitutes one widely utilized option. We review the application of ^{27}Al NMR experimentation to aluminosilicate glasses, emphasizing practical aspects of performing MAS and triple-quantum MAS NMR experiments, as well as options for data analysis to extract ^{27}Al NMR parameters and quantifying AlO_p populations. Despite that parts of the text apply generally to MAS NMR targeting half-integer spins as structural probes in both crystalline and amorphous materials, the focus remains on ^{27}Al NMR applications to aluminosilicate glasses, whose basic structural features are outlined together with a survey of the most central research problems in the field. The current update of our previous article from 2015 [Edén, Annu. Rep. NMR Spectrosc. 86 (2015) 237–331] also reviews the principles of heteronuclear NMR experimentation and its applications to study both the short ($\lesssim$0.3 nm) and medium ($\approx$0.3–1 nm) range structure of aluminosilicate glasses. Moreover, the present article also extends our previous account by reviewing some of the most recent work in the area up to the beginning of year 2020, where several sections are significantly modified and expanded. The presentation is intended for a broad audience, encompassing both experienced researchers in solid-state NMR and/or glass structures, as well as to beginners in either area.

Keywords: Glass structure, Network polymerization, Cation field strength, Al/Si intermixing, BO/NBO distribution, Half-integer spin quadrupolar nuclei, Practical advice on ^{27}Al NMR experimentation, Estimation of ^{27}Al NMR parameters, MQMAS, Heteronuclear NMR

1. Introduction

Owing to the importance of aluminosilicate (AS) glasses for both Earth Sciences (for instance as melt-models of magma) and Material Sciences, as well as for numerous industrial applications, vast efforts have been spent to improve the understanding of this family of amorphous materials. The pioneering investigations targeting the local structure of AS glasses by magic-angle-spinning (MAS) NMR were performed during the 1980s [1–7], along with other spectroscopic techniques, notably so Raman spectroscopy [8–11]. They emanated into the "conventional" structural model outlined herein in Section 3. The early MAS NMR studies primarily utilized ^{29}Si, whereas ^{27}Al NMR experimentation was a complement that offered less transparent and reliable—yet important—information for advancing the structural insight.

The initially modest role of ^{27}Al NMR stems from the spin-5/2 nature of the nuclide, which renders MAS NMR experimentation challenging owing to the inherently low spectral resolution stemming from NMR signal broadenings due to the quadrupolar interactions, in turn hampering the extraction of accurate information (see Section 5). These obstacles are compounded by the amorphous character of AS glasses. Yet, standard MAS application is often sufficient for NMR-signal discrimination and quantification the various (potentially coexisting) AlO_4, AlO_5, and AlO_6 polyhedra, thereby offering important insight into the Al speciation. Thanks to the pioneering work towards improving the understanding of MAS NMR on half-integer spins and their utilization to address structural problems in oxide-based materials (with dominant contributions from Lippmaa et al. [12–20])—as well as the concurrent MAS NMR hardware improvements in terms of faster spinning rates and high-field superconducting magnets—facilitated ^{27}Al NMR studies of glasses significantly. Yet, the research field remained narrow until the mid 1990s when high-resolution NMR of half-integer spins became tractable by standard MAS NMR probeheads [21, 22].

The present article expands on our previous review on AS glasses from 2015 [23], which focussed primarily on their short-range ($\lesssim$0.3 nm) structure, to also encompass accounts on heteronuclear NMR methods that may unveil the medium-range ($\approx$ 0.3–1 nm) structural features (Section 11). Most of sections 2–10 of Ref. [23] have only undergone relatively minor updates, whereas Section 12 is modified significantly and expanded; this particularly concerns discussions on the structural role of the minor but non-negligible AlO_5 and AlO_6 moieties, along with their bearings on the AS glass properties. Yet, besides also reviewing the most recent progress in the area of ^{27}Al NMR on AS glasses up to early year 2020, the main spirit and contents of the previous article remain: (*i*) We outline the basic structural features of (alumino)silicate glasses in Section 3. This text is written for an audience without expertise on glasses, whereas Section 12 provides a revised and more detailed structural picture, towards which ^{27}Al and ^{17}O MAS NMR have played major roles. (*ii*) With the theoretical presentation on quadrupolar nuclei kept at a minimum (Section 5), we discuss fundamental aspects of direct relevance for *performing* ^{27}Al MAS NMR experiments (Sections 6–8 and 11.3).

The present text also aims at introducing routine ^{27}Al MAS NMR from a *practical* viewpoint; this includes advice on the recording of quantitative NMR spectra by single-pulse excitation, along with instrumental experimental techniques such as spin-locking (Section 8.5), triple-quantum MAS (3QMAS) [21, 22] (Section 9.2), while Section 11.3 outlines the experimental optimization of cross-polarization (CP) [24–26] in the context of ^{27}Al NMR and AS glasses. Moreover, options for extracting quantitative ^{27}Al NMR parameters and $\{AlO_4, AlO_5, AlO_6\}$ populations from AS glasses are discussed (Section 10), where we also highlight the distinctions between ^{27}Al NMR applications to well-ordered structures and glasses, whose amorphous nature yields distributions of the NMR observables (Section 5). Finally, Section 13 concludes by providing a summary and outlook of the field.

For further reading on the NMR theory and technique-development of half-integer spins, we particularly recommend the deep but concise theoretical/methodological account on quadrupolar nuclei by Jerschow [27], the comprehensive text on theory/applications of MQMAS by Goldbourt and Madhu [28], and the overview on MAS NMR methodology for half-integer spins by Fernandez and Pruski [29]. For applications of both homonuclear and heteronuclear NMR experimentation on other oxide-based glasses, we refer the reader to our previous review [30].

2. Nomenclature and NMR conventions

2.1 Notation for glass compositions and structural concepts

The nomenclature and notation employed in the present article overall adhere to that of our previous review on oxide-based glasses [30]. Glass compositions are expressed by their oxide constituents specified in mol%, or stoichiometric formulations with n_E denoting the stoichiometric amount of the element/oxide E, whereas x_E denotes its molar (atomic) fraction.

Generally, the symbol x is dedicated for *fractions*, whose entity is specified by a subscript and defined in its context of usage. The coordination number of the central atom/ion α with respect to another species β is denoted by a superscript following the element symbol. For instance, $Al^{[4]}$ denotes Al^{3+} in fourfold coordination by O, i.e., a (pseudo)tetrahedral AlO_4 group.

The *cation field strength* (CFS) of an ion M^{z+} with charge z and radius R is defined by z/R^2. Hence, large CFS values are encountered for small and highly charged cations. Sometimes the closely related cation potential (z/R) is encountered in the literature. Note that the radius R depends on the cation coordination number; Shannon provides tabulated values of R [31].

2.2 NMR frequencies and chemical shifts

The Zeeman Hamiltonian for an arbitrary site of the nuclear spin species "S" with spin-quantum number S and located in a strong external magnetic field $\mathbf{B}_0$ is given by

$$H_Z = -\gamma B_0 S_z = 2\pi\nu_0 S_z, \tag{1}$$

where B_0 is the field magnitude in Tesla (T) [32]. From this definition follows that the Larmor frequency $\omega_0 = 2\pi\nu_0$ is *negative* for nuclei with a positive spin magnetogyric ratio γ (as for ^{27}Al) [32, 33]. While the NMR Hamiltonians involve angular frequencies (denoted ω) in units of rads^{-1}, we will consistently express frequencies (ν) in Hz, each being related to its corresponding angular frequency by $\omega = 2\pi\nu$.[a] When also (only) accounting for the isotropic chemical shift δ_{iso}^{j} of a given site j of species S, the resonance frequency becomes [32, 34]

$$\nu_j = \nu_0(1 + \delta_{iso}^{j}). \tag{2}$$

In general, we will identify "S" with ^{27}Al, but whenever required, the spin species is specified by a superscript, e.g., δ^{Al} for a general ^{27}Al "shift" and δ_{iso}^{Al} for the *isotropic* chemical shift.

The NMR spectra presented herein employ a shift scale in ppm, where δ increases from *right to left* such that a "shielded" ("deshielded") nucleus resonates at a lower (higher) chemical shift. Any shift is defined from

$$\delta = \frac{\nu - \nu_{ref}}{\nu_{ref}}, \tag{3}$$

[a] Herein, we strictly use frequencies in units of Hz, where we have cleaned up some inconsistencies in the ω/ν notation of our previous article [23], notably so in Sections 5 and 7.

where ν_{ref} is the resonance frequency of the reference compound that defines $\delta = 0$. All ^{27}Al shifts herein are reported relative to 1 M Al^{3+}(aq) as the standard reference.

NMR spectra in *frequency units* are herein displayed with *positive* frequency-values appearing to the *right* and negative values to the left of $\nu = 0$ [32, 33]; compared with NMR spectra recorded on most spectrometers, this implies performing a sign-reversal of all frequencies such that "positive values become negative", and *vice versa.* Note that the centre frequency $\nu = 0$ of the NMR spectrum corresponds to the position of the *spectrometer reference frequency* (ν_{rf}) [32], often also referred to as the "carrier" or "transmitter" frequency. Carefully distinguish between the stumblingly similar labels ν_{ref} and ν_{rf} that defines $\delta = 0$ and $\nu = 0$ on the shift and frequency scales, respectively [32].

2.3 Magic-angle spinning

Unless stated otherwise, all NMR experimentation discussed below involves glass powders undergoing MAS, meaning that the sample holder (rotor) rotates at the constant frequency ν_r, while subtending the "*magic angle*" $\beta_m = \arctan(\sqrt{2}) \approx 54^{\circ}$ relative to the $\mathbf{B}_0$ direction [35–37]. The "rotor period" τ_r constitutes the time interval for the rotor to make one complete revolution and is given by the inverse of the MAS frequency: $\tau_r = \nu_r^{-1}$. At sufficiently fast MAS, the orientation-*dependent* (*anisotropic*) part of the NMR interaction that transforms as a second-rank spherical tensor is averaged out, such as the chemical shift anisotropy (CSA) and homonuclear/heteronuclear dipolar interactions. This occurs because β_m is a root of $P_2(\beta) = (3\cos^2\beta - 1)/2$, meaning that $P_2(\beta_m) = 0$, and implying that only the orientation-*in*dependent (*isotropic*) component of the chemical shifts and through-bond (J) interactions remain. Refs. [34, 38] provide gentle introductions to tensors in NMR and the orientation-dependence of the spin interactions in the solid-state.

Unless the MAS rate exceeds the magnitude of the anisotropic NMR interaction of a given spin-site j ($\nu_r > |\nu^j_{aniso}|$), a manifold of *spinning sidebands* (ssb) is observed. The *centreband* ($k = 0$) appears at the isotropic chemical shift frequency ($\nu_j^{(0)} = \nu^j_{iso}$) and is flanked by all other sidebands that appear at frequencies $\nu_j^{(k)} = \nu_j^{(0)} + k\nu_r$ ($k = \pm1, \pm 2, \pm3, \ldots$) [34, 37], where the index k is called the *sideband order*.

3. Basic structural features of (alumino)silicate glasses

This section introduces the basic concepts and nomenclature for describing aluminosilicate structures, including the current structural model of AS glasses that has prevailed for several decades. Some deviations from this conventional view are reported over the last 20 years, as commented on in Section 3.3.3 and discussed further in Section 12.

3.1 Amorphous SiO_2 and silicate glasses

Silicon remains predominantly in tetrahedral coordination in both silicates and aluminosilicates. The structure of amorphous (vitreous) SiO_2 is a three-dimensional (3D) network of corner-shared SiO_4 tetrahedra, where each *bridging oxygen* (BO) atom at the corner of one tetrahedron links to a neighbouring silicate group; see Fig. 1. Such a tetrahedron may be denoted $SiO_{4/2}$ to stress that each BO atom connects two SiO_4 groups. Si is a *network former* and the SiO_4 tetrahedron represents the basic building block of all silicate phases, crystalline as well as amorphous, meaning that they all share the feature of *short-range* order. The primary distinction between crystals and glasses is the lack of "long-range" ($>$ 1 nm) structural order of the latter, as reflected already at the medium-range scale ($\approx$0.3–1 nm) by random

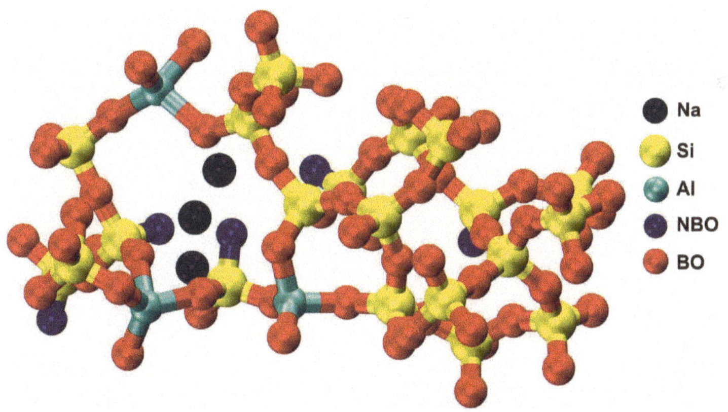

Fig. 1 A structural fragment observed from a Si-rich and highly polymerized Na_2O–Al_2O_3–SiO_2 glass, constituting of a network of Q^n_{Si} and Q^4_{Al} tetrahedra, interconnected by BO atoms. All primary AS glass features (**1**)–(**3**) of Section 3.3.2 are obeyed: only $AlO_{4/2}$ groups are observed with the absence of direct $Al^{[4]}$–O–$Al^{[4]}$ linkages, while all NBO ions are located at the Q^n_{Si} groups. The structural fragment was obtained from an atomistic molecular dynamics (MD) simulation.

SiO_4–SiO_4 inter-connectivity patterns and *distributions* of Si–O distances, as well as Si–O–Si bond angles among the tetrahedral building blocks.

Silicate glasses involve at least one electropositive metal cation (M^{z+}), usually a monovalent alkali (M^{+}) or a divalent alkaline-earth (M^{2+}) ion, added as a carbonate ($M_{(2)}CO_3$) or oxide ($M_{(2)}O$) in the melt-quench glass preparation procedure. Such a cation is referred to as a *network modifier*, as for each *unit* (+1) charge of the modifier, one Si–O–Si bond of the glass network is broken and a *nonbridging oxygen* (NBO) ion is introduced at the SiO_4 group. Hence, for a divalent modifier ion, two BO → NBO conversions occur, which may be expressed schematically as

$$\text{Si}-\text{O}-\text{Si} + (M^{2+}\cdots\text{O}^{2-}) \rightarrow \text{Si–O}^{-}\cdots M^{2+}\cdots{}^{-}\text{O–Si}, \quad (4)$$

where "··· " indicates the electrostatic M^{2+}–NBO interaction that is much weaker than the predominantly covalent bonds of the Si–O–Si bridges associated with the BO atoms. Note that the positive network-modifier cations provide local charge-balance to the negative NBO species, meaning that both the M^{z+} and NBO moieties are closely associated in the glass structure, as shown in Fig. 1. If only counting bonds to network-forming atoms (here Si), the BO and NBO species correspond to $O^{[2]}$ and $O^{[1]}$ coordinations, respectively (see Fig. 1).

The symbol "Q_{Si}^{n}" is a commonly adopted shorthand notation for a $[SiO_n(O^-)_{4-n}]$ moiety with n BO atoms and $4 - n$ NBO ions around the central atom of the tetrahedron [39], whose identity is indicated by a subscript (here: Si). Fig. 2 illustrates the various basic building motifs of silicate structures. A modified silicate glass generally incorporates a *set* $\{Q_{Si}^{n}\}$ of

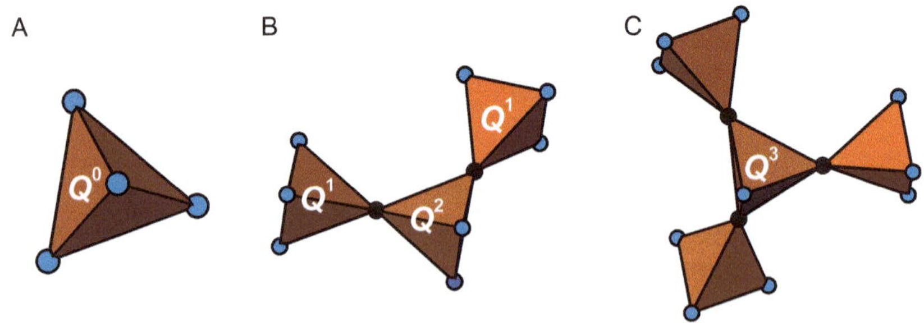

Fig. 2 Illustration of the Q_E^n nomenclature for a tetrahedrally coordinated atom species "*E*". Only oxygen atoms are indicated, with bridging and nonbridging entities depicted in *black* and *blue colour*, respectively. (A) An "isolated" Q^0 tetrahedron. (B) A Q^2 group linked to two Q^1 terminal tetrahedra. A central (C) Q^3 and (D) Q^4 tetrahedron connected to Q^1 groups.

coexisting groups with the respective fractional *populations* $\{x^n_{Si}\}$, the latter obeying the normalization condition

$$\sum_{n=0}^{4} x^n_{Si} = 1. \quad (5)$$

As each BO → NBO conversion is accompanied by a $Q^n_{Si} \rightarrow Q^{n-1}_{Si}$ transformation and a concurrent glass-network depolymerization, the network topology is reduced for increasing modifier content: the 3D "tectosilicate" structure of amorphous silica, built solely from Q^4_{Si} motifs, converts into a "sheet-like" (phyllosilicate) structure where Q^3_{Si} groups dominate. As more M^{z+} cations are added, rings and chains of interconnected Q^2_{Si} moieties form that characterize "metasilicates", whereas eventually only dimers (Q^1_{Si}) and monomers (Q^0_{Si}) remain in the "disilicate" and "orthosilicate" structures, respectively.

Some comments on the "Q^n_{Si} concept" and its relationship to the glass structure are warranted: there is a widespread misconception in the glass community that the "silicate speciation" informs about the *medium*-range glass structure. This is in our opinion incorrect, because the "silicate speciation"—i.e., the set of fractional populations $\{x^n_{Si}\}$ of the set $\{Q^n_{Si}\}$—only provides statistics on the BO/NBO distribution around Si, i.e., $\{x^n_{Si}\}$ represents *statistics on the short-range structure*.[b] Notably, distinct-n Q^n_{Si} groups only differ in their *numbers* of Si–BO and Si–NBO bonds, each one inarguably being a *short*-range property which is only relevant for defining the structure across <0.2 nm. Indeed, knowledge about the silicate speciation offers no insight into *how* these groups are *interconnected* in the glass network (other than constraining its average topology; see below). In contrast, each silicate group connectivity [30] represents a *medium*-range structural feature, i.e., *which* $Q^n_{Si}-Q^m_{Si}$ *pairs* (and larger silicate group aggregates) are present in the glass network.

We stress that in contrast to a crystalline silicate structure—usually built from *one* Q^n_{Si} species with n fixed—a given silicate glass network always involves nonnegligible amounts of *at least two* tetrahedral groups with distinct/variable numbers of BO atoms. Thus, without experimental data on the precise $\{x^n_{Si}\}$ distribution, one may only discuss the *average* glass-network topology, which is characterized by the *silicate network connectivity* [39, 41, 42],

$$\overline{N}^{Si}_{BO} = 2(4 - r_{Si}) = \sum_{n=0}^{4} n x^n_{Si}, \quad (6)$$

[b] In direct analogy to the "O speciation", which informs about the proportions of BO and NBO species in the glass structure. Here we disregard the $O^{[3]}$ coordinations (oxygen triclusters) [40], which may be present in significant amounts in (dense) AS glass networks; see Section 12.1.

where $\overline{N}_{BO}^{Si}$ is the *average* number of BO atoms at the tetrahedra of the $\{Q_{Si}^n\}$ ensemble, whose respective populations $\{x_{Si}^n\}$ may often be obtained experimentally by ^{29}Si NMR [30]. $\overline{N}_{BO}^{Si}$ is related to the glass *network polymerization* (or condensation) degree r_{Si} [39, 41, 42],

$$r_{Si} = n_O^{Si}/n_{Si}, \tag{7}$$

which represents the mean number of O atoms per Si atom in the structure (with each BO and NBO counting as 1/2 and 1 O atom, respectively), whereas n_O^{Si} is the stoichiometric amount of O that form one or multiple Si–O bonds. For a silicate glass where Si constitutes the sole network former (as is assumed at the moment), n_O^{Si} equals the total stoichiometric amount of O in the glass ($n_O^{Si} \equiv n_O$), meaning that r_{Si} is readily calculated from Eq. (7) along with the glass composition. The sub/superscript "Si" (e.g.,$r_{Si}, \overline{N}_{BO}^{Si}$) is then superfluous and is only included here for the sake of clarity and for subsequent generalizations to AS glasses (Section 3.4). For glasses involving two or several network formers (as in aluminosilicate, phosphosilicate or borosilicate glasses), the n_O^{Si}, r_{Si} and $\overline{N}_{BO}^{Si}$ parameters are not known a priori.

Carefully distinguish between the two entities $\overline{N}_{BO}^{Si}$ (silicate network connectivity) and r_{Si} (silicate network polymerization), which represents the (average) numbers of *oxygen* atoms and *bridging oxygen* atoms per Q_{Si}^n group, respectively. There are several options for conveying the parameters $\overline{N}_{BO}^{Si}$ and r_{Si}; in the Geoscience community, the complementary terminology "NBO/T" is normally employed, thereby focussing instead on specifying the complementary entity $\overline{N}_{NBO}^{Si}$ (in our notation). Note that

$$\overline{N}_{BO}^{Si} + \overline{N}_{NBO}^{Si} = 4. \tag{8}$$

3.2 Al_2O_3 and aluminate phases

As opposed to the well-defined tetrahedral coordination of Si in silicates, Al may be coordinated by four, five or six O atoms, i.e., up to three distinct Al-centred $\{AlO_4, AlO_5, AlO_6\}$ polyhedra may coexist in an oxide-based structure; Fig. 23 of Section 12.1 depicts each such structural moiety. Six-coordinated Al is most common in simple crystalline Al-bearing compounds, such as $Al(OH)_3$ and salts like $Al(NO_3)_3$ and "alum", $KAl(SO_4)_2 \cdot 12H_2O$ [39, 43]. Alumina occurs in several polymorphs. The most stable form is α-Al_2O_3, whose structure is built from face-shared $AlO_{6/4}$

octahedra. However, numerous metastable forms of alumina exist: for instance, the disordered structure of γ-Al_2O_3 comprises AlO_4 and AlO_6 polyhedra in the approximate proportions 3:7 [44–47], possibly with minor AlO_5 contributions inferred by ^{27}Al MAS NMR [46–48], whereas all three $\{Al^{[4]}, Al^{[5]}, Al^{[6]}\}$ coordinations coexist in ρ-Al_2O_3 [44]. $Al^{[4]}$ is the prevailing Al coordination in the vast class of ordered aluminosilicate structures, encompassing minerals, zeolites and many ceramics [39, 43]. Tetrahedrally coordinated Al is also most common in $MO_{z/2}$–Al_2O_3 "aluminate phases", with M^{z+} typically representing a divalent alkaline-earth (where Ca^{2+} is most common) or a trivalent rare-earth cation, such as La^{3+} or Y^{3+}.

The structural role of Al in amorphous phases varies depending on the nature and contents of the other glass constituents. Notably, despite that AlO_6 octahedra are the most prominent building blocks of crystalline Al-based structures (together with AlO_4), significant fractions ($>$ 10%) of $Al^{[6]}$ coordinations are rarely encountered in *amorphous* Al-bearing phases where—besides the dominating tetrahedral coordination—AlO_5 merely constitutes the second most abundant species. While a vast majority of the research community nowadays advocates the presence of AlO_5 species in AS glasses, a disputed matter in earlier literature was whether ^{27}Al NMR signals observed ~30 ppm stem from distorted AlO_4 tetrahedra sharing an $O^{[3]}$ moiety [40] rather than from AlO_5 polyhedra; see Section 12.4. The structural role of the $Al^{[5]}$ and $Al^{[6]}$ species and their bearings for the glass properties are discussed in Sections 12.3.1 and 12.5–12.6.

Although it is not possible to produce vitreous alumina by conventional melt-quench techniques, thin films of amorphous Al_2O_3 have been prepared and characterized by ^{27}Al MAS NMR, revealing a structure involving all three $\{AlO_4, AlO_5, AlO_6\}$ polyhedral types, with AlO_4 and AlO_5 groups being most abundant [49, 50]. Likewise, the networks of $MO_{z/2}$–Al_2O_3 aluminate glasses (or gels) with $M^{z+} = \{Ca^{2+}, La^{3+}, Y^{3+}\}$ are primarily built by AlO_4 groups, but with significant contributions of AlO_5 (major) and AlO_6 (minor) polyhedra [51–55].

3.3 Aluminosilicate glasses

3.3.1 Basic nomenclature

Onwards, we consider a general aluminosilicate glass composition involving one or several network-modifying cations (M^{z+}), with the ternary $MO_{z/2}$–Al_2O_3–SiO_2 system representing the archetypical case of modified amorphous aluminosilicates. Such a network comprises SiO_4 and AlO_4 tetrahedra that are corner-shared by BO atoms, as depicted by Fig. 1. These moieties will be denoted $Q^n_{Si}(m\text{Al})$ and $Q^n_{Al}(m\text{Si})$, respectively, with the sub/superscript labels having identical meanings as for (modified) silicate glasses. Out of these n bridges, there are m **Si**–O–Al and m **Al**–O–Si bonds at each $Q^n_{Si}(m\text{Al})$ and $Q^n_{Al}(m\text{Si})$ moiety, respectively, implying the occurrence of $n - m$ corresponding Si–O–Si and Al–O–Al homoatomic bridges. Whenever the identity of nearest-neighbouring TO_4 groups is immaterial, the shorter labels Q^n_{Si} and Q^n_{Al} will be employed for the respective SiO_4 and AlO_4 groups.

Relative to the $SiO_{4/2}$ moiety, a $TO_{4/2}$ tetrahedron involving a *trivalent* cation T (e.g., Al^{3+}, B^{3+}) carries one excess negative charge-unit that requires nearby cations (M^{z+}) for retaining local charge-neutrality. The presence of $[AlO_{4/2}]^-$ and $[BO_{4/2}]^-$ groups strongly impacts Al and/or B bearing glass structures, as examined further in Section 12.

3.3.2 Conventional structural picture of aluminosilicate glass networks

The structural description outlined below has prevailed for decades and centres around three features that hold well for glasses based on low-CFS M^{z+} cations, in practice encompassing all alkali- and alkaline-earth cations except for the high-CFS Mg^{2+} ion (yet, with the deviations from the structural model becoming nonnegligible already for Ca^{2+}). All of features (**1**)–(**3**) discussed below are direct consequences of the excess negative charge of the Q^4_{Al} (i.e., $[AlO_{4/2}]^-$) groups. *Provided* that the AS glass composition simultaneously adhere to the requirements

$$(\mathbf{A})\ n_{Al}/n_M \leq z, \tag{9}$$

and

$$(\mathbf{B})\ n_{Al}/n_{Si} \leq 1, \tag{10}$$

then the AS glass structure exhibits the following three features:

(1) Both Si and Al are tetrahedrally coordinated ($Si^{[4]}$, $Al^{[4]}$). Note that $n_{Al} = zn_M$ [c.f. Eq. (9)] represents the minimum M^{z+} content required for balancing the entire Al speciation as $AlO_{4/2}$ tetrahedra; higher-coordination AlO_5 and AlO_6 polyhedra form whenever Eq. (9) is not obeyed [7, 56]. Moreover, since the $AlO_{4/2}$ content in the glass network restricts its net amount of NBO ions, the latter may be estimated after accounting for the fraction of the total modifier reservoir that is consumed for balancing the $[AlO_{4/2}]^-$ groups. The dual role of a M^{z+} cation as a network modifier or charge *compensator* of $AlO_{4/2}$ groups has bearings for glasses comprising several distinct cation types, where a given species may assume a preferential structural role: in Na_2O–CaO–Al_2O_3–SiO_2 glasses, for instance, the monovalent Na^+ cations naturally compensate the charges of $[AlO_{4/2}]^-$ tetrahedra, whereas Ca^{2+} mainly acts as a modifier [9, 57, 58].

(2) The "Loewenstein Al avoidance" rule [59] implies that $Al^{[4]}$–O–$Al^{[4]}$ bridges are absent to avoid local negative charge accumulation in the structure. There is generally a strong preference for Si–O–$Al^{[4]}$ linkages, such that the network comprises only Si–O–$Al^{[4]}$ and Si–O–Si bridges. When the Al content exceeds that of Si, i.e., if the requirement Eq. (10) is violated, the Loewenstein rule *cannot* be obeyed and higher-coordination AlO_5/AlO_6 polyhedra form [2, 5, 6].[c]

(3) AlO_4 tetrahedra avoid NBO ions and tend to occupy the most polymerized Q^n (highest n) sites in the glass network [2–6, 9–11, 60–62], in practice translating into the sole presence of $Q^4_{Al}(4Si)$ groups. For Si-rich and highly polymerized AS networks, the strong propensity for Si–NBO contacts relative to Al–NBO implies that all NBO species are located at the SiO_4 tetrahedra [60–62]. When the network-modifier content is increased, the additional NBO ions initially enter at the Q^n_{Si} groups before being accommodated by the Q^n_{Al} tetrahedra. Note that charge-neutralization of a Q^n_{Al} group consumes $5 - n$ positive charges.

The structural features (**1**)–(**3**) imply that for fully polymerized tectosilicate glasses with $n_{Al} = n_{Si} = zn_M$, the 3D framework structure is formally devoid of NBO ions and built from strictly alternating $Q^4_{Si}(4Al)$ and $Q^4_{Al}(4Si)$ tetrahedra [2–6, 9]. This scenario remains for glass compositions with $n_{Al} = zn_M$ but arbitrary molar ratios $n_{Al}/n_{Si} < 1$, with the distinction that Si–O–Si linkages must exist, giving a glass network comprising $Q^4_{Al}(4Si)$ groups and a set of $\{Q^n_{Si}(mAl)\}$ tetrahedra for which n varies. The constraint $0 \leq m \leq n \leq 4$ implies the *potential* existence of up to 15 $Q^n_{Si}(mAl)$ structural species; in practice, however, a significantly smaller subset is observed in AS glasses [2, 3, 6].

3.3.3 Limitations of the model

The compliance of the "real" AS glass structure to the "conventional" structural model tends to relax gradually as the CFS of the network-modifier

[c] We note that condition (**B**) [Eq. (10)] is a necessary but not sufficient prerequisite for the absence of $Al^{[4]}$–O–$Al^{[4]}$ linkages. Strictly, a minimum number of BO atoms must be available to exclusively enable Si–O–Si and Si–O–Al bond formation, meaning that the NBO content of the glass cannot be too large. When also accounting for the fractional populations of BO (x_{BO}) and NBO (x_{NBO}) species (with $x_{BO} + x_{NBO} = 1$) and assuming the sole presence of Si–NBO (as opposed to $Al^{[p]}$–NBO) bonds, condition (**B**) becomes $n_{Al}/n_{Si} \leqslant 1 - x_{NBO}$.

grows [2, 6, 62–66]. For increasing CFS of M^{z+}, the glass structure becomes randomized, as reflected both in the SiO_4/AlO_4-ordering of the network and the NBO-partitioning among these TO_4 groups, as well as in growing AlO_5 and AlO_6 populations. Over the past two decades, violations of properties (**1**) and (**2**) are well documented for AS glasses based on mono/divalent cations, whereas those involving trivalent rare-earth (RE^{3+}) cations manifest strong violations of all three properties (**1**)–(**3**); we defer more thorough discussions to Section 12, and here limit ourselves to conveying the general trends observed for "tectosilicate" AS glass compositions. Despite that they *formally* exhibit full charge-balance of all AlO_4 tetrahedra by the M^{z+} cations, several ^{27}Al MAS NMR studies of alkaline-earth-based AS glasses reveal minor fractional populations ($\lesssim$7%) of AlO_5 groups [62, 67–69] that remain even for compositions with $n_{Al}/n_M < z$ [62, 68, 70]. Also at odds with expectations, the O speciation involves a few percent of NBO species in tectosilicate glasses [71].

Moreover, while Loewenstein's rule [feature (**2**)] is a well-established paradigm for crystalline AS phases whose compositions obey Eq. (10) (such as zeolites and minerals [39, 43]), minor deviations thereof are reported for charge-balanced $CaO–Al_2O_3–SiO_2$ glasses exhibiting $n_{Al} \lesssim n_{Si}$ [65, 66]. As discussed further in Section 12.3.3, only feature (**3**) remained unquestioned up to very recently [72–75]: while direct and unambiguous experimental proof of $Al^{[4]}$–NBO bonds existed solely for highly fragmented aluminate or Si-poor AS networks, only circumstantial evidence pointed towards $Q^n_{Al} \rightarrow Q^{n-1}_{Al}$ depolymerization processes occurring in Si-rich AS glass networks [6, 60–62].

3.4 Split-network representation of aluminosilicate glasses

A network-forming element F (e.g., Al, B, Ge, Si and P) usually adopts tetrahedral coordination, meaning that the network consists of FO_4 tetrahedra. Whenever more than one network former is present in the glass, the BO/NBO-partitioning between them is nonuniform, i.e., the NBO ions are not distributed statistically among the various network formers $\{F, F', F'', \ldots\}$ according to their relative stoichiometric amounts $\{n_F, n_{F'}, n_{F''}, \ldots\}$. We have already highlighted the prevailing accommodation of NBO ions by SiO_4 groups, whereas the AlO_4 counterparts normally avoid contacts with NBO species.

The precise BO/NBO-partitioning among the network formers is generally unknown and experimental data are required for detailed insight. Unfortunately, the experimental assessment of the BO/NBO distribution is far from trivial for glasses incorporating multiple network formers (see Section 12). Yet, the nuclear charge and electronegativity of each network former F allow for *qualitative* predictions of the relative propensity of each FO_4 tetrahedral type to coordinate BO and NBO species [30, 76].

It is sometimes useful to envisage the glass network as comprising a set of *fictive* "*subnetworks*", each one associated with one FO_4 species. Fig. 3 provides an example. We recently introduced this "*split-network*" picture for a convenient bookkeeping of the average network polymerization (r_F) and accompanying mean number of BO ($\overline{N}_{BO}^{F}$) and NBO ($\overline{N}_{NBO}^{F}$) species per FO_4 group of *each individual* network-forming species $\{F, F', F'', \ldots\}$ in a multicomponent oxide glass [42, 76]. The following *split-network representation* of the glass composition provides directly the polymerization degree of each subnetwork and conveys the most basic insight into the BO/NBO-partitioning among the various network-forming species [42, 76]:

$$n_F F\text{O}_{r_F} \cdot n_{F'} F'\text{O}_{r_{F'}} \cdot n_{F''} F''\text{O}_{r_{F''}} \cdots \tag{11}$$

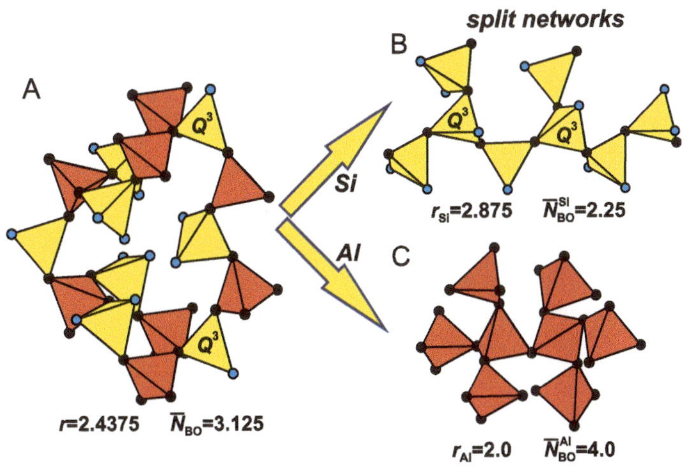

Fig. 3 (A) An aluminosilicate network built from eight SiO_4 (*yellow colour*) and eight AlO_4 (*red colour*) tetrahedra, with *black* and *blue colours* representing BO and NBO species, respectively. Aluminium is exclusively present as Q^4_{Al} units, whereas there are six Q^2_{Si} and two Q^3_{Si} tetrahedra. (B, C) "Split-network" representation [42] involving two fictive networks of (B) SiO_4 and (C) AlO_4 tetrahedra. Note the very different polymerization degrees (r_{Si} and r_{Al}) and average number of BO atoms ($\overline{N}_{BO}^{Si}$ and $\overline{N}_{BO}^{Al}$) of the Al and Si networks. These parameters are related by Eqs (12) and (13) to their respective average values r and $\overline{N}_{BO}$ in (A). *Reproduced from M. Edén, The split-network analysis for exploring composition-structure correlations in multi-component glasses: I. Rationalizing bioactivity-composition trends of bioglasses, J. Non-Cryst. Solids 357 (2011) 1595–1602, with permission from Elsevier.*

Each $\{r_F, r_{F'}, r_{F''}, \ldots\}$ parameter is defined according to Eq. (7), e.g., $r_F = n_O^F/n_F$. Locating the entire set of values becomes the central task. Provided that all parameters—but one—are known, that remaining entity is readily calculated by the split-network procedure, as described in detail in Ref. [42] and exemplified in the AS context below.

In the following, we illustrate the application of the split-network formalism to a ternary glass of stoichiometry $n_M MO_{z/2}$–$n_{Al}AlO_{3/2}$–$n_{Si}SiO_2$, where we assume that AlO_4 accounts for the entire Al speciation; see Refs. [42, 76] for generalizations to scenarios where AlO_5 and AlO_6 are also present.[d] Noteworthy, the *average* network polymerization degree (r) and mean number of BO atoms ($\overline{N}_{BO}$) per TO_4 group ($T = \{F, F', F'', \ldots\}$) in the *entire* network—i.e., the respective average values over the entire $\{SiO_4, AlO_4\}$ ensemble in the glass—are in direct analogy with Eqs (6) and (7) defined by [41, 42]

$$r = n_O/(n_{Si} + n_{Al}), \tag{12}$$

$$\overline{N}_{BO} = 2(4 - r), \tag{13}$$

respectively, whereas the average number of NBO ions per tetrahedron is given by

$$\overline{N}_{NBO} = 2r - 4 = 4 - \overline{N}_{BO}. \tag{14}$$

The expressions Eqs (12), (13) and (14) are due to Stevels [41], who introduced the parameters R, X and Y that in our notation correspond to r, $\overline{N}_{NBO}$, and $\overline{N}_{BO}$, respectively. Note that each of them is *readily calculated* from the (known) *glass composition.*

The split-network representation of the AS glass is

$$n_{Si}SiO_{r_{Si}} \cdot n_{Al}AlO_{r_{Al}}, \tag{15}$$

where the values of $\{r_{Si}, r_{Al}\}$ must be located, and from which those of $\{\overline{N}_{BO}^{Si}, \overline{N}_{BO}^{Al}\}$ may be calculated *via* Eq. (6). The parameter-sets of the sub-network ensemble are related to their respective averages, r and $\overline{N}_{BO}$, according to [42, 76, 77]

$$r = x_{Si} r_{Si} + x_{Al} r_{Al}, \tag{16}$$

[d] If several $\{M_1^{z_1}, M_2^{z_2}, M_3^{z_3}, \ldots\}$ network-modifying cations are present, they are treated collectively by using their total stoichiometric amount ($n_M = n_1 + n_2 + n_3 + \ldots$) and effective charge $z_M = (z_1 n_1 + z_2 n_2 + z_3 n_3 + \ldots)/n_M$, meaning that the modifier ensemble is replaced by a single (fictive) cation M^{z_M}.

$$\overline{N}_{BO} = x_{Si}\overline{N}_{BO}^{Si} + x_{Al}\overline{N}_{BO}^{Al}, \tag{17}$$

where x_{Si} and x_{Al} are the molar fractions of Si and Al, respectively, out of all *network-forming* cations: $x_T = n_T/(n_{Si} + n_{Al})$, with $T = \{Si, Al\}$.

Now given that most (if not all!) NBO species are located at the SiO_4 groups, we assume the sole presence of Q_{Al}^4 groups in the network, i.e., that $r_{Al} = 2$ and $\overline{N}_{BO}^{Al} = 4$. This provides the following expression for r_{Si}:

$$r_{Si} = \frac{1}{2}(zn_M - n_{Al} + 4). \tag{18}$$

If instead assuming the presence of Al–NBO bonds and consequently an arbitrary average number $\overline{N}_{BO}^{Al}$ of BO atoms at the AlO_4 groups, we have $r_{Al} = 4 - \overline{N}_{BO}^{Al}/2$. Then, the mean polymerization degree of the "silicate subnetwork" becomes [77]

$$r_{Si} = \frac{1}{2}\left(zn_M + n_{Al}\left[\overline{N}_{BO}^{Al} - 5\right] + 4\right). \tag{19}$$

Now both parameters of the split-network representation of Eq. (15) are defined.

Fig. 3 illustrates the separation of an AS network into two *fictive* Si and Al subnetworks that exhibit distinctly different polymerization degrees. Despite that the average polymerization $r = 2.44$ in Fig. 3A *could* imply a structure built predominantly by Q_{Si}^3 and Q_{Al}^3 units, there are only two such moieties (both are Q_{Si}^3) out of the total 16 tetrahedra. The Al-associated network comprises solely Q_{Al}^4 groups, according to the "conventional" AS glass-network model. This underscores the limited ability of any average structural parameter (e.g., r or $\overline{N}_{BO}$) to provide *detailed* insight into the underlying $\{Q_T^n\}$ speciations: because the BO/NBO-partitioning is *uneven* among the various $\{Q_{Si}^n\}$ and $\{Q_{Al}^n\}$ species, the entities r and $\overline{N}_{BO}$ may at best *constrain* its underlying "subnetwork" parameters *via* Eqs (16) and (17). As commented in Refs. [42, 76], the "split-network" strategy may be applied to amorphous as well as (mixtures of) crystalline phases, where we stress that the "split networks" is a *mind construction*, which does neither relate to, nor imply, a physical phase-separation of the glass.

We have utilized the split-network formalism for numerous purposes, encompassing composition-structure analyses of Na_2O–CaO–B_2O_3–SiO_2–P_2O_5 glasses [78, 79], the rationalization of composition-bioactivity relationships in Na_2O–CaO–SiO_2–P_2O_5 glasses [42, 80], as well as for designing new compositions thereof with fixed silicate network connectivity ($\overline{N}^{\mathrm{Si}}_{\mathrm{BO}}$) but variable $n_{\mathrm{P}}/n_{\mathrm{Si}}$ molar ratios [81, 82], along with structural analyses of aluminoborosilicate glasses and glass-wool fibres [76]. Split-network concepts were also employed to (semi-quantitatively) predict ^{29}Si chemical shifts in La_2O_3–Al_2O_3–SiO_2 glasses that in conjunction with experimental ^{29}Si MAS NMR data provided insight into both the SiO_4/AlO_4 intermixing in the AS networks and the BO/NBO-partitioning among its Q^n_{Si} and Q^n_{Al} constituents [77]. At odds with the conventional structural model of AS glasses, the results suggested a pronounced network disorder, as mirrored in strong violations of Loewenstein's rule, as well as $\mathrm{Al}^{[4]}$–NBO bonds being present. Both predictions were subsequently proven by more advanced ^{27}Al-based NMR experimentation, as discussed further in Section 12.

4. Chemical shifts of $^{27}AlO_p$ groups in oxide-based materials

Reasonably accurate predictions of chemical shifts require either density functional theory (DFT) calculations [83–85] or machine learning concepts [86] (as very recently applied to AS glasses by Charpentier et al.). Here we discuss a crude and simple approach, which despite being oversimplified usually offers a sufficient predictive power for qualitative assessments of ^{27}Al chemical shift trends in oxide-based materials. The ^{27}Al isotropic shift of an AlO_p moiety in such structures is effectively dictated by paramagnetic shifts and correlates approximately with the electron density at the nucleus: when the *electron density* is *increased*, ^{27}Al *deshields*, i.e., $\delta^{\mathrm{Al}}_{\mathrm{iso}}$ *increases*. Hence, if two distinct local structural fragments are compared, the one predicting the highest electron density at ^{27}Al is expected to provide the largest $\delta^{\mathrm{Al}}_{\mathrm{iso}}$-value. In the following, $\delta^{[p]}_{\mathrm{iso}}$ denotes the isotropic chemical shift of an $^{27}\mathrm{Al}^{[p]}$ site. Tabulated ^{27}Al chemical shifts are provided by MacKenzie and Smith [43, 87].

In general, the ^{27}Al isotropic chemical shift depends on the same structural factors as those of ^{29}Si and ^{31}P [4, 6, 20, 39, 43, 88–90]. Consequently, according to the well-established trends for ^{29}Si, $\delta^{[4]}_{\mathrm{iso}}$ of an $^{27}AlO_4$ group depends on the average inter-tetrahedral Al–O–T bond angle ($\bar{\theta}_{\mathrm{Al-O-}T}$) and average Al–O distance ($\bar{d}_{\mathrm{Al-O}}$), where $T=\{\mathrm{Si, Al}\}$: ^{27}Al deshields by a few ppm when either $\bar{\theta}_{\mathrm{Al-O-}T}$ is decreased or $\bar{d}_{\mathrm{Al-O}}$ is increased [20, 39, 43, 88–90]. However, in practice, these local structural parameters yield little information from glass powders, as the subtle shift-alterations are obscured by the (*i*) structural disorder that produces distributions of the NMR parameters (see Section 5.4), and primarily that (*ii*) ^{27}Al is a quadrupolar nucleus affected by large quadrupolar interactions that swamps all minor shift-changes (Section 5.3). Hence, only local structural modifications accompanied by significant electron density alterations are discernible in ^{27}Al MAS NMR spectra from disordered structures.

In oxide-based materials, the strongest electron density alteration at the ^{27}Al nucleus occurs for a change in Al coordination number: as illustrated by several of the figures that follow (Figs 8, 9, and 26), various coexisting $Al^{[p]}$ coordinations are often readily identified by the ^{27}Al MAS NMR spectrum. The ^{27}Al isotropic chemical shift is reduced by around 30 ppm for each increase in coordination number, $Al^{[p]} \rightarrow Al^{[p+1]}$. In AS glasses, $\delta_{iso}^{[p]}$-values of 65 ppm, 35 ppm and 5 ppm are typically observed for the respective $^{27}AlO_4$, $^{27}AlO_5$ and $^{27}AlO_6$ polyhedra, the precise values depending on the molar n_{Al}/n_{Si} ratio and glass-network polymerization (Section 12.2). The further away from the ^{27}Al nucleus that a given local structural modification occurs, the less altered is the electron density at the nucleus and the lower becomes the ^{27}Al chemical shift displacement. Yet, the $\delta_{iso}^{[p]}$ value depends on the following structural features:

(i) As anticipated from the correlation between the electron density and the chemical shift, the (Pauling) electronegativity of the ligand atom around Al affects $\delta_{iso}^{[p]}$: for instance, each O → N replacement at an AlO_4 group typically *increases* $\delta_{iso}^{[4]}$ by around 10–12 ppm [43, 91–93]. The ^{27}Al deshielding effects from nearest-neighbouring N atoms are evident from the NMR spectra of Fig. 15 in Section 9.2 that were collected from oxynitride AS glasses with increasing N contents [93]. While $Al^{[4]}$ reluctantly forms bonds to F, O→F substitutions at AlO_6 octahedra *decreases* $\delta_{iso}^{[6]}$ by a few ppm per F atom, altogether amounting to ≈ 20 ppm lower ^{27}Al chemical shifts of $^{27}AlF_6$ groups relative to $^{27}AlO_6$ [94].

(ii) ^{29}Si and ^{31}P chemical shifts depend strongly on the relative numbers of BO/NBO species located at the Q_{Si}^n (or Q_P^n) tetrahedron, with δ_{iso}^{Si} increasing by around 8–12 ppm for each $Q_{Si}^n \rightarrow Q_{Si}^{n-1}$ conversion [2, 4, 95–97]. Although early ^{27}Al NMR studies on crystalline aluminate phases indicated the absence of any correlation between $\delta_{iso}^{[4]}$ and the number of BO atoms (n) at the Q_{Al}^n tetrahedron [98, 99], the databank of ^{27}Al isotropic chemical shifts from phyllosilicate minerals, which comprise sheets of interconnected Q_{Al}^3(3Si) groups, reveal about 10 ppm higher shifts (typically ∼ 65–75 ppm) than those of Q_{Al}^4(4Si) groups (∼ 55–70 ppm) in 3D framework structures, such as zeolites [39, 43]. Moreover, several MAS NMR studies of Al-bearing aluminate and aluminosilicate glasses manifest weak but consistent ^{27}Al deshielding effects for increasing network-modifier content of the glass [6, 51, 62, 72]. These relatively minor ^{27}Al chemical shift effects naturally stem from the lower tendency of AlO_4 groups to accommodate NBO ions relative to SiO_4 [6, 51, 60–62], and is presumably the main reason why there is little discussion in the literature [6, 51, 62, 72] about ^{27}Al shifts in relation to the condensation degree n of Q^n_{Al} groups (yet, see Sections 11.5.1 and 12.3.3).

(iii) Each $^{27}Al^{[p]}$–O–**Al** $\rightarrow$ $^{27}Al^{[p]}$–O–***X*** substitution of Al for a more electronegative species X in the *second* coordination sphere of Al *reduces* its ^{27}Al isotropic chemical shift by ≈3–8 ppm, with the shielding effect growing according to Al < Si < P. For instance, the $\delta^{[4]}_{iso}$ shifts of $^{27}AlO_4$ groups are typically 70–80 ppm, 60–70 ppm and 35–45 ppm in aluminate, aluminosilicate, and aluminophosphate glasses, respectively, with the precise value depending on the number of Al/Si/P neighbours in the second coordination sphere of Al [4, 6, 39, 43, 88].

5. NMR features of half-integer spin quadrupolar nuclei

5.1 Quadrupolar nuclei and electric field gradients

^{27}Al has spin $S = 5/2$, thereby constituting a *quadrupolar nucleus* with half-integer spin, for which spin quantum numbers $S = q/2$ with $q = 3, 5, 7, 9$ are encountered across the Periodic Table [27, 100, 101]. Other common half-integer spin nuclei utilized in the context of AS glasses are ^{23}Na ($S = 3/2$), ^{17}O ($S = 5/2$), and ^{45}Sc ($S = 7/2$).

As illustrated in Fig. 4, a spin-S nucleus has $2S + 1$ Zeeman energy levels with corresponding z-angular momentum components m in the range $\{\pm 1/2, \pm 3/2, \ldots, \pm S\}$. There are $2S$ single-quantum coherences ($\pm$1QC)—often referred to as "transitions"—where we focus on the (-1)QC associated with a $|m\rangle \rightarrow |m-1\rangle$ transition, which is detected on modern NMR spectrometers [32, 33]. The $|1/2\rangle \rightarrow |-1/2\rangle$ "transition"—as also shared with $S = 1/2$ nuclei—is called the "*central transition*" (CT) [27, 100, 101]. All other single-quantum transitions are referred to as *satellite transitions* (STs). For $S = 5/2$ nuclei (e.g., ^{27}Al), there are two *pairs* of "inner" and "outer" satellite transitions that will be book-kept together as "$ST_{3/2}$" and "$ST_{5/2}$", respectively, with the subscript denoting the magnitude of the higher-m quantum number of the Zeeman state-pairs involved in the transition; see Fig. 4. Every coherence order $M \leq 2S$ may be generated in an ensemble of (isolated) spin-S nuclei. All orders $M > 1$ are called *multiple-quantum coherences* [32], where

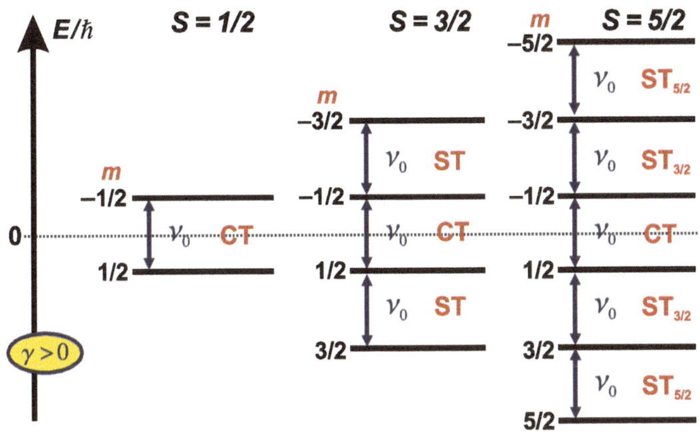

Fig. 4 Zeeman energy level diagrams for $S = 1/2$, $S = 3/2$ and $S = 5/2$ nuclei. Note that all half-integer spins share the $m = \pm 1/2 \rightarrow m = \mp 1/2$ transition; for cases of $S > 1/2$, this is referred to as the central transition (CT), whereas all other $m \rightarrow m \pm 1$ transitions are called satellite transitions (STs). A $S = 5/2$ nucleus has two pairs of distinct STs that are labelled $ST_{3/2}$ ($1/2 \rightarrow 3/2$ and $-1/2 \rightarrow -3/2$) and $ST_{5/2}$, ($5/2 \rightarrow 3/2$ and $-3/2 \rightarrow -5/2$). The diagram is shown for a nucleus with $\gamma > 0$, for which negative (positive) m-values are associated with $E > 0$ ($E < 0$).

for instance ± 3QC are associated with transitions between the two $ST_{3/2}$ Zeeman levels ($|\pm 3/2\rangle \Leftrightarrow |\mp 3/2\rangle$) in Fig. 4.

Spin-1/2 nuclei, such as ^{1}H, ^{13}C, ^{29}Si and ^{31}P, only experience NMR interactions of magnetic origin, such as those involving the external field (B_0), the radio-frequency (rf) pulses (B_1), and the direct/indirect spin–spin couplings stemming from the magnetic moments of neighbouring nuclei [32, 34]. All these interactions also affect quadrupolar nuclei. However, owing to its uneven nuclear charge distribution, a quadrupolar nucleus possesses an electric quadrupole moment (eQ) that also interacts with a *variation* in the electric field—i.e., the *electric field gradient* (EFG)—in its vicinity. This interaction is characterized by a *quadrupolar coupling constant* C_Q (in units of Hz),

$$C_Q = (eq) \cdot (eQ)/h = e^2qQ/h, \tag{20}$$

and an asymmetry parameter η of the electric field gradient (EFG) tensor,

$$\eta = (V_{yy} - V_{xx})/V_{zz}, \text{ with } 0 \leq \eta \leq 1, \tag{21}$$

where e is the elementary charge and $V_{zz} = eq$ is the largest component of the EFG tensor, whose principal values $\{V_{xx}, V_{yy}, V_{zz}\}$ are ordered according to $V_{zz} \geq V_{xx} \geq V_{yy}$ [34]. Note that another ordering ($V_{zz} \geq V_{yy} \geq V_{xx}$) and accompanying distinct definition of η is common in the literature [27–29, 100, 101].

The pair of $\{C_Q, \eta\}$ values reflects the local symmetry of the charge distribution around the quadrupolar site. The EFG tensor is traceless ($V_{xx} + V_{yy} + V_{zz} = 0$), meaning that its orientational average is zero and that C_Q vanishes for perfectly symmetric (e.g., cubic, octahedral or tetrahedral) charge distributions. However, such ideal geometries are very rarely encountered in oxide-based inorganic structures. For instance, the structural disorder of AS glasses manifests in sizable ^{27}Al quadrupolar coupling constants, with typical (average) C_Q-values in the 5–10 MHz range.

For spin-1/2 nuclei, the Larmor frequency of the Zeeman Hamiltonian [Eq. (1)] is several orders of magnitude larger than any other NMR interaction frequency. It follows that only the corresponding Hamiltonian operator parts that commute with S_z may significantly affect the eigenvalues of the total spin Hamiltonian and thereby the NMR observables. Consequently, each NMR Hamiltonian is treated as a perturbation to the Zeeman counterpart, where only S_z-commuting operators are retained; this is referred to as the "high-field approximation" or "Zeeman truncation". We recently provided a gentle but thorough introduction to the subject [102, 103]. In contrast, the presence of very strong quadrupolar interactions have bearings for NMR applications: this is hinted by the more elaborate perturbation approach required in formulating the quadrupolar Hamiltonian under high-field conditions. The *strength* of the quadrupolar interaction is mirrored in the *quadrupolar frequency*,

$$\nu_Q = \frac{C_Q}{2S(2S-1)}, \tag{22}$$

which typically conforms to the 0.1–1 MHz range. Such values are sufficiently small that perturbation theory with a sole consideration of the Hamiltonian eigenvalues is sufficient (i.e., the set of Zeeman eigenstates remains a valid eigen-basis set [27, 37, 102]), yet sufficiently large that the Zeeman-truncated quadrupolar Hamiltonian must invoke a *second*-order correction term $H_Q^{(2)}$:

$$H_Z + H_Q \approx H_Z + H_Q^{(1)} + H_Q^{(2)}, \tag{23}$$

where $H_Q^{(1)} \sim \nu_Q$ and $H_Q^{(2)} \sim \nu_Q^2/\nu_0$ [37]. Eq. (23) is a good approximation whenever $\nu_0 \gtrsim 100\nu_Q$. The following two subsections discuss the features and practical consequences of each *first*- and *second*-order quadrupolar Hamiltonian term. More thorough treatments of the quadrupolar interactions are provided in several texts [28, 32, 100, 101], notably in Jerschow's excellent review [27].

5.2 First-order quadrupolar interaction

The *first-order* Hamiltonian

$$H_Q^{(1)} = \nu_Q^{(1)}(\Omega)\frac{1}{2}\{3S_z^2 - S(S+1)\hat{\mathbf{1}}\}, \tag{24}$$

represents the interaction between the nuclear quadrupole moment and the EFG tensor for a quadrupolar site in one (stationary) "crystallite" orientation Ω.[e] $\hat{\mathbf{1}}$ is the unity operator. Eq. (24) assumes high-field NMR applications where the Larmor frequency ν_0 of the Zeeman Hamiltonian far exceeds the *first-order quadrupolar frequency* $\nu_Q^{(1)}(\Omega)$ [27, 28, 37, 100, 101]. This frequency is directly proportional to ν_Q (and thereby also to C_Q),

$$\nu_Q^{(1)}(\Omega) = \nu_Q f(\Omega) \tag{25a}$$

$$= \frac{C_Q\{3\cos^2\beta - 1 + \eta\cos 2\alpha\sin^2\beta\}}{4S(2S-1)}, \tag{25b}$$

but also depends on the relative orientation between the V_{zz} EFG component and the $\mathbf{B}_0$ vector direction, as encoded by the function $f(\Omega) = (3\cos^2\beta - 1 + \eta\cos 2\alpha\sin^2\beta)/2$. We will not discuss the theoretical background of the orientation-dependence (see Refs. [27, 28, 101] for details), but will merely focus on its practical consequences. For a given C_Q-value, the denominator of Eq. (25b) implies that the first-order quadrupolar effects diminish when the spin number S is increased. Note that C_Q is uniquely defined, but the expression for ν_Q differs in the literature.

While only amounting to <1% of the ^{27}Al Larmor frequency at reasonably high magnetic fields of $B_0 \geq 9.4$ T, the quadrupolar frequency remains 2–3 orders of magnitude larger than any other NMR interaction for ^{27}Al. As follows from Eq. (25b) and illustrated by Fig. 5, the NMR signals from the STs are generally separated from the CT by a few hundred kHz for typical C_Q-values of 5–10 MHz for ^{27}Al in an aluminosilicate glass. Moreover, the distinct orientations of the various crystallites broaden the NMR signals observed from powders. Owing to the second-rank characteristics of the first-order quadrupolar interaction, it may, somewhat loosely, be visualized as a huge chemical shift anisotropy; the only essential difference is that the

[e] In the absence of a clear/concise terminology for amorphous materials, we use the (somewhat misleading) term "crystallite" for *any* (~μm-sized) fragment of a powder, regardless of whether the underlying structure is ordered or disordered.

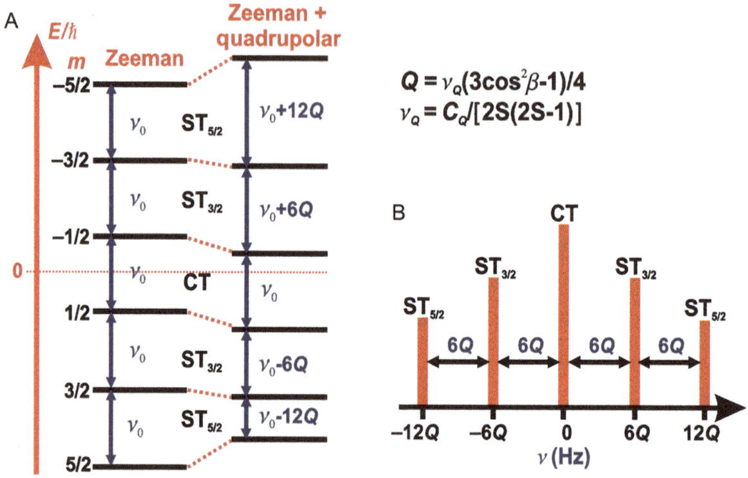

Fig. 5 (A) The energy-shifts of the Zeeman levels induced by the first-order quadrupolar interaction, relevant for a $S = 5/2$ site in a single crystallite in a strong magnetic field. The EFG tensor is assumed to be axially symmetric ($\eta = 0$), with the V_{zz} component subtending the angle β with respect to $\mathbf{B}_0$. (B) A schematic NMR spectrum, with each peak identified with its respective CT/ST transition; the frequency axis (in Hz) is expressed as multiples of the constant $Q = \nu_Q(3\cos^2\beta - 1)/4$ that is proportional to the quadrupolar frequency [Eq. (22)]. Note that the CT resonance is independent of ν_Q. The spectrometer reference frequency coincides with the Larmor frequency.

CSA Hamiltonian depends linearly on the operator S_z while $H_Q^{(1)}$ features a S_z^2-dependence. The periodic time-modulation induced by MAS on the CSA results in a manifold of spinning sidebands (Section 2.3). However, even the fastest MAS rates $\nu_r \approx 130$ kHz that are routinely accessible today do not permit a complete averaging of the first-order ^{27}Al quadrupolar interaction, in practice leading to an extensive ssb formation across the MAS NMR spectrum [17, 19], as depicted by Fig. 6.

The orientation-dependence leads to resonance-broadening from powders, except for any symmetric $\pm m \rightarrow \mp m$ transition that *remains independent of* $\nu_Q^{(1)}$ [27, 28, 101], with the CT being the most important example.[f] The invariance of the CT to the quadrupolar interaction to first-order perturbation theory is of pivotal importance for practical NMR applications to

[f] This property follows directly from the operator form of $H_Q^{(1)}$ in Eq. (24): only unity operators $\hat{\mathbf{1}}^{m,-m}$ appear when $H_Q^{(1)}$ is decomposed as a sum of fictitious spin-1/2 operators [104, 105] that connect symmetric $|m\rangle\langle -m|$ eigenstates of S_z: $H_Q^{(1)} = \sum_{m=1/2}^{S} \omega_Q^{(1)}\{3m^2 - S(S+1)\}\hat{\mathbf{1}}^{m,-m}$.

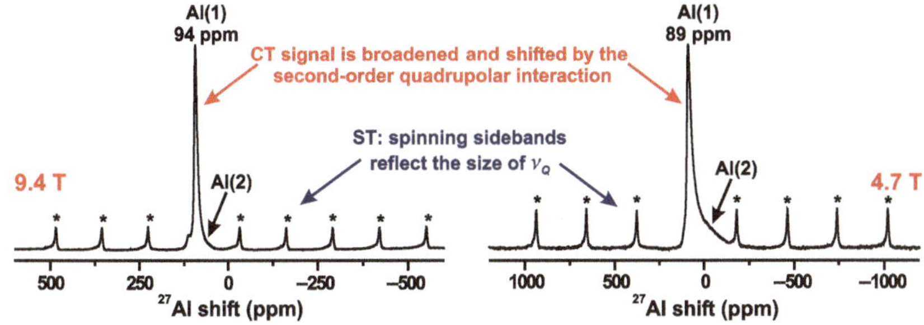

Fig. 6 ^{27}Al MAS NMR spectra recorded at two distinct magnetic fields of $B_0 = 9.4$ T (*left panel*) and $B_0 = 4.7$ T (*right panel*) from the oxynitride "S phase" $Ba_2Al_2Si_{10}N_{14}O_4$ [106, 107]. The spectra are zoomed around the CT-signal region. The structure comprises two nonequivalent Al sites, Al(1) and Al(2), the latter featuring $C_Q = 4.2$ MHz. Although the former site is associated with a small coupling constant of $C_Q = 2.0$ MHz, it is sufficiently large to produce noticeably different NMR peak-maximum positions in spectra obtained at distinct external magnetic fields, where both peaks appear lower than the isotropic chemical shift of $\delta_{iso} = 97$ ppm. Note that the NMR spectra are presented over the same frequency-span. The MAS rates were 13.492 kHz and 14.596 kHz for the data collected at 9.4 T and 4.7 T, respectively. *Asterisks* mark spinning sidebands from the first-order quadrupolar interaction.

half-integer spins: this property renders data interpretation feasible from ^{27}Al sites that experience large quadrupolar interactions of $\nu_Q \sim 1$ MHz that otherwise would be intractable. Compare for instance with the $S = 1$ nucleus ^{14}N that lacks a central transition but is also ≈100% naturally abundant and its $\pm$ 1Q coherences evolve under similarly-sized quadrupolar frequencies as ^{27}Al. The ^{14}N NMR signal broadening inflicted by $H_Q^{(1)}$ in static powders—or the pronounced ssb formation with very weak ssb intensities under MAS conditions—makes ^{14}N NMR very challenging and it has not found widespread usage until more recently.

5.3 Second-order quadrupolar interaction

5.3.1 General features

The *second-order* quadrupolar Hamiltonian of a stationary EFG orientation Ω is complicated [27, 28, 37], but may be expressed schematically according to:

$$H_Q^{(2)} \sim \frac{\nu_Q^2}{\nu_0} \left\{ \begin{array}{l} A_{00} S_z \left[3S_z^2 - S(S+1)\hat{\mathbf{1}}\right] + A_{20}(\Omega) S_z \left[-12S_z^2 + 8S(S+1)\hat{\mathbf{1}} - 3\hat{\mathbf{1}}\right] + \\ A_{40}(\Omega) S_z \left[-34S_z^2 + 18S(S+1)\hat{\mathbf{1}} - 5\hat{\mathbf{1}}\right] \end{array} \right\}. \quad (26)$$

Its essentials are (*i*) the ν_Q^2/ν_0 dependence that renders the second-order interaction frequency much smaller (~kHz) than its first-order counterpart $\nu_Q^{(1)}$;[g] (*ii*) $H_Q^{(2)}$ comprises *both isotropic* ($\sim A_{00}$) and *anisotropic* (Ω-dependent) contributions to the NMR observables that MAS is unable to eliminate completely, because the anisotropy transforms as a sum of a second [$\sim A_{20}(\Omega)$] and fourth [$\sim A_{40}(\Omega)$] rank tensors [13, 17, 27, 28, 37]. These features are examined further below. To emphasize its key properties, we express the net second-order quadrupolar frequency (in Hz) as follows:

$$\nu_Q^{(2)}(\Omega) = \nu_{\text{iso}}^{Q} + \nu_{\text{aniso}}^{Q}(\Omega) \sim \frac{C_Q^2 f(S)}{\gamma B_0 [S(2S-1)]^2}, \tag{27}$$

where *f*(*S*) is a function of the spin quantum number *S*. The consequences of each ν_{iso}^{Q} and $\nu_{\text{aniso}}^{Q}(\Omega)$ term is discussed in the remaining of this section.

While the second-order quadrupolar effects may convey important structural information, they often present significant hurdles for the identification and quantification of the unique resonances from the structure. We briefly comment on the important consequences of the right-hand side of Eq. (27) that summarizes how $\nu_Q^{(2)}$ scales with the spin parameters and the external field B_0: the second-order quadrupolar terms grow *quadratically* with the quadrupolar coupling constant, while for a fixed C_Q-value, they attenuate as either *S* or B_0 is increased. Consequently, NMR experimentation targeting quadrupolar nuclei is markedly facilitated by utilizing high magnetic fields $B_0 \gtrsim 14$ T, particularly for nuclides with low magnetogyric ratios γ and/or low spin numbers ($S=1$ and $S \leq 3/2$), for which the second-order effects become substantial even for modest quadrupolar coupling constants of a few MHz. Note, however, that albeit the denominator of Eq. (27) suggests an inverse $S^2(2S-1)^2$ dependence of $\nu_Q^{(2)}(\Omega)$, once the factor *f*(*S*) of the nominator is also accounted for, an approximate scaling of $\sim (S^{-3} + S^{-2})$ against *S* is observed in practice.

5.3.2 Anisotropic part

The *anisotropic* frequency $\nu_{\text{aniso}}^{Q}(\Omega)$ broadens the CT NMR signal, as well as those observed from all STs. The explicit expression for ν_{aniso}^{Q} is not considered here, but we note that it depends on the precise MAS/static

[g] Note the *quadratic* dependence on ν_Q in Eq. (26). This notation must not be confused with the perturbation orders specified within parentheses for the frequencies $\nu_Q^{(1)} = \nu_Q f(\Omega)$ [Eq. (25b)] and $\nu_Q^{(2)}$.

experimental condition of the powdered sample, as well as on *which* 1QC that is considered; CT, $ST_{3/2}$ or $ST_{5/2}$. The effects of MAS on the second-rank component, $\sim A_{20}(\Omega)$ in Eq. (26), is formally identical to the scenarios of the CSA or first-order quadrupolar interactions. However, the fourth-rank contribution [$\sim A_{40}(\Omega)$] only becomes *scaled,*[h] meaning that a sizable broadening remains, regardless of the MAS rate [17, 43]. Altogether, this leads to a CT MAS NMR peak being $\sqrt{92/7} \approx 3.6$ times narrower compared with its static powder counterpart [39, 108], yet remaining several kHz wide, as illustrated in Fig. 7. The second-order quadrupolar Hamiltonian is therefore a nuisance as it partially spoils the favourable property that the CT is devoid of quadrupolar broadenings to first-order perturbation theory. The anisotropy

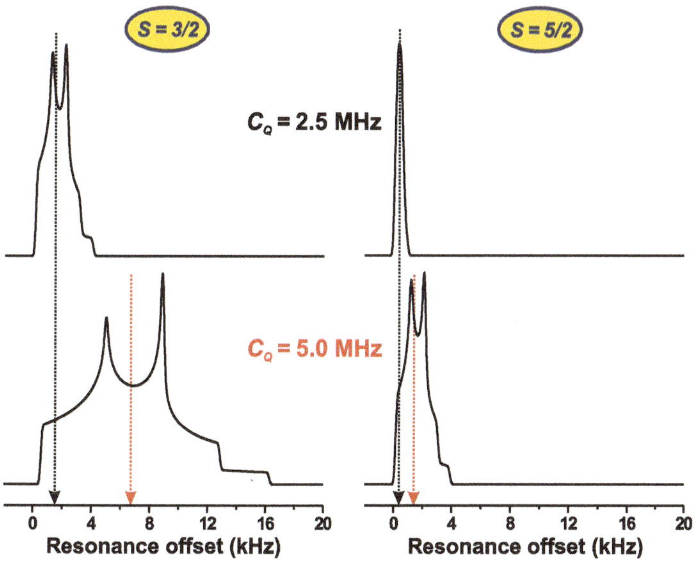

Fig. 7 Simulated CT peakshapes from $S = 3/2$ (*left panel*) and $S = 5/2$ (*right panel*) quadrupolar sites associated with an asymmetry parameter of $\eta = 0.50$ and quadrupolar coupling constants of 2.5 MHz (*top row*) and 5.0 MHz (*bottom row*). The centre-of-gravity (COG) resonance frequencies from quadrupolar sites with $C_Q = 2.5$ MHz and $C_Q = 5.0$ MHz are indicated with black and red dotted arrows, respectively. Throughout, the isotropic *chemical* shift is at the zero frequency, meaning that the isotropic *quadrupolar* shift coincides with the COG frequency coordinate. Note that the second-order quadrupolar broadening scales inversely with S but quadratically with C_Q.

[h] The scaling is given by $P_4(\beta_m) = (35\cos^4\beta_m - 30\cos^2\beta_m + 3)/8 \approx -0.39$. The negative sign implies that the MAS NMR powder spectrum appears mirrored around its centre-of-gravity frequency/shift relative to the spectrum observed from a static powder.

of the second-order quadrupolar Hamiltonian is sufficient to severely compromise the MAS NMR spectral resolution from structures comprising several nonequivalent quadrupolar sites, particularly so for amorphous structures. Therefore, intense efforts were spent during the 1980s and 1990s to find techniques capable of eliminating the second-order broadenings (see Section 9).

Both the second- and fourth-rank contributions to $\nu_{\text{aniso}}^{\text{Q}}(\Omega)$ become periodically modulated by the action of MAS and generate spinning sidebands of the CT NMR signals [37, 39]. However, for ^{27}Al in AS glasses at moderately high magnetic fields and reasonably fast MAS ($\nu_r > 15$ kHz), these CT-stemming spinning sidebands are much weaker than their centreband, because the magnitude of $\nu_{\text{aniso}}^{\text{Q}}(\Omega)$ typically only amounts to ≈20–30 kHz. For $\nu_r > 40$ kHz, they vanish altogether, such that the entire CT-signal intensity becomes concentrated in the centreband (this statement must not be confused with the CT peak-*broadening*, which persists).

Yet, for $S = 5/2$ nuclei (e.g., ^{27}Al and ^{17}O), the $\text{ST}_{3/2}$ NMR-peak width only amounts to ≈30% of its CT counterpart, thereby offering significant resolution improvements, provided that a sufficient NMR signal-to-noise ratio (S/N) may be achieved for the weak ssb peaks [19, 109, 110]; see Fig. 6. Another nice property of the ST allows for determining the ^{27}Al chemical shift and quadrupolar interaction parameters, as reviewed in Section 10.2. Table 1 lists the relative second-order isotropic shifts and broadening factors for various spin quantum numbers and $\pm$ 1Q transitions. Note that relative to the CT, the second-order broadening is *only substantially* lower for the $\text{ST}_{3/2}$ of $S = 5/2$ nuclei, as opposed to its $\text{ST}_{5/2}$ counterpart, or the scenario of any other spin-S transition (besides the practically less relevant $\text{ST}_{5/2}$ transition for $S = 9/2$). In contrast, the $\text{ST}_{5/2}$ NMR signals of $S = 5/2$ present a very significant broadening that is 1.8 and 6.3 times larger than that of the CT and $\text{ST}_{3/2}$ transition, respectively. This feature makes the detection of the $\text{ST}_{5/2}$ sidebands stemming from the MAS-modulation of $\nu_{\text{Q}}^{(1)}(\Omega)$ very difficult to observe in MAS NMR spectra, unless the magic-angle setting is very precise and the S/N is high; see Fig. 6.

5.3.3 Isotropic part

Each transition of a half-integer spin exhibits an *isotropic* second-order *quadrupolar* shift, often referred to as the *quadrupolar-induced shift* (QIS). As for $\nu_{\text{aniso}}^{\text{Q}}(\Omega)$, its precise value depends on the particular $|m\rangle \rightarrow |m'\rangle$ transition considered, onwards expressed as $m \rightarrow m'$. The general expression for an arbitrary $m \rightarrow m-1$ (i.e., -1Q) transition of a spin-S is [19]

Table 1 Fractional population, nutation frequency, and second-order isotropic shift/anisotropic broadening of each spin-S -1Q transition.

S	m[a]	F_m[b]	$\nu_{nut}^{[m,\, m-1]}/\nu_1$[c]	$\delta_{iso}^{Q}[m, m-1]/\delta_{iso}^{Q}[CT]$[d]	$\nu_{aniso}^{Q}[m, m-1]/\nu_{aniso}^{Q}[CT]$[e]
3/2	1/2	0.400	2	1	1
	3/2	0.600	$\sqrt{3} \approx 1.73$	−2.000	−0.889
5/2	1/2	0.257	3	1	1
	3/2	0.457	$2\sqrt{2} \approx 2.83$	−0.125	0.292
	5/2	0.286	$\sqrt{5} \approx 2.24$	−3.500	−1.833
7/2	1/2	0.190	4	1	1
	3/2	0.357	$\sqrt{15} \approx 3.87$	0.400	0.622
	5/2	0.286	$2\sqrt{3} \approx 3.46$	−1.400	−0.511
	7/2	0.166	$\sqrt{7} \approx 2.65$	−4.400	−2.400
9/2	1/2	0.152	5	1	1
	3/2	0.291	$2\sqrt{6} \approx 4.90$	0.625	0.764
	5/2	0.255	$\sqrt{21} \approx 4.58$	−0.500	0.056
	7/2	0.194	4	−2.375	−1.125
	9/2	0.109	3	−5.000	−2.778

[a]Refers to the -1Q transition $m \rightarrow m-1$.
[b]Fractional signal intensity present in the CT ($1/2 \rightarrow -1/2$) or each ST_m ($m \rightarrow m-1$ and $-m+1 \rightarrow -m$) out of the total -1QC signal intensity: $F_m \equiv F_{m,\, m-1} + F_{-m+1,\, -m,}$, with $F_{m,\, m-1}$ given by Eq. (37).
[c]Ratio between the nutation frequency for selective excitation of the given transition (Eq. 43) relative to the nonselective scenario [Eq. (38)].
[d]The isotropic second-order quadrupolar shift relative to the CT: $\delta_{iso}^{Q}[m,\ m-1]/\delta_{iso}^{Q}[1/2,\ -1/2]$. A negative value means a shift towards *higher* δ-values (note that $\delta_{iso}^{Q}[1/2,\ -1/2] < 0$ throughout). The data were reproduced from Samoson [19].
[e]The anisotropic second-order quadrupolar broadening of transition $m \rightarrow m-1$ relative to that of the CT: $\nu_{aniso}^{Q}[m,\ m-1]/\nu_{aniso}^{Q}[1/2,\ -1/2]$. A negative value implies a mirrored powder pattern relative to that of the CT. The data were reproduced from Samoson [19].

$$\nu_{iso}^{Q}[m, m-1] = \frac{3C_Q^2(1+\eta^2/3)[3+9m(m-1)-S(S+1)]}{40\nu_0 S^2(2S-1)^2} \text{ [in Hz]}, \quad (28)$$

which for the central ($1/2 \rightarrow -1/2$) transition becomes

$$\nu_{iso}^{Q} = \frac{3C_Q^2(1+\eta^2/3)[3-4S(S+1)]}{160\nu_0 S^2(2S-1)^2} \text{ [in Hz]}, \quad (29)$$

$$\delta_{\text{iso}}^{\text{Q}} = \nu_{\text{iso}}^{\text{Q}} / \nu_0 \text{ [in ppm]}. \tag{30}$$

The isotropic character of $\nu_{\text{iso}}^{\text{Q}}$ means that it is *independent on orientation* and the expression holds both for static powders and those undergoing MAS. Hence, in the presence of significant second-order quadrupolar interactions, the *centre-of-gravity* (CG) peak position in the NMR spectrum of any $m \rightarrow m - 1$ transition is given by the *sum* of the *isotropic chemical* (δ_{iso}) and *isotropic second-order quadrupolar* ($\delta_{\text{iso}}^{\text{Q}}$) shifts:

$$\delta^{\text{CG}} = \delta_{\text{iso}} + \delta_{\text{iso}}^{\text{Q}}. \tag{31}$$

For the CT, $\delta_{\text{iso}}^{\text{Q}}$ is *negative* (regardless of the spin quantum number *S*), meaning that the isotropic quadrupolar shift displaces the centre-of-gravity shift (δ_{CG}) to a ppm-value that is *lower* than δ_{iso}; see Figs 6 and 7.

The following three important consequences of Eq. (31) are worth stressing: (*i*) The shift value at the maximum peak amplitude (δ_{max}) does *not* coincide with the isotropic chemical shift; (*ii*) Consequently, the axes of the NMR spectra from quadrupolar nuclei are herein labelled with "shift"—as opposed to "chemical shift"—as these entities are *not* equal owing to the presence of second-order quadrupolar shifts. (*iii*) When expressed in frequency units, the quadrupolar-induced shift in frequency units ($\nu_{\text{iso}}^{\text{Q}}$) depends inversely on ν_0 [Eq. (29)], meaning that $\delta_{\text{iso}}^{\text{Q}} = \nu_{\text{iso}}^{\text{Q}} / \nu_0$ scales as B_0^{-2} [see Eq. (27) and (30)]. Hence, the "shift scale" [see Eq. (3)] that was introduced to *apparently* remove the linear field-dependence of the chemical shift interaction is of much less utility when dealing with quadrupolar nuclei. More generally, Eq. (27) reveals that *both* the width and the position of the NMR peak from a quadrupolar spin-site depends directly on the exact magnetic field (B_0-value) used to record the NMR spectrum, as is evident from Fig. 6.

The separate extraction of δ_{iso} and $\delta_{\text{iso}}^{\text{Q}}$ [Eq. (31)] by MAS NMR spectra analysis is not trivial and will be reviewed in Section 10. Due to these problems, much of the early literature before ~1990 on ^{27}Al MAS NMR applications to glasses reported the "centre of gravity" or "peak-maximum position" rather than the (desirable) isotropic chemical shift. The frequent use of low magnetic fields ($B_0 < 9.4$ T) in those studies implied that $\delta_{\text{iso}}^{\text{Q}}$ often exceeded 30 ppm, thereby leading to significant differences between δ^{CG} and δ_{iso}. As pointed out in the milestone paper by Lippmaa et al. in 1986: "*Ignoring the second-order quadrupolar shifts ... would introduce errors exceeding one-tenth of the whole ^{27}Al chemical shift range in aluminosilicates, large enough to swamp most structural effects in the ^{27}Al MAS NMR spectra*" [20].

5.4 Distributions of chemical shift and quadrupolar parameters in glasses

The lack of long-range structural order in AS glasses translates into variations of *T*–O bond-lengths and *T*–O–*T* bond angles among the various TO_4 sites with $T=\{Si, Al\}$. Consequently, the *distribution* of such structural parameters of each SiO_4 and AlO_4 ensemble (as well as AlO_5/AlO_6 polyhedra, if present) translates into spreads of their corresponding chemical shifts and $\{C_Q, \eta\}$ parameters. For a spin-1/2 nucleus, the observed broad and Gaussian-shaped peaks stem exclusively from chemical shift distributions. For instance, the ^{29}Si resonance from vitreous silica is a Gaussian peak with full-width at half maximum (FWHM) height of 11–12 ppm, which may be contrasted with the FWHM $\approx$ 0.3–0.5 ppm typically resulting for well-ordered SiO_2 polymorphs [39, 43]; when recorded at a well-shimmed magnet, the ^{29}Si NMR peak widths of crystalline silicates are limited only by spin–spin (T_2) relaxation and strains/imperfections of the microcrystals that may cause a minor spread of resonance frequencies.

As discussed further in Section 10.3, NMR peak deconvolution by numerical simulations may be employed for extracting the underlying average NMR parameters and their distribution widths. The chemical-shift distribution is usually modelled by a Gaussian distribution, resulting in an average value $\bar{\delta}_{iso}$ (henceforth denoted "δ_{iso}" for simplicity) and an FWHM of the NMR peak. Yet, this assumption has been questioned [111].

Yet, quadrupolar nuclei manifest further complications owing to the *additional* distribution of quadrupolar parameters, where we will solely consider CT MAS NMR responses. From Eqs (27) and (31) follow that for increasing quadrupolar coupling constants, the NMR signal broadens and concurrently shifts towards lower ppm values by δ_{iso}^{Q}. Consequently, the largest C_Q values over the distribution manifest as a tail of the NMR peakshape, as illustrated by the ^{27}Al NMR spectra displayed in Fig. 8 that were recorded from various AS phases associated with increasing degree of structural disorder. The spectrum in (A) was obtained from a perfectly ordered structure, SrSiAlD [112] that exhibits the characteristic second-order powder pattern. Fig. 8B shows the ^{27}Al NMR spectrum from a layer mineral chlorite [113], which consists of sheets of interconnected SiO_4/AlO_4 groups alternating with layers comprising MgO_6/AlO_6 octahedra. Despite that the Al/Si ordering in the tetrahedral sheets obeys Loewenstein's rule and chlorite is formally an *ordered* phase, long-range order is absent due to stacking disorder of the various layers. Notably, the presence of (weak) disorder is revealed by the "tail" of the $^{27}AlO_6$ resonance towards

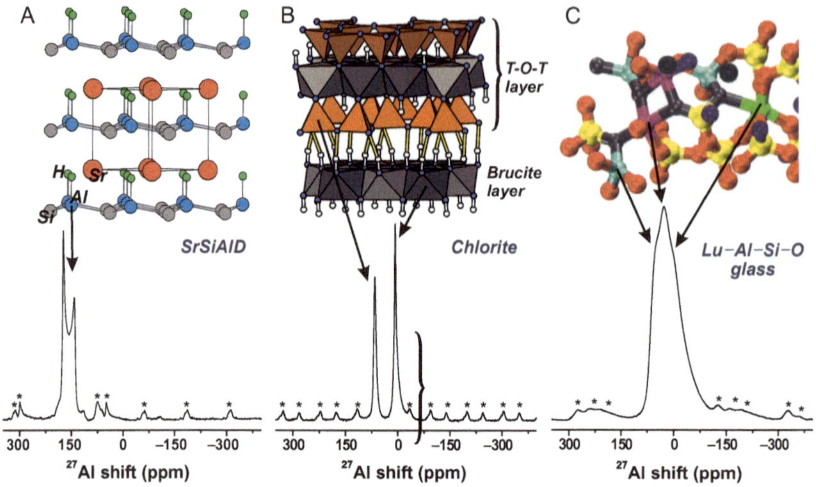

Fig. 8 ^{27}Al MAS NMR spectra recorded from (A) SrSiAlD with one crystallographic ^{27}Al site ($C_Q = 7.5$ MHz; $\eta \approx 0$) [112]; (B) chlorite, a synthetic phyllosilicate mineral of composition $(Mg_{9.0}Al_{3.48})(Si_{5.0}Al_{2.53})O_{20}(OH)_{16}$ that comprises $Al^{[4]}$ and $Al^{[6]}$ coordinations [113]; (C) a $Lu_{0.94}AlSiO_{4.91}$ glass involving three {$Al^{[4]}$, $Al^{[5]}$, $Al^{[6]}$} coordinations [72, 114]. Note the alterations of the ^{27}Al resonances for increasing structural disorder, i.e., when progressing from the perfectly ordered SrSiAlD structure that reveals a characteristic second-order quadrupolar-broadened NMR powder pattern, to the weakly disordered octahedral layers of the chlorite mineral that exhibit a "tailed" $^{27}Al^{[6]}$ resonance, to the fully amorphous AS glass structure. All NMR spectra were recorded at $B_0 = 9.4$ T with MAS rates of (A) 13.0 kHz, (B) 11.0 kHz, and (C) 21.0 kHz. Note that all spectra are presented over identical ppm-ranges; the narrower NMR signals from the chlorite mineral stems from its lower quadrupolar products (<4 MHz).

lower shifts. Moreover, as opposed to the partially disordered chlorite mineral that predominantly exhibits a distribution of quadrupolar coupling constants, the full effects of distributions in *both* chemical shifts and quadrupolar parameters on the ^{27}Al MAS NMR spectrum are witnessed in Fig. 8C, which was recorded from a Lu-based AS *glass* [72, 114]. While the spectral resolution is sufficient for evidencing the coexistence of $\{AlO_4, AlO_5, AlO_6\}$ polyhedra in the glass structure, the ^{27}Al resonances are markedly broader than those observed from the perfectly ordered or partially disordered structures in Fig. 8A and B.

A rigorous modelling of the quadrupolar-parameter distributions is not trivial. We first note that $C_{Q\eta}^2 = C_Q^2(1 + \eta^2/3)$ enters Eq. (29), as well as in many other $H_Q^{(2)}$-deriving expressions (not explicitly shown herein). This feature complicates the disentangling of C_Q and η, analogously to the unambiguous separation of δ_{iso} and δ_{iso}^Q in Eq. (31). For well-ordered structures, C_Q and η can often be determined separately, for instance by fitting the

NMR line-shape to numerically calculated NMR spectra. Note that the peakshape of the CT resonance depends on η [27, 100, 101], as may be appreciated by comparing the spectra in Figs 8A and 7 that illustrate the cases of $\eta = 0$ and $\eta = 0.5$, respectively. The ill-defined quadrupolar NMR lineshapes associated with disordered structures (such as AS glasses) implies that only the *quadrupolar product* ($C_{Q\eta}$) is experimentally accessible. It is defined from the quadrupolar coupling constant C_Q and the asymmetry parameter η of the EFG tensor as follows:

$$C_{Q\eta} = C_Q\sqrt{1 + \frac{\eta^2}{3}}. \tag{32}$$

We remind that specific values of quadrupolar constants/products discussed herein from glasses concern each respective *average* value over its associated distribution, i.e., strictly implying $\bar{C}_Q$ and $\bar{C}_{Q\eta}$, respectively. Moreover, the range of definition of the asymmetry parameter ($0 \leq \eta \leq 1$) [34] provides the following constraint of the quadrupolar coupling constant:

$$0.866C_{Q\eta} \leq C_Q \leq C_{Q\eta}. \tag{33}$$

Various models have been proposed to handle the $\{C_Q, \eta\}$ parameter distributions. One option involves generating normal distributions of two EFG tensor components of the $\{V_{xx}, V_{yy}, V_{zz}\}$ triplet, whereas the third one is deduced by the constraint $V_{xx} + V_{yy} + V_{zz} = 0$, from which each distribution of C_Q and η is generated [115]. Alternatively, η is (arbitrarily) assumed to be fixed [44, 116]. Simpler approaches (albeit not physically sound, as discussed by de Lacaillerie [117]), assume a Gaussian distribution of C_Q, while the asymmetry parameter is either kept constant [118–121] or distributed independently [122]. The currently most popular distribution is that of Czjzek et al. [123], first implemented for analyzing Mössbauer spectra and later introduced to NMR on quadrupolar nuclei by Massiot et al. [124, 125]. Fig. 9 illustrates how the convolution of two independent Gaussian and Czjzek distributions in δ_{iso} and $C_{Q\eta}$, respectively, naturally account for the characteristic ^{27}Al MAS NMR peakshapes observed from an amorphous material. However, while independent $\{C_Q, \eta\}$ distributions are normally assumed, they are not universally applicable, for instance if (a priori unknown) correlations exist between the chemical shift and EFG parameters.

The Czjzek distribution is implemented in several public software packages for fitting MAS NMR spectra [125, 127–129]. The Czjzek model was recently extended to separate the contributions from the local coordination

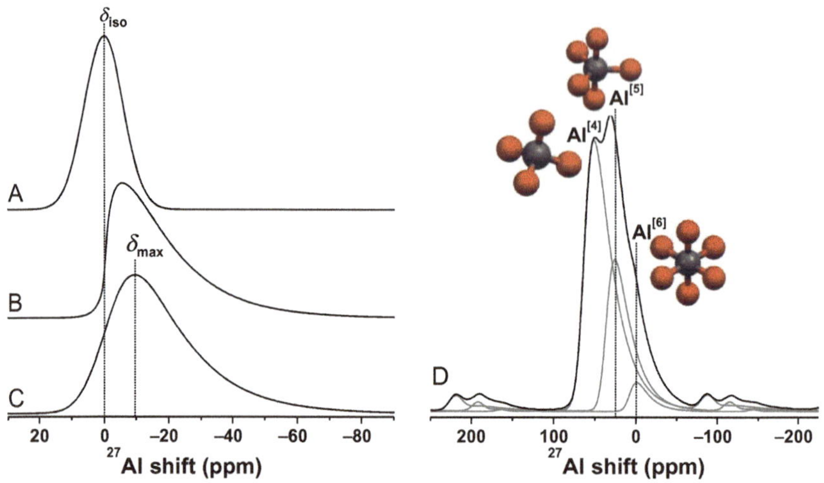

Fig. 9 (A)–(C) Numerically simulated MAS NMR peakshapes from an ^{27}Al site ($\delta_{iso} = 0$ ppm at $B_0 = 14.1$ T), illustrating the effects from distributions over (A) a Gaussian isotropic chemical shifts with a FWHM = 12 ppm or (B) a Czjzek distribution of quadrupolar products [123–125], with $\bar{C}_{Q\eta} = 6.5$ MHz. (C) The results of (A) and (B) combined. Note the shift of peak maxima of (B) and (C) relative to each other and (A). (D) Experimental ^{27}Al MAS NMR spectrum from a Sc_2O_3–Al_2O_3–SiO_2 glass [126], shown together with a spectral deconvolution result revealing $Al^{[4]}$, $Al^{[5]}$ and $Al^{[6]}$ coordinations in the structure. The relative area beneath each peak is proportional to its site population, corresponding to 65% AlO_4, 31% AlO_5, and 4% AlO_6 [126].

environments and those of longer-range effects to the effective $C_{Q\eta}$-value [130]. Notably, de Lacaillerie et al. [117] demonstrated that the Czjzek distribution produces an average asymmetry parameter of $\bar{\eta} \approx 0.61$, which is close to that of $\eta \approx 0.50$ [131–133] sometimes invoked as a fixed value together with a distribution of C_Q in amorphous materials.

6. Sensitivity considerations for ^{27}Al NMR

The *receptivity* ($\mathcal{R}_S$) of a spin-S nuclide with natural abundance a_S (in %) and magnetogyric ratio γ_S scales according to [100]

$$\mathcal{R}_S \sim a_S \gamma_S^3 S(S+1). \tag{34}$$

Tabulated data of a_S, γ_S and $\mathcal{R}_S$ are available in Refs. [100, 134] for various nuclides. However, worth stressing is that the receptivity concept is formulated from solution NMR standpoints: it does not account for other important NMR properties such as quadrupolar interactions or longitudinal (T_1) relaxation times in the solid-state and conclusions drawn only from receptivity-values may sometimes be very misleading. In practice, it is the NMR signal strength from a given sample that matters—i.e., the NMR signal "*sensitivity*" that $\mathcal{R}_S$ aims at predicting.

For experimentation focussing on CT detection, it should be remembered that only a *fraction* of the total -1QC NMR signal from a spin-S nucleus resides in the CT and that this fraction reduces as S is increased [19, 135–137]:

$$F_{1/2,-1/2} = \frac{3(S+1/2)}{4S(S+1)}. \qquad (35)$$

Now there is yet another caveat: the value of $F_{1/2,\,-1/2}$ in Eq. (35) is of little practical use, as it is only realized under so-called *nonselective* rf excitation conditions, whereas if the CT-signal is *selectively excited* using weak rf pulses (see Section 7), a factor of $(S+1/2)$ lower NMR signal intensity is observed [19, 135–137]:

$$\mathcal{F}^{\mathrm{CT-sel}}_{1/2,-1/2} = \frac{3}{4S(S+1)} \ [\text{CT-selective excitation}]. \qquad (36)$$

Consequently, the factor $S(S+1)$ cancels when Eqs (34) and (36) are combined and the *receptivity becomes independent on S if only the CT is detected*. The *fractional* signal *intensity* ($F_{m,m-1}$) present in the $m \rightarrow m-1$ transition out of the total -1QC signal intensity is given by [135–137]

$$F_{m,m-1} = \frac{3[S(S+1) - m(m-1)]}{2S(S+1)(2S+1)}, \qquad (37)$$

with the normalization condition $\sum_{m=-S+1}^{S} F_{m,m-1} = 1$. Note that the *total* satellite transition intensity $F_m \equiv F_{m,m-1} + F_{-m+1,-m}$, is a sum of two contributions, each evaluated *via* Eq. (37). For instance, $F_{3/2} = F_{3/2,1/2} + F_{-1/2,-3/2}$ constitutes that of $ST_{3/2}$ (see Fig. 4 and Section 5.1). The fractional intensities F_m resulting for various spin numbers and transitions are summarized in Table 1.

Besides originating from the frequent occurrence of Al in crystalline and amorphous aluminosilicates and other materials of great technological and geological importance, the very widespread use of ^{27}Al NMR stems from the high receptivity *and* sensitivity of ^{27}Al, thanks to its combination of 100% natural abundance (n.a.), a relatively high magnetogyric ratio

($\gamma_{Al} = 6.976 \cdot 10^7$ rad $s^{-1}T^{-1}$; 26% of γ_H), and a moderately high spin-quantum number ($S = 5/2$). Moreover, these favourable properties are usually not spoilt by too extensive quadrupolar interactions, because sufficiently high local symmetries are observed for the various AlO_p polyhedra in oxide-based materials, thereby alleviating huge second-order quadrupolar broadenings potentially arising from the sizable ^{27}Al electric quadrupolar moment ($Q = 14.66$ fm^2).

The problem of poor NMR receptivity of low-γ nuclei with half-integer spin is frequently exacerbated by the accompanying sensitivity reduction from large second-order quadrupolar broadenings [see Eq. (27)]. It is instructive to compare the sensitivity of ^{27}Al with other spin-5/2 nuclei, e.g., ^{17}O and ^{25}Mg that feature $a_S = 0.04\%$ and $a_S = 10\%$, respectively. Their low abundances constitute significant challenges, but even when exploiting high ^{17}O/^{25}Mg isotopic enrichment, they offer lower sensitivity than ^{27}Al due to the cubic dependence of the receptivity on the magnetogyric ratio in Eq. (34); the ratios $\gamma_{Al}/\gamma_O = 1.92$ and $\gamma_{Al}/\gamma_{Mg} = 4.26$ *alone* translate in $\approx$7 and $\approx$77 times higher receptivity of ^{27}Al relative to ^{17}O and ^{25}Mg, respectively. ^{25}Mg is very challenging for NMR owing to its high quadrupolar moment ($Q = 19.94$ fm^2) combined with the unfavourable C_Q^2/γ scaling of the $\nu_{\text{aniso}}^{Q}(\Omega)$-stemming broadenings of the CT resonances. In contrast, ^{17}O NMR on AS glasses (typically prepared with ^{17}O enrichment to 20–40%) is nowadays routinely utilized with good S/N ratio. Here the very low ^{17}O quadrupole moment ($Q = 2.56$ fm^2) is advantageous: for crystalline and amorphous aluminosilicates, quadrupolar coupling constants typically conform to the 2–6 MHz range [43], yielding significantly narrower CT ^{17}O NMR peaks (relative to those of ^{27}Al) due to the lower second-order quadrupolar broadenings. Nevertheless, despite these favourable properties of ^{17}O, practical MAS NMR data collection from glasses is generally much more time-consuming than for ^{27}Al.

We next illustrate the problems with NMR on nuclei situated in low-symmetry environments *and* exhibiting sizable electric quadrupole moments, by comparing the feasibility of detecting $^{27}AlO_p$ polyhedra in oxide-based materials with their $^{45}ScO_p$ and $^{139}LaO_p$ analogues. The $S = 7/2$ character and $\approx$100% n.a. of both the ^{45}Sc and ^{139}La nuclides make them *formally* more receptive than ^{27}Al, although not for experimentation targeting the CT. Moreover, their higher quadrupolar moments of $|Q| = 22.0$ fm^2 (^{45}Sc) and $Q = 20.0$ fm^2 (^{139}La) may induce very substantial quadrupolar coupling constants spanning tens of MHz. Consequently, while ^{45}Sc MAS NMR was only recently demonstrated on glasses [126, 138, 139], the few current ^{139}La

MAS applications involved either highly symmetric La environments in ordered/crystalline structures [140–144], or NMR on static powders of metallocenes [145] or ceramics and other oxide-based crystalline phases [146, 147], for which very high ^{139}La C_Q-values around 50 MHz and 30–90 MHz were reported, respectively. While MacKenzie *et al.* observed weak ^{139}La NMR signal intensities from precursor gels used to prepare $LaAl_{11}O_{18}$ [53], no ^{139}La MAS NMR results are to our knowledge reported from glasses, presumably owing to a substantial CT-signal broadening stemming from very large C_Q-values and the modest magnetogyric ratio of ^{139}La (which is close to that of ^{17}O). Moreover, significant distributions of both $C_{Q\eta}$ and δ_{iso} parameters are expected to deteriorate the NMR signal resolution and sensitivity further.

7. Rf pulses and quadrupolar interactions

When an rf pulse of magnitude B_1 is applied to an ensemble of spin-1/2 nuclei, each spin rotates around the rf field vector at a rate given by the *nutation frequency* $\nu_{nut} = \nu_1$,

$$\nu_1 = |\gamma| B_1 / (2\pi) \tag{38}$$

that gives a net rotation (flip) angle of

$$\beta = \nu_1 \tau_p \text{ [in rad]}, \tag{39}$$

for a pulse of duration τ_p [32, 34]. As illustrated by Fig. 10, the detected NMR signal intensity grows according to

$$I(\tau_p) = I_0 \sin\{2\pi\nu_1\tau_p\}, \tag{40}$$

where its maximum I_0 reflects the relative abundance of the given site in the structure. Eq. (39) assumes that any spin evolution due to other NMR interactions may be neglected during the rf pulse. This approximation holds well for strong pulses ($\nu_1/2\pi \gtrsim$ 70 kHz) applied near resonance ($\nu_{rf} \approx \nu_j$; see Eq. (2) and Section 2.2), because the rf nutation frequency is much larger than any other spin-1/2 interaction.[i]

Unfortunately, in general this simple picture does not hold for quadrupolar nuclei, due to the dominating (first-order) quadrupolar frequency ν_Q that typically implies $\nu_Q \gtrsim \nu_1$ or $\nu_Q \gg \nu_1$. Strictly, it is the

[i] Except for the Zeeman interaction that, however, becomes irrelevant when using the customary spin dynamics description in the rotating frame [32] .

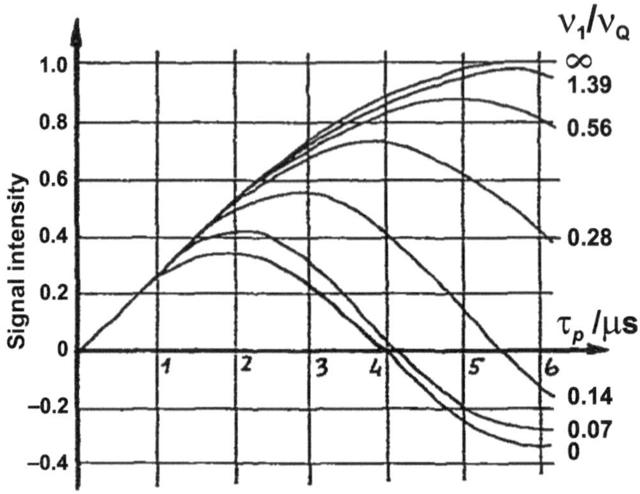

Fig. 10 Calculated CT-signal intensity observed from an $S = 5/2$ site in a single crystallite for variable pulse-lengths τ_p and ratios ν_1/ν_Q [Eq. (40)]. The simulations assumed a nutation frequency $\nu_1 = 41.67$ kHz. The cases $\nu_1/\nu_Q \sim \infty$ and $\nu_1/\nu_Q \sim 0$ correspond to nonselective and selective excitation, respectively. In the "linear" excitation regime of very short pulses, the signal buildup is independent of ν_Q. *Adapted from D. Fenzke, D. Freude, T. Frölich, J. Haase, NMR intensity measurements of half-integer quadrupolar nuclei, Chem. Phys. Lett. 111 (1984) 171–175, with permission from Elsevier.*

magnitude of the frequency $\nu_Q^{(1)}(\Omega)$ that should be compared with ν_1, but here we ignore the complications of $\nu_Q^{(1)}(\Omega)$ varying among different crystallite orientations in the powder, as well as its time-dependence under MAS conditions. The effective nutation frequency of a quadrupolar spin during an rf pulse exhibits a complicated dependence on *both* ν_1 and ν_Q, which cannot be predicted unless ν_Q is a priori known [16, 18, 148, 149]. Notably, in the case of several quadrupolar "sites" with widely differing quadrupolar frequencies, each NMR signal intensity no longer faithfully reflects the respective site abundance in the structure and the NMR spectrum is not "quantitative". The spin dynamics is only easily predicted in two limiting cases, for which the *rf nutation frequency is independent on* ν_Q:

(i) $\nu_1 \gg \nu_Q$: This limit is referred to as *nonselective* excitation, because all -1QC—associated with the CT and each ST_m—are uniformly excited with an identical rf nutation frequency as in the spin-1/2 case, i.e., all of Eqs (38)–(40) hold (Fig. 10). However, nonselective rf irradiation may rarely be arranged for solid samples and virtually never for glasses, owing to the spread in ν_Q (see Section 5.4), except for nuclei with low electric quadrupolar moments, e.g., ^{2}H ($S = 1$), ^{6}Li ($S = 1$), and ^{7}Li ($S = 3/2$), or quadrupolar sites in perfectly symmetric environments, such as ^{23}Na in the cubic structure of NaCl. Nevertheless, the "nonselective" regime is exploited routinely for rf power calibrations on aqueous solutions of salts comprising quadrupolar nuclei, such as Al^{3+}(aq) for ^{27}Al. For the latter, it is the *only* approach to determine ν_1 accurately, as even the most symmetric ^{27}Al environments encountered in solids are associated with ν_Q-values in the order of tens of kHz that are sufficiently large to deteriorate the accuracy of the ν_1 estimate. Consequently, we strongly advice against using solid Al-bearing samples for rf pulse calibrations!

(ii) $\nu_1 \ll \nu_Q$: This so-called *selective* excitation regime implies that the rf field only excites *one* transition, which in practice always constitutes the CT: "*CT-selective excitation*" [18, 135–137]. Provided that $\nu_1 \ll \nu_Q^j$ is obeyed for each quadrupolar spin-site j in a multisite structure (where "$\ll$" implies that the magnitudes of ν_1 and ν_Q^j are separated by at least one order of magnitude), all are uniformly excited and the resulting relative NMR intensities quantitatively reflect the various site abundances. However, the *CT nutation frequency* during the rf pulse is a factor of $(S + 1/2)$ higher than ν_1 [18, 136, 150, 151];

$$\nu_{\text{nut}}^{\text{CT}} = (S + 1/2)\nu_1 \ [\text{if } \nu_1 \ll \nu_Q], \tag{41}$$

meaning that the magnetization vector associated with the CT from a spin-S ensemble rotates $S + 1/2$ times faster than the rate observed at a "nonselective" condition, as illustrated by Fig. 10 for $S = 5/2$. With the maximum attainable signal intensity from Eq. (36), the CT-signal depends on τ_p according to [c.f. Eq. (40)]:

$$I_{\text{CT}}(\tau_p) = \frac{3}{4S(S+1)} \sin\left\{2\pi(S + 1/2)\nu_1\tau_p\right\}. \tag{42}$$

Eq. (41) is a special case of the more general expression

$$\nu_{\text{nut}}^{[m,m-1]} = \sqrt{S(S+1) - m(m-1)}\,\nu_1 \ [\text{if } \nu_1 \ll \nu_Q] \tag{43}$$

for the nutation frequency associated with selective rf excitation of the $m \to m - 1$ transition. Table 1 lists the values of $\nu_{\text{nut}}^{[m,\,m-1]}$ for each S and given transition. For details, we refer to the theoretical treatments on 1QC excitation of half-integer spins presented in Refs. [136, 150, 151].

Nonselective rf excitation is practically irrelevant for solid-state NMR experimentation, and CT-selective excitation may be problematic for samples exhibiting quadrupolar sites with widely different quadrupolar products and/or spread of resonance frequencies (a particular concern for ^{27}Al NMR applications to glasses). The most general approach to simultaneously excite and detect *all* -1Q transitions *quantitatively* is to exploit strong rf powers (i.e., $\nu_1 \approx 100$ kHz) but utilizing very *short* ($\sim$0.3–0.5 μs) pulses that obey [19, 20, 39]

$$\tau_p \leq \frac{1}{12(S+1/2)\nu_1}. \tag{44}$$

Despite that the actual spin nutation frequencies are generally unknown and differ among distinct 1Q transitions and structural sites, in this so-called *linear excitation regime*, the NMR response becomes to a good approximation independent on ν_Q; see Fig. 10. Consequently, the NMR signal from each transition/quadrupolar site grows proportionally to the respective site abundances and fractional intensities $F_{m,\,m-1}$. See Refs. [19, 20, 39, 136, 150, 151] for details.

8. Practical advice for recording quantitative ^{27}Al MAS NMR spectra

Here we summarize the relevant experimental precautions/considerations for recording NMR spectra from half-integer spins, either by utilizing CT-selective excitation or short and strong rf pulses. Each of them is discussed in its dedicated subsection, including the pivotal role of the B_0 magnitude and the importance of the choice of MAS rate, the magic-angle setting, rf pulse calibrations, and the selection of relaxation delays. Although the discussion focusses on ^{27}Al NMR in AS glasses, most of the concerns/precautions apply generally to experimentation of other half-integer spins in inorganic materials, such as ^{11}B, ^{17}O, and ^{23}Na that often exhibit comparable quadrupolar frequencies as those of ^{27}Al.

8.1 Magnetic field and MAS rate

Given that the CT NMR signal is only affected by the quadrupolar interaction to second-order perturbation theory (Section 5.3), it becomes the primary target of concern when selecting the experimental MAS NMR conditions. This applies regardless of whether *only* focussing on the CT *via* selective excitation, or if aiming at a quantitative excitation of *all* distinct -1Q coherences from each nonequivalent quadrupolar site in the structure. Eq. (27) conveys that for large C_Q-values, the use of high external magnetic fields and fast MAS conditions minimize the residual quadrupolar broadening while concentrating the CT intensity in the centreband.

Considering the 4–10 MHz range of (average) ^{27}Al quadrupolar coupling constants typically encountered in AS glasses, we recommend performing the NMR experiments at $B_0 \geq 9.4$ T and $\nu_r \geq 20$ kHz to ensure MAS NMR data that allows for a meaningful analysis (Section 10.3). Lower magnetic fields emphasize second-order effects even for ^{27}Al sites with relatively small C_Q-values of a few MHz, as may be verified from the MAS NMR spectra of a *crystalline* oxynitride "S phase" [106, 107] shown in Fig. 6: despite that both ^{27}Al environments exhibit low coupling constants of $C_Q \leq 4.2$ MHz, as manifested by narrow CT MAS NMR peaks observed both at 9.4 T and 4.7 T, the broadening associated with the larger C_Q value of Al(2) remains evident at the lower magnetic field. For glasses revealing markedly higher ^{27}Al quadrupolar couplings, the bearings from insufficient MAS-narrowing of the CT resonance is well documented in earlier reviews from the 1990s [39, 100, 109, 110]: when the MAS rate is *decreased*, the ^{27}Al CT NMR peak generally *narrows*, while its maximum moves towards *higher* shifts [7, 152]. These seemingly counter-intuitive observations originate from the inability of the sample spinning to narrow the NMR responses of the largest-C_Q ^{27}Al sites, whose resonances appear either as a broad "baseline" around the narrower CT-signal-portion, or escape detection altogether. Consequently, the de-facto observed $C_{Q\eta}$-distribution becomes skewed towards the smaller couplings, with a concomitant displacement of the CT resonance towards higher shifts [7, 152]. These problems of overestimated (chemical) shifts, accompanied by an underestimation of both the $C_{Q\eta}$-values and the ^{27}Al site populations, plagued all early MAS NMR reports from AS glasses, which were typically performed at $B_0 \leq 9.4$ T and using modest MAS rates $\lesssim 15$ kHz.

For ^{27}Al MAS NMR on AS glasses, we recommend MAS rates $\nu_r > 20$ kHz if working at fields of 9.4–11.7 T, whereas even modest MAS rates around 15–20 kHz at $B_0 \geq 14.1$ T provide significantly better NMR signal resolution and more reliable parameter sets. Given the high sensitivity of ^{27}Al, utilizing 3.2 mm or 2.5 mm rotors that allows for MAS rates up to ≈ 25 kHz and ≈ 35 kHz, respectively, are usually ideal. The drawback with small rotors, however, is the sensitivity-loss, which is generally immaterial when the signal detection is limited to solely the CT, but makes NMR acquisitions time-consuming for attaining sufficient S/N of the much weaker ST spinning sidebands, or when dealing with more advanced homo/heteronuclear experimentation.

8.2 Magic-angle calibration

Besides needing the shortest possible delay between the excitation pulse and the start of the NMR signal sampling (the "dead-time"; typically, 5–7 μs), a high accuracy ($\lesssim 0.05°$ deviation) of the magic-angle setting is of pivotal importance for a faithful recording of the full spinning-sideband manifold stemming from the ST-coherence evolution under large first-order quadrupolar interactions, such as those of Fig. 11. As for the case of the CT, the peak-width of a ST ssb is *ideally* broadened solely by the MAS-surviving fourth-rank component $[A_{40}(\Omega)]$ of the second-order quadrupolar interaction [Eq. (26)]. However, since the CT is invariant to $H_Q^{(1)}$, this property holds regardless of the precise rotor-axis inclination angle β_r (relative to $\mathbf{B}_0$), whereas each ST ssb broadens by $\approx \nu_Q(3\cos^2\{\Delta\beta\} - 1)$ for a deviation $\Delta\beta$ from the exact magic angle $\beta_m = \arctan\left(\sqrt{2}\right) \approx 54.7356°$ [37]. Hence, by virtue of the large ν_Q magnitude ($\sim 10^5$–10^6 Hz), even minute deviations $\beta_r \neq \beta_m$ give large ($\approx$kHz) broadening effects that make the weak ssb peaks disappear within the NMR spectral noise. This broadening mechanism applies to any spatial NMR interaction portion that transforms as a second-rank tensor, such as the CSA and the $A_{20}(\Omega)$ part of $H_Q^{(2)}$ [Eq. (26)]. Fortunately, the relatively small size (kHz range) of these anisotropies sets much lower demands on the precise β_r-value. Consequently, the sole observation of CT-stemming signals in MAS and 3QMAS [21, 28] NMR spectra is relatively forgiving for having a precise magic-angle setting, in contrast to experiments that acquire the full ssb manifold from large-C_Q sites and rely on information from the STs, such as satellite transition spectroscopy (SATRAS) [116, 153–156] or satellite transition magic-angle spinning (STMAS) experimentation [157–159] (Section 9.2 reviews the 3QMAS/STMAS NMR techniques).

At MAS rates of 5–10 kHz, the "magic-angle" condition is readily arranged by monitoring the response of the rotational echoes from the STs of ^{79}Br ($S = 3/2$) in a powder of KBr [161]. This is performed by varying the position of the "stator" (which holds the rotor in position by flowing air or nitrogen gas), and thereby the inclination angle between the rotor axis and $\mathbf{B}_0$ [161, 162]. The periodic MAS-modulation of the quadrupolar frequency makes rotational echoes forming at every multiple of the rotor period ($n\tau_r$). Such echoes would appear perpetually in the hypothetical case of a perfect magic-angle setting and the absence of NMR signal relaxation, but deviations of the rotor axis from β_m extinct the rotational echoes for long times. Observation of echoes for $\gtrsim$12 ms permits locating the exact magic

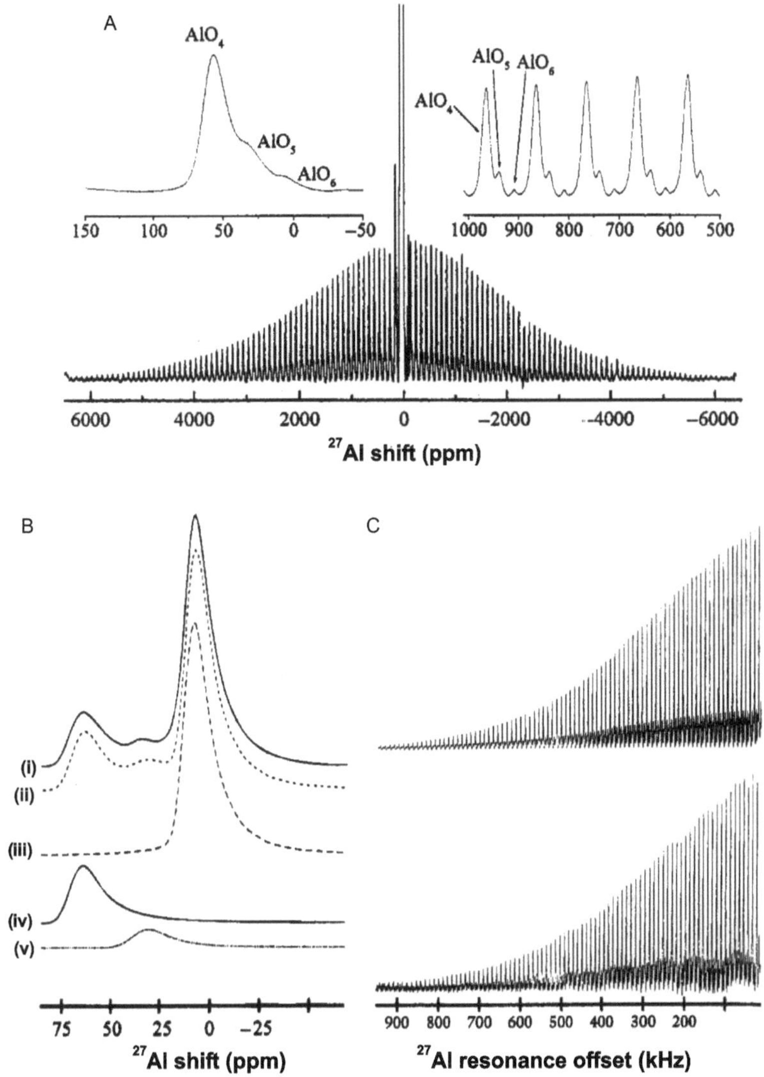

Fig. 11 Examples of SATRAS ^{27}Al NMR applications. (A) Spectrum from an aluminoborate glass of composition $0.20Na_2O–0.30Al_2O_3–0.50B_2O_3$ shown together with zooms around the CT-signal region (*left*) and a selected range of the ST sidebands (*right*) [160]. NMR peaks associated with the {AlO_4, AlO_5, AlO_6} polyhedra are indicated. (B, C) Spectra recorded from an amorphous Al_2O_3 specimen [156]. Part (B) is a zoom around the CT region of the NMR spectrum, where (i) displays the experimental result (solid trace) that is shown together with its (ii) best-fit simulation and each signal component of (iii) AlO_6, (iv) AlO_4 and (v) AlO_5 polyhedra. (C) Left part of the ST sideband manifold, with the upper and lower panel representing simulated and experimental data, respectively. *Part (A) was adapted with permission from L. Züchner, J.C.C. Chan, W. Müller-Warmuth, H. Eckert, Short-range order and site connectivities in sodium aluminoborate glasses: I. Quantification of local environments by high-resolution ^{11}B, ^{23}Na, and ^{27}Al solid-state NMR, J. Phys. Chem. B 102 (1998) 4495–4506. Copyright (1998) American Chemical Society. Parts (B) and (C) were adapted with permission from G. Kunath-Fandrei, T.J. Bastow, J.S. Hall, C. Jäger, M.E. Smith, Quantification of aluminum coordinations in amorphous aluminas by combined central and satellite transition magic angle spinning NMR spectroscopy, J. Phys. Chem. 99 (1995) 15138–15141. Copyright (1995) American Chemical Society.*

angle within deviations of $<0.05°$. This allows for the observation of many ST ssb orders in single-pulse excited MAS NMR spectra that is sufficient for numerical fitting of the CT-signal region (Section 10.3). A rough initial magic-angle adjustment is accomplished in real time by observing repeatedly acquired time-domain signal transients while adjusting the stator position. A subsequent fine-tuning must zoom in the signal-portion $\gtrsim 8$ ms, typically requiring signal averaging of 32–128 signal transients to allow observation of the weakest rotational echoes. We refer to Taylor for additional information about the magic-angle setting [162].

If stable spinning below $\nu_r \approx 10$ kHz is not feasible, such as for small (outer) rotor diameters $\leqslant 2.5$ mm, the FWHM of the first-order spinning sideband observed in the ^{79}Br MAS NMR spectrum from KBr offers a better β_m-calibration criterion: then, a good magic-angle setting translates into FWHM $\lesssim 200$ Hz.

^{23}Na NMR on $NaNO_3$ is also suitable for locating β_m; while allowing for more accurate angle-settings due to the higher quadrupolar coupling constant ($C_Q = 337$ kHz) compared with ^{79}Br in KBr ($C_Q \approx 100$ kHz), the calibration is more time-consuming owing to the slower T_1-relaxation relative to the KBr case, for which relaxation delays of ≈ 0.25 s is readily utilized, only being limited by rf heating effects of the MAS probehead.

Depending on the MAS probehead design, rotor insertion/removal may deteriorate the magic-angle setting, sufficiently much for precluding STMAS NMR implementations [159], whereas this is unproblematic if the subsequent structural analysis only requires a limited number (<20) of ST sidebands. Notably, the precise angle between the rotor axis and $\mathbf{B}_0$ may change slightly depending on the particular MAS rate used, implying that if high precision is required, the calibration should be performed at the ν_r value to be used in the subsequent experimentation [159]. Since the stator position stability varies between MAS probeheads, it is recommended to first gain familiarity with each one by verifying the absence of any significant alteration of the stator position on either multiple rotor switches, or on remounting the probehead after its removal from the magnet, as well as exploring any significant dependence on the magic-angle setting on the particular MAS rate.

For AS glasses, second-order quadrupolar broadenings always dominate the ^{27}Al NMR peak widths. Before starting any ^{27}Al MAS NMR acquisition of the targeted sample—which may require many hours for achieving sufficient S/N to allow for detailed analysis of the $ST_{3/2}$ ssb manifold (Section 10.2)—we recommend verifying that the $ST_{3/2}$ peaks-widths amount to $\approx 30\%$ of their CT counterparts; this only applies to $S = 5/2$ nuclei (see Table 1) and ignores the NMR-peak broadening from chemical shift dispersion in glasses. If a noticeable resolution improvement is not observed, the stator is likely off its optimal position. The only exception when the stator may be perfectly positioned but $ST_{3/2}$ sidebands remains broad is if the sample exhibits molecular dynamics, which is sometimes relevant for ^{27}Al NMR on microporous materials (e.g., zeolites), where motions of physisorbed water molecules (or remaining templating molecules) may accelerate spin–spin (T_2) relaxation [163].

8.3 T_1 relaxation concerns

As highlighted previously [110, 164], a common misconception is that the mere size of ν_Q ensures rapid NMR relaxation of quadrupolar nuclei. However, a fast spin-lattice (T_1) relaxation requires *molecular motions* to modulate the NMR interaction frequency at a rate comparable to the Larmor frequency; this does not apply to rigid inorganic materials, where minute paramagnetic impurities are often mainly responsible for driving relaxation [15, 165, 166]. Similar misbeliefs apply for ^{1}H: the very *rapid* relaxation frequently observed from organic molecules stem predominantly from molecular segments that exhibit fast internal rotations (such as CH_3 and NH_3 moieties) coupled with ^{1}H spin diffusion. Yet, protons within inorganic structures devoid of significant internal motions generally relax much faster than notoriously slowly relaxing spin-1/2 nuclei in glasses (e.g., ^{29}Si, ^{31}P, ^{13}C and ^{89}Y), normally allowing for using ^{1}H relaxation delays <10 s. Nevertheless, cases do exist that require very long relaxation delays (pulse delays; recycle delays). One example is the Si–O^{1}H moieties present in very low amounts in vitreous SiO_2, for which we have encountered $T_1 \approx 190$ s, which if using 90° excitation pulses would require relaxation delays of 16 minutes in the NMR signal averaging to ensure a "quantitative" ^{1}H NMR spectrum [167].

Consequently, in Al-bearing materials that only comprise ppm-levels of paramagnetic species—for instance α-Al_2O_3 [164] and many N-bearing ceramic phases [106, 107, 168]—relaxation delays of several minutes may be required for complete relaxation after a 90° excitation pulse. Also, ^{23}Na MAS NMR on sodium salts commonly utilized for NMR (e.g., $NaNO_3$, Na_2SO_4, and Na_2SO_3) normally demands delays of around 1 min to admit complete recovery of the longitudinal CT magnetization. Luckily, ^{27}Al relaxation in AS glasses is somewhat faster, generally requiring $\lesssim$30 s for complete recovery after application of CT-selective 90° pulses, whereas < 5 s are typically sufficient when recording NMR spectra by short (hard) rf pulses; here the requirement of flip angles < 10° in Eq. (44) for $S = 5/2$ helps limiting the relaxation delays to speed up the NMR signal averaging. Paramagnetic doping with Mn^{2+}, Fe^{3+} or Gd^{3+} in amounts of $\approx$0.1 wt% is often employed for accelerating the T_1 relaxation, thereby reducing the experimental relaxation delays to $<$10 s for CT-selective NMR measurements.

Notably, in stark contrast to scenarios where molecular motions *are present*, the ^{27}Al ST resonances usually reveal *larger* T_1-values than their CT counterparts [110, 169]. Owing to the weak ST ssb intensities and associated difficulties in assessing their relaxation properties, the choice of relaxation delays in the recording of the MAS NMR spectrum by "hard" rf pulses is sometimes not obvious. Moreover, the multilevel Zeeman structure of half-integer spins (see Fig. 4)—with the various populations being coupled and thereby affecting each other [135, 170] —renders the mere *assessment* of the T_1-relaxation of each 1Q transition nontrivial, as discussed recently by Yesinowski [170].

If working with small quantities of glasses with low Al contents (a few at-%) for the purpose of NMR parameter extraction by iterative numerical fitting (see Section 10.3), a trade-off must eventually be made between ensuring "quantitative" signals and achieving a sufficiently high S/N to preclude additional uncertainties arising at the NMR data-fitting stage. Here our somewhat pragmatic recommendation is simply to select the longest relaxation delay that permits achieving sufficiently high S/N of the $ST_{3/2}$ peaks in the MAS spectrum to avoid unnecessary numerical fitting errors.

8.4 Rf pulse calibrations

8.4.1 Initial calibrations on Al^{3+} solution

Besides for shift referencing, Al^{3+}(aq) serves the purpose of calibrating rf nutation frequencies [Eq. (38)] by the standard procedure of performing a "nutation array", i.e., conducing a series of NMR acquisitions at a fixed nutation frequency and progressively lengthened excitation pulses; in our experience, the ν_1-value is then preferably calculated from the 360° pulse-width according to $\nu_1 = \tau_{360}^{-1}$ (as opposed to using τ_{90} or τ_{180}). Albeit more time-consuming and not feasible for samples providing weak NMR signals, the most accurate rf power calibrations result by recording (finely sampled) nutation arrays out to large flip angles (beyond 1440°), followed by Fourier transformation of the array of integrated signal intensities with respect to the pulse-length τ_p; this yields a "spectrum" with the (various) nutation frequencies peaking at the respective ν_{nut} values for the given rf power value.

For the subsequent solid-state NMR work, we recommend to locate (*i*) one *high* nutation frequency to be employed for hard-pulse excitation, obeying $\nu_1 \geq 70$ kHz and preferably $\nu_1 \approx 100$ kHz; (*ii*) one suitable *low* power level for CT-selective pulse application, selected as $\nu_1^{\text{soln}} \approx 2\nu_r/[3(S + 1/2)]$, where ν_r is the MAS rate intended for the solid-state experimentation (clarified below). For example, ^{27}Al NMR at $\nu_r = 24$ kHz requires $\nu_1^{\text{soln}} \approx 5.3$ kHz, as then the ^{27}Al CT nutation frequency *approximately* obeys $\nu_{\text{nut}}^{\text{CT}} \approx 3\nu_1^{\text{soln}} \approx 16$ kHz [Eq. (41)].

8.4.2 Spin-lock array

All remaining rf power calibrations are carried out on the solid sample, initiated by the recording of an array of *spin-lock* MAS NMR experiments. Fig. 12A displays the pulse scheme. Spin-locking starts with application of a CT-selective 90° rf pulse of phase ϕ, followed by a longer rf pulse of fixed duration (τ_{SL}) that should be rotor-synchronized, i.e., $\tau_{\text{SL}} = m\tau_r$ with $m = 1, 2, \ldots$. Here, $\tau_{\text{SL}} \approx 2$–5 ms is typical for ^{27}Al NMR on AS phases. The rf phase of the spin-lock pulse must be orthogonal to the 90° pulse, i.e., $\phi + 90°$ (or $\phi + \pi/2$ in radians). Then, the CT magnetization becomes aligned with the $\mathbf{B}_1$ field vector, thereby remaining "locked" along the $\phi + 90°$ direction of the rotating-frame transverse plane. With the aim of finding the $\nu_{\text{nut}}^{\text{CT}}$-value that provides the best spin-locking (thereby minimizing the NMR signal losses) at the spinning frequency to be utilized in the subsequent MAS NMR experimentation with CT-selective excitation pulses, an array is performed where the nutation frequency is varied in fine steps over the range $0 \lesssim \nu_{\text{nut}}^{\text{CT}} \lesssim 1.5\nu_r$. By positioning the spectrometer reference frequency (ν_{rf}) at the centre-of-gravity of the NMR signal region, resonance offsets are minimized during spin-locking. This precaution is also relevant for any other experimental protocol discussed below, notably so for those discussed in Section 11.

Rf pulse application to half-integer spins is well-known to provide a reduced CT-signal intensity due to magnetization-leakage out to the STs [171–176]: unless *either* truly nonselective *or* selective pulses may be accomplished to spin-lock the entire magnetization vector or its CT component, respectively, the spin-lock is usually imperfect. The aim is to arrange sufficiently strong rf pulses that remain (essentially) CT-selective but permit accurate manipulations of the CT magnetization vector despite the complications of resonance spreads from chemical shifts and second-order quadrupolar interactions. This may necessitate relatively high MAS rates (> 20 kHz). Further details of the spin-locking properties of half-integer spins are given in Refs. [171–177].

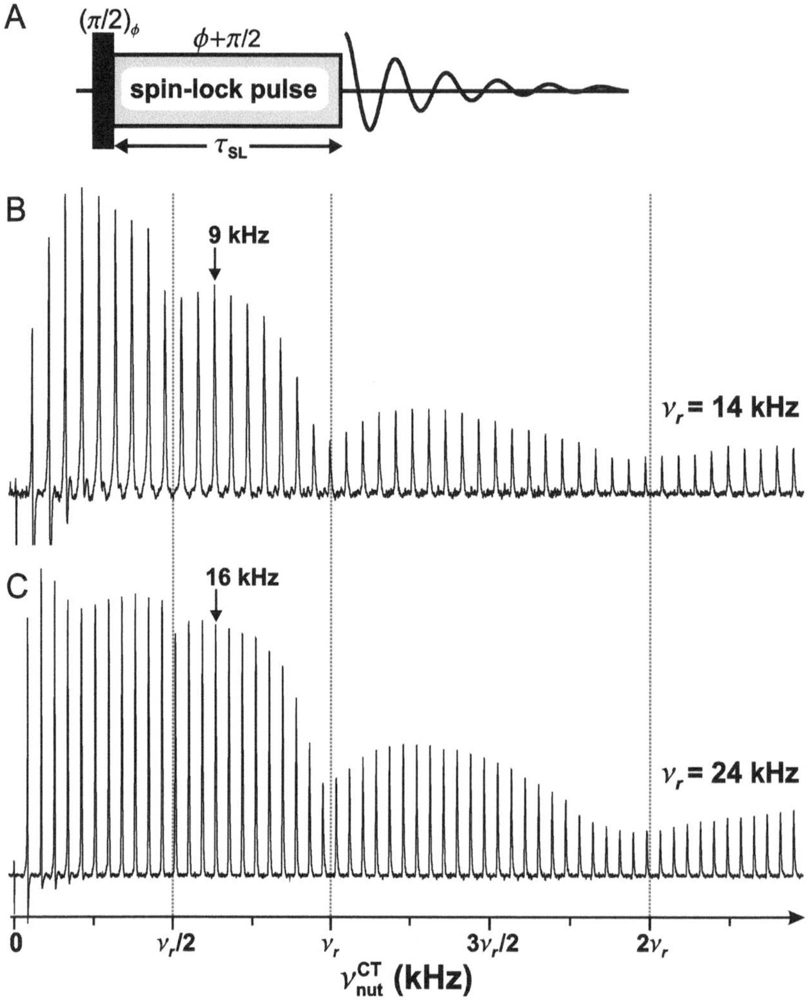

Fig. 12 (A) Rf pulse diagram for a spin-lock NMR experiment. (B, C) Experimental spin-lock ^{27}Al MAS NMR results recorded from two La AS glasses at MAS rates of (B) 14.00 kHz ($B_0 = 9.4$ T) and (C) 24.00 kHz ($B_0 = 14.1$ T). The spin-lock interval was kept fixed at $\tau_{SL} = 5.0$ ms while ν_{nut}^{CT} was varied. Note the diminished NMR signal intensities near each R^3 condition [Eq. (45)] marked by dotted lines, and the overall intensity-reduction as ν_{nut}^{CT} is increased. Each arrow mark the nutation frequency suggested to be used for CT-selective rf pulses (in subsequent experimentation) at each respective MAS rate: these values provide the best compromise between retaining CT selectivity (for minimizing NMR signal losses in multiple-pulse applications) and employing a sufficiently strong rf power to allow for accurate manipulations of the CT magnetization.

The CT-signal losses are minimized for truly selective spin-locking pulses, *except* when $\nu_{\text{nut}}^{\text{CT}}$ is close to a *rotary resonance recoupling* (R^3) condition [178], meaning that

$$\nu_{\text{nut}}^{\text{CT}} = n\nu_r, n = \{1/2, 1, 2\} \ [\text{R}^3 \text{ conditions}]. \quad (45)$$

When the CT nutation frequency is a (sub)-multiple of the spinning frequency, the spin-lock efficiency is drastically reduced [see Fig. 12B and C] because various anisotropic interactions become "*recoupled*"—i.e., they are no longer fully suppressed by the MAS and their effects are reintroduced. By fulfilling Eq. (45), the CSA, heteronuclear dipolar, and/or second-order quadrupolar interactions may be recoupled [176–179]. The $n = 1/2$ case of Eq. (45) is referred to as the "2Q-HORROR" condition [180], as it selectively reintroduces homonuclear through-space interactions that may be exploited to excite double-quantum coherences (2QC) among the CTs of spin-pairs [181–184]. Notably, even during one CT-selective pulse operating at/near an R^3 condition, the undesirable recoupled interactions may interfere with the NMR signal excitation and distort the NMR peakshape. As revealed by Fig. 12B and C, the $n = 1/2$ case of $\nu_{\text{nut}}^{\text{CT}} \approx 7$ kHz and $\nu_{\text{nut}}^{\text{CT}} \approx 12$ kHz, respectively, does not lead to as detrimental signal losses as the other R^3 conditions. The longer the spin-lock period τ_{SL}, the easier the identification of each R^3 match becomes.

Performing the spin-lock array is primarily for *locating* the R^3 conditions so that they may be *avoided* (unless they are to be exploited for interaction-recoupling [176, 177, 181, 182, 184–186]), as well as for identifying the $\nu_{\text{nut}}^{\text{CT}}$-value that *minimizes* the CT NMR signal losses during the subsequent CT-selective pulse application [176, 177, 181, 184]. Fig. 12B and C reveal that $\nu_{\text{nut}}^{\text{CT}} \approx 2\nu_r/3$ provides the best compromise between retaining CT-selectivity to minimize the NMR-signal losses and using a sufficiently large nutation frequency to achieve accurate manipulations of the CT magnetization.[j]

Although decent spin-locking is also observed in the range *below* the 2Q-HORROR condition $\nu_{\text{nut}}^{\text{CT}} < \nu_r/2$, such low rf powers strongly restrict the frequency/shift-range across which spin-locking is successful. Hence,

[j] This is our reason for recommending to independently estimate $\nu_1^{\text{soln}} \approx 2\nu_r/[3(S + 1/2)]$ already at the initial rf power calibrations on the aqueous solution. However, to avoid too low nutation frequencies for samples featuring large resonance-spreads ($\gtrsim$10 kHz) when using MAS rates below 20 kHz, it is better to select $\nu_{\text{nut}}^{\text{CT}}$ to be slightly larger than the spinning frequency ($\nu_{\text{nut}}^{\text{CT}} \approx 1.25\nu_r$) to ensure sufficiently accurate manipulations of the CT magnetization. Similar precautions apply for high-field MAS NMR experimentation ($B_0 > 14.1$ T) where resonance-offset effects become magnified.

the regime $\nu_{nut}^{CT} < \nu_r$ is generally too narrow for accurately controlling the magnetization associated with the broad ^{27}Al NMR resonances from AS glasses, as is for instance reflected in Fig. 12B and C where parts of the NMR signals appear out of phase. Despite that the spin-locking performance is formally distinct from that of applying CT-selective pulses to rotate the CT magnetization, the variations in the NMR CT-signal intensity profile *versus* ν_{nut}^{CT} matches well the corresponding relative NMR intensities observed on application of a series of CT-selective pulses at the given rf power level.

We warmly recommend to utilize spin-lock arrays *routinely* for identifying the most favourable rf nutation frequency of the CT-selective pulses [184]. Whenever feasible, our group employs this procedure prior to *any* detailed rf power calibration, encompassing widely spanning types of NMR experimentation (that is not restricted to ^{27}Al NMR and AS glasses): (*i*) To locate suitable cross-polarization (CP) experimental conditions involving half-integer spins (see Section 11.3) [93, 187, 188]; (*ii*) To establish optimum conditions for homonuclear recoupling of spin-1/2 nuclei [189–191] or half-integer spins [182–184, 192–195], as well as for (*iii*) both homonuclear and heteronuclear spin-1/2 applications at fast MAS (>35 kHz), e.g., for optimizing CP and locating favourable proton decoupling conditions [189, 191].

8.4.3 Rf nutation array

After selecting the rf power to be used for the CT-selective pulses from the set of spin-lock NMR experiments, an rf nutation array is performed to precisely determine τ_{90}^{CT}. A ν_{nut}^{CT}-value in the 16–25 kHz range—that results for typical MAS rates between 24–35 kHz together with the choice $\nu_{nut}^{CT} \approx 2\nu_r/3$—is usually sufficiently large to ensure accurate CT magnetization manipulations for systems exhibiting resonance-spreads up to 10–15 kHz. However, we recommend to check for proper pulse-operation by verifying (*i*) the absence of NMR peakshape distortions relative to those recorded by short and strong pulses; (*ii*) that the signal may be inverted properly by using a CT-selective 180° pulse; (*iii*) that no significant $ST_{3/2}$ ssb formation is observed (weak excitation of the 1–2 ST sideband orders closest to the CT-signal region is usually unavoidable). If any of these simple tests fails, the rf pulses are *either too strong*, and thereby not truly CT-selective, such that the NMR signal buildup does not obey Eq. (42) simultaneously for all quadrupolar sites, *or* the pulses are *too weak* relative to the resonance offsets. For AS glasses revealing very large spreads of resonance frequencies, quantitative excitation may be precluded, necessitating strong-pulse excitation.

8.5 Useful extensions of single-pulse experiments

8.5.1 Saturation recovery

Unless one utilizes relaxation delays ensuring a fully recovered spin ensemble for the given sample and rf excitation condition, we strongly recommend any NMR acquisition that aims at optimizing experimental parameters to start by a "saturation-recovery block", i.e., a series of strong pulses for saturating the spin system by equalizing the Zeeman populations, followed by a relaxation delay (τ_{relax}); see Fig. 13. When appended by a CT-selective or "hard" excitation pulse, this scheme represents the well-known saturation-recovery experiment for measuring T_1. Besides being the *only* practical option for ensuring reproducible results when optimizing pulse-widths or other experimental parameters, such a pulse-block also allows for a quick testing/optimization of parameters, as $\tau_{relax} \sim 2$ s is usually sufficient for work on ^{27}Al in glasses; yet, a suitable τ_{relax}-value must first be located and verified to give sufficiently quantitative results so that meaningful parameter optimizations may be performed on a particular sample. The same precautions apply when utilizing the CT-signal enhancement schemes discussed below.

8.5.2 CT-signal enhancement techniques

A very useful approach for speeding up parameter optimizations is to exploit CT-signal enhancement techniques that starts by either inverting or equalizing the ST_m populations before exciting the CT magnetization by a single pulse [151, 198–200]. Several options exist for MAS experimentation, including frequency sweeps [200, 201], shaped pulses [202, 203] or rotor assisted population transfer [197] (RAPT); the latter builds on the fast

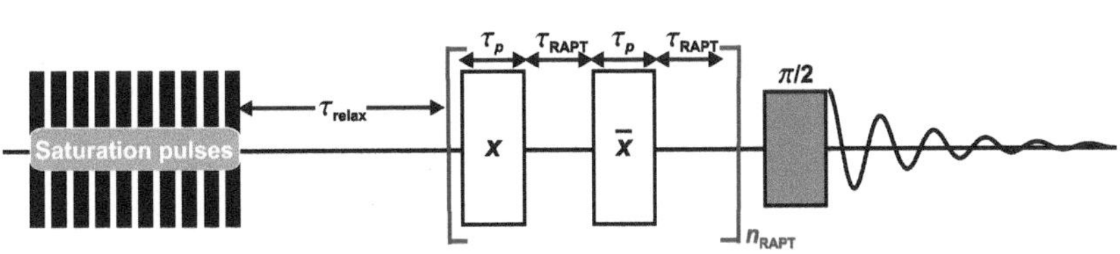

Fig. 13 Rf pulse diagram for a RAPT sequence [196, 197] applied before a CT-selective 90° pulse (*grey rectangle*) for enhancing the CT NMR-signal intensity. The scheme starts with a pulse-block for signal saturation followed by a relaxation delay τ_{relax}. The basic RAPT block of two pulses and delays of equal duration $\tau_{RAPT} = \tau_p$ is repeated n_{RAPT} times. The x and $\bar{x}$ letters denote the respective rf phases of the pulses, i.e., $\phi = 0$ and $\phi = 180°$. Note that neither the timings nor rf amplitudes are drawn to scale: the strong RAPT pulses last for around 0.6–1 μs, whereas the CT-selective pulse (~ 10 μs) operates at a much lower rf power.

amplitude modulation [196] (FAM) technique, originally introduced for 3QMAS applications (Section 9.2) and subsequently developed further into different incarnations [204–206]. For a spin-S nucleus, such CT polarization enhancement techniques may (ideally) provide factors of $2S$ (for ST population inversion) or $S + 1/2$ (for ST saturation) more intense NMR signals relative to that of a CT-selective 90° pulse [151, 197–200]. Since the optimal parameters of any such signal enhancement technique depend on the quadrupolar frequency (that will vary among distinct spin sites), they should *not* be employed for acquisitions aiming at precise quantifications of site populations or $\{\delta_{\text{iso}}, C_{Q\eta}\}$ parameter distributions. Yet, substantial NMR intensity and/or peakshape distortions only result for samples with multiple sites whose $C_{Q\eta}$-values differ widely (by several MHz). We therefore suggest to exploit signal enhancement protocols prior to *any* experimental scheme building around detection of CT magnetization of half-integer spins S (e.g., see [205, 207]), encompassing heteronuclear $S \rightarrow I$ polarization-transfer schemes, such as CP [187, 188, 208].

Thanks to its straightforward setup, we prefer using the RAPT protocol of Yao et al. [197]. It builds around a pair of short pulses $\{\beta_p, \bar{\beta}_p\}$ with equal durations τ_p but opposite phases. The pulses are interleaved with delays, each lasting for $\tau_{\text{RAPT}} = \tau_p$. The pulse-pair is repeated n_{RAPT} times such that the output RAPT scheme becomes

$$\left[\beta_p - \tau_{\text{RAPT}} - \bar{\beta}_p - \tau_{\text{RAPT}}\right]_{n_{\text{RAPT}}}, \tag{46}$$

which is preceded by a saturation-recovery block and followed by a CT-selective $\pi/2$ pulse, as depicted in Fig. 13. With quite effortless optimization, giving typical parameters of $\tau_{\text{RAPT}} \sim$0.6–0.9 μs and $n_{\text{RAPT}} \sim$20–30, one reaches CT-signal enhancement factors around 1.7–2.0 for ^{27}Al in AS glasses under fast MAS conditions, thereby offering a nearly fourfold reduction in the experimental time. Yet, these enhancements are lower than that ≈3 ideally expected for a complete $ST_{3/2}$-population equalization; the primary reasons for the lower yields from disordered materials are their wide $C_{Q\eta}$-distributions and the typical use of relatively short relaxation delays (≲2 s), which normally do not permit a full recovery of thermal equilibrium that is a prerequisite for achieving the ideal enhancement factor (recall that the ST magnetization usually recover slower than the CT [169]). Extended RAPT protocols may provide higher population transfers [204–206].

9. High-resolution techniques

9.1 Gist of the methodology: DOR and DAS

Provided that the various AlO_p populations in a given AS structure do not differ widely, MAS alone achieves sufficient NMR spectra resolution for identifying and quantifying the set $\{x_{Al}^{[p]}\}$; see Section 10.3. However, this is not applicable to the majority of AS glass compositions for which $Al^{[4]}$ coordinations dominate. The second-order quadrupolar broadening $\nu_{aniso}^{Q}(\Omega)$ [Eq. (27)] limits the resolution (e.g., see Figs 7 and 8) and since the first MAS NMR demonstrations to quadrupolar spins in the early 1980s, significant research has been devoted to reach comparable resolution for $S > 1/2$ nuclei as that accomplished by MAS for spins-1/2 in powders. Although being distinctly different in terms of concrete implementations, all such hitherto proposed options revolve around the same idea: sample the NMR time-signal such that the spin evolution effectively becomes independent of *both* the second-rank $[A_{20}(\Omega)]$ and fourth-rank $[A_{40}(\Omega)]$ contributions to Eq. (26) [27–29, 100].

One of the first high-resolution techniques was *double rotation* (DOR) NMR [209–212]. DOR implements a *simultaneous* sample rotation around *two* axes (relative to $\mathbf{B}_0$) to eliminate the broadening from each $A_{20}(\Omega)$ and $A_{40}(\Omega)$ term by selecting an axis-pair $\{\beta_m^{(2)}, \beta_m^{(4)}\}$ that arranges $P_2(\beta_m^{(2)}) = 0$ and $P_4(\beta_m^{(4)}) = 0$. Here $\beta_m^{(2)} \equiv \beta_m = 54.74°$ is the "regular magic angle", whereas $\beta_m^{(4)}$ represents either 30.56° or 70.12°. DOR constitutes the *conceptually* most elegant solution to the high-resolution problem by extending the single-axis MAS idea to encompass two rotation axes. Despite having the decisive advantage of providing high-resolution NMR spectra in real time, DOR suffers from a number of practical drawbacks: it requires a sophisticated DOR probehead where one small rotor spins *inside* of a larger one. Thus, arranging stable spinning is a formidable task and the NMR signal-sensitivity is nonoptimal owing to a low "filling factor". Moreover, the "outer" rotor must be large enough to accommodate the "inner" one, implying that its maximum spinning rate is restricted to a few kHz; this unfortunately leads to an excessive ssb formation across the CT-centreband region. Owing mainly to these cumbersome technical obstacles, DOR has not gained widespread use. Nevertheless, a number of impressive experimental demonstrations were presented for estimating various structural parameters [213–216], notably for gaining connectivity information using through-space/bond spin–spin couplings among half-integer spins [214, 217–219].

Another high-resolution strategy for half-integer spins in powders, *dynamic angle spinning* (DAS) [209, 212], was devised simultaneously as DOR. DAS experimentation also requires a specialized probehead that allows for rapid switching and stabilization of the rotor between two rotation axes during the experiment. In contrast to DOR, however, DAS is a 2D NMR technique, utilizing that each $A_{20}(\Omega)$ and $A_{40}(\Omega)$ factor [Eq. (26)] depends on the precise rotor inclination angle β_r; the protocol may be expressed schematically as $t_1(\beta_r = 37.38°)$—(rotor reorientation)—$t_2(\beta_r = 79.19°)$, with z-magnetization stored while the rotor-axis position is altered. With this combination of β_r orientations (one out of several options [27, 209, 220]), the phase accumulated by the -1QC evolution at the time-point t_2 becomes free of both second/fourth-rank tensor contributions from the second-order quadrupolar interaction, leading to the formation of an "echo" with solely the isotropic frequency component encoded [209]. The echo-signal position shifts concurrently with the incrementing t_1-value. Besides involving a 2D NMR acquisition—a drawback shared by all subsequently introduced high-resolution analogues—application of DAS is restricted to systems with weak homonuclear interactions [209, 220]. This aspect may be problematic for ^{27}Al-rich glasses. Indeed, although quantitative (distributions of) isotropic chemical shifts, quadrupolar products and site populations are reported from crystalline and amorphous (alumino)silicate, borate, and borosilicate phases by using both ^{17}O and ^{11}B DAS/DOR NMR [212, 221–224], we are unaware of ^{27}Al NMR applications thereof to AS glasses.

9.2 MQMAS

As it only demands standard MAS NMR equipment, the *multiple-quantum MAS* (MQMAS) technique of Frydman et al. [21, 22] revolutionized the area of probing half-integer spin quadrupolar nuclei in inorganic materials; besides intense developments of the technique itself by numerous groups, the amount of structural reports targeting half-integer spins as structural probes in general—and ^{27}Al in particular—has grown enormously since the introduction of MQMAS in 1995. MQMAS NMR not only improves spectral resolution but also permits the determination of NMR parameters. Stebbins et al. provided the first ^{27}Al 3QMAS application to disordered materials, encompassing an AS glass of the anorthite composition ($CaAl_2Si_2O_8$) [225]. The technique is nowadays routinely employed for probing ^{27}Al environments in AS glasses and other disordered solids.

MQMAS shares the 2D NMR character of DAS but instead of switching the rotor-axis position, echo formation is accomplished by *altering* the *coherence order* so as to correlate MQC evolution during t_1 with that of CT1QC during t_2; see Fig. 14. For spin-5/2 nuclei, the relevant combination is $\{p(t_1), p(t_2)\} = \{+3, -1\}$, i.e., the +3QC evolution should be correlated with that of the detected −1QC during t_2. While the use of higher-order MQC (e.g., 5QC for ^{27}Al) is advantageous for better signal-separation among distinct quadrupolar sites [231–233], it also accentuates the Achilles heel of the method, namely the modest signal-sensitivity stemming from magnetization losses associated with the MQC filtering [22, 234]. The efficiency of both the MQC excitation/conversion processes benefit from using the highest possible rf nutation frequency [232, 234], in practice requiring $\nu_1 \gtrsim 80$ kHz [Eq. (28)]. To date, Frydman's original triple-quantum version (3QMAS) remains as the most utilized option relative to its higher-order coherence-counterparts. Options for remedying the primary weakness of

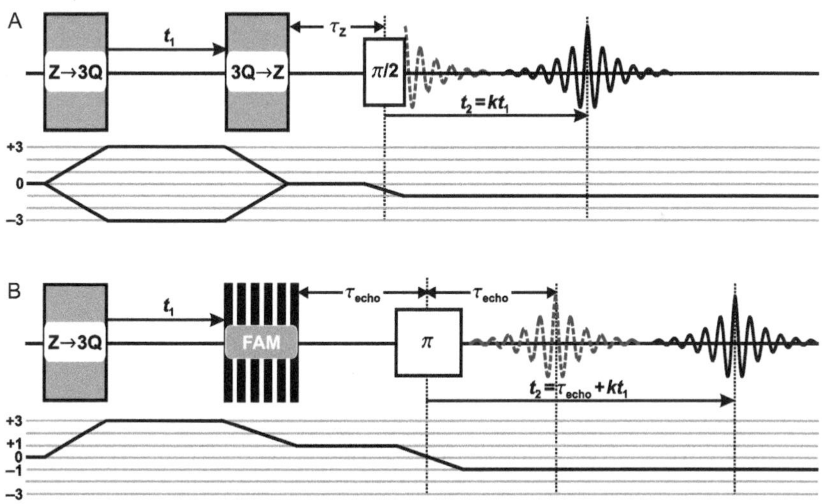

Fig. 14 Schematic rf pulse diagrams of the (A) Z-filter [226] and (B) shifted-echo [227] 3QMAS NMR experiments relevant for $S = 5/2$ (i.e., ^{27}Al) applications, shown together with the coherence-transfer pathways [228] beneath. The shifted-echo sequence involves the FAM scheme [196] for performing the +3QC → +1QC conversion [229]. Note that the echo-interval τ_{echo} should be rotor-synchronized: $\tau_{echo} = n\tau_r$, $n = 0, 1, 2, \ldots$. The time-signals at $t_1 = 0$ and $t_1 > 0$ are depicted by *dotted grey* and *solid black* traces, respectively, with the top of the echo shifting in time by kt_1, where $k = 19/12$ for $S = 5/2$. The missing initial part of the full echo-signal in (A) requires the use of a hypercomplex acquisition [228, 230] for obtaining absorptive 2D NMR spectra with frequency-sign discrimination along the indirect dimension.

the inefficient MQC generation/reconversion were explored intensively for about a decade after the birth of MQMAS; e.g., see Refs. [196, 200, 235–241].

3QMAS NMR spectra are normally "sheared" to produce an "isotropic" (high-resolution) indirect spectral dimension (ν_1), revealing one narrow NMR peak for each chemically distinct site, while the "anisotropic" dimension (ν_2) comprises the respective broad signal as detected in a regular 1D MAS NMR acquisition. However, while the isotropic dimension of the 3QMAS spectrum from a disordered/amorphous material is devoid of quadrupolar broadenings, the chemical shift dispersion remains intact. The $\delta_{\text{iso}}/C_{Q\eta}$-distributions manifest in 3QMAS spectra by broad ridges extending along both spectra dimensions, as illustrated in Fig. 15. Bodart presented expressions accounting for their appearance [242], and the distribution widths may be estimated by using numerical simulations [52], as discussed further in Section 10.4. Besides allowing for estimating the average δ_{iso} and $C_{Q\eta}$ parameters of each $^{27}AlO_p$ moiety in an AS glass, the enhanced resolution in 3QMAS 2D NMR spectra significantly assists the unambiguous detection of the (usually) weak NMR signals from the minor $^{27}AlO_5$ and $^{27}AlO_6$ species, whose resonances are often swamped by the much more intense $^{27}AlO_4$ counterpart; unless the higher-coordination $Al^{[p]}$ species are present to significant amounts ($\gtrsim$10%), they may escape detection in standard 1D MAS NMR spectra; see Fig. 15.

A plethora of distinct but related 3QMAS implementations exist [27–29]. Most of them are modifications of two prototypes: the Z-filter [226, 243] and the shifted-echo [227, 244, 245] protocols shown in Fig. 14A and B, respectively. Both schemes improve on Frydman's original two-pulse implementation [21, 22] and start with 3QC excitation by applying a strong rf pulse ($\approx$ 180° for $S = 5/2$ applications [232, 234]), followed by the 3QC evolution interval (t_1). The Z-filter scheme involves simultaneous evolution of $\pm$3Q coherences (thereafter converted into z-magnetization), whereas the shifted-echo protocol selects solely the (echo-forming) +3Q coherence that is subsequently transferred to +1QC by using a FAM pulse-block [196] (see Eq. (46) and Section 8.5.2). Either 3QMAS scheme also invokes CT-selective pulses, constituting a 90° readout pulse for the Z-filter scheme in Fig. 14A, and a 180° pulse for the echo segment in Fig. 14B. The detected echo displaces according to $t_2 = kt_1$, with the value of k depending both on the evolved coherence order and the spin number S; $k = 19/12$ for ^{27}Al 3QMAS.

To avoid a substantial ssb formation along the ν_1 spectral dimension stemming from the MQC conversion process [246], it is beneficial to

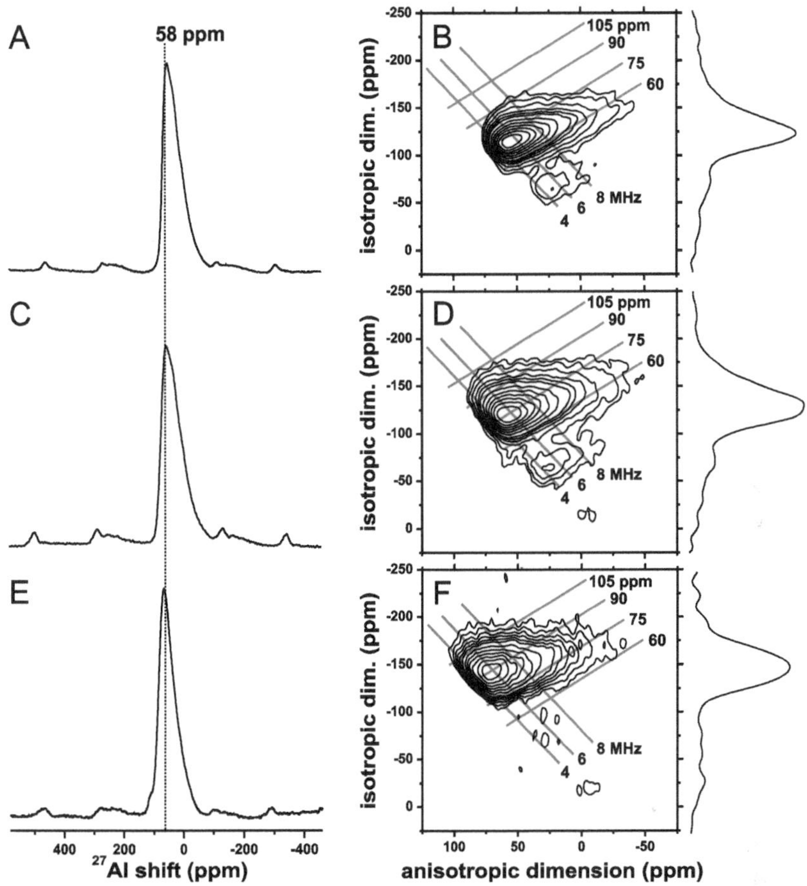

Fig. 15 Experimental MAS (*left column*) and 3QMAS (*right column*) ^{27}Al NMR spectra recorded at 9.4 T from La–Al–Si–O–(N) glasses of variable compositions [93]. Note the deshielding of $^{27}AlO_4$ when the N-content of the glass is increased according to (A, B) 0%, (C, D) 17%, and (E, F) 36%, implying the formation of "mixed" $Al(O, N)_4$ tetrahedra [percentages represent $100 \times n_N/(n_N + n_O)$]. The *lines* with positive slopes in the 3QMAS spectra mark displacement directions originating from the $C_{Q\eta}^{[4]}$-distribution (at the as-indicated constant $\delta_{iso}^{[4]}$-values), whereas lines with negative slopes show displacements resulting from the $\delta_{iso}^{[4]}$-spreads (at the as-indicated fixed $C_{Q\eta}^{[4]}$-values); the ^{27}Al deshielding from the Al–N bonds are witnessed by an overall displacement of the 2D peak towards the *upper left* corner of the 3QMAS spectrum. Note that $^{27}AlO_5$ resonances are only evidenced from the 3QMAS NMR spectra, whereas they are swamped by the much more intense $^{27}AlO_4$ signals in the 1D MAS counterparts.

rotor-synchronize the t_1-evolution ($t_1 = n\tau_r$, $n = 0, 1, 2, \ldots$) such that the spectral window and MAS rate are equal and all sidebands superimpose on their associated centreband [247] (see Section 10.2). This precaution reduces the experimental time and provides more reliable NMR parameter analyses, because the 2D NMR peakshape becomes identical to that observed for infinitely fast MAS. However, rotor-synchronization is often precluded for ^{27}Al work on AS glasses comprising multiple $^{27}Al^{[p]}$ coordinations whose resonance-spread along the ν_1-dimension typically exceeds the MAS rate. We then recommend to minimize the number of sidebands by incrementing t_1 by a submultiple of the rotor period; $t_1 = n\tau_r/m$, where m is the smallest integer that allows for accommodation of all NMR CT-signals.

For ^{27}Al applications to AS glasses, the shifted-echo 3QMAS version offers decisive advantages over its Z-filter counterpart: (*i*) It yields a superior signal-to-noise (S/N) ratio thanks to its significantly more efficient 3QC reconversion implementation. (*ii*) The echo-segment shifts the entire NMR signal into the acquisition window, thereby boosting the signal-sensitivity further by a factor of $\sqrt{2}$ [227, 248]. (*iii*) It offers a two-fold reduced experimental time by circumventing the need of a States-type hypercomplex acquisition [230] that is required for the Z-filter 3QMAS scheme. A single 2D NMR data-set acquired by the scheme in Fig. 14B followed by a complex 2D Fourier transformation yields purely absorptive 2D NMR spectra with frequency-sign discrimination along the indirect dimension. The rotor-synchronized echo-interval τ_{echo} of Fig. 14B must be selected to exceed the "life-time" of the t_2-detected 1QC signal. The shifted-echo experiment is useful whenever the -1QC signal damps rapidly from the interference among the multitude of components emanating from the structural disorder, whereas the (irreversible) signal-decay stemming from T_2-relaxation is negligible during the echo-interval τ_{echo}; this applies nearly always to ^{27}Al in rigid diamagnetic solids. We recommend to utilize the shifted-echo scheme for ^{27}Al 3QMAS work on AS glasses, where normally $\tau_{echo} \approx 0.4$–0.8 ms.

Another high-resolution 2D NMR protocol, STMAS, was introduced by Gan [157, 158]. It offers higher signal-sensitivity as it only involves 1QC evolution (as in DAS) but requires a very high accuracy in the magic-angle setting (Section 8.2), as it partially invokes evolution of ST-coherence evolution that is affected by the first-order quadrupolar interactions [157–159]. Despite that the signal-sensitivity of 3QMAS NMR is generally inferior to other high-resolution options (DOR, DAS, STMAS), it remains the method of choice for studying both ordered and disordered structures. There are several experimental studies comparing the relative merits (such as signal-sensitivity and resolution) among various high-resolution techniques [222, 224, 249]—including evaluations of different MQMAS incarnations [229, 240, 245, 250]—as well as a large number of thorough reviews on theory and methodological advances, where we single out Refs. [27–29, 43, 101, 159].

10. Determining ^{27}Al NMR parameters and AlO_p populations

10.1 Obtaining (δ_{iso}, $C_{Q\eta}$) from CT MAS spectra at multiple fields

The isotropic *chemical* shift and $C_{Q\eta}$ parameters of a given half-integer spin-site are not directly available from the CT NMR peakshape that only provides the sum over the two δ_{iso} and δ_{iso}^{Q} components of Eq. (31). The two desirable contributions of the CG shift (δ^{CG}) may nevertheless be separated by exploiting their distinctly different dependencies on the external magnetic field, where δ_{iso} is independent on B_0 while δ_{iso}^{Q} scales as B_0^{-2}. By combining δ^{CG} results from NMR spectra recorded at several (distinct) magnetic fields, the $B_0 = 0$ intercept in a plot of δ^{CG} against B_0^{-2} provides the δ_{iso}-value, whereas δ_{iso}^{Q} is proportional to the slope of the line. However, with only two unknown parameters, the set $\{\delta_{iso}, C_{Q\eta}\}$ is readily obtained from Eqs (31) and (54) sampled at two B_0-values:

$$\delta_{iso} = \frac{\delta_{I}^{CG}(B_0^{I})^2 - \delta_{II}^{CG}(B_0^{II})^2}{(B_0^{I})^2 - (B_0^{II})^2}, \tag{47a}$$

$$C_{Q\eta} = \pm\frac{\gamma S B_0^{I} B_0^{II}}{2\pi}\sqrt{\frac{160(2S-1)(\delta_{I}^{CG} - \delta_{II}^{CG})}{3(2S+3)\left[(B_0^{I})^2 - (B_0^{II})^2\right]}}, \tag{47b}$$

where δ_{I}^{CG} and δ_{II}^{CG} denote the centre-of-gravity shifts of the NMR peak observed at the two respective fields B_0^{I} and B_0^{II}.

This early procedure for estimating the chemical shift and quadrupolar parameters has been superseded by easier and more accurate alternatives that only demand data at one magnetic field.

10.2 Obtaining (δ_{iso}, $C_{Q\eta}$) by combining CT and $ST_{3/2}$ shifts

10.2.1 Gist of the procedure

We start by conveying the idea to estimate the isotropic chemical shift and the quadrupolar coupling product $C_{Q\eta}$ by combining the information from the CG frequencies of the CT and $ST_{3/2}$ MAS NMR peaks from a $S = 5/2$ site, as introduced by Samoson et al. [19, 20]. Here we describe our implementation of the procedure in detail. Strictly, it requires knowledge of the exact CG shift of the NMR peak from each quadrupolar "site", but as the various $^{27}AlO_p$ resonances observed from AS glasses partially overlap, we must assume that the frequency/shift at the *maximum* spectral *amplitude* is a sufficiently good approximation of the real CG peak-position. The validity of this approximation improves when B_0 is increased, owing to the gradual suppression of the second-order quadrupolar broadening and the accompanying NMR signal "tail" (see Fig. 8).

Onwards focusing on the CG shifts of the central and the "inner" satellite transitions, Eq. (31) implies

$$\delta_{\mathrm{CT}}^{\mathrm{CG}} = \delta_{\mathrm{iso}} + \delta_{\mathrm{CT}}^{\mathrm{Q}} \tag{48}$$

$$\delta_{\mathrm{ST}}^{\mathrm{CG}} = \delta_{\mathrm{iso}} + \delta_{\mathrm{ST}}^{\mathrm{Q}} \tag{49}$$

where we here identify $\mathrm{ST} \equiv \mathrm{ST}_{3/2}$. We remind that the $ST_{3/2}$ of $S = 5/2$ nuclei is markedly less affected by the second-order quadrupolar interaction, manifested in second-order quadrupolar *isotropic* shift only amounting to $-1/8$ that of the CT (see Table 1). The CG of the $ST_{3/2}$ *centreband* is denoted $\delta_{\mathrm{ST}}^{(0)}$ when expressed in ppm relative to the standard reference [see Eq. (3)], or $\nu_{\mathrm{ST}}^{(0)}$ when expressed in Hz relative to the spectrometer reference frequency. While the $ST_{3/2}$ centreband signal is located close to the *isotropic chemical* shift δ_{iso}, it generally overlaps with that from the CT and $\delta_{\mathrm{ST}}^{(0)}$ cannot be extracted directly from the NMR spectrum. Yet, one may utilize the other $ST_{3/2}$ spinning sidebands to estimate the frequency of the centreband $\nu_{\mathrm{ST}}^{(0)}$, exploiting that the frequency position $\nu_{\mathrm{ST}}^{(k)}$ of any sideband order k is related to $\nu_{\mathrm{ST}}^{(0)}$ by $k\nu_r$ [34]. Then the identification $\delta_{\mathrm{iso}} = \delta_{\mathrm{ST}}^{\mathrm{CG}}$ is a good approximation of the isotropic chemical shift for $S = 5/2$ sites with small/moderately large ($\lesssim$4 MHz) values of C_Q [19, 20, 39]. Note that the chemical shift estimation based on $\delta_{\mathrm{ST}}^{\mathrm{CG}}$ is eight times more accurate than the corresponding approximation $\delta_{\mathrm{iso}} \approx \delta_{\mathrm{CT}}^{\mathrm{CG}}$ that is generally poor unless C_Q is (unusually) small ($\lesssim$0.5 MHz).

10.2.2 Step-by-step protocol

The following experimental parameters are required in the subsequent calculations: (*i*) The spectrometer reference frequency ν_{rf}, (*ii*) the MAS rate ν_r (that is assumed to be positive) used to record the NMR spectrum, and (*iii*) the standard reference frequency ν_{ref}. We here only consider the estimate of one $S=5/2$ site of the structure; if several sites are present, the procedure below is repeated for each of them:

(1) Determine the CG frequency of the CT peak ($\nu_{\mathrm{CT}}^{\mathrm{CG}}$), and select a set $\{k,\ \nu_{\mathrm{ST}}^{(k)}\}$ of CG frequencies (or peak maxima) of the $\mathrm{ST}_{3/2}$ spinning sidebands. Note that the sideband order k is defined such that the first ssb appearing to the right of the CT peak is $\nu_{\mathrm{ST}}^{(1)}$, while that to the left of the CT peak is $\nu_{\mathrm{ST}}^{(-1)}$. The literature suggests using the low-order $k=\{\pm1,\ \pm2\}$ sidebands [19, 20, 39]. However, as they often overlap with the CT-sidebands, we rather recommend using the higher orders $k=\pm3,\ \pm4,\ \ldots$.

(2) For each ssb frequency $\nu_{\mathrm{ST}}^{(k)}$, obtain the corresponding estimate of $\nu_{\mathrm{ST}}^{(0)}$ by subtracting the value $k\nu_r$ (this operation is called "folding"):

$$\nu_{\mathrm{ST}}^{(0)}=\nu_{\mathrm{ST}}^{(k)}-k\nu_r. \tag{50}$$

Despite that Eq. (50) is exact when considering the narrow sidebands generated by the CSA interaction for spins-1/2 [34], the orientation-dependent second-order broadening results in minor variations of the NMR peakshapes among distinct $\mathrm{ST}_{3/2}$ ssb orders [109, 251]. Moreover, as each estimate of $\nu_{\mathrm{ST}}^{(0)}$ from $\nu_{\mathrm{ST}}^{(k)}$ is also associated with other experimental uncertainties, at least four distinct sidebands should be taken into accounted (the more, the better), preferentially selected symmetrically to include both k and $-k$ orders. Calculate the mean value of the set $\{\nu_{\mathrm{ST}}^{(0)}\}$.

(3) Convert each $\nu_{\mathrm{CT}}^{\mathrm{CG}}$ and $\nu_{\mathrm{ST}}^{(0)}$ frequency into its corresponding shift by using Eq. (3), i.e., $\delta_{\mathrm{ST}}^{(0)}=(\nu_{\mathrm{ST}}^{(0)}-\nu_{\mathrm{ref}})/\nu_{\mathrm{ref}}$, and likewise for $\delta_{\mathrm{CT}}^{\mathrm{CG}}$. Onwards, we identify $\delta_{\mathrm{ST}}^{\mathrm{CG}}\equiv\delta_{\mathrm{ST}}^{(0)}$.

(4) A reasonable estimate of the isotropic *chemical* shift is obtained from the following approximation,

$$\delta_{\mathrm{iso}}\approx\delta_{\mathrm{ST}}^{\mathrm{CG}}=\delta_{\mathrm{ST}}^{(0)}, \tag{51}$$

where the residual error stems from the second-order isotropic quadrupolar shift according to Eq. (49). Yet, a better estimate of δ_{iso} results by combining the CG frequencies from both CT and $\mathrm{ST}_{3/2}$ transitions [19, 20, 39]:

$$\delta_{\mathrm{iso}}\approx\delta_{\mathrm{ST}}^{\mathrm{CG}}-\frac{1}{18}\left(\delta_{\mathrm{ST}}^{\mathrm{CG}}-\delta_{\mathrm{CT}}^{\mathrm{CG}}\right). \tag{52}$$

The factor of 1/18 stems from Engelhardt and Michel [39], and differ from that (1/9) initially recommended in Refs. [19, 20].

(5) With δ_{iso} known, the isotropic second-order quadrupolar shift δ_{CT}^{Q} (in ppm) is readily obtained from Eq. (48). Convert this value into units of Hz:

$$\nu_{CT}^{Q} = \nu_0 \delta_{CT}^{Q}. \tag{53}$$

(6) Calculate the quadrupolar product from

$$C_{Q\eta} = 2S(2S-1)\sqrt{\frac{40\nu_0 \nu_{CT}^{Q}}{3[3-4S(S+1)]}}. \tag{54}$$

If η may be determined separately, Eq. (32) provides the pair of values $\{C_Q, \eta\}$.

10.3 Deconvolution of 1D MAS NMR spectra by numerical simulations

The fractional population $x_{Al}^{[p]}$ of each $^{27}AlO_p$ polyhedron and its parameters characterizing the accompanying distributions of isotropic chemical shifts ($\delta_{iso}^{[p]}$) and quadrupolar products ($C_{Q\eta}^{[p]}$) may be extracted from the respective MAS NMR peakshape by iterative fitting of numerically simulated NMR spectra to the experimental data until the root-mean-square deviation is minimized. Such a procedure normally yields significantly more accurate $\{\delta_{iso}^{[p]}, C_{Q\eta}^{[p]}\}$ estimates than the two approaches outlined above.

One option aims at reproducing the entire $ST_{3/2}$ sideband envelope by numerical simulations, as first implemented by Jakobsen et al. [153, 154]. The approach was developed further by Jäger et al. [116, 155, 156], who termed it "SATRAS" (satellite transition spectroscopy) and provided the first ^{27}Al NMR applications to structurally disordered phases, such as mullite, [155], χ-Al_2O_3 [156], and alumino-borophosphate glasses [116]; see Fig. 11. SATRAS is best suited for structures involving one crystallographically/chemically unique quadrupolar site with modest $C_{Q\eta}$-value. It is in practice demanding for application to amorphous phases, as recording of the entire $ST_{3/2}$ ssb manifold requires a highly accurate magic-angle setting and use of very large spectral windows that reduces the S/N ratio. High-quality and *quantitative* NMR acquisitions become time-consuming for AS glasses.

Nowadays, it is most common to extract the quadrupolar parameters by focussing on the signal region around the CT and the low-order $ST_{3/2}$ sidebands in ^{27}Al MAS NMR spectra, or solely targeting the CT-signals if the data were acquired by using CT-selective pulses. Each NMR peakshape is deconvoluted into its components by iterative fitting using numerical simulations. Fig. 9D shows an example from a Sc_2O_3–Al_2O_3 –SiO_2 glass [126], which reveals contributions from all three $\{Al^{[4]}, Al^{[5]}, Al^{[6]}\}$ coordinations. Each peak-component is associated with a quartet of best-fit parameters $\{x_{Al}^{[p]}, \delta_{iso}^{[p]}, C_{Q\eta}^{[p]}, W_{iso}^{[p]}\}$, where $W_{iso}^{[p]}$ is the FWHM of the (as-assumed) Gaussian chemical-shift distribution. A multitude of additional deconvolutions are provided from CaO–Al_2O_3–SiO_2 glasses [62] in Fig. 24 of Section 12.

Several public programs exist for deconvoluting MAS NMR spectra, e.g., DMFit [125, 251], QuadFit [127], and "*EGdeconv*" (EASY-GOING deconvolution) [128]. All of them implement a Czjzek distribution of $C_{Q\eta}$. Besides differing in the user interface and required operating system, the main distinctions between the various deconvolution/fitting options is their assumptions underlying the generation of the NMR spectrum from each NMR parameter set and crystallite orientation: DMFit and QuadFit implements analytical solutions, whereas *EGdeconv* accounts rigorously for the given experimental conditions by invoking numerically exact simulations that integrates the time-dependent Schrödinger equation to calculate the spin dynamics, as utilized in several general-purpose simulation platforms, such as GAMMA [252], SIMPSON [253], and SPINEVOLUTION [254]. Deconvolutions by DMFit accounts well (albeit approximately) for the ssb formation by combining different analytically derived expressions [251, 255], whereas QuadFit only operates in the limit of infinitely fast MAS [127]. When fitting ^{27}Al NMR spectra recorded from sites with large $C_{Q\eta}$-values (which applies to AS glasses) at moderately large fields, such an assumption introduces systematic errors in the best-fit parameters, because the anisotropy $\nu_{aniso}^{Q}(\Omega)$ is generally sufficiently large to produce nonnegligible CT spinning sidebands.

Following previous developments of efficient protocols for spin dynamics calculations [256, 257] and orientational averaging/spectra interpolation [258–260], we employ a program developed in our laboratory for deconvoluting 1D MAS NMR spectra [72, 126, 195] (courtesy of K. Okhotnikov and B. Stevensson), which bears some similarities with the *EGdeconv* package [128]. The program incorporates *restricted* parameter *ranges* (as opposed to *fixed* parameters) in the numerical fitting: this is useful for precluding physically unreasonable results and ensuring consistency in the handling of a large series of AS glasses, for instance for monitoring dependencies of the NMR parameters on a glass composition variable [72, 126, 195]. Surprisingly, among the free deconvolution software available to date, only *EGdeconv* appears to offer options for restricted parameter fitting.

Out of the possibilities reviewed herein, we believe that a simultaneous fitting of the CT peakshape together with a few of the $ST_{3/2}$ sidebands to be the most convenient option for extracting the parameter set $\left\{x_{\mathrm{Al}}^{[p]}, \delta_{\mathrm{iso}}^{[p]}, C_{Q\eta}^{[p]}, W_{\mathrm{iso}}^{[p]}\right\}$ of each $^{27}\mathrm{Al}^{[p]}$ species in an AS glass. However, given the large number of a priori unknown and correlated fitting parameters in multisite structures, the reliability of the numerical fitting improves significantly if complementary information is used to constrain the parameter optimizations; we therefore recommend to first arrange independent estimates of $\{\delta_{\mathrm{iso}}^{[p]}, C_{Q\eta}^{[p]}\}$ from a combined $CT/ST_{3/2}$ analysis (Section 10.2). The primary obstacle is the strong correlation between the two broadening factors, i.e., the quadrupolar product and the chemical shift distribution width ($W_{\mathrm{iso}}^{[p]}$). They become much easier to disentangle if MAS NMR spectra are available from two distinct external magnetic fields, where low (high) B_0-values provide more reliable best-fit parameters of $C_{Q\eta}^{[p]}$ ($W_{\mathrm{iso}}^{[p]}$), respectively.

10.4 Analysis of 3QMAS NMR spectra

3QMAS NMR offers several options for determining the parameter-triplet $\{x_{\mathrm{Al}}^{[p]}, \delta_{\mathrm{iso}}^{[p]}, C_{Q\eta}^{[p]}\}$. Worth noting, however, is the inherent nonquantitative character of MQMAS experimentation: the *M*QC excitation/reconversion processes depend strongly on the frequency $\nu_{Q}^{(1)}(\Omega, t)$, which generally result in distorted 2D peakshapes. As shown by the excitation profile in Fig. 16, 3QMAS spectra tend to emphasize signals from quadrupolar sites with modest values of $C_{Q\eta}$ (yet not too low, i.e., $\lesssim 3$ MHz), whereas ^{27}Al sites exhibiting large quadrupolar products are partially suppressed [22, 52, 234, 235]. Hence, for disordered structures, the extracted average $C_{Q\eta}$ (δ_{iso}) values tend to be somewhat underestimated (overestimated) [117, 261]. The extent of such problems depend strongly on the particular 3QMAS pulse-scheme employed. For instance, the Z-filter protocol in Fig. 14A is markedly more susceptible to the precise quadrupolar products than its shifted-echo counterpart in (B) that implements FAM [196] for 3QC reconversion; this is another impetus for employing the shifted-echo scheme for experimentation on AS glasses.

By estimating the centre-of-gravity 2D coordinates $\{\delta_1^{\mathrm{CG}}, \delta_2^{\mathrm{CG}}\}$ of the *sheared* 3QMAS spectrum, a straightforward extraction of the parameter set $\{\delta_{\mathrm{iso}}, C_{Q\eta}\}$ of a given $S = 5/2$ (^{27}Al) site is available from the expressions

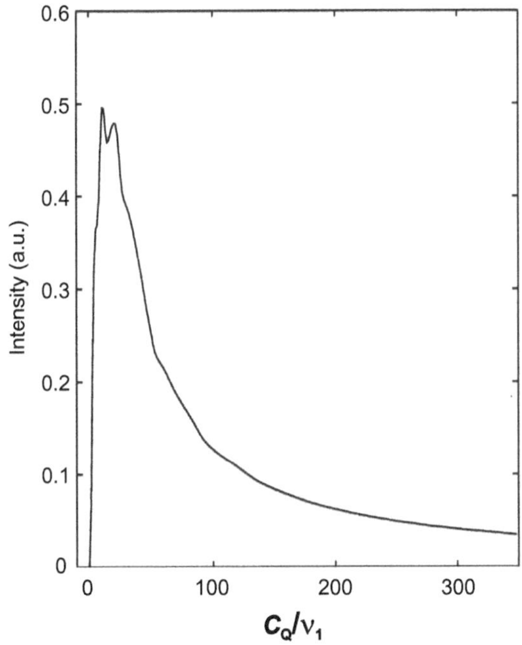

Fig. 16 Simulated NMR intensity from the Z-filter 3QMAS experiment of Fig. 14A for increasing C_Q/ν_1 ratios. The calculations assume a powder comprising $S = 5/2$ nuclei, and 180° and 60° rf pulses for 3QC excitation and reconversion, respectively. *Reproduced with permission from D. Iuga, S. Simon, E. de Boer, A.P.M. Kentgens, A nuclear magnetic study of amorphous and crystalline lanthanum-aluminates, J. Phys. Chem. B 103 (1999) 7591–7598. Copyright (1999) American Chemical Society.*

$$\delta_{\text{iso}} = -\frac{12}{27}\delta_1^{\text{CG}} + \frac{10}{27}\delta_2^{\text{CG}} \tag{55a}$$

$$\delta_{\text{iso}}^{\text{Q}} = \frac{4}{9}\delta_1^{\text{CG}} + \frac{17}{27}\delta_2^{\text{CG}}, \tag{55b}$$

where $C_{Q\eta}$ is derived from $\delta_{\text{iso}}^{\text{Q}}$ via Eqs (53) and (54). For disordered structures, each resulting $\{\delta_{\text{iso}}, C_{Q\eta}\}$ pair yields the respective mean value over a distribution. Application of Eqs (55a) and (55b)—the "C_z" convention of Man [262] that was introduced by Frydman [21, 22]—assumes a sweep width $sw = 1/\Delta t_1$ for the (sheared) isotropic dimension, where Δt_1 is the *physical* Δt_1-incrementation (dwell-time) of the 3QC evolution (see Fig. 14). Moreover, the *unaltered* frequencies along the indirect dimension—i.e., those resulting directly after a two-dimensional FT, without sign-reversal to consider/correct for that the +3QC and −3QC frequencies have opposite signs—are used as input to Eqs (55a) and (55b). Further details about the axes-labelling of MQMAS spectra and NMR parameter extraction are provided in several reviews [27, 28, 262, 263].

The DMFit [125] and *EGdeconv* [129] softwares allow for deconvoluting 3QMAS spectra—albeit both give approximate calculations, as the spin dynamics during the rf pulses is not (fully) accounted for—thereby allowing for determination of the entire $\{x_{\rm Al}^{[p]}, \delta_{\rm iso}^{[p]}, C_{Q\eta}^{[p]}, W_{\rm iso}^{[p]}\}$ parameter-quartet from each $^{27}Al^{[p]}$ coordination in AS glasses. Fig. 17 displays experimental and modelled 3QMAS ^{27}Al NMR spectra from $CaO–Al_2O_3–SiO_2$ glasses, as reported by Massiot et al. [69]. While numerical fitting of the hard-pulse-excited MAS NMR spectrum (Section 10.3) generally provides the most accurate determinations of AlO_p populations, deconvoluting 3QMAS spectra offer generally good estimates thereof, provided that all experimental conditions are emulated, including the 3QC excitation/reconversion dynamics. By employing a 3QC excitation profile (such as that shown in Fig. 16), the $C_{Q\eta}$-dependence of each best-fit $\{x_{\rm Al}^{[p]}\}$ result may be corrected for *approximately* by a weighting factor [264], although this strategy (as implemented) does not capture the effects from the $C_{Q\eta}$ *distribution*. Numerically exact

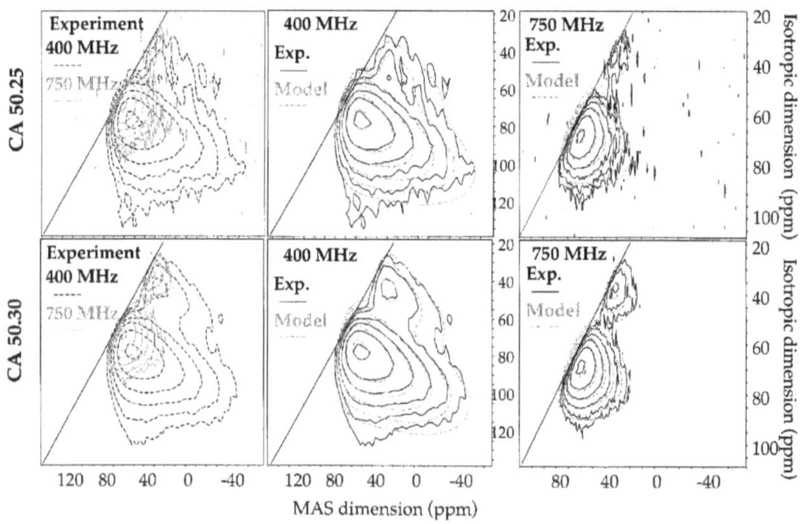

Fig. 17 Experimental and numerically simulated ^{27}Al 3QMAS NMR spectra from $CaO–Al_2O_3–SiO_2$ glasses of composition $0.25CaO–0.25Al_2O_3–0.50SiO_2$ (CA50.25) and $0.20CaO–0.30Al_2O_3–0.50SiO_2$ (CA50.30) [69]. Data were obtained both at $B_0 =$ 9.4 T (400 MHz ^{1}H frequency) and at $B_0 =$ 17.6 T (750 MHz ^{1}H frequency). *Reproduced from D.R. Neuville, L. Cormier, D. Massiot, Al environment in tectosilicate and peraluminous glasses: a ^{27}Al MQ-MAS NMR, Raman and XANES investigation, Geochim. Cosmochim. Acta 68 (2004) 5071–5079, with permission from Elsevier.*

3QMAS NMR simulations that consider *all* events of the 3QMAS scheme (including the rf pulses) [253, 254, 265] yield the most accurate 3QMAS-derived NMR parameters. However, to date such approaches are used sparsely (if at all) for 3QMAS spectra deconvolutions. Yet, given the rapid enhancement of CPU speeds, as well as advances in numerical simulation protocols coupled with the use of parallel computing [266], it is likely only a matter of time until such fitting strategies are utilized routinely within public simulation/deconvolution software.

An alternative route to determine $\{\delta_{iso}, C_{Q\eta}\}$-distributions is to employ an "inversion" procedure that maps the regular 2D $\{\delta_1, \delta_2\}$ spectra representation onto the $\{\delta_{iso}, C_{Q\eta}\}$ space, from which the average values and distribution widths are directly accessible. In contrast to NMR parameter extraction by fitting experimental data to numerical simulations that are inevitably biased to some *assumed* distribution model (see Section 5.4), an inversion procedure is advantageous in being model-free. Moreover, it may also readily reveal if correlations exist between the $\delta_{iso}/C_{Q\eta}$ distributions. Spectra inversion was introduced to NMR by Zwanziger [221] in the context of DAS, and has been employed extensively by Charpentier et al. to determine $\delta_{iso}/C_{Q\eta}$-distributions from $^{17}O/^{23}Na/^{27}Al$ 3QMAS data of various amorphous systems [131, 132, 267]. By using ^{27}Al 3QMAS inversion complemented by results from molecular dynamics (MD) simulations, Angeli et al. investigated the Al–O–Si bond angle distribution in (Na/Ca)–Si–Al–O glasses, inferring that $\bar{\theta}_{Al-O-Si}$ decreased when Si was substituted for Ca at a fixed Na content of 25 mol% [268].

11. Heteronuclear NMR techniques and their application

Double-resonance heteronuclear NMR experimentation involving two *distinct* spin species I and S, with the respective spin quantum numbers I and S, may be exploited for probing not only the short-range AS glass structure, but also its *medium*-range order (≈0.3–1 nm). Typical problems include examining the preferential interbond connectivities among the various glass-network formers (e.g., the Si/Al ordering in AS glasses), and how the network modifiers distribute around the network groups. Depending on the pair of nuclides involved, structural features are accessible across a length-scale up to ≈0.7 nm. Below, we outline the most commonly employed heteronuclear NMR techniques and review their application to structural studies of AS glasses, where ^{27}Al constitutes one of the targeted nuclei of the I–S pair.

11.1 Cross polarization

The rf pulse scheme for a cross-polarization (CP) experiment [24, 25] is depicted in Fig. 18: it extends an S-spin-locking (Section 8.4.2) segment into a double-resonance MAS NMR experiment by simultaneously irradiating the I-spins [26], whereupon the S-spin signal is detected. The analogue of τ_{SL} is called the *contact interval* (τ_{CP}), during which the *heteronuclear* I–S *through-space dipolar* interaction becomes *recoupled* (restored), such that I-spin magnetization is transferred to their S neighbours. For CP under MAS conditions (CPMAS), the magnetization-transfer process occurs whenever any of the following *modified Hartmann-Hahn* conditions [25, 26, 269–279] is fulfilled:

$$\nu_1(\mathrm{I}) \pm \nu_1(\mathrm{S}) = n\nu_r, \quad n = \pm 1, \pm 2. \tag{56}$$

Here each nutation frequency of the I- and S-spins, $\nu_1(\mathrm{I})$ and $\nu_1(\mathrm{S})$, depends on its respective spin quantum number I and S, as well as on the precise experimental condition and the value of the quadrupolar frequency (if either I or S is a quadrupolar nucleus). For the S-spins, $\nu_1(\mathrm{S})$ is given by

$$\nu_1(\mathrm{S}) \equiv \nu_1^{\mathrm{S}} = |\gamma_{\mathrm{S}}|B_1^{\mathrm{S}}/(2\pi) \text{ [if } S = 1/2 \text{ or } \nu_1^{\mathrm{S}} \gg \nu_{\mathrm{Q}}], \tag{57a}$$

$$\equiv \nu_{\mathrm{nut}}^{\mathrm{CT}}(\mathrm{S}) = (S + 1/2)\nu_1^{\mathrm{S}} \text{ [if } \nu_1^{\mathrm{S}} \ll \nu_{\mathrm{Q}}]. \tag{57b}$$

Hence, for spins $S = 1/2$, or $S = q/2$ with $q = \{3, 5, 7, 9\}$ in the *nonselective* regime, the relevant nutation frequency is given by Eq. (57a), whereas for an

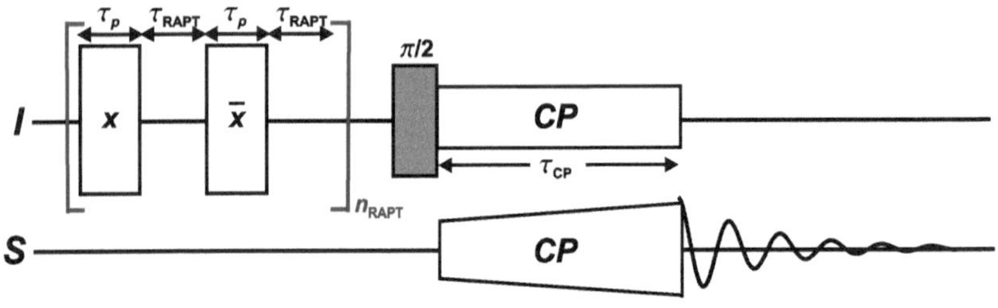

Fig. 18 Rf pulse scheme for an I → S cross-polarization (CP) experiment under MAS conditions [24, 25]. During the contact interval (τ_{CP}), magnetization is transferred from the I-spin species to S-spins in their close proximity. The figure shows the case where I is a half-integer spin quadrupolar nucleus (e.g., ^{27}Al), so that an enhanced the I/S-spin signal intensity may be achieved by preceding the CP stage by the RAPT protocol [196, 197] (optional, see Fig. 13), and resulting in the RAPT-CP scheme [187]. The detected S-spin may represent either a $S = 1/2$ (e.g., ^{1}H or ^{23}Si) or $S > 1/2$ spin (e.g., ^{17}O, ^{11}B). Note the "ramp" of the S-spin pulse around the modified Hartmann-Hahn condition (Eq. 56), which stabilizes and enhances the magnetization transfers [280].

half-integer spin S in the *CT-selective* regime, the CT nutation frequency [$\nu_{\mathrm{nut}}^{\mathrm{CT}}(\mathrm{S})$ of Eq. (57b)] should obey Eq. (56); see Section 8.4.2. Analogous expressions apply to the I-spin species, i.e., $\nu_1(\mathrm{I})$ is either equal to ν_1^{I} or to $\nu_{\mathrm{nut}}^{\mathrm{CT}}(\mathrm{I})$ depending on the spin quantum number I and the experimental conditions. Note that the amplitudes of the S-spin NMR signals are associated with opposite signs for the $\nu_1(\mathrm{I}) + \nu_1(\mathrm{S})$ and $\nu_1(\mathrm{I}) - \nu_1(\mathrm{S})$ Hartmann-Hahn conditions of Eq. (56): these respective "2Q" and "ZQ" conditions may therefore interfere destructively [270, 274, 277–279].

The rate of the I→S magnetization transfer, i.e., how rapidly the S-spin NMR signal forms, depends on the value of the I–S dipolar coupling constant (b_{IS}), which exhibits an inverse cubic relationship to the internuclear distance r_{IS}; $b_{\mathrm{IS}} \sim r_{\mathrm{IS}}^{-3}$ [26, 34]. For *short* contact intervals, the S-spin NMR signal intensity is proportional to $(b_{\mathrm{IS}}\tau_{\mathrm{CP}})^2$. Hence, by arranging a small value of τ_{CP}, *solely* S-spin sites in close proximity (within a few hundred pm) to an I-spin neighbour are detected [26, 30]. Moreover, implementing progressively lengthened τ_{CP} periods informs about the relative I–S internuclear "dipolar contacts" in the structure [26, 271, 272, 278]. Here, a "strong" dipolar contact implies a large number of I-spins *nearby* the given S site, whereas a "weak" dipolar contact implies few(er) I-spin neighbours and/or long internuclear I–S distances.

Since it constitutes a double-resonance extension of spin-locking, CPMAS unfortunately inherits the complexities and CT NMR signal losses when dealing with quadrupolar nuclei (Section 8.4.2). Such obstacles accentuate when *both* I and S represent half-integer spin species, such as for $^{27}\mathrm{Al}/^{17}\mathrm{O}$ [73, 75, 208] or $^{11}\mathrm{B}/^{27}\mathrm{Al}$ [274–277, 281]. Consequently, despite that CPMAS was initially introduced as an S-signal *enhancement* technique [24, 25]—and is indeed utilized as such in organic/biological solids when exploiting $^{1}\mathrm{H}\rightarrow^{13}\mathrm{C}$ or $^{1}\mathrm{H}\rightarrow^{15}\mathrm{N}$ magnetization transfers—its implementations to amorphous inorganic phases that involve at least one half-integer spin quadrupolar nucleus are plagued by very low magnetization-transfer efficiencies (typically 1–10% of those observed for spin-1/2 pairs). This feature primarily stems from a strong NMR signal damping ($T_{1\rho}$ relaxation [282]) during spin-locking of quadrupolar nuclei, thereby imposing an upper limit of the feasible contact period that may be used in the CPMAS experiment. Hence, regardless if applied to a crystalline or to an amorphous network structure, CPMAS mainly serves as a source of qualitative I–S *proximity* information, although it may be analyzed for extracting semi-quantitative internuclear distances [271, 272, 278].

11.2 CPMAS applications to aluminosilicate glasses

11.2.1 $^{27}Al\leftrightarrow{}^{1}H$ CPMAS experimentation

Most CPMAS applications to crystalline and amorphous aluminosilicate phases involve either the spin-pairs ^{29}Si and ^{1}H (e.g., see [283, 284]) or ^{27}Al and ^{1}H [121, 285–293] . Out of the latter, several reports utilized magnetization transfers *from* ^{1}H (i.e., $^{1}H\rightarrow{}^{27}Al$) [121, 285, 286, 288, 289] along the lines for recording $^{13}C/^{15}N$ NMR spectra from organic molecules. Such $^{1}H\rightarrow{}^{27}Al$ CPMAS experimentation enables a surface-*selective* probing of the $\{Al^{[p]}\}$ speciation of initially anhydrous AS glasses upon their exposure to aqueous solutions, as demonstrated by Tsomaia et al. [288] for Na_2O–Al_2O_3–SiO_2 glasses subjected to acidic solutions with pH 2. While directly (single-pulse) excited ^{27}Al MAS NMR spectra revealed the sole presence of AlO_4 tetrahedra in the *bulk* of the glass (as for the pristine specimen), $^{1}H\rightarrow{}^{27}Al$ CPMAS NMR evidenced a prevalence of AlO_6 groups at the leached AS glass *surface* [288].

We next focus on reviewing the utilization of the reverse magnetization transfer i.e., $^{27}Al\rightarrow{}^{1}H$ CPMAS [290–293], as first introduced for studying AlOH motifs in zeolites [290]. While $^{1}H\rightarrow{}^{27}Al$ CPMAS reveals the alterations of the Al speciation at the glass surface, the $^{27}Al\rightarrow{}^{1}H$ counterpart has proven to be extremely valuable for probing the local proton environments in *hydrous* AS glasses. Here the work of Xue and Kanzaki [291–293] on various alkali/alkaline-earth bearing AS glasses has provided a particularly detailed picture of the various coexisting OH moieties in the bulk structure of hydrous ($\lesssim$2 wt% H_2O) AS glasses involving either Na^+ or K^+ as the network modifier [291], or glasses that simultaneously incorporate Ca^{2+} and Mg^{2+} [292, 293]. From an array of NMR experimental protocols encompassing directly excited ^{1}H NMR spectra, homonuclear ^{1}H correlation NMR experiments, 3QMAS ^{27}Al NMR, as well as various $^{27}Al\rightarrow{}^{1}H$ CPMAS options, Xue and Kanzaki inferred the presence of three main groups of OH environments, encompassing (*i*) "free OH groups", i.e., those associated solely with the network-modifier cations and resonating around $\delta_H = 1.2$ ppm; (*ii*) SiOH (silanol) groups with variable degrees of H bonding ($\delta_H \gtrsim 5$ ppm), as well as (*iii*) $Al^{[p]}$–OH moieties whose ^{1}H NMR signals appear in the 2–3 ppm range [292, 293]. The latter involve primarily $Al^{[4]}$–OH, although minor amounts of $Al^{[5]}$–OH and $Al^{[6]}$–OH motifs were also detected, as evidenced by the coupling of 3QMAS with $^{27}Al\rightarrow{}^{1}H$ CPMAS experimentation; see Fig. 19. ^{1}H resonances appear along the horizontal dimension of such a 2D NMR spectrum, yet *only* from proton sites in close proximity to $^{27}Al^{[p]}$, whose respective ^{1}H NMR peaks are resolved along the vertical (isotropic) dimension.

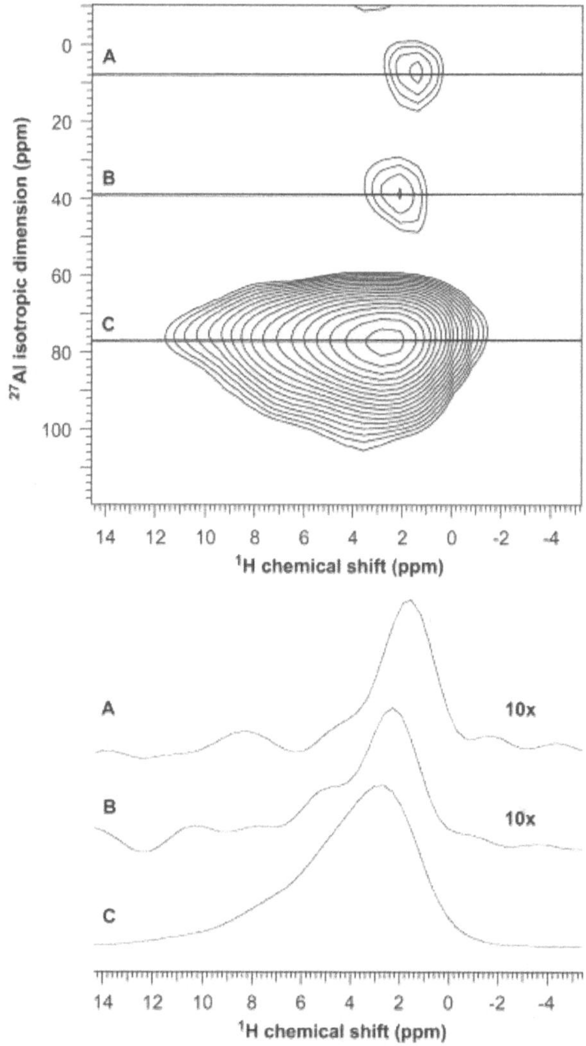

Fig. 19 *Top panel*: 2D NMR spectrum recorded from an hydrous CaO–MgO–Al_2O_3–SiO_2 glass (1.7 wt% H_2O) at $B_0 = 9.4$ T and 15.00 kHz MAS [292], and using a combination of 3QMAS [21] and $^{27}Al \rightarrow {}^{1}H$ CP [290] to obtain a high-resolution (vertical) ^{27}Al spectral dimension revealing $^{27}Al^{[4]}$ (dominating), $^{27}Al^{[5]}$, and $^{27}Al^{[6]}$ resonances correlated with the respective ^{1}H NMR signals from Al–O^{1}**H** groups that appear along the horizontal dimension. *Bottom panel*: slices along the ^{1}H dimension taken through each (A) $^{27}Al^{[6]}$, (B) $^{27}Al^{[5]}$, and (C) $^{27}Al^{[4]}$ resonance. *Reproduced from X. Xue, M. Kanzaki, Al coordination and water speciation in hydrous aluminosilicate glasses: direct evidence from high-resolution heteronuclear ^{1}H-^{27}Al correlation NMR, Solid State Nucl. Magn. Reson. 31 (2007) 10–27, with permission from Elsevier.*

11.2.2 $^{27}Al \rightarrow ^{29}Si$ CPMAS experimentation

Once considering options for CPMAS applications to probe the *bulk* structure of *anhydrous* AS glasses, utilization of *at least one* quadrupolar nucleus becomes mandatory, because ^{27}Al, ^{17}O, as well as essentially all M^{z+} cations, possess half-integer spin. Hence, ^{29}Si is typically the sole spin-1/2 nuclide.[k] Nonetheless, the repetition rate of the I → S CP NMR signal averaging depends on the spin-lattice relaxation properties of the I-spins [24, 25]. Consequently, for structures devoid of rapid proton mobility, it may be advantageous to instead exploit the typically much faster T_1-relaxation of ^{27}Al in $^{27}Al \rightarrow ^{29}Si$ CP experiments to accelerate the acquisition of ^{29}Si NMR spectra [187] (where ^{29}Si usually exhibits very slow T_1 relaxation in rigid solids; see Section 8.3). Such $^{27}Al \rightarrow ^{29}Si$ CPMAS NMR experiments require long contact intervals (10^1–10^2 ms) to ensure magnetization transfers also from (very) distant ^{27}Al sites. While the $^{27}Al \rightarrow ^{29}Si$ CPMAS concept has been demonstrated on crystalline AS phases [187, 295, 296], we are only aware of one application to AS-based *glasses* (but without much added structural insight) [93]. Similar ideas of using $^{7}Li \rightarrow ^{29}Si$ CP to accelerate the recording of ^{29}Si MAS NMR spectra were utilized for Li_2O–SiO_2 glass-ceramics [297]. Likewise, although $^{27}Al \rightarrow ^{31}P$ CPMAS reports exist for the significantly more NMR signal sensitive $^{31}P/^{27}Al$ pair (where $\gamma_P \approx 2\gamma_{Si}$, thereby offering a $2^3 = 8$ fold better S/N for ^{31}P relative to ^{29}Si; see Section 6) in the context of aluminophosphate glasses [298–301], they remain much fewer than those targeting crystalline phases (e.g., see Refs. [271, 272, 278]).

Besides the more favourable T_1-relaxation properties of ^{27}Al relative to $X = \{^{29}Si, ^{31}P\}$, large $^{27}Al \rightarrow X$ CPMAS NMR experimental time-savings are offered by implementing RAPT or other CT ^{27}Al NMR signal enhancement techniques (see Section 8.5.2), as well by "ramping" the amplitude of the X-spin rf pulse [280] (Fig. 18), as demonstrated in the context of CPMAS from $I > 1/2$ nuclei in Refs. [187, 302]. Further signal-sensitivity boosts are offered by coupling the RAPT-CP [187] protocol with the Carr-Purcell-Meiboom-Gill (CPMG) technique [303, 304] during the ^{29}Si NMR signal detection [294, 295].

Moreover, $^{27}Al \rightarrow ^{17}O$ CPMAS has been utilized for supporting results from dipolar-dephasing NMR experiments towards an unambiguous proof of direct ^{27}Al–NBO bonding in AS glasses [73] (see Sections 11.5.1 and

[k] Exceptions are, for instance, the spin-1/2 nuclides ^{109}Ag and ^{89}Y, where $^{27}Al \rightarrow ^{89}Y$ CPMAS has been applied to AS glasses [294] .

12.3.3). Yet, $^{27}Al \rightarrow ^{17}O$ and $^{27}Al \rightarrow ^{29}Si$ CPMAS experimentation is hitherto only exploited to a minor extent of their full potential: particularly for the $^{27}Al/^{29}Si$ spin-pair, we anticipate further utilizations for examining SiO_4–AlO_p connectivities and Si/Al intermixing in AS glasses. Notably, however, when it comes to (chemical) shift-correlation experimentation, there are better alternatives to the I → S CP-based heteronuclear correlation (HETCOR) experiment [30, 305], such as using the dipolar-based heteronuclear multiple-quantum coherence (D-HMQC) protocol [185, 306, 307]. This technique is particularly useful for spin-pairs involving *at least one* quadrupolar species (I or S), as for instance utilized for $^{1}H/^{27}Al$ [306–308], $^{1}H/^{23}Na$ [309] $^{27}Al/^{31}P$ [301, 306, 307], $^{23}Na/^{27}Al$ [310], $^{23}Na/^{31}P$ [306], and $^{11}B/^{31}P$ [79, 301]. However, the D-HMQC method has also proven valuable for applications targeting the more straightforward $I = 1/2$ scenarios of (for instance) $^{13}C/^{15}N$ [307] and $^{1}H/^{31}P$ [311, 312].

11.3 Protocol for implementing CPMAS experiments

Along the lines of Section 8—and particularly connecting to the spin-lock experimentation of Section 8.4.2—we here outline the practical implementation of the I → S CPMAS rf pulse scheme shown in Fig. 18, assuming that I and S feature $I > 1$ and $S = 1/2$, respectively. Its first step involves selecting the MAS rate, which must be made as a compromise between using the lowest possible MAS rate to ensure a reasonably efficient spin-locking of the half-integer spin quadrupolar nucleus, yet being sufficiently rapid to (*i*) provide a decent resonance-offset bandwidth of both the I and S species (the bandwidth scales roughly linearly with each $\nu_1(S)$ and $\nu_1(I)$ nutation frequency); (*ii*) avoiding overlap between the $\nu_1(S) + \nu_1(I)$ and $\nu_1(S) - \nu_1(I)$ CPMAS conditions that otherwise interfere destructively and compromises the I→S magnetization transfer [270, 274, 277–279] (see Section 11.1), as well as (*iii*) for suppressing the spinning sidebands of the second-order quadrupolar interaction (see Section 5.3.2). The latter two concerns are particularly relevant if *also* S possesses half-integer spin. For cases of I≡^{27}Al and AS glasses exhibiting nonnegligible $Al^{[5]}/Al^{[6]}$ populations, MAS rates of $\nu_r \lesssim$ 10 kHz that typically provide the most efficient magnetization transfers are insufficient for ensuring uniform magnetization transfers across the entire span of $^{27}Al^{[p]}$ resonance frequencies at (moderately) high magnetic fields $B_0 \gtrsim 9.4$ T. For such cases, however, MAS rates ≈20 kHz provide decent resonance-offset bandwidths at the expense of overall (markedly) lower magnetization-transfer efficiencies.

Throughout, we assume that all I-spin sites manifest large quadrupolar frequencies so that the CT-selective regime applies (Section 7) and $\nu_{\text{nut}}^{\text{CT}}(\text{I})$ is given by Eq. (57b). Once the MAS rate is selected, the CPMAS optimization proceeds as follows, with the first three steps only involving (sequential) rf application on either the I- *or* S-spin species:

(1) For each of I and S, the rf carrier frequency is positioned such that it coincides with the centre-of-gravity of the I (S) NMR signals in the spectrum. This precaution is crucial due to the low rf amplitudes involved and thereby compromised I and S resonance-offset bandwidths (see above).

(2) The wisest choice of the I-spin CT nutation frequency normally coincides with that providing an optimum spin-locking, i.e., $\nu_{\text{nut}}^{\text{CT}}(\text{I}) \approx 2\nu_r/3$. Hence, the rf amplitude obeying this condition is first located by performing a spin-lock array of the I-spins, as described in Section 8.4.2. Note that the thereby obtained value of $\nu_{\text{nut}}^{\text{CT}}(\text{I})$ is often a suitable choice *both* for the spin-lock rf pulse during CP and for the preceding CT-selective 90° pulse (see Fig. 18). The duration of the latter is calibrated according to Section 8.4.3.

(3) A suitable Hartmann-Hahn condition is selected out of the possible options of Eq. (56), where the "2Q condition" $\nu_{\text{nut}}^{\text{CT}}(\text{I}) + \nu_1(\text{S}) = \nu_r$ is usually most efficient [274, 277], provided that the MAS rate exceeds the frequency-spread of the S-spins (Δ_S), such that $\Delta_\text{S} < \nu_1(\text{S}) \approx \nu_r/3$. If that is not the case, the second-best choices are typically either $\nu_1(\text{S}) = 2\nu_r - \nu_{\text{nut}}^{\text{CT}}(\text{I})$ [i.e., $\nu_{\text{nut}}^{\text{CT}}(\text{I}) + \nu_1(\text{S}) = 2\nu_r$ of Eq. (56)] or $\nu_1(\text{S}) = \nu_{\text{nut}}^{\text{CT}}(\text{I}) + \nu_r$. However, it is important to *also* avoid any rotary resonance (including 2Q-HORROR) condition for the S-spins, i.e., to ensure that $\nu_1(\text{S}) \neq n\nu_r$ with $n = \{1/2, 1, 2\}$; see Section 8.4.2. This is accomplished by also performing a spin-locking NMR experiment for the S-spin species, which also provides information about the relative spin-lock efficiency at each potential Hartmann-Hahn condition of Eq. (56). Analogous concerns about the I-spin resonance spread (Δ_I) relative to $\nu_{\text{nut}}^{\text{CT}}(\text{I})$ also applies in step (2). However, for the as-assumed scenario of $\{I > 1,\ S = 1/2\}$, it is usually advantageous to arrange the largest possible $\nu_1(\text{S})/\Delta_\text{S}$ ratio to improve the CPMAS NMR performance, whereas experiments with $\nu_{\text{nut}}^{\text{CT}}(\text{I}) > \nu_r$ degrade the magnetization-transfer efficiency. However, the situation complexifies if *also* $S > 1$; then a decision must be made whether to arrange $\nu_{\text{nut}}^{\text{CT}}(\text{I}) \approx 2\nu_r/3$ *or* $\nu_{\text{nut}}^{\text{CT}}(\text{S}) \approx 2\nu_r/3$ in step (2).

(4) All remaining optimization stages involve double-resonance application with the CPMAS protocol of Fig. 18: First, an educated guess of τ_{CP} is required, which is ideally the value that provides the strongest S-spin NMR signal. Note that τ_{CP} depends on b_{IS}, and thereby on both the γ_I and γ_S values of the nuclides involved, as well as on the I–S internuclear distance. The NMR relaxation during spin-locking ($T_{1\rho}$) is usually rapid for ^{27}Al and other half-integer spins. Hence, towards selecting a suitable τ_{CP} value, we propose to first perform a series of NMR acquisitions with progressively increasing spin-lock periods using the protocol of Fig. 12A at the preselected ν_{nut}^{CT}(I)-value from stage (3). In the absence of further guidance, we suggest choosing $\tau_{CP} \approx 5$ ms and $\tau_{CP} \approx 10$ ms for ^{27}Al→^{17}O and ^{27}Al→^{29}Si CPMAS optimizations on AS glass specimens, respectively.

(5) Once each targeted nutation frequency of ν_{nut}^{CT}(I) and ν_1(S) is located, the latter is kept fixed, while an array of variable I-spin rf amplitudes is performed to locate the ν_{nut}^{CT}(I)-value that provides the strongest S-spin signal intensity. Ideally, this optimization should result in $\nu_{nut}^{CT}(I) \approx 2\nu_r/3$ of stage (2), thereby ensuring that the spin-locking of the quadrupolar nucleus remains near-optimal.

(6) Once the pair $\{\nu_{nut}^{CT}(I), \nu_1(S)\}$ of "optimal" nutation frequencies obeying Eq. (56) is deduced, a linear "ramp" of the rf amplitude is introduced (Fig. 18), such that the nutation frequency is swept between $\nu_1(S) \pm \nu_\Delta$. Typically, the optimum ν_1(S)-range is narrow ($\nu_\Delta/\nu_1(S) \lesssim 0.05$), meaning that the *span* of nutation frequencies is less than 10% of the centre value ν_1(S).

(7) Finally, an array of CPMAS NMR experiments with variable contact periods is performed, using the optimal set of $\{\nu_{nut}^{CT}(I), \nu_1(S), \nu_\Delta\}$ parameters obtained in the previous steps of the optimization procedure.

11.4 Dipolar-dephasing NMR experiments

11.4.1 Principles

An S{I} "dipolar-dephasing" experiment probes S–I proximities without involving direct magnetization transfers, but the spatial contact between a detected S-spin site and its surrounding I-spins is merely manifested by an NMR signal *attenuation (dephasing)* upon a temporary recoupling of the I–S dipolar interaction. This is achieved by applying rf pulses to the (non-observed) I-spins during a given *dipolar recoupling period* (τ_{rec}) [313–315]. The notation "S{I}" implies that the S-spin NMR signal is *detected*, whereas the rf pulses for recoupling the I–S dipolar interactions are applied to I. The most commonly employed dipolar-dephasing techniques are rotational-echo double resonance (REDOR) [313], rotational-echo adiabatic-passage double resonance (REAPDOR) [316], and transfer of populations in double resonance (TRAPDOR) [317–319]. Below we outline the essence of these NMR techniques and review their use in the context of ^{27}Al NMR applied to AS glasses.

Both the REDOR (Fig. 20A) and REAPDOR (Fig. 20B) NMR protocols recouple the dipolar I–S interactions by using a series of rotor-synchronized 180° rf pulses, but they differ in that the pulse-train is applied

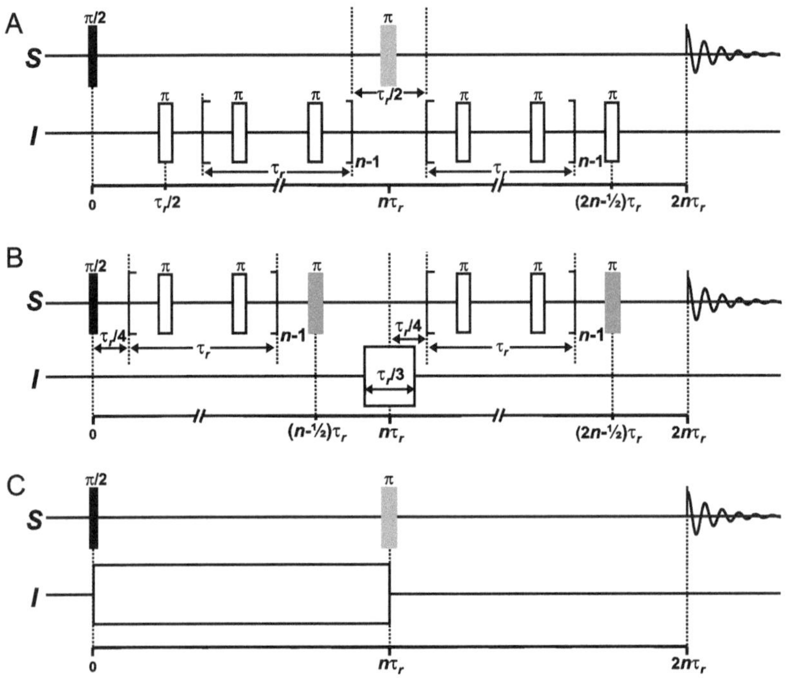

Fig. 20 Schematic rf pulse protocols for S{I} dephasing NMR experiments based on (A) REDOR [313], (B) REAPDOR [316], and (C) TRAPDOR [317, 318], where applications to spin-pairs with $S = 1/2$ and $I > 1$ are assumed. *Solid black rectangles* and *open/grey rectangles* represent 90° ($\pi/2$) and 180° (π) rf pulses, respectively. The *grey rectangles mark* pulses with fixed positions, regardless of the choice of the duration of the recoupling period; they are responsible for refocusing S-spin isotropic chemical shifts. In (A) and (B), the number of 180° pulses (*open rectangles*) for dipolar recoupling is incremented in blocks of four pulses that together span two rotor periods, thereby giving a recoupling/dephasing period of $\tau_{rec} = 2n\tau_r$, where $n = 1, 2, 3, \ldots$. The *white rectangle* in (C) depicts the strong TRAPDOR dephasing pulse, which is incremented in completed rotor periods: $\tau_{rec} = n\tau_r$. Note that the TRAPDOR experiment may be utilized when both spin quantum numbers I and S are half-integer: then, the S-spin 180° rf pulse must be CT-selective.

to the I- and S-spins, respectively. Moreover, the REAPDOR NMR experiment (Fig. 20B) involves a strong "adiabatic-passage" rf pulse applied to the I-spins for $\approx \tau_r/3$, which along with the 180° pulse-train on the S-spins is responsible for recoupling the I–S dipolar interactions [314, 316]. Both methods are suitable for probing heteronuclear proximities in I–S pairs involving *one* spin-1/2, which for the REDOR (REAPDOR) experimentation should be selected to constitute the I (S) spin species. For a REDOR implementation, this ensures a minimal rf irradiation of the detected $S > 1/2$ nuclei. We note that Eckert's group has also introduced an alternative S{I} REDOR version [315, 320], which only differs from the REAPDOR protocol of Fig. 20B in that one 180° rf pulse is applied to the $I > 1/2$ nucleus instead of the adiabatic-passage pulse. Here we do not consider this REDOR incarnation further.

Each REDOR, REAPDOR, and TRAPDOR protocol involves recording an array of NMR experiments with a progressively lengthened recoupling/dephasing interval $\tau_{\rm rec}$, which for REDOR and REAPDOR is sampled at any even integer multiple of the rotor period ($\tau_{\rm rec} = 2n\tau_r$) [313, 314]; see Fig. 20(A, B). In contrast, TRAPDOR employs a strong continuous rf pulse to the I-spins (with $I > 1/2$) throughout $\tau_{\rm rec} = n\tau_r$, where $n = 1, 2, \ldots$ [317–319] (Fig. 20C), similarly to the adiabatic-passage pulse of the REAPDOR scheme [314, 316]. The thereby recoupled I–S dipolar interactions lead to a diminished (*dephased*) S-spin NMR signal [$S(\tau_{\rm rec})$] relative to that observed in a separately acquired spin-echo (reference) NMR experiment, denoted $S_0(\tau_{\rm rec})$ [313–315]. Here a *low* (*high*) NMR signal attenuation is equivalent to a *large* (*small*) dipolar contact, i.e., a *high* (*low*) value of $b_{\rm IS}$.

11.4.2 Extraction of dipolar second moments

In an amorphous structure, quantitative information about the I–S contacts is available from the Van Vleck *dipolar second moment* [321],

$$M_2(\mathrm{S-I}) = \frac{\mu_0^2\hbar^2\gamma_\mathrm{I}^2\gamma_\mathrm{S}^2}{64\pi^4 N_\mathrm{S}}\sum_{j=1}^{N_\mathrm{S}}\sum_{k=1}^{N_\mathrm{I}}\left[r(\mathrm{S}_j - \mathrm{I}_k)\right]^{-6}\ [\text{unit}: \mathrm{s}^{-2} = \mathrm{Hz}^2], \tag{58a}$$

$$= N_\mathrm{S}^{-1}\sum_{j=1}^{N_\mathrm{S}}\sum_{k=1}^{N_\mathrm{I}}\left[b(\mathrm{S}_j - \mathrm{I}_k)\right]^2, \tag{58b}$$

which is proportional to the sum over the inverse sixth power of the interatomic distance of each S_j–I_k pair. Here, μ_0 is the permeability of vacuum, while N_I and N_S represent the total numbers of I and S nuclei in the glass, respectively (which are obtained from the glass stoichiometry).

As illustrated in Fig. 21A and B, the value of M_2(S–I) may be extracted by fitting the *initial* S-spin signal dephasing from an S{I} REDOR/REAPDOR NMR experiment to the following expression [275, 315, 320, 322, 323]:

$$\Delta S/S_0 \equiv [S_0(\tau_{\text{rec}}) - S(\tau_{\text{rec}})]/S_0(\tau_{\text{rec}}) = \frac{A}{I(I+1)} M_2(\text{S}-\text{I})\tau_{\text{rec}}^2. \quad (59)$$

Here, $A = 1$ for REDOR NMR applications with $I = 1/2$ [275, 315, 322, 323], while A must be determined empirically by separate experiments for

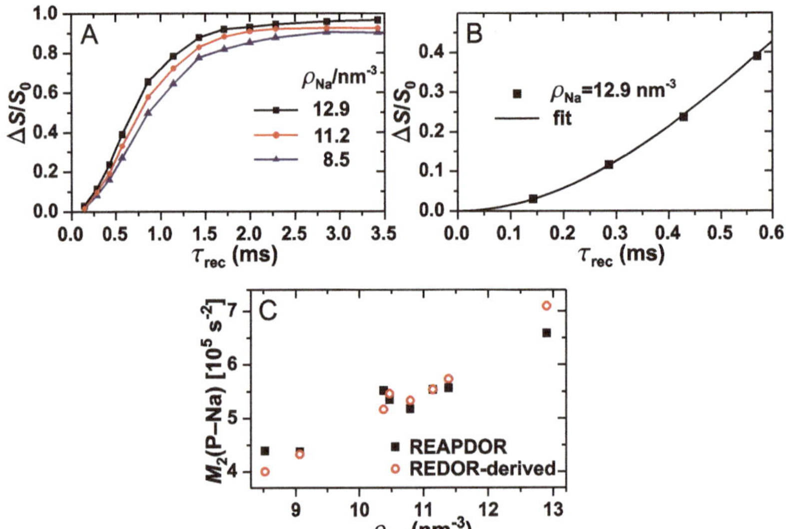

Fig. 21 (A) $^{31}P\{^{31}Na\}$ REAPDOR NMR dephasing curves ($\Delta S/S_0$) plotted against the recoupling/dephasing period (τ_{rec}). The data were recorded at $B_0 = 9.4$ T and $\nu_r = 14.00$ kHz from Na_2O–CaO–SiO_2–P_2O_5 glasses with increasing Na content, as specified by the number density ρ_{Na} [323]. (B) Zoom of the initial region of one dephasing curve, where the parabola represents the best-fit results to Eq. (59), which was used to extract the dipolar second moment $M_2(^{31}P–^{23}Na)$. (C) REAPDOR-derived $\{M_2(^{31}P–^{23}Na)\}$ data plotted against ρ_{Na}; they are contrasted with the results obtained by a similar $^{31}Na\{^{31}P\}$ REDOR NMR data analysis on the same glasses, yielding the set $\{M_2(^{23}Na–^{31}P)\}$, where each value was subsequently converted into its $M_2(^{31}P–^{23}Na)$ counterpart (REDOR-derived) by using Eq. (60) [323]. The two independent $\{M_2(^{31}P–^{23}Na)\}$ data sets agree very well, thereby demonstrating that the more readily implemented and analyzed $^{31}Na\{^{31}P\}$ REDOR NMR experiment may be utilized be obtain the same information as *if* $^{31}P\{^{31}Na\}$ REAPDOR was employed.

implementations involving $I > 1/2$ [275, 315, 320, 322, 324], as well as for REAPDOR (see below and Ref. [323]). The $\{\tau_{rec}, \Delta S/S_0\}$ data-set used for the numerical fitting must be restricted to short dephasing periods, because Eq. (59) is a Taylor expansion (truncated at the first term) of an exact spin dynamics solution for an isolated *pair* of spins-1/2 [325], where the bearings from homonuclear I–I and S–S dipolar interactions are not accounted for [326]. Yet, the approximation of the S-spin dephasing dynamics for SI_n clusters expressed as a superposition of n individual S–I_k spin-pair responses is adequate in the limit $\Delta S/S_0 \rightarrow 0$. This analysis approach, however, tends to *underestimate* the M_2(S–I) value in that *the lower the largest $\Delta S/S_0$-value included in the fitting, the higher the best-fit dipolar second moment* (e.g., see the best-fit results in Ref. [323] that were obtained by using progressively increasing upper values of $\Delta S/S_0$). Hence, whenever possible, we recommend restricting the numerical fitting to the range $\Delta S/S_0 \lesssim 0.1$ [82, 323, 324], which is lower than previous recommendations of $\Delta S/S_0 \lesssim 0.2$ [275, 315, 320] or $\Delta S/S_0 \lesssim 0.3$ [322, 326, 327]. However, here a compromise must be made between the increased best-fit-data uncertainties from using a strongly restricted $\{\tau_{rec}, \Delta S/S_0\}$ set (of typically 2–4 data points) and the inevitable systematic underestimation of the dipolar second moment resulting by including larger values $\Delta S/S_0 > 0.1$.

The determination of quantitative dipolar second moments by using Eq. (59) was until recently restricted solely to S{I} REDOR NMR applications. Yet, Stevensson et al. [323] demonstrated its applicability also for REAPDOR analyses (Fig. 21), which otherwise demand time-consuming—yet approximate and model-dependent—numerical spin dynamics simulations for I–S proximity information (e.g., see Refs. [299, 300, 315, 328, 329]). However, the REAPDOR-dephasing analysis *via* Eq. (59) first requires a "calibration" of the A parameter by performing an independent NMR experiment (under identical conditions) on a crystalline phase whose M_2(S–I) value is known [323]. Such an independent calibration procedure is also required for S{I} REDOR NMR implementations involving $I > 1/2$ [275, 315, 320, 322, 324], but then enables M_2(S–I) determinations even for spin systems where *both* species I and S possess half-integer spins. Hence, the REDOR and TRAPDOR NMR protocols constitute the *only* options available to date for such scenarios. However, the TRAPDOR technique may only provide *qualitative* internuclear distance information, as its NMR signal dephasing dynamics differs inherently from the parabolic $\Delta S/S_0$-dependence ($\sim \tau_{rec}^2$) on the recoupling/dephasing period given by Eq. (59) [319].

11.4.3 Two dipolar second moments at the price of one

Given the very limited options of S{I} and I{S} dephasing experiments available for certain I–S pairs, the following property of the corresponding M_2(S–I) and M_2(I–S) dipolar second moments is very useful [323]:

$$f_S n_S S(S+1) M_2(\mathrm{S-I}) = f_I n_I I(I+1) M_2(\mathrm{I-S}). \tag{60}$$

Here n_I and n_S are the stoichiometric amounts of the I and S nuclei in the sample, respectively, while f_I and f_S are the corresponding (fractional) isotopic abundances. Eq. (60) has the following important consequence: *Both* second moments M_2(S–I) or M_2(I–S) are available from *one sole* heteronuclear S{I}*or* I{S} NMR experiment, which is selected as the most favourable/straightforward one to implement [323]. Fig. 21C illustrates a good agreement observed between the sets of second moments $\{M_2(^{31}\mathrm{P}–^{23}\mathrm{Na})\}$ from Na_2O–CaO–SiO_2–P_2O_5 glasses obtained either directly by fitting ^{31}P{^{31}Na} REAPDOR NMR data to Eq. (59), or extracted from the $\{M_2(^{23}\mathrm{Na}–^{31}\mathrm{P})\}$ results of ^{23}Na{^{31}P} REDOR NMR experiments fed through Eq. (60).

In some cases, Eq. (60) enables determining dipolar second moments that may be very difficult—or even impossible—to obtain experimentally by other means. For instance, M_2(Na–B$^{[3]}$) and M_2(Na–B$^{[4]}$) data are currently very challenging to extract by applying rf pulses to ^{11}B due to the very different quadrupolar coupling constants of ^{11}B$^{[3]}$ ($C_Q \approx 2.5$ MHz) and ^{11}B$^{[4]}$ ($C_Q < 0.8$ MHz); see Section 7. Yet, such M_2(Na–B$^{[p]}$) results were recently derived for several Na_2O–B_2O_3–SiO_2 glasses from the corresponding M_2(B$^{[3]}$–Na) and M_2(B$^{[4]}$–Na) data obtained from a ^{11}B{^{23}Na} REDOR NMR experiment [324].

The M_2(S–I) $\leftrightarrow$ M_2(I–S) conversion strategy of Eq. (60) has thus far been exploited to study glass structures when using ^{23}Na along with either ^{31}P, ^{29}Si, and ^{11}B [82, 324]. Yet, it is by no means restricted to spin-pairs involving ^{23}Na. For instance, it could greatly assist the probing of relative bonding preferences between ^{29}Si and each {^{27}Al$^{[4]}$, ^{27}Al$^{[5]}$, ^{27}Al$^{[6]}$} species in AS glasses where M_2(Si–Al$^{[p]}$) data are accessible from the corresponding ^{27}Al{^{29}Si} REDOR-derived $\{M_2(\mathrm{Al}^{[p]}–\mathrm{Si})\}$ results. Although an implementation of the latter may require AS glasses with isotopic ^{29}Si enrichment (due to the low n.a. of ^{29}Si), combining its outcome with Eq. (60) yields the full set of six $\{M_2(\mathrm{Al}^{[p]}–\mathrm{Si})\}$ and $\{M_2(\mathrm{Si}–\mathrm{Al}^{[p]})\}$ values for the three coexisting Al$^{[p]}$ coordinations in the structure. In contrast, the alternative of using ^{29}Si{^{27}Al} REAPDOR NMR is only applicable to (the nonetheless majority of) AS glasses that feature negligible Al$^{[5]}$/Al$^{[6]}$ populations, as it may only yield the *net* second moment, M_2(Si–Al) = M_2(Si–Al$^{[4]}$) + M_2(Si–Al$^{[5]}$) + M_2(Si–Al$^{[6]}$), because REAPDOR NMR does not admit Al-coordination-*selective* dipolar recoupling, nor does any other currently available alternative NMR scheme.

11.5 Application of dipolar-dephasing experiments to aluminosilicate glasses

11.5.1 Short-range structural probing

Dipolar-dephasing NMR experimentation with ^{17}O detection has recently proven to be a powerful information source about the various BO and NBO environments in multicomponent AS-based glasses. Their utility should be viewed in relation to what routine ^{17}O (3Q)MAS NMR may offer, which is reviewed below.

Depending on the identity and CFS of the glass-network modifier (M^{z+}) in ternary AS glasses, the two main groups of ^{17}O$^{[1]}$ and ^{17}O$^{[2]}$ resonances are often separated by standard MAS NMR at $B_0 \geq 9.4$ T, whereas if they are not, application of 3QMAS [21, 22] will in general *not* improve the spectral resolution further, as discussed in Ref. [74]. Owing to the distinctly different quadrupolar coupling constants between the ^{17}O$^{[1]}$ and ^{17}O$^{[2]}$ sites, work at low magnetic fields ($B_0 < 9.4$ T) may offer exceptions of improved NMR signal-separation of ^{17}O$^{[1]}$ and ^{17}O$^{[2]}$ by 3QMAS, e.g., see Ref. [330]. For AS glasses where the $\{AlO_p\}$ speciation is dominated by AlO_4, all ^{17}O$^{[2]}$ resonances from the various Si/Al$^{[4]}$–O–Si/Al$^{[4]}$ bridges are typically not resolved by 1D MAS but sometimes when exploiting 3QMAS, such as for Na-bearing glasses [57, 331, 332], as opposed to their Ca, Mg or RE-based counterparts, where overlap among ^{17}O$^{[1]}$ and ^{17}O$^{[2]}$ resonances is often an additional complication at low or moderately large magnetic fields ($B_0 \lesssim 11.7$ T) [57, 66, 70, 71, 74, 332, 333]. However, ^{17}O (3Q) MAS NMR generally fails to resolve the multitude of distinct ^{17}O$^{[1]}$–Si/Al$^{[p]}$ and Si/Al$^{[p]}$–^{17}O$^{[2]}$–Si/Al$^{[q]}$ resonances *within* each respective group of NBO and BO motifs in AS glasses that *also* feature nonnegligible Al$^{[5]}$ and Al$^{[6]}$ populations (such as RE-based AS glasses; see Section 12) and/or incorporate additional network formers (e.g., B and P). The resolution-enhancements by employing 3QMAS relative to MAS alone appear to decrease for increasing CFS of the network-modifier cations [74].[1]

[1] Here we even ignore the further complications from the presence of ^{17}O$^{[3]}$ structural moieties that coordinate variable constellations of Si/Al$^{[p]}$ neighbours (see Section 12). The existence of such motifs has hitherto resisted unambiguous identification in AS glasses, but their ^{17}O resonances are expected to overlap severely with those from ^{17}O$^{[2]}$ [74, 85, 334–340] .

As discussed further in Section 12.3.3, $^{17}O\{Al\}$ TRAPDOR experimentation was employed to unambiguously prove the presence of Al–NBO bonding in RE AS glasses [73]. Such NMR experimentation was expanded significantly by Stebbins et al. [75], who demonstrated that application of $^{17}O\{Al\}$ TRAPDOR NMR on AS glasses—as well as $^{17}O\{^{27}Al\}$ and $^{17}O\{^{11}B\}$ TRAPDOR experimentation on aluminoborosilicate (ABS) glasses—enabled assignment of the various (otherwise overlapping) $^{17}O^{[1]}$ and $^{17}O^{[2]}$ resonances. Notably, their experiments allowed for (roughly) estimating the corresponding populations of the $^{17}O^{[1]}$–Al/Si and Si/Al–O–Si/Al structural motifs. For instance, each $^{17}O\{^{27}Al\}$ and $^{17}O\{^{11}B\}$ dephasing experiment informs directly that the given ^{17}O site involves (at least) one bond to ^{27}Al and ^{11}B, respectively, whereas those resonances with an insignificant dephasing in both NMR experiments must stem from $^{17}O^{[1]}$–Si and Si–^{17}O–Si motifs [75]. We expect further developments and applications of such ^{17}O-based dipolar-dephasing experimentation to glasses, notably so their coupling with high-resolution techniques such as 3QMAS [21, 22].

Dipolar-dephasing NMR experiments are of course not restricted to probe direct bonds between ^{17}O and the network-*forming* species in glasses. They may also inform about the network-modifier constellations around the various BO/NBO sites. Given that the local charge requirement of a given $O^{[p]}$ site in an AS glass depends on the particular constellation of its directly bonded Si/$Al^{[p]}$ neighbours, the degree of ^{17}O–M^{z+} dipolar contact may indirectly convey information about the Si/$Al^{[p]}$ constellation around a BO atom. For instance, the average *number* of M^{z+} cations around the various $^{17}O^{[2]}$ moieties increase according to Si–O–Si < Si–O–Al < Al–O–Al (e.g., see Ref. [74] for concrete numbers in the context of RE_2O_3–Al_2O_3–SiO_2 glasses), meaning that the extent of $^{17}O\{M\}$ NMR dephasing increases concurrently. Fig. 22 shows an example of the identification of the Al–$^{17}O^{[2]}$–Al resonance in a Sc_2O_3–Al_2O_3–SiO_2 glass by $^{17}O\{^{45}Sc\}$ TRAPDOR NMR [74]—and thereby direct evidence for violation of the Loewenstein Al avoidance [59]: the ^{17}O resonance-region between 0 and 125 ppm stems from Si–^{17}O–Si, Si–^{17}O–Al, and Al–^{17}O–Al linkages. However, only two NMR signals are hinted in the ^{17}O MAS NMR spectrum, where that ≈80 ppm was attributed to Al–^{17}O–Al bridges by using DFT calculations [74]. Nonetheless, independent support for this NMR peak assignment are given by the $^{17}O\{^{45}Sc\}$ TRAPDOR NMR results of Fig. 22: besides the as expected strong ^{17}O NMR signal attenuation from the $^{17}O^{[1]}$ sites, a significant dephasing is also observed for the ^{17}O NMR

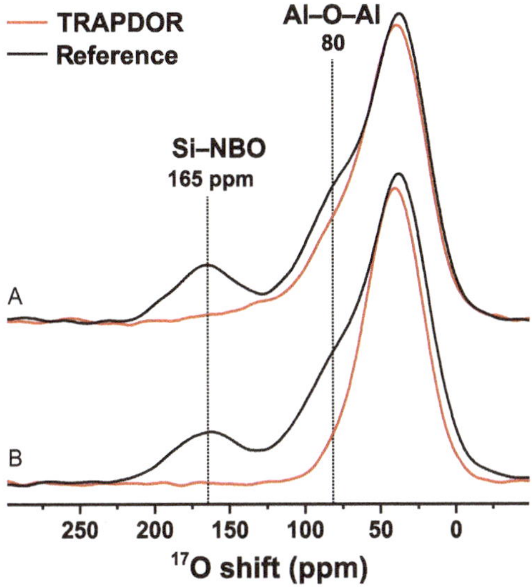

Fig. 22 $^{17}O\{^{45}Sc\}$ TRAPDOR [318] NMR spectra (red traces) recorded at $B_0 = 14.1$ T from a Sc-based glass with molar composition $17Sc_2O_3$–$28Al_2O_3$–$55SiO_2$ for two dephasing periods of (A) $\tau_{rec} = 2.50$ ms, and (B) τ_{rec}=5.00 ms [74]. The difference between the *red* and *black* (reference) spectra reflects the extent of ^{17}O–^{45}Sc dipolar contacts among the different $^{17}O^{[p]}$ sites. Note the strong dephasing observed for the resonances around 165 ppm and 80 ppm, which stem from Si–$^{17}O^{[1]}$ and Al–$^{17}O^{[2]}$–Al moieties, respectively. *Reproduced with permission from A. Jaworski, B. Stevensson, M. Edén, The bearings from rare-earth (RE = La, Lu, Sc, Y) cations on the oxygen environments in aluminosilicate glasses: a study by solid-state ^{17}O NMR, molecular dynamics simulations, and DFT calculations, J. Phys. Chem. C 120 (2016) 13181–13198. Copyright (2016) American Chemical Society.*

peak ≈80 ppm, as anticipated from the comparatively strong $^{17}O^{[2]}$–Sc contacts of the Al–O–Al linkages (see above). Analogous dipolar-dephasing experiments involving, for example, $^{17}O\{^{23}Na\}$ and $^{17}O\{^{7}Li\}$ might be useful for exploring the various BO/NBO environments in Na_2O–Al_2O_3–SiO_2 and Li_2O–Al_2O_3–SiO_2 glasses, respectively.

11.5.2 Medium-range structural probing

We next consider the utilization of dipolar-dephasing experiments to probe the *intermediate*-range order in AS-based glasses. While there are several reports targeting Si/Al contacts in (well-ordered) zeolite networks by using both $^{29}Si\{^{27}Al\}$ REAPDOR [341, 342] and TRAPDOR [272, 343] NMR, analogous experimentation to *amorphous* aluminosilicates are very sparse [328, 329]. Bradtmüller el al. [328] recently reported a study on an alumino(boro)silicate glass series where both the relative amounts of the Al/B network formers and the Na^+/Mg^{2+} modifiers were varied: $^{29}Si\{^{27}Al\}$ REAPDOR NMR application to AS glasses manifested a decelerated ^{29}Si NMR signal dephasing when Mg was substituted for Na, which was attributed to the emergence of $Al^{[4]}$–O–$Al^{[4]}$ and/or $Al^{[4]}$–NBO motifs at the expense of the overall dominating Si–O–Al linkages [328]. Furthermore, $^{11}B\{^{27}Al\}$ REDOR NMR experiments on ABS glasses revealed $B^{[3]}$–O–$Al^{[4]}$ bonds in the structure, with the dipolar contacts increasing concurrently with the Al content, but being independent on the presence of Na or Mg in the glass [328]. Overall, the $^{11}B\{^{27}Al\}$ REDOR NMR dephasing data from the Na/Mg-based ABS glasses were consistent with a statistical intermixing of the various {Si, $B^{[3]}$, $B^{[4]}$, $Al^{[4]}$} network formers, except for the near absence of $B^{[4]}$–O–$Al^{[4]}$ linkages [328] (as expected from the Loewenstein Al avoidance rule [59]; see Section 12.3.2).

Moreover, Schaller et al. [344] probed the structural role of P when present as a minor component (4 mol% P_2O_5) in Na_2O–Al_2O_3–SiO_2 glasses by $^{31}P\{^{27}Al\}$ TRAPDOR NMR, identifying both isolated ortho- and pyro-phosphate moieties, along with PO_4 groups participating in the glass network by forming P–O–Al bonds.

Concerning the glass-network former/modifier association in AS glasses, Le Losq et al. [345] presented a combined molecular dynamics (MD) simulation and Raman/NMR study of AS glasses with variable Na^+ and K^+ contents, encompassing $^{27}Al\{^{23}Na\}$ and $^{23}Na\{^{27}Al\}$ REDOR NMR dephasing data for glasses with increasing K^+-for-Na^+ substitution. Their results suggested a preference for Na^+ (relative to K^+) to charge-compensate the $[AlO_4]^-$ groups [345]. Altogether, the NMR data were argued to support an AS glass structural picture where the modifier ions cluster to form "ion channels", along the modified random network (MRN) model proposed earlier for $M_{(2)}O$–SiO_2 glasses [346, 347]. A similar structural scenario was proposed for mixed-cation CaO–MgO–Al_2O_3–SiO_2 glasses from circumstantial experimental evidence and MD simulations [348, 349].

12. Aluminosilicate glass systems and unconventional structural features

12.1 Overview

Here we provide a survey of ^{27}Al MAS NMR reports on specific AS glass systems and discuss various "unconventional" structural features that imply significant deviations from the prevailing AS glass model outlined in Section 3; some of them remain controversial and are subject to debate.

The structural fragment shown in Fig. 23 will serve for introducing the "unconventional" structural properties not conforming to the prevailing view. It was obtained by an atomistic MD simulation of an Y-based AS glass of stoichiometry $Y_{0.56}Al_{0.40}SiO_{3.44}$ [72] and manifests the following remarkable features, despite that the glass composition exhibits $n_{Al}/n_{RE} = 0.71 < 3$ and $n_{Al}/n_{Si} = 0.40 < 1$, such that both Eqs (9) and (10) are readily obeyed:

(1) There are significant contributions from AlO_5 and AlO_6 polyhedra, which besides the more commonly encountered corner-sharing also share *edges*.

(2) The glass network breaks Loewenstein's rule by comprising directly connected AlO_4 tetrahedra.

(3) In contrast with the Na-based AS glass in Fig. 1, one AlO_4 tetrahedron in Fig. 23 coordinate NBO ions.

(4) $O^{[3]}$ species connecting *three* SiO_4/AlO_p polyhedra are present. Lacy argued against AlO_6 groups as (significant) building blocks of AS glasses and introduced the term "oxygen tricluster" for an O atom that interlink three *tetrahedral* moieties [40], suggesting it as an alternative charge-compensation mechanism to the network modifiers M^{z+}.

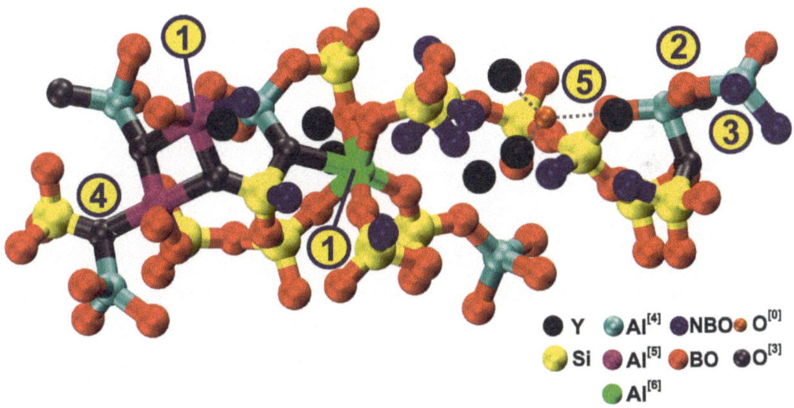

Fig. 23 A structural fragment from an atomistic MD-derived Y_2O_3–Al_2O_3–SiO_2 glass model of composition $Y_{0.56}Al_{0.40}SiO_{3.44}$ [72, 350]. The fragment is representative for RE AS glasses, which in contrast with their Na AS counterparts (see Fig. 1) manifest several features at odds with the conventional view of AS glass structure: (**1**) abundant AlO_5 and AlO_6 polyhedra that frequently connect to other polyhedra by edge-sharing; (**2**) direct $Al^{[4]}$–O–$Al^{[4]}$ linkages that violates Loewenstein's rule; (**3**) AlO_4 groups with NBO ions ($Al^{[4]}$–$O^{[1]}$ bonds) are encountered. Moreover, besides the BO/NBO species, (**4**) oxygen triclusters ($O^{[3]}$) and (**5**) free O^{2-} ions ($O^{[0]}$) are present. Each of features (**1**)–(**5**) are discussed throughout Section 12.

The neutral oxygen triclusters have also been suggested to induce NBO formation in tectosilicate glasses [71]. Note that although Fig. 23 reveals several distinct types of "$O^{[3]}$" species, none of them interconnect three *tetrahedra*: onwards, "$O^{[3]}$" denotes a central O atom in *any* constellation of interconnected SiO_4/AlO_p polyhedra, while the symbol "$O_T^{[3]}$" refers specifically to Lacy's tricluster definition. $O_T^{[3]}$ species are well-known to exist in several ordered aluminate phases, such as grossite ($CaAl_4O_7$), as well as aluminosilicates, e.g., mullite ($Al_{4+2x}Si_{2-2x}O_{10-x}$; $0.2 \lesssim x \lesssim 0.6$), but their presence in AS glasses still awaits *direct* experimental proof. Convincing (albeit not unambiguous) evidence for the occurrence of $O^{[3]}$ species in amorphous systems is hitherto only provided for amorphous Al_2O_3 by using ^{17}O 3QMAS NMR [340], and a $CaAl_2O_4$ glass composition by using heteronuclear $^{27}Al\{^{17}O\}$ double-resonance NMR experimentation [337].

(5) Besides the $O^{[p]}$ ($p = 1, 2, 3$) moieties, MD simulations of RE AS glasses (see Fig. 23) also predict another controversial entity in glass structures—"free O^{2-} ions", i.e., $O^{[0]}$ species that solely coordinate modifier cations [351–353]. This implies the formation of M–$O^{[0]}$–M type motifs that are intimately related to tendencies of modifier clustering. As for the triclusters, *direct* experimental proof for free O^{2-} ions in AS glasses has hitherto remained elusive, but their presence has been conjectured from circumstantial evidence in various silicate-based glasses, [114, 344, 354, 355]. Recently their existence was ascertained in highly fragmented ($\overline{N}_{BO} < 2$) "invert" Ca/Mg-SiO_2 glasses, [356] where they indeed are *expected*. See Sawyer et al. [353] for a recent discussion on the subject.

We discuss these "unexpected" glass features (**1**)–(**5**) further in the following subsections. Notably, given that the Y_2O_3–Al_2O_3–SiO_2 glass structure manifests all of them, it is significantly more disordered than its Na-based AS counterpart of Fig. 1. Although the structural fragment of Fig. 23 stems from a MD-generated *model*, we observe a very good agreement between the predictions from MD simulations and NMR experiments concerning the Al and O speciations: all of features (**1**)–(**3**) are fully corroborated by experiments [72, 77, 114, 126, 195] (whereas the presence of $O^{[0]}$ and $O^{[3]}$ species in AS glasses still awaits direct experimental proof).

12.2 CaO–Al_2O_3–SiO_2 glasses

A vast majority of all structural reports on AS glasses by solid-state NMR and other characterization options targeted alkali, alkaline-earth, or mixed-modifier systems. Thanks to its unusually wide glass-forming region that spans the entire compositional range between each CaO–Al_2O_3 (60–70 mol% Al_2O_3) and CaO–SiO_2 binary, the ternary CaO–Al_2O_3–SiO_2 system is the hitherto most widely investigated and well-characterized AS glass family. Here we summarize the main findings from the most comprehensive investigations that explored large compositional ranges and established the general trends.

Engelhardt et al. provided the first thorough MAS NMR investigation over a large compositional space of the CaO–Al_2O_3–SiO_2 system [3]. The low MAS rates (< 5 kHz) employed in their ^{27}Al NMR experimentation, however, precluded observation of anything but $Al^{[4]}$ coordinations in the glasses, even for "peraluminous" glass compositions ($n_{Al}/n_{Ca} > 2$) where the Ca^{2+} content is insufficient for balancing the entire Al speciation as AlO_4. They suggested that the excess of Al formed neutral $[Al^{3+}(AlO_2^-)_3]$ clusters or $O_T^{[3]}$ species [3]. Engelhardt et al. also demonstrated a roughly linear correlation between the average ^{29}Si chemical shift and the ^{27}Al NMR peak maximum, implying that $^{27}AlO_4$ groups exhibit a very similar shift dependence on composition/structure factors as ^{29}Si (Section 4). These inferences were for various M^+/M^{2+} AS glasses strengthened and developed further by Oestrike [4, 5], Merzbacher [6], and their coworkers, amounting in the following conclusions (note that NMR peak maxima rather than $\delta_{iso}^{[4]}$ data were reported): $\delta_{iso}^{[4]}$ increases concomitantly and roughly linearly with the molar fraction $x_{Al} = n_{Al}/(n_{Si} + n_{Al})$, as well as for increasing n_M/n_{Al} ratio [3–6].

The Al speciation of the CaO–Al_2O_3–SiO_2 system was particularly well explored in the more recent seminal work by Neuville et al.: they employed numerical fitting of ^{27}Al MAS and 3QMAS NMR data (see Figs 17 and 24 for examples) and gained complementary information from Raman and X-ray absorption spectroscopies [62, 69, 357]. They established the correlations between the glass composition and the relative populations of the various $^{27}AlO_p$ groups, as well as with their isotropic chemical shifts [62, 69]. A few percent of AlO_5 groups were observed in essentially all glasses, with up to 7% in tectosilicate compositions [69]. These estimates were somewhat higher than the first report of $\approx$ 1% by Stebbins et al. from a $CaAl_2Si_2O_8$ glass [67]. As also noted previously for "peraluminous" ($n_{Al}/n_{Ca} > 2$) Ca AS glasses that violate Eq. (9) [7], Neuville et al. reported that for a fixed SiO_2 content, the AlO_5 populations grew markedly for increasing n_{Al}/n_{Ca} ratio, but they levelled out for $n_{Al}/n_{Ca} \gtrsim 4$ [i.e., $n(Al_2O_3)/n(CaO) \gtrsim 2$], resulting in the highest value $x_{Al}^{[5]} \approx 0.28$ observed around 50 mol% SiO_2 [62]. No AlO_6 groups were detected, except for minor amounts ($x_{Al}^{[6]} \lesssim 0.02$) in very Al-rich glass compositions with $n_{Al}/n_{Ca} > 2$ [69].

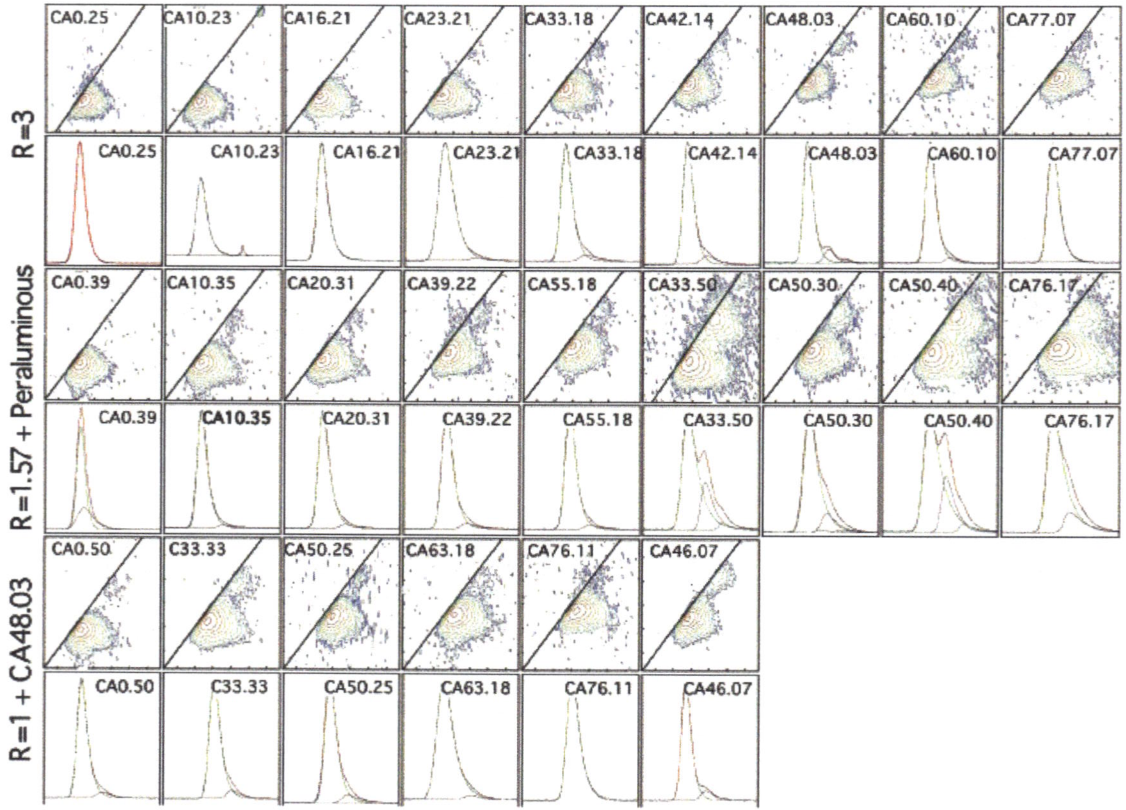

Fig. 24 Experimental ^{27}Al MAS and 3QMAS NMR spectra recorded from $CaO–Al_2O_3–SiO_2$ glasses of variable compositions, where CA*x*.*y* corresponds to $(1 - x - y)CaO–xAl_2O_3–ySiO_2$, whereas $R = n(CaO)/n(Al_2O_3)$ [62]. The best-fit results of the 1D MAS spectra are also displayed together with the peak components from each observed AlO_p moiety. *Reproduced from D.R. Neuville, L. Cormier, D. Massiot, Al Coordination and speciation in calcium aluminosilicate glasses: effects of composition determined by ^{27}Al MQ-MAS NMR and Raman spectroscopy, Chem. Geol. 229 (2006) 173–185, with permission from Elsevier.*

The trends in $^{27}AlO_4$ chemical shifts observed by Neuville et al. from NMR data at high magnetic fields and fast MAS [62, 69] generally consolidated the inferences from early ^{27}Al MAS NMR studies [3–6], demonstrating that $\delta_{iso}^{[4]}$ increases linearly when either (*i*) the SiO_2 content of the glass is increased at a fixed ratio $R = n\ (CaO)/n(Al_2O_3)$, or when (*ii*) R grows at constant amount of SiO_2. Neuville et al. studied glass compositions along joins of three distinct R-values of 1, 1.57 and 3 [62]; see Fig. 24. For increasing Al_2O_3 content in the peraluminous region, $\delta_{iso}^{[4]}$ remained nearly invariant, which was attributed to the sole presence of $Q_{Al}^4(4Si)$ groups while the "excess" of Al formed high-coordination species [69]. For Si-rich (> 45 mol%) glasses, $\delta_{iso}^{[4]}$ increased linearly and independently of R, whereas in the Si-poor (< 20 mol% SiO_2) "percalcic" region ($n_{Al}/n_{Ca} < 2$) where solely AlO_4 species exist, different slopes of $\delta_{iso}^{[4]}$ *versus* $x(SiO_2)$ were observed when R was varied. This was interpreted as the sole presence of fully polymerized tetrahedral moieties (Q_{Al}^4) in percalcic glasses with > 45 mol% SiO_2, whereas progressive $Q_{Al}^n \rightarrow Q_{Al}^{n-1}$ conversions occur when the silica content is decreased, with $\{Q_{Al}^3, Q_{Al}^2\}$ groups dominating in highly depolymerized networks of low SiO_2 contents (< 20 mol%) [62].

12.3 Consequences of the CFS for the structural (Dis)Order

The field strength of each cation in a glass affects significantly its degree of structural (dis)order. Albeit the Si^{4+} and Al^{3+} network formers exhibit the highest CFS values, they are common to all AS glasses: in practice, the charge and size of each network-modifying cation M^{z+} control the structural *details* associated with a given (fixed) $\{M$, Si, Al$\}$ composition, i.e., how well they conform to the "conventional" glass description (Section 3). Hence, the AS glass structure and its physical/chemical properties are effectively dictated *both* by the relative $\{M$, Si, Al$\}$ contents [*via* Eqs (9) and (10)] and the CFS of the network modifier(s). Generally, the structural disorder and the M^{z+} CFS increase concurrently [2, 6, 63–66, 139, 358], with the following consequences: (**1**) The relative populations of AlO_5/AlO_6 polyhedra elevate; (**2**) the adherence to Loewenstein's rule relaxes and the disorder of the SiO_4/AlO_4 intermixing is increased; (**3**) although prevailing, the preference for Si–NBO bonds weakens, i.e., the number of Al–NBO contacts grows.

Each "structural disorder" aspect (**1**)–(**3**) are discussed further below, together with the possibilities/limitations of ^{27}Al NMR for their accurate probing.

12.3.1 The AlO_p speciation and its implications for mechanical properties

We first consider the concomitant growth of the AlO_5/AlO_6 contents with the M^{z+} CFS [62–64, 67–70, 72, 114, 139, 261, 358]. The latter is maximized for small and highly charged cations. The smaller size of Mg^{2+} relative to Ca^{2+} is reflected by overall higher amounts of AlO_5 groups in $MgO–Al_2O_3–SiO_2$ or "mixed-cation" $CaO–MgO–Al_2O_3–SiO_2$ glasses compared with their Ca AS counterparts of comparable compositions, whereas the latter manifest higher AlO_5/AlO_6 populations than alkali-metal-based glasses [56, 68, 70, 139, 330, 333, 358–361].

Consequently, the significant field strength of any *trivalent* RE^{3+} species induces very substantial structural disorder effects and the features (**1**)–(**3**) manifest the strongest for $RE_2O_3–Al_2O_3–SiO_2$ glasses [63, 64, 72–74, 93, 114, 126, 139, 195, 261, 362] . Large amounts of AlO_5 polyhedra generally persist across all RE AS glass compositions, encompassing those adhering to *both* conditions Eqs (9) and (10). Moreover, the AlO_5 population grows markedly for increasing CFS along the series $La^{3+} < Y^{3+} < Lu^{3+} < Sc^{3+}$ (due to the ion-radius contraction) with typical contributions amounting to ~8%, ~20%, ~25% and ~30% out of the total Al speciation, respectively [72, 114, 126, 195, 363]. The fractional AlO_6 population manifests a similar trend but remains $\lesssim$5% throughout all glass compositions obeying both Eqs (9) and (10) [114, 126, 195, 261].

Compared with their $M_{(2)}O–Al_2O_3–SiO_2$ analogues, RE AS glasses exhibit enhanced physicochemical properties, such as improved corrosion resistance and higher glass-transition temperature as well as enhanced Vickers hardness (H_V; "microhardness") [114, 363–366]. Such trends have traditionally been attributed to the relatively strong RE–O bonds. However, from a strong observed correlation between H_V and the *average* coordination number of Al in a large set of $RE_2O_3–Al_2O_3–SiO_2$ (RE={La, Y, Lu, Sc}) glasses of widely spanning compositions, we recently proposed a new explanation for the well-known microhardness enhancement of RE AS glasses when the RE^{3+} CFS is increased [72, 363]: it originates from the concurrent growth of the AlO_5 and AlO_6 populations, and their enhanced network cross-linking effects through relatively strong $Al^{[5]}–O–Si^{[4]}/Al^{[4]}$ bonds (also see Section 12.6).[m] Such a structure-strengthening mechanism may possibly also control other physical properties, such as the glass

[m] While the effects are in principle even stronger for the AlO_6 octahedra, their abundance is significantly lower.

transition temperature, and particularly the elastic modulus (which often correlates strongly with H_V).

Although demonstrated and discussed in the context of RE_2O_3 –Al_2O_3–SiO_2 glasses [72, 363], the microhardness-boosting effects of the high-coordination $Al^{[5]}$ species (as well as $Al^{[6]}$; see above) are doubtlessly relevant for *any* oxide-based glass structure that manifests abundant AlO_5 and AlO_6 polyhedra, encompassing Mg^{2+}-bearing AS glasses (see above), and notably so binary Al_2O_3–SiO_2 glasses (whose structures reveal significant AlO_5 and AlO_6 contents; see Section 12.4). Indeed, as discussed further below, very recent advances towards the combination of a high microhardness and crack resistance of glasses concerned systems such as MgO–Al_2O_3–SiO_2 [367], RE_2O_3–Al_2O_3 [368], Ta_2O_5–Al_2O_3 [369], Al_2O_3–SiO_2 [370], and Li_2O–Al_2O_3–B_2O_3 [371]—all of which typically manifest significant AlO_5/AlO_6 populations (although not quantified in the reports cited above, except for Refs. [369, 371]). Hence, the favourable H_V results reported in Refs. [367–371] strongly corroborate the inferences of Refs. [72, 363] about the high-coordination $Al^{[5]}/Al^{[6]}$ species boosting the glass microhardness.

Notably, the presence of AlO_5/AlO_6 polyhedra in the AS glass structures involving the high-CFS Mg^{2+} cation naturally explain the high H_V-values observed by Ke et al. [367] for a 11MgO–x Al_2O_3–(89– x)SiO_2 glass series with a low and constant amount of 11 mol% MgO along with increasing $0.46 \leqslant n_{Al}/n_{Si} \leqslant 1.54$. Throughout their glass series, large AlO_5 populations are anticipated by the violation of condition (**A**) [Eq. (9)], while $x_{Al}^{[5]}$ is moreover boosted by the concurrent increase in n_{Al}/n_{Si}, such that condition (**B**) [Eq. (10)] was *also* not obeyed for the Al-richest glasses. While the authors of Ref. [367] reported a consistently lower microhardness of each MgO–Al_2O_3–SiO_2 glass relative to its Al_2O_3–SiO_2 counterpart of Ref. [370] with a similar n_{Al}/n_{Si} molar ratio, they observed an asymptotic elevation of H_V for the Al-rich Mg-bearing glasses towards that of the Al_2O_3–SiO_2 counterpart. Ke et al. attempted to rationalize the composition–microhardness trend by considering solely the Mg^{2+} species (which may act either as network modifiers or network formers), yet ignoring the potential role of the AlO_5/AlO_6 moieties [72, 363] by assuming the sole presence of AlO_4 groups in the MgO–Al_2O_3–SiO_2 glasses [367].

We merely believe that the observed microhardness values of the MgO–Al_2O_3–SiO_2 glasses [367] compared with the Al_2O_3–SiO_2 counterparts [370] are readily explained by the distinct AlO_5/AlO_6 contents in each respective AS glass structure; the latter aspect was not addressed in Refs. [367, 370], but the high-coordination Al populations are doubtlessly larger in the binary Al_2O_3–SiO_2 glass (which is devoid of M^{2+} ions to stabilize the $[AlO_4]^-$ tetrahedra). Notably, this case constitutes an excellent illustration of where application of *routine* ^{27}Al MAS NMR could readily settle whether the speculations of Ref. [367], or those herein, would best rationalize the observed H_V alterations of Al-rich Al_2O_3–SiO_2 glasses upon a modest incorporation of Mg^{2+}.

The recent reports on Al-bearing glasses that *simultaneously* manifest high microhardness *and* crack resistance [367–371] are noteworthy, because these two properties are difficult to combine, in that a high H_V-value of the glass is normally accompanied by its strong tendency of cracking when a load is applied. Nonetheless, a synergistic increase of the microhardness and the crack resistance for a concurrent elevation of the Al content (and thereby—by the glass design—also of the AlO_5/AlO_6 contributions to the structure) were reported for binary AS glasses [370], as well as for the ternary counterparts incorporating high-CFS Mg^{2+} [367] or RE^{3+} [368] cations. In the absence of any detailed structural study, these composition–property trends remain largely unexplained. Nonetheless, from the glass-series design alone, the AlO_5/AlO_6 abundances are without doubt increasing concomitantly with both the microhardness and crack resistance properties (see above), strongly suggesting their involvement not only for H_V [72, 363] but also for the crack resistance, as indeed proposed by Rosales-Sosa et al. [370] and Januchta et al. [371] in the contexts of Al_2O_3–SiO_2 and Li_2O–Al_2O_3–B_2O_3 glasses, respectively.

12.3.2 Al/Si intermixing

Several studies have shown that Loewenstein's rule is only partially obeyed in AS glass networks that involve network modifiers with low or moderately large field strength (e.g., Na^+ and Ca^{2+}) [65, 66, 195, 264]. The degree of $Al^{[4]}$–O–$Al^{[4]}$ bonding emphasizes when the M^{z+} CFS is increased, as reflected in around 6%, 7%, 15%, and 35% deviations from a perfect Al–O–Si ordering observed for AS glasses with $n_{Al}/n_{Si} \approx 1$ that incorporate Na^+, Li^+, Ca^{2+}, and Mg^{2+}, respectively [65, 66, 70]. (Here a 100% deviation implies that the $Al^{[4]}$–O–$Al^{[4]}$ population matches that of a statistical Al/Si intermixing.) Note that a high NBO content of the glass naturally boosts the occurrence of $Al^{[4]}$–O–$Al^{[4]}$ fragments due to the strong propensity for Si–NBO bond formation, which limits the number of Si–BO contacts of the $\{Q_{Si}^{n<4}\}$ groups in the depolymerized glass network [61, 195, 348, 349].

Unfortunately, routine ^{27}Al (3Q)MAS NMR offers limited insight into the detailed Al/Si ordering across the glass network, as the ^{27}Al chemical shift dependence on the parameters n and m of a $Q^n_{Al}(m\text{Si})$ moiety is too weak compared with the large ^{27}Al resonance-widths stemming from the quadrupolar broadenings (which for AS glasses is exacerbated further by the structural disorder). Some information has been gained by fitting the isotropic projections of ^{27}Al 3QMAS NMR spectra into signals from distinct $Q^n_{Al}(m\text{Si})$ groups [264], or deconvolution of essentially featureless ^{29}Si MAS NMR peakshapes into an *ad hoc* assumed underlying set of $\{Q^n_{Si}(m\text{Al})\}$ components to probe the Al/Si intermixing [65, 329, 348, 372–374]. Hence, given that neither of the standard ^{29}Si MAS and ^{27}Al (3Q)MAS NMR tools offers sufficient spectral resolution to address this task accurately and reliably for AS glass networks, notably not when *both n* and *m* distributions remain *a priori* unknown, we advice against attempting anything but *qualitative* ^{29}Si MAS NMR analyses, such as those of Refs. [6, 77, 375]. Currently, ^{17}O 3QMAS NMR provides superior insight: the ^{17}O chemical shift is a sensitive probe to distinguish the relative numbers of Al/Si atoms in the first coordination shell, i.e., to assess the ensemble of $\{T\text{–O–}T'\}$ motifs [66, 331, 332, 376–378]. However, it requires expensive ^{17}O NMR isotopic enrichment, and often fails to provide sufficient resolution and/or accurate signal intensities when applied to multicomponent AS glasses, as discussed further in Section 11.5.1.

While the short-range ^{27}Al chemical shift information alone does not inform about spatial proximities among various AlO_p polyhedra, such insight may be gained by exploiting the homonuclear ^{27}Al–^{27}Al through-space dipolar interaction that depends on the inverse cube of the Al–Al interatomic distance—analogously with the I–S heteronuclear interactions (see Section 11). The ^{27}Al–^{27}Al interaction is thus emphasized for *directly linked* $^{27}AlO_p$–$^{27}AlO_q$ groups. Hence, recording a double-quantum–single-quantum (2Q–1Q) correlation NMR spectrum that involves double-quantum (2Q) filtering of the CT-signals in a spin-S *pair* provides direct information about the various coexisting $^{27}AlO_p$–$^{27}AlO_q$ pairs in the structure [181–183, 186, 194, 379] (This technique must not be confused with that of MQMAS [21, 22] which utilizes *single*-spin multiple-quantum coherences).

The indirect "2Q dimension" of a 2Q–1Q 2D NMR experiment encodes the $^{27}AlO_p$–$^{27}AlO_q$ connectivities, whose 2Q NMR shift-coordinate $\delta^{pq}_{2Q} = \delta^{[p]} + \delta^{[q]}$ represents the sum of the respective correlated $\delta^{[p]}$ and $\delta^{[q]}$ shifts that appear along the direct "1Q dimension" of the 2D NMR spectrum; see Fig. 25. The NMR signal from a $^{27}AlO_p$–$^{27}AlO_p$ group connectivity aligns with the "auto-correlation diagonal" ($\delta^{pp}_{2Q} = 2\delta^{[p]}$).

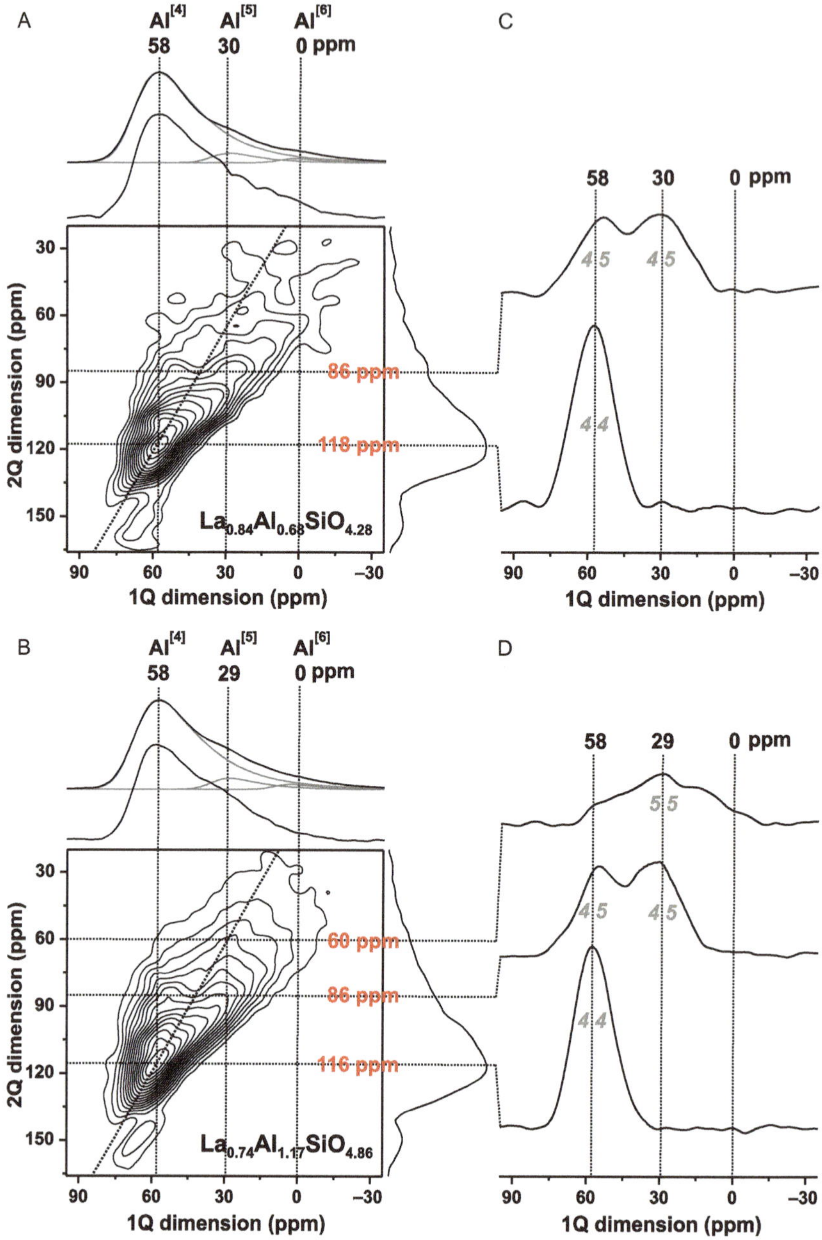

Fig. 25 (A, B) 2Q–1Q ^{27}Al NMR correlation spectra recorded at B_0 =14.1 T from two La-based AS glasses with compositions (A) $La_{0.84}Al_{0.68}SiO_{4.28}$ and (B) $La_{0.74}Al_{1.17}SiO_{4.86}$. The 2D spectra are shown together with projections along both spectral dimensions. Each respective directly excited (by CT-selective pulses) 1D MAS NMR spectrum is displayed topmost of each panel, where the results of its deconvolution into each $^{27}AlO_4$, $^{27}AlO_5$ and ^{27} AlO_6 signal component is also shown (*grey traces*). (C, D) Slices along the 1Q dimension taken through the as-indicated 2Q shifts in (A) and (B); they reveal the primary $^{27}AlO_p$–$^{27}AlO_q$ connectivities (labelled by *pq*) contributing to the given 2Q shift value. *Reproduced from A. Jaworski, B. Stevensson, B. Pahari, K. Okhotnikov, M. Edén, Local structures and Al/Si ordering in lanthanum aluminosilicate glasses explored by advanced ^{27}Al NMR experiments and molecular dynamics simulations, Phys. Chem. Chem. Phys. 14 (2012) 15866–15878, with permission from the Royal Chemical Society.*

Fig. 25 provides two examples of 2Q–1Q NMR correlations in the context of La_2O_3 –Al_2O_3–SiO_2 glasses [195]. Both reveal $Al^{[4]}$–O–$Al^{[4]}$ and $Al^{[4]}$–O–$Al^{[5]}$ associations in the glass structures, with a dominance of the former, thereby unambiguously revealing significant violations of the Loewenstein Al avoidance in these glasses. Such 2Q–1Q correlation experimentation has also been applied for probing $Al^{[p]}$–O–$Al^{[q]}$ connectivities in La_2O_3–Al_2O_3 [183], Al_2O_3–SiO_2 [329], Na_2O–Al_2O_3–SiO_2–P_2O_5 [380] and MgO–Al_2O_3–B_2O_3 [381] glasses. We guide the reader to our previous reviews [30, 186, 379] for further information about NMR techniques utilizing homonuclear dipolar interactions among half-integer spins and their application to structural studies.

Each ^{17}O and 2Q–1Q ^{27}Al NMR option has its pros and cons. 3QMAS ^{17}O NMR offers an overall richer information content than 2Q–1Q correlation ^{27}Al NMR experiments—as it often, yet not always (see Section 11.5.1)—unveils information about the Al–O–Al, Al–O–Si, as well as Si–O–Si structural fragments (along with the various $O^{[1]}$–Si/Al contacts). However, 2Q–1Q ^{27}Al NMR exhibits a superior NMR signal resolution from $^{27}Al^{[p]}$–O–$^{27}Al^{[q]}$ linkages of *distinct* Al coordination numbers $\{p, q\}$, which generally remain unresolved in 3QMAS ^{17}O NMR spectra (Section 11.5.1). Such signal-separation advantages are even more pronounced for ^{11}B (at magnetic fields $B_0 > 9.4$ T), where 2Q–1Q correlation experimentation readily separates all three $^{11}B^{[3]}$–O–$^{11}B^{[3]}$, $^{11}B^{[3]}$–O–$^{11}B^{[4]}$ and $^{11}B^{[4]}$–O–$^{11}B^{[4]}$ resonances in multicomponent borosilicate-based glasses [79, 382, 383]. In contrast, signals from the corresponding $^{17}O^{[2]}$ sites are normally unresolved by 3QMAS ^{17}O NMR [384–388].[n] We conclude that for glasses comprising network formers with multiple coordination states (such as B and Al), their direct detection by 2Q–1Q NMR—or alternatively, the more straightforwardly implemented "NOESY-like" [389] 1Q–1Q correlation experimentation [192, 390–396] —are significantly more informative about the homonuclear $F^{[p]}$–O–$F^{[q]}$ linkages than their indirect observation *via* ^{17}O NMR.

Notably, along the lines of the Loewenstein Al avoidance of $Al^{[4]}$–O–$Al^{[4]}$ motifs in AS glasses, a similar "B avoidance" of the $B^{[4]}$–O–$B^{[4]}$

[n] We comment that although 3QMAS ^{17}O NMR spectra do not unambiguously resolve the signals among all $B^{[3]}$–^{17}O–$B^{[3]}$, $B^{[3]}$–^{17}O–$B^{[4]}$ and Si–^{17}O–$B^{[3]}$/$B^{[4]}$ motifs, their projections along the "isotropic dimension" often suggest the presence of such (overlapping) peak components of the main B–^{17}O–B and Si–^{17}O–B resonances. Complete sets of F–$O^{[2]}$–F' motifs with $\{F, F'\} = \{B^{[3]}, B^{[4]}, Si\}$ have indeed been extracted by deconvolutions of 3QMAS projections [385, 386]. However, such analyses require assumptions and the best-fit results are potentially associated with large uncertainties.

counterparts has often been assumed to operate in borosilicate glasses [397–399]. Yet, recent 2Q–1Q ^{11}B NMR results evidence not only that $B^{[4]}$–O–$B^{[4]}$ motifs *exist* in borosilicate-based glasses—they are in fact comparatively *abundant* [79, 383], even for glasses incorporating the low-CFS Na^+ cation [383]. Indeed, by comparing results in Refs. [65, 66, 383], it appears that the degree of "Al avoidance" is stronger in Na_2O–Al_2O_3–SiO_2 glasses than the "B avoidance" counterpart in Na_2O–B_2O_3–SiO_2 glasses of comparable compositions, thereby translating into an overall larger number of $B^{[4]}$–O–$B^{[4]}$ linkages relative to $Al^{[4]}$–O–$Al^{[4]}$. Apparently, directly bonded $[BO_4]^-$–$[BO_4]^-$ moieties are more readily stabilized by the network-modifier cations than their $[AlO_4]^-$–$[AlO_4]^-$ counterparts. However, the comparatively more abundant $B^{[4]}$–O–$B^{[4]}$ bridges likely reflects the presence of $B^{[3]}$ coordinations in the borosilicate glass networks (besides $Si^{[4]}$ and $B^{[4]}$), which naturally restricts the number of BO linkages available relative to the AS scenario that manifests $Si^{[4]}/Al^{[4]}$ (and often also higher-coordination $Al^{[5]}/Al^{[6]}$) species.

12.3.3 BO/NBO distribution among tetrahedral groups

As for the weak dependence of the ^{27}Al chemical shift on the distinct number of ^{27}Al–O–**Si/Al** bridges [i.e., the *m*-value of $Q^n_{Al}(m\text{Si})$] discussed in Section 12.3.2, ^{27}Al (3Q)MAS NMR cannot reliably distinguish $Q^n_{Al}(m\text{Si})$ tetrahedra with distinct number of BO atoms (n). This hampers the probing of the BO/NBO-partitioning among the SiO_4/AlO_4 network groups by standard ^{27}Al NMR experimentation. If ^{27}Al NMR is to address this important question, advances are required to improve the understanding of the precise relationships between the parameter n and the ^{27}Al chemical shift of the $Q^n_{Al}(m\text{Si})$ moiety, as well as NMR-technique developments. Here, density functional theory (DFT) calculations of ^{27}Al chemical shifts are currently the most promising route to improved insight [58, 83–85].

^{17}O MAS and 3QMAS NMR are normally the methods of choice for addressing the BO/NBO distribution among the glass-network formers, as well as that between the various network modifiers [57, 60, 61, 71]. However, besides the interpretation of the frequently observed $^{27}AlO_4$ deshielding trends for increasing network-modifier content as $Q^n_{Al} \rightarrow Q^{n-1}_{Al}$ depolymerizations [6, 51, 62, 72]—in direct analogy with the well-established responses of the ^{29}Si chemical shift (see Section 5 and Ref. [30])—direct and unambiguous evidence for Al–NBO bonds were until recently only available from ^{17}O MAS NMR applied to Si-poor and highly fragmented Ca AS glass networks [60, 61]. The overwhelming propensity for Si–NBO bond formation relative to that of Al–NBO in $M_{(2)}O$–Al_2O_3 –SiO_2 glasses based on cations with small (such as those of the alkali metals) or moderately large (e.g., Ca^{2+}) field strengths implies the absence of lower-connectivity $Q_{Al}^{n<4}$ groups in Si-rich ($\gtrsim$40 mol% SiO_2) glass networks of reasonably high polymerization ($\overline{N}_{BO} \gtrsim 2.5$) [60, 61].

Yet, again the high charge and CFS of the RE^{3+} cations may randomize the NBO-partitioning among the tetrahedral network species to a much higher extent than that encountered in a M^+/M^{2+} bearing glass. MD simulations of RE_2O_3–Al_2O_3–SiO_2 glasses with $0.30 \leqslant n_{Al}/n_{Si} \leqslant 1.8$ has suggested that $Al^{[4]}$–NBO contacts constitute significant fractions (20–40%) of all $O^{[1]}$ anions (with the latter accounting for 15–37% of all O species) [72–74]. We recently presented unambiguous experimental evidence for a significant Al–$O^{[1]}$ bonding in RE AS glasses by exploiting $^{17}O\{^{27}Al\}$ TRAPDOR and $^{27}Al{\rightarrow}^{17}O$ CPMAS NMR experimentation [73]; see Section 11.5.1. Notwithstanding that the experimentally derived Al–NBO populations were almost three times lower than those of the MD-derived glass models, they amount to 7–12% of all $O^{[1]}$ moieties and 1–4% out of the entire O reservoir, across wide ranges of n_{Al}/n_{Si} ratios and NBO contents [73, 74]. Hence, their appearance seems to be a general structural feature of RE_2O_3–Al_2O_3–SiO_2 glasses.

^{17}O NMR experiments evidence that the degree of Al–NBO bonding elevates concomitantly with increasing Al and/or NBO contents of the AS glass [60, 61, 73, 74, 349], as well as when faster melt-cooling is employed in the glass preparation [400]. Consequently, while the discrepancies between the experimental and modelled Al–$O^{[1]}$ populations [72–74] *partially* stem from the orders-of-magnitude more rapid melt-cooling rate employed by the MD simulations (as suggested in Ref. [75]), we believe that they *mainly* reflect an imperfect performance of the force fields employed to model the various cation–BO and cation–NBO interactions. Notably, MD simulations of Ca- or Na-based AS glasses by Pedone et al. [401] (using a shell-model-based force field that accounts for polarization effects) did not reveal Al–NBO bonding, as expected. We infer that Pedone's force field gave more accurate predictions of the (absent) Al–NBO populations in the Ca/Na AS glasses than that used for the RE counterparts in Refs. [72–74]. Indeed, the predictions of the relative degree of Al–NBO bonding offered by various force fields were very recently compared in the context of Na_2O–K_2O–Al_2O_3–SiO_2 glasses [402]: large variations were observed, where only the shell-model force field yielded an excellent precision (i.e., that Al–NBO contacts are essentially absent) [402]. To conclude, we believe that a (significant) reduction in the cooling rate of the MD calculations is not a practically sensible approach, but enhancing the force fields is merely the way to go. Indeed, an improved shell-model-based force field [403] that permitted computationally tractable melt-cooling rates of 5–10 K/ps, provided a superior performance in reproducing experimentally observed phosphate speciations $\{Q_P^n\}$ in Na_2O–CaO–SiO_2–P_2O_5 glasses [81] as compared to employing inferior force fields combined with prohibitively time-consuming calculations to implement much slower melt-cooling [404, 405].

12.4 Al_2O_3–SiO_2 glasses: Oxygen triclusters or AlO_5?

By virtue of their absence of modifier cations to charge-balance the $[AlO_4]^-$ groups, the parent Al_2O_3–SiO_2 glass system presents several intriguing features, where the charge-neutralizing mechanism of each AlO_p moiety is the central question. The preparation of homogenous Al_2O_3–SiO_2 glasses is problematic by traditional melt-quench techniques due to large regions of liquid immiscibility. Single-phase glasses are reported for compositions with $\lesssim$10 mol% Al_2O_3 [406, 407] and ~60–67 mol% Al_2O_3 [152, 408], whereas outside of these ranges, phase-separation occurs into Si-rich and Al-rich domains [152, 406].

Risbud et al. [406] reported the first ^{27}Al and ^{29}Si NMR investigation of the Al_2O_3–SiO_2 system. The authors concluded the presence of all three $\{Al^{[4]}, Al^{[5]}, Al^{[6]}\}$ coordinations in their glasses, which incorporated 6–37 mol% Al_2O_3. The existence of AlO_5 polyhedra was inferred from an NMR signal ≈30 ppm, appearing in-between the typical shifts ≈60 ppm and ≈0 ppm associated with AlO_4 and AlO_6 polyhedra, respectively (see Section 4). Other reports supporting this NMR peak assignment argued for its very similar shift with those observed from $^{27}AlO_5$ groups in crystalline AS phases, such as the Al_2SiO_5 polymorph andalusite.

The results of Risbud et al. [406] were later questioned from two standpoints, the first being the coexistence of all three $\{AlO_4, AlO_5, AlO_6\}$ polyhedral types in Al_2O_3–SiO_2 glasses. Sato et al. [152, 409] provided evidence for a strong dependence of the precise $\{AlO_p\}$ speciation on the melt-quench rate utilized in the glass preparation, claiming that all $\{Al^{[4]}, Al^{[5]}, Al^{[6]}\}$ species exist *only* for very fast cooling rates (10^5-10^6 K/s), whereas standard rates of 10^2-10^3 K/s result only in AlO_4 and AlO_6 moieties. However, their conclusions were also questioned in subsequent work [407, 408] that overall supported the results of Risbud et al. For a series of Al_2O_3–SiO_2 glasses with variable compositions, Sen and Youngman [407] concluded from ^{27}Al and (3Q)MAS ^{17}O NMR that $O_T^{[3]}$ species are responsible for charge-balancing the AlO_4 groups in glasses incorporating $\lesssim$0.6 mol% Al_2O_3, whereas all three $Al^{[p]}$ coordinations coexist in glass structures that exhibit larger Al_2O_3

contents, with the AlO_4/AlO_6 and AlO_4/AlO_5 pairs of polyhedra dominating in the approximate $x(Al_2O_3)$ range of 0.6–4 mol% and $\gtrsim$4 mol% Al_2O_3, respectively [407]. Fig. 26 shows a selection of ^{27}Al MAS NMR spectra from Al_2O_3–SiO_2 glasses.

The main controversy of the pioneering work of Risbud et al. [406], however, was the assignment of what became coined as "the 30 ppm peak". While the recent mainstream opinion supports the original interpretation of AlO_5 groups as its origin in ^{27}Al NMR spectra of (any) AS glass, this resonance has also been attributed either to (*i*) Lacy's triclusters, as suggested by Schmücker et al. [410–412], or to (*ii*) severely distorted AlO_4 moieties [44, 119], where strong deviations from tetrahedral symmetry may give anomalously shielded and/or strongly second-order quadrupolar-displaced ^{27}Al resonances outside of the typical shift range of $^{27}AlO_4$ groups (see Section 4). McManus et al. [121] contrasted the ^{27}Al NMR parameters observed from

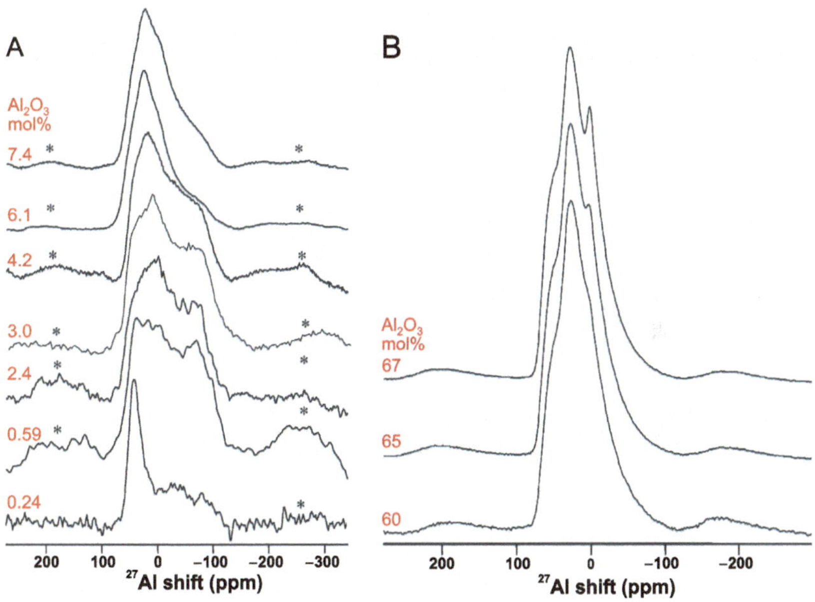

Fig. 26 ^{27}Al MAS NMR spectra recorded at $B_0 = 11.7$ T and $\nu_r \approx 25$ kHz from Al_2O_3–SiO_2 glasses with increasing Al_2O_3 content, as indicated at the bottom of each spectrum. Asterisks mark spinning sidebands. *The spectra in the left panel were reproduced with permission from S. Sen, R.E. Youngman, High-resolution multinuclear NMR structural study of binary aluminosilicate and other related glasses, J. Phys. Chem. B 108 (2004) 7557–7564. Copyright (2004) American Chemical Society. The spectra in the right panel were reproduced with permission from R. Weber, S. Sen, R.E. Youngman, R.T. Hart, C.J. Benmore, Structure of high alumina content Al_2O_3-SiO_2 composition glasses, J. Phys. Chem. B 112 (2008) 16726–16733. Copyright (2008) American Chemical Society.*

$\{Al^{[4]}, Al^{[5]}, Al^{[6]}\}$ coordinations of various amorphous Al_2O_3–SiO_2 phases with those of the $Al^{[4]}$–$O_T^{[3]}$ moieties in mullite and the AlO_5 groups in crystalline aluminosilicate phases. They concluded that albeit the *average* $\{\delta_{iso}, C_{Q\eta}\}$ values extracted from Al_2O_3–SiO_2 glasses are not easily reconciled with *either* $Al^{[4]}$–$O_T^{[3]}$ or AlO_5 species, when also considering the distribution widths of the $\{\delta_{iso}, C_{Q\eta}\}$ parameters, the attribution to *either* structural entity is possible [121]. Notably, as discussed further below, the various NMR spectra assignments of "the 30 ppm peak" may not necessarily be mutually exclusive: Al_2O_3–SiO_2 glasses may display a continuum of structural motifs with variable Al/O atom-positions/distances between the more categorical limits of $Al^{[4]}$–$O_T^{[3]}$ and AlO_5 moieties [121, 334].

12.5 Oxygen triclusters and AlO_5/AlO_6 polyhedra are closely related structural motifs

Note that in the MD-derived structural fragment of the Y_2O_3–Al_2O_3–SiO_2 glass shown in Fig. 23, *both* $O^{[3]}$ and AlO_5 moieties *not only coexist*, but are also *closely associated local structural motifs* in the common edge-sharing of AlO_5 (or AlO_6) polyhedra in RE AS structures [72, 74, 350]; such high-coordination AlO_p moieties share edges among themselves, as well as with tetrahedral groups [72, 74, 350]. That there may be an intimate structural relationship between $Al^{[5]}$ and $O^{[3]}$ species was suggested earlier by Poe et al. [413] for rationalizing pressure effects in AS melts. Similar inferences of direct $Al^{[5]}$–$O^{[3]}$ contacts were also reached from computer modelling of AS glasses based on either Na [414], Sr [415], or mixed-cation Ca/Mg [349] AS glasses, as well as RE-based aluminate [416] glasses. Altogether, the results of Refs. [72, 74, 349, 350, 413–416] suggest that the intimate relationship between $O^{[3]}$ and $Al^{[5]}$/$Al^{[6]}$ structural motifs is likely a *general* property of Al-bearing glass structures that involve $Al^{[5]}$/$Al^{[6]}$ species.

Notably, although Lacy argued against significant amounts of AlO_6 groups in AS glasses for reasons of spatial constraints—instead proposing the neutral triclusters as a more likely alternative—he did suggest the possibility of edge-shared AlO_4/AlO_6 polyhedra, as well as making the following statement "*...due to the formation of tri-clusters, oxygen may be made available for incorporation in octahedral groups*" [40]. Such features are indeed manifested by the structural fragment of Fig. 23, although due to the much higher abundance of AlO_5 polyhedra, $Al^{[5]}$–$O^{[3]}$ motifs are in practice more important for controlling AS glass/melt properties than their $Al^{[6]}$–$O^{[3]}$ counterparts.

Interestingly, the members of the seemingly most distinct AS glass systems—RE_2O_3–Al_2O_3–SiO_2 glasses with highly charged network modifiers, and their Al_2O_3–SiO_2 counterparts devoid of modifiers—incidentally reveal similar NMR spectra (compare Figs 8, 9 and 26) and doubtlessly share several structural features that control their physical properties; see Section 12.3.1.

12.6 Structural role of AlO_5/AlO_6

The identification and detection of the high-coordination AlO_5 and AlO_6 polyhedra is a central task for ^{27}Al NMR. However, although the AlO_5/AlO_6 groups doubtlessly originate from local charge-balance considerations of the glass network, their precise structural role and formation mechanisms still remain unclear: they are often referred to as "excess Al" and even classified as "network modifiers", then implying all $Al^{[p]}$ species *not* present as $Al^{[4]}$, and thereby not considered to participate in the AS glass network [8, 9]. Furthermore, some workers in the field view the $Al^{[5]}$/$Al^{[6]}$ species on equal footing with the electropositive M^{z+} cations of the alkali and alkaline-earth metals.

Yet, the substantial CFS of Al^{3+} and the high Al–O bond strength imply a distinctly different structural role of the AlO_5/AlO_6 moieties than of the network modifiers. There is a very *strong preference* for the M^{z+} cations (i.e., the "network modifiers") to coordinate NBO rather than BO [57, 350, 415, 417]. Consequently, in all but very NBO-poor glasses, the electropositive cations coordinate primarily (but *not* exclusively) NBO anions, whereas BO atoms ($O^{[2]}$ and $O^{[3]}$) account for nearly all O species at the AlO_5, and AlO_6 polyhedra [74, 350, 415]. Indeed, MD simulations of RE_2O_3–Al_2O_3–SiO_2 glasses reveal an *even lower* propensity for $Al^{[5]}$–NBO and $Al^{[6]}$–NBO bond formation than for $Al^{[4]}$–NBO [74, 350], whereas the preferences of the network modifiers (i.e., the RE^{3+} cations) to coordinate NBO are an order of magnitude larger than those for $Al^{[5]}$–NBO or $Al^{[6]}$–NBO bonding [350]. Also see Section 12.3.3.

Hence, as commented in Refs. [293, 414, 418], the AlO_5, and AlO_6 moieties are merely to be considered as glass-network *formers*. Indeed, their presence strengthens (rather than weakens) the AS glass/melt structure by an enhanced cross-linking [72, 363], as for instance reflected in a concurrently elevated density, microhardness, and crack resistance of the AS glass for increasing AlO_5/AlO_6 populations [72, 358, 363] (Section 12.3.1), as well as in a decrease of the configurational entropy of the corresponding melt [414]. Recent discussions on the (non)correlation between the AlO_5/AlO_6 populations and the NBO content is given by Stebbins et al. [351, 418–420].

13. Concluding remarks and outlook

After outlining how ^{27}Al MAS and 3QMAS NMR may be implemented and analyzed to probe the Al speciation in AS glasses, we discussed the utility of these techniques, as well as their relative merits and shortcomings. Moreover, we reviewed the basic local structural features of AS glasses and how they relate both to the glass composition and the field strength of the glass-network-modifier cations. As its predecessor [23], the present review mainly focussed on the *short*-range ($\lesssim$0.3 nm) structural features primarily reflected in the cation/oxygen coordinations. While even the understanding of the *local* structures of multicomponent AS glasses is far from complete, the accurate probing of *medium*-range structural aspects—i.e., to address structural motifs across the 0.3–1 nm length-scale—remains the major future challenge. As we hope to have conveyed in Sections 11 and 12, the utilization of distance/geometry-dependent homo/heteronuclear dipolar interactions involving ^{27}Al is a promising avenue, but the engineering and application of such MAS NMR techniques for glasses are still in their infancy; e.g., see [29, 30, 101, 186, 315, 379] for further details.

To address structural problems of AS glasses, it is common to utilize multinuclear NMR that combines information from ^{27}Al with that of ^{29}Si and (notably) ^{17}O; this expands the structural insight and provides a mutual validation of the various pieces of independent information needed for solving the complex puzzle of an AS glass structure. Here, a very promising approach is the exploitation of DFT to first calculate NMR parameters from atom coordinates of MD-generated glass models, followed by feeding the $\{\delta_{iso}, C_{Q\eta}\}$ data to simulate (3Q)MAS NMR spectra [58, 83–85]. This strategy is particularly rewarding when evaluated directly against experimental results.

Moreover, *modelling* by atomistic MD simulations is a rich source of information, particularly so for unravelling the medium-range glass structure, as well as for gathering insight into not only its "exotic" features that has hitherto resisted unambiguous and direct experimental detection, such as $O^{[0]}$ and $O^{[3]}$ species (see Section 12.1), but also the nowadays experimentally confirmed $Al^{[4]}$–O–$Al^{[4]}$ (Section 12.3.2) and $Al^{[4]}$–NBO (Section 12.3.3) motifs. Provided that accurate force fields are employed (see Section 12.3.3), MD-derived models of oxide-based glasses comprise a wealth of structural information. They have proven particularly powerful when evaluated directly against constraints on the medium-range glass structure from advanced homonuclear or heteronuclear NMR experiments that provide means to consolidate the modelled data [79, 82, 324, 383, 421]. At present, however, such tandem MD/NMR applications to AS glasses are very sparse [73, 195, 345], but they are anticipated to grow over time.

Although ^{27}Al is a highly receptive nucleus (see Section 6) and readily admits routine 1D NMR experimentation on glasses incorporating only a few at% of Al, most of the nonstandard NMR techniques that may inform about the medium-range glass structure—e.g., the heteronuclear $^{27}Al \rightarrow ^{29}Si$ CPMAS experiments described in Section 11, and the homonuclear 2Q–1Q correlation ^{27}Al NMR counterpart discussed in Section 12.3.2—only leave a minor fraction of the full NMR signal available for detection. Hence, the NMR signal-sensitivity is in practice a major limitation for the application of several potentially very informative experimental protocols. Yet, over the past ≈15 years, NMR coupled with dynamic nuclear polarization (DNP) have emerged. In such experiments, a tens-to-hundreds-fold stronger NMR signal is achieved by impregnating the sample with radicals and next exploiting the much stronger polarization of the unpaired electrons *via* magnetization transfers to the nuclei (analogously with CP; Section 11.1). Such DNP/NMR experimentation on inorganic materials have thus far focused on surface-selective studies of crystalline materials for applications in catalysis, then targeting relatively unreceptive nuclei such as ^{29}Si and ^{13}C [422, 423], and notably ^{17}O at natural abundance [424, 425]. However, although not yet applied to ^{27}Al for structural studies of AS glasses, some recent ^{27}Al DNP/NMR work is reported on γ-Al_2O_3 [426, 427], Al-bearing mesoporous silica [428], as well as amorphous aluminosilicates for catalytic purposes [429, 430]. We expect a substantial future expansion of such DNP/NMR investigations to also encompass ^{27}Al and (particularly) ^{17}O in AS glasses. Besides significantly widening the feasible glass-composition range amenable for experiments, DNP/NMR experimentation would be of utmost value for examining minor O species in (AS) glasses, notably so "free O^{2-}anions" ($O^{[0]}$; see Section 12.1), as well as elucidating the relative propensities of the various $\{Al^{[4]}, Al^{[5]}, Al^{[6]}\}$ coordinations to coordinate the BO and NBO species.

We conclude by noting that since its introduction the early 1980s, ^{27}Al MAS NMR application to AS glasses has advanced significantly and become an ubiquitous routine tool, but further developments are required concerning new NMR methodology for more efficiently investigating the medium-range glass structure, as well as for answering hitherto open questions about the local structure.

Acknowledgment

I thank Baltzar Stevensson for help with preparing the manuscript.

References

[1] B.H. W.S. de Jong, C.M. Schramm, V.E. Parziale, Polymerization of silicate and aluminate tetrahedra in glasses, melts, and aqueous solutions—IV. Aluminum coordination in glasses and aqueous solutions and comments on the aluminum avoidance principle., Geochim. Cosmochim. Acta 47 (1983) 1223–1236.

[2] J.B. Murdoch, J.F. Stebbins, I.S.E. Carmichael, High-resolution ^{29}Si NMR study of silicate and aluminosilicate glasses: the effect of network-modifying cations, Am. Mineral. 70 (1985) 332–343.

[3] G. Engelhardt, M. Nofz, F. Forkel, F.G. Wishmann, M. Mägi, A. Samoson, E. Lippmaa, Structural studies of calcium aluminosilicate glasses by high resolution solid state ^{29}Si and ^{27}Al magic angle spinning nuclear magnetic resonance, Phys. Chem. Glasses 26 (1985) 157–165.

[4] R. Oestrike, W.-H. Yang, R.J. Kirkpatrick, R.L. Hervig, A. Navrotsky, B. Montez, High-resolution ^{23}Na, ^{27}Al and ^{29}Si NMR spectroscopy of framework aluminosilicate glasses, Geochim. Cosmochim. Acta 51 (1987) 2199–2209.

[5] R. Oestrike, R.J. Kirkpatrick, ^{27}Al and ^{29}Si MASS NMR spectroscopy of glasses in the system anorthite-diopside-forsterite, Am. Mineral. 73 (1988) 534–546.

[6] C.I. Merzbacher, B.L. Sheriff, J.S. Hartmann, W.B. White, A high-resolution ^{29}Si NMR and ^{27}Al NMR study of alkaline earth aluminosilicate glasses, J. Non-Cryst. Solids 124 (1990) 194–206.

[7] R.K. Sato, P.F. McMillan, P. Dennison, R. Dupree, A structural investigation of high alumina glasses in the CaO-Al_2O_3-SiO_2 system via Raman and magic angle spinning nuclear magnetic resonance spectroscopy, Phys. Chem. Glasses 32 (1991) 149–156.

[8] B.O. Mysen, D. Virgo, C.M. Scarfe, Relations between the anionic structure and viscosity of silicate melts—a Raman spectroscopic study, Am. Mineral. 65 (1980) 690–710.

[9] B.O. Mysen, D. Virgo, I. Kushiro, The structural role of aluminium in silicate melts; a Raman spectroscopic study at 1 atmosphere, Am. Mineral. 66 (1981) 678–791.

[10] B.O. Mysen, D. Virgo, F. Siefert, Relationships between properties and structure of aluminosilicate melts, Am. Mineral. 70 (1985) 88–105.

[11] F. Domine, B. Piriou, Raman spectroscopic study of the SiO_2-Al_2O_3-K_2O vitreous system: distribution of silicon second neighbors, Am. Mineral. 71 (1986) 38–50.

[12] D. Müller, W. Gessner, H.-J. Behrens, G. Scheler, Determination of the aluminium coordination in aluminium-oxygen compounds by solid-state high-resolution ^{27}Al NMR, Chem. Phys. Lett. 79 (1981) 59–62.

[13] E. Kundla, A. Samoson, E. Lippmaa, High-resolution NMR of half-integer quadrupolar nuclei in rotating solids, Chem. Phys. Lett. 83 (1981) 229–232.

[14] M.D. Meadows, K.A. Smith, R.A. Kinsey, M. Rothiger, R.P. Skarjune, E. Oldfield, High resolution solid-state NMR of quadrupolar nuclei, Prog. Nat. Acad. Sci. USA 79 (1982) 1351–1355.

[15] E. Oldfield, R.A. Kinsey, K.A. Smith, J.A. Nichols, R.J. Kirkpatrick, High-resolution NMR of inorganic solids. Influence of magnetic centers on magic-angle sample-spinning lineshapes in some natural aluminosilicates, J. Magn. Reson. 51 (1983) 325–329.

[16] D. Fenzke, D. Freude, T. Frölich, J. Haase, NMR intensity measurements of half-integer quadrupolar nuclei, Chem. Phys. Lett. 111 (1984) 171–175.

[17] A. Samoson, E. Kundla, E. Lippmaa, High resolution MAS-NMR of quadrupolar nuclei in powders, J. Magn. Reson. 49 (1982) 350–357.
[18] A. Samoson, E. Lippmaa, Excitation phenomena and line intensities in high-resolution NMR powder spectra of half-integer quadrupolar nuclei, Phys. Rev. B. 28 (1983) 6567–6570.
[19] A. Samoson, Satellite transition high-resolution NMR of quadrupolar nuclei in powders, Chem. Phys. Lett. 119 (1985) 29–32.
[20] E. Lippmaa, A. Samoson, M. Mägi, High-resolution ^{27}Al NMR of aluminosilicates, J. Am. Chem. Soc. 86 (1986) 1730–1735.
[21] L. Frydman, J.S. Harwood, Isotropic spectra of half-integer quadrupolar spins from bidimensional magic-angle-spinning NMR, J. Am. Chem. Soc. 117 (1995) 5367–5368.
[22] A. Medek, J.S. Harwood, L. Frydman, Multiple-quantum magic-angle-spinning NMR: a new method for the study of quadrupolar nuclei in solids, J. Am. Chem. Soc. 117 (1995) 12779–12787.
[23] M. Edén, ^{27}Al NMR studies of aluminosilicate glasses, Annu. Rep. NMR Spectrosc. 86 (2015) 237–331.
[24] A. Pines, M.G. Gibby, J.S. Waugh, Proton-enhanced NMR of dilute spins in solids, J. Chem. Phys. 59 (1973) 569–590.
[25] E.O. Stejskal, J. Schaefer, J.S. Waugh, Magic-Angle spinning and polarization transfer in proton-enhanced NMR, J. Magn. Reson. 28 (1977) 105–112.
[26] W. Kolodziejski, J. Klinowski, Kinetics of cross-polarization in solid state NMR: a guide for chemists, Chem. Rev. 102 (2002) 613–628.
[27] A. Jerschow, From nuclear structure to the quadrupolar NMR interaction and high-resolution spectroscopy, Prog. Nucl. Magn. Reson. Spectrosc. 46 (2005) 63–78.
[28] A. Goldbourt, P.K. Madhu, Multiple-quantum magic-angle spinning: high-resolution solid-state NMR of half-integer spin quadrupolar nuclei, Ann. Rep. NMR Spectr. 54 (2005) 79–153.
[29] C. Fernandez, M. Pruski, Probing quadrupolar nuclei by solid-state NMR spectroscopy: recent advances, Top. Curr. Chem. 306 (2012) 119–188.
[30] M. Edén, NMR studies of oxide-based glasses, Annu. Rep. Prog. Chem. Sect. C Phys. Chem. 108 (2012) 177–221.
[31] R.D. Shannon, Revised effective ionic-radii and systematic studies of interatomic distances in halides and chalcogenides, Acta Cryst. A32 (1976) 751–767.
[32] M.H. Levitt, Spin Dynamics. Basics of Nuclear Magnetic Resonance, second ed., Wiley, Chicester, 2007.
[33] M.H. Levitt, The signs of frequencies and phases in NMR, J. Magn. Reson. 126 (1997) 164–182.
[34] M. Edén, Computer simulations in solid state NMR: I. Spin dynamics theory, Concepts Magn. Reson. A 17 (2003) 117–154.
[35] E.R. Andrew, A. Bradbury, R.G. Eades, Removal of dipolar broadening of nuclear magnetic resonance spectra of solids by specimen rotation, Nature 183 (1959) 1802–1803.
[36] I.J. Lowe, Free induction decays of rotating solids, Phys. Rev. Lett. 2 (1959) 285–287.
[37] M.M. Maricq, J.S. Waugh, NMR in rotating solids, J. Chem. Phys. 70 (1979) 3300–3316.
[38] M. Edén, Computer simulations in solid state NMR: III. Powder averaging, Concepts Magn. Reson. A 18 (2003) 24–55.
[39] G. Engelhardt, D. Michel, High-Resolution Solid-State NMR Spectroscopy of Silicates and Zeolites, Wiley, Chicester, 1987.
[40] E.D. Lacy, Aluminium in glasses and melts, Phys. Chem. Glasses 4 (1963) 234–238.

[41] J.M. Stevels, Neue Erkenntnisse über die Struktur des Glases, Philips Techn. Rundsch. 9/10 (1960) 337–349.
[42] M. Edén, The split network analysis for exploring composition-structure correlations in multi-component glasses: I. Rationalizing bioactivity-composition trends of bioglasses, J. Non-Cryst. Solids 357 (2011) 1595–1602.
[43] K.J.D. MacKenzie, M.E. Smith, Multinuclear Solid-State NMR of Inorganic Materials, Pergamon Press, Amsterdam, 2002.
[44] R.H. Meinhold, R.C.T. Slade, R.H. Newman, High field MAS NMR, with simulations of the effects of disorder on linehshape, applied to thermal transformations of alumina hydrates, Appl. Magn. Reson. 4 (1993) 121–140.
[45] M.-H. Lee, C.-F. Cheng, V. Heine, J. Klinowski, Distribution of tetrahedral and octahedral Al sites in gamma alumina, Chem. Phys. Lett. 265 (1997) 673–676.
[46] A. Düvel, E. Romanova, M. Sharifi, D. Freude, M. Wark, P. Heitjans, M. Wilkening, Mechanically induced phase transformation of γ-Al_2O_3 into α-Al_2O_3. Access to structurally disordered γ-Al_2O_3 with a controllable amount of pentacoordinated Al sites, J. Phys. Chem. C 115 (2011) 22770–22780.
[47] L. Samain, A. Jaworski, M. Edén, D.M. Ladd, D.-K. Seo, F.J. Garcia-Garcia, U. Häussermann, Structural analysis of highly porous γ-Al_2O_3, J. Solid State Chem. 217 (2014) 1–8.
[48] L.M. Perander, Z.D. Zujovic, T. Groutso, M.M. Hyland, M.E. Smith, L.A. O'Dell, J.B. Metson, Characterization of metallurgical-grade aluminas and their precursors by ^{27}Al NMR and XRD, Can. J. Chem. 85 (2007) 889–897.
[49] S.K. Lee, S.B. Lee, S.Y. Park, Y.S. Yi, C.W. Ahn, Structure of amorphous aluminium oxide, Phys. Rev. Lett. 103 (2009) 095501.
[50] V. Sarou-Kanian, A.N. Gleizes, P. Florian, D. Samélor, D. Massiot, C. Vahlas, Temperature-dependent 4-, 5- and 6-fold coordination of aluminum in MOCVD-grown amorphous alumina films: a very high field ^{27}Al-NMR study, J. Phys. Chem. C 117 (2013) 21965–21971.
[51] P.F. McMillan, W.T. Petuskey, B. Coté, D. Massiot, C. Landron, J.-P. Coutures, A structural investigation of CaO-Al_2O_3 glasses via ^{27}Al MAS-NMR, J. Non-Cryst. Solids 195 (1996) 261–272.
[52] D. Iuga, S. Simon, E de Boer, A.P.M. Kentgens, A nuclear magnetic study of amorphous and crystalline lanthanum-aluminates, J. Phys. Chem. B 103 (1999) 7591–7598.
[53] K.J.D. MacKenzie, M. Schmücker, L. Mayer, Evolution of crystalline aluminates from hybrid gel-derived precursors studied by XRD and multinuclear solid-state NMR III. Lanthanum hexaluminate, $LaAl_{11}O_{18}$, Themochim. Acta 335 (1999) 73–78.
[54] M.C. Wilding, C.J. Benmore, P.F. McMillan, A neutron diffraction study of yttrium and lanthanum-aluminate glasses, J. Non-Cryst. Solids 297 (2002) 143–155.
[55] J.A. Tangeman, B.L. Phillips, P.C. Nordine, J.K.R. Weber, Thermodynamics and structure of single- and two-phase yttria-alumina glasses, J. Phys. Chem. B 108 (2004) 10663–10671.
[56] P.F. McMillan, R.J. Kirkpatrick, Al coordination in magnesium aluminosilicate glasses, Am. Mineral. 77 (1992) 898–900.
[57] S.K. Lee, S. Sung, The effect of network-modifying cations on the structure and disorder in peralkaline Ca-Na aluminosilicate glasses: O-17 3QMAS NMR study, Chem. Geol. 256 (2008) 326–333.
[58] E. Gambuzzi, A. Pedone, M.C. Menziani, F. Angeli, D. Caurant, T. Charpentier, Probing silicon and aluminium chemical environments in silicate and aluminosilicate glasses by solid state NMR spectroscopy and accurate first-principles calculations, Geochim. Cosmochim. Acta 125 (2014) 170–185.
[59] W. Loewenstein, The distribution of aluminium in the tetrahedra of silicates and aluminates, Am. Mineral. 39 (1954) 92–96.

[60] J.R. Allwardt, S.K. Lee, J.F. Stebbins, Bonding preferences of non-bridging O atoms: evidence from ^{17}O MAS and 3QMAS NMR on calcium aluminate and low-silica Ca-Aluminosilicate glasses, Am. Mineral. 88 (2003) 949–954.
[61] S.K. Lee, J.F. Stebbins, Disorder and the extent of polymerization in calcium silicate and aluminosilicate glasses: O-17 NMR results and quantum chemical molecular orbital calcualtions, Geochim. Cosmochim. Acta 70 (2006) 4275–4286.
[62] D.R. Neuville, L. Cormier, D. Massiot, Al Coordination and speciation in calcium aluminosilicate glasses: effects of composition determined by ^{27}Al MQ-MAS NMR and Raman spectroscopy, Chem. Geol. 229 (2006) 173–185.
[63] J.T. Kohli, J.E. Shelby, J.S. Frye, A structural investigation of Yttrium aluminosilicate glasses using ^{29}Si and ^{27}Al magic angle spinning nuclear magnetic resonance, Phys. Chem. Glasses 33 (1992) 73–78.
[64] T. Schaller, J.F. Stebbins, The structural role of Lanthanum and Yttrium in aluminosilicate glasses: a ^{27}Al and ^{17}O MAS NMR study, J. Phys. Chem. B 102 (1998) 10690–10697.
[65] S.K. Lee, J.F. Stebbins, The degree of aluminium avoidance in aluminosilicate glasses, Am. Mineral. 84 (1999) 937–945.
[66] S.K. Lee, J.F. Stebbins, Al-O-Al and Si-O-Si sites in framework aluminosilicate glasses with Si/Al=1: quantification of framework disorder, J. Non-Cryst. Solids 270 (2000) 260–264.
[67] J.F. Stebbins, S. Kroeker, S.K. Lee, T.J. Kiczenski, Quantification of five- and six-coordinated aluminium in aluminosilicate and fluoride-containing glasses by high-field ^{27}Al NMR, J. Non-Cryst. Solids 275 (2000) 1–6.
[68] M.J. Toplis, S.C. Kohn, M.E. Smith, I.J.F. Poplett, Fivefold coordinated aluminium in tectosilicate glasses observed by triple-quantum MAS NMR, Am. Mineral. 85 (2000) 1556–1560.
[69] D.R. Neuville, L. Cormier, D. Massiot, Al environment in tectosilicate and peraluminous glasses: a ^{27}Al MQ-MAS NMR, Raman and XANES investigation, Geochim. Cosmochim. Acta 68 (2004) 5071–5079.
[70] S. Lee, H. Kim, J. Kim, K. Mun, S. Ryu, Extent of disorder in magnesium aluminosilicate glasses: insights from ^{27}Al and ^{17}O NMR, J. Phys. Chem. C 120 (2016) 737–749.
[71] J.F. Stebbins, Z. Xu, NMR evidence for excess non-bridging oxygen in an aluminosilicate glass, Nature 390 (1997) 60–62.
[72] S. Iftekhar, B. Pahari, K. Okhotnikov, A. Jaworski, B. Stevensson, J. Grins, M. Edén, Properties and structures of RE_2O_3-Al_2O_3-SiO_2 (RE=Y, Lu) glasses probed by molecular dynamics simulations and solid-state NMR: the roles of aluminum and rare-earth ions for dictating the microhardness, J. Phys. Chem. C 116 (2012) 18394–18406.
[73] A. Jaworski, B. Stevensson, M. Edén, Direct ^{17}O NMR experimental evidence for Al-NBO bonds in Si-Rich and highly polymerized aluminosilicate glasses, Phys. Chem. Chem. Phys. 17 (2015) 18269–18272.
[74] A. Jaworski, B. Stevensson, M. Edén, The bearings from rare-earth (RE=La, Lu, Sc, Y) cations on the oxygen environments in aluminosilicate glasses: a study by solid-state ^{17}O NMR, molecular dynamics simulations, and DFT calculations, J. Phys. Chem. C 120 (2016) 13181–13198.
[75] M. LaComb, D. Rice, J.F. Stebbins, Network oxygen sites in calcium aluminoborosilicate glasses: results from ^{17}O{^{27}Al} and ^{17}O{^{11}B} double resonance NMR, J. Non-Cryst. Solids 447 (2016) 248–254.
[76] M. Edén, P. Sundberg, C. Stålhandske, The split network analysis for exploring composition-structure correlations in multi-component glasses: II. Multinuclear NMR studies of alumino-borosilicates and glass-wool fibers, J. Non-Cryst. Solids 357 (2011) 1587–1594.

[77] S. Iftekhar, E. Leonova, M. Edén, Structural characterization of lanthanum aluminosilicate glasses by ^{29}Si solid-state NMR, J. Non-Cryst. Solids 355 (2009) 2165–2174.
[78] Y. Yu, M. Edén, Structure-composition relationships of bioactive borophosphosilicate glasses probed by multinuclear ^{11}B, ^{29}Si, and ^{31}P solid-state NMR, RSC Adv. 6 (2016) 101288–101303.
[79] Y. Yu, B. Stevensson, M. Edén, Medium-range structural organization of phosphorus-bearing borosilicate glasses revealed by advanced solid-state NMR experiments and MD simulations: consequences of B/Si substitutions, J. Phys. Chem. B 121 (2017) 9737–9752.
[80] Y. Yu, R. Mathew, M. Edén, Quantitative composition-bioactivity relationships of phosphosilicate glasses: bearings from the phosphorus content and network polymerization, J. Non-Cryst. Solids 502 (2018) 106–117.
[81] R. Mathew, B. Stevensson, A. Tilocca, M. Edén, Toward a rational design of bioactive glasses with optimal structural features: composition-structure correlations unveiled by solid-state NMR and MD simulations, J. Phys. Chem. B 118 (2014) 833–844.
[82] R. Mathew, B. Stevensson, M. Edén, Na/Ca intermixing around silicate and phosphate groups in bioactive phosphosilicate glasses revealed by heteronuclear solid-state NMR and molecular dynamics simulations, J. Phys. Chem. B 119 (2015) 5701–5715.
[83] T. Charpentier, The PAW/GIPAW approach for computing NMR parameters: a new dimension added to NMR of solids, Solid State Nucl. Magn. Reson. 40 (2011) 1–20.
[84] T. Charpentier, M.C. Menziani, A. Pedone, Computational simulations of solid state NMR spectra: a new era in structure determination of oxide glasses, RSC Adv. 3 (2013) 10550–10578.
[85] A. Pedone, E. Gambuzzi, G. Malavasi, M.C. Menziani, First-principles simulations of the ^{27}Al and ^{17}O solid-State NMR spectra of the $CaAl_2Si_3O_{10}$ glass, Theor. Chem. Acc. 131 (2012) 1147.
[86] Z. Chaker, M. Salanne, J.-M. Delaye, T. Charpentier, NMR shifts in aluminosilicate glasses *via* machine learning, Phys. Chem. Chem. Phys. 21 (2019) 21709–21725.
[87] M.E. Smith, Application of ^{27}Al NMR techniques to structure determination in solids, Appl. Magn. Reson. 4 (1993) 1–64.
[88] E. Hallas, U. Sternberg, New aspects of the semiempirical theory of ^{27}Al-chemical shifts and its application to glasses, Mol. Phys. 68 (1989) 315–326.
[89] S.C. Kohn, C.M.B. Henderson, R. Dupree, Al-Si ordering in leucite group minerals and ion-exchanged analogues: an MAS NMR study, Am. Mineral. 82 (1997) 1133–1140.
[90] Y. Liu, H. Nekvasil, J. Tossell, Explaining the effects of T-O-T bond angles on NMR chemical shifts in aluminosilicates: a natural bonding orbital NBO and natural chemical shielding NCS analysis, J. Phys. Chem. A 109 (2005) 3060–3066.
[91] M.E. Smith, Observation of mixed $Al(O,N)_4$ structural units by ^{27}Al magic angle spinning NMR, J. Phys. Chem. 96 (1992) 1444–1448.
[92] J.J. Fitzgerald, S.D. Kohl, G. Piedra, Observation of four-coordinate aluminum oxynitride $AlO_{4-x}N_x$ environments in AlON solids by MAS ^{27}Al NMR at 14 T, Chem. Mater. 6 (1994) 1915–1917.
[93] E. Leonova, A.S. Hakeem, K. Jansson, B. Stevensson, Z. Shen, J. Grins, S. Esmaeilzadeh, M. Edén, Nitrogen-rich La-Si-Al-O-N oxynitride glass structures probed by solid state NMR, J. Non-Cryst. Solids 354 (2008) 49–60.
[94] P.J. Dirken, J.B.H. Jansen, R.D. Schuiling, Influence of octahedral polymerization on ^{23}Na and ^{27}Al MAS NMR in alkali fluoroaluminates, Am. Mineral. 77 (1992) 718–724.
[95] E. Lippmaa, M. Mägi, A. Samoson, G. Engelhardt, A.-R. Grimmer, Structural studies of silicates by solid-state high-resolution ^{29}Si NMR, J. Am. Chem. Soc. 102 (1980) 4889–4893.

[96] A.-R. Grimmer, M. Mägi, M. Hähnert, H. Stade, A. Samoson, W. Weiker, E. Lippmaa, High resolution solid state ^{29}Si nuclear magnetic resonance spectroscopic studies of binary alkali silicate glasses, Phys. Chem. Glasses 25 (1984) 105–109.
[97] H. Maekawa, T. Maekawa, K. Kawamura, T. Yokokawa, The structural groups of alkali silicate glasses determined by ^{29}Si MAS-NMR, J. Non-Cryst. Solids 127 (1991) 53–64.
[98] D. Müller, W. Gessner, A. Samoson, E. Lippmaa, G. Scheler, Solid-state aluminium-27 nuclear magnetic resonance chemical shift and quadrupole coupling data for condensed AlO_4, J. Chem. Soc. Dalton Trans. (1986) 1277–1281.
[99] D. Müller, W. Gessner, A. Samoson, E. Lippmaa, G. Scheler, Solid-state ^{27}Al NMR studies of polycrystalline aluminates of the system CaO-Al_2O_3, Polyhedron 5 (1986) 779–785.
[100] M.E. Smith, E.R.H. van Eck, Recent advances in experimental solid state NMR methodology for half-integer spin quadrupolar nuclei, Prog. Nucl. Magn. Reson. Spectrosc. 34 (1999) 159–201.
[101] S.E. Ashbrook, M.J. Duer, Structural information from quadrupolar nuclei in solid state NMR, Concepts Magn. Reson. A 28 (2006) 183–248.
[102] M. Edén, Zeeman truncation in NMR. I. The role of operator commutation, Concepts Magn. Reson. 43A (2014) 91–108.
[103] M. Edén, Zeeman truncation in NMR. II. Time averaging in the rotating frame, Concepts Magn. Reson. 43A (2014) 109–126.
[104] A. Wokaun, R.R. Ernst, Selective excitation and detection in multilevel spin systems: application of single transition operators, J. Chem. Phys. 67 (1977) 1752–1758.
[105] S. Vega, Fictitious spin 1/2 operator formalism for multiple quantum NMR, J. Chem. Phys. 68 (1978) 5518–5527.
[106] S. Esmaeilzadeh, J. Grins, Z. Shen, M. Edén, M. Thiaux, A study of sialon S-phases $M_2Al_xSi_{12-x}N_{16-x}O_{2+x}$, M=Ba and $Ba_{0.9}Eu_{0.1}$ by X-ray single crystal diffraction, X-ray powder diffraction and solid state nuclear magnetic resonance, Chem. Mater. 16 (2004) 2113–2120.
[107] M. Edén, J. Grins, K. Jansson, Z. Shen, Al and O substitutions in Ba S-phases, $Ba_2Al_xSi_{12-x}N_{16-x}O_{2+x}$: a TEM, XRD and solid-state NMR study, Solid State Sci. 10 (2008) 50–60.
[108] D. Freude, J. Haase, J. Klinowski, T.A. Carpenter, G. Roniker, NMR line shifts caused by the second-order quadrupolar interaction, Chem. Phys. Lett. 119 (1985) 365–367.
[109] L.B. Alemany, Critical factors in obtaining meaningful fast MAS NMR spectra of non-integral spin quadrupolar nuclei. A review with particular emphasis on ^{27}Al MAS NMR of catalysts and minerals, Appl. Magn. Reson. 4 (1993) 179–201.
[110] A.P.M. Kentgens, A practical guide to solid-state NMR of half-integer quadrupolar nuclei with some applications to disordered systems, Geoderma 80 (1997) 271–306.
[111] J. Mahler, A. Sebald, Deconvolution of ^{29}Si magic-angle spinning nuclear magnetic resonance spectra of silicate glasses revisited—some critical comments, Solid State Nucl. Magn. Reson. 5 (1995) 63–78.
[112] T. Björling, D. Noréus, K. Jansson, M. Andersson, E. Leonova, M. Edén, U. Hålenius, U. Häussermann, SrAlSiH: a polyanionic semiconductor hydride, Angew. Chem. Int. Ed. 44 (2005) 7269–7273.
[113] Å. Zazzi, T.K. Hirsch, E. Leonova, A. Kaikkonen, J. Grins, H. Annersten, M. Edén, Structural investigations of natural and synthetic chlorite minerals by X-ray diffraction, Mössbauer spectroscopy and solid-state nuclear magnetic resonance, Clays Clay Miner. 54 (2006) 252–265.
[114] S. Iftekhar, J. Grins, P.N. Gunawidjaja, M. Edén, Glass formation and structure-property-composition relations of the RE_2O_3-Al_2O_3-SiO_2 (RE=La, Y, Lu, Sc) systems, J. Am. Ceram. Soc. 94 (2011) 2429–2435.

[115] D. Coster, A.L. Blumenfeld, J.J. Fripiat, Lewis acid sites and surface aluminum in aluminas and zeolites: a high-resolution NMR study, J. Phys. Chem. 98 (1994) 6201–6211.
[116] C. Jäger, G. Kunath, P. Losso, G. Scheler, Determination of distributions of the quadrupolar interaction in amorphous solids by ^{27}Al satellite-transition spectroscopy, Solid State Nucl. Magn. Reson. 2 (1993) 73–82.
[117] J.-B. d'Espinose de Lacaillerie, C. Fretigny, D. Massiot, MAS NMR spectra of quadrupolar nuclei in disordered solids: the Czjzek model, J. Magn. Reson. 192 (2008) 244–251.
[118] B.L. Phillips, R.J. Kirkpatrick, G.L. Hovis, ^{27}Al, ^{29}Si, and ^{23}Na MAS NMR study of an Al, Si ordered alkali feldspar solid solution series, Phys. Chem. Miner. 16 (1988) 262–275.
[119] M.P.J. Peeters, A.P.M. Kentgens, A ^{27}Al MAS, MQMAS and off-resonance nutation NMR study of aluminium containing silica-based Sol-Gel materials, Solid State Nucl. Magn. Reson. 9 (1997) 203–217.
[120] G.L. Hoatson, D.H. Zhou, F. Fayon, D. Massiot, R.L. Vold, ^{93}Nb magic angle spinning NMR study of perovskite relaxor ferroelectrics $(1 - x)$Pb$(Mg_{1/3}Nb_{2/3})O_3$-xPb$(Sc_{1/2}Nb_{1/2})O_3$, Phys. Rev. B. 66 (2002) 224103.
[121] J. McManus, S.E. Ashbrook, K.J.D. MacKenzie, S. Wimperis, ^{27}Al multiple-quantum MAS and ^{27}Al{^{1}H} CPMAS NMR study of amorphous aluminosilicates, J. Non-Cryst. Solids 282 (2001) 278–290.
[122] P. Mustarelli, R. Riccardi, S. Scotti, M. Villa, Profile analysis of quadrupolar broadened nuclear-magnetic-resonance spectra of glasses, Phys. Chem. Glasses 32 (1991) 129–131.
[123] G. Czjzek, J. Fink, F. Götz, H. Schmidt, J.M.D. Coey, J.-P. Rebouillat, A. Liénard, Atomic coordination and the distribution of electric field gradients in amorphous solids, Phys. Rev. B. 23 (1981) 2513–2530.
[124] B. Bureau, G. Silly, J.Y. Buzaré, C. Legein, D. Massiot, From crystalline to glassy gallium fluoride materials: an NMR study of ^{69}Ga and ^{71}Ga quadrupolar nuclei, Solid State Nucl. Magn. Reson. 14 (1999) 181–190.
[125] D. Massiot, F. Fayon, M. Capron, I. King, S. Le Calvé, B. Alonso, J.-O. Durand, B. Bujoli, Z. Gan, G. Hoatson, Modelling one- and two-dimensional solid-state NMR spectra, Magn. Reson. Chem. 40 (2002) 70–76.
[126] B. Pahari, S. Iftekhar, A. Jaworski, K. Okhotnikov, K. Jansson, B. Stevensson, J. Grins, M. Edén, Composition-property-structure correlations of scandium aluminosilicate glasses revealed by multinuclear ^{45}Sc, ^{27}Al and ^{29}Si solid state NMR, J. Am. Ceram. Soc. 95 (2012) 2545–2553.
[127] T.F. Kemp, M.E. Smith, QuadFit—a new cross-platform computer program for simulation of NMR line shapes from solids with distributions of interaction parameters, Solid State Nucl. Magn. Reson. 35 (2009) 243–252.
[128] D.L.A.G. Grimminck, B.J.W. Polman, A.P.M. Kentgens, W.L. Meerts, *EASY-GOING deconvolution*: combining accurate simulation and evolutionary algorithms for fast deconvolution of solid-state quadrupolar NMR spectra, J. Magn. Reson. 211 (2011) 114–120.
[129] D.L.A.G. Grimminck, B. van Meerten, M.H.W. Verkuijlen, E.R.H. van Eck, W.L. Meerts, A.P.M. Kentgens, *easy-going deconvolution*: automated mqmas nmr spectrum analysis based on a model with analytical crystallite excitation efficiencies, J. Magn. Reson. 228 (2013) 116–124.
[130] G. Le Caër, B. Bureau, D. Massiot, An extension of the Czjzek model for the distributions of electric field gradients in disordered solids and an application to NMR spectra of ^{71}Ga in chalcogenide glasses, J. Phys. Condens. Matter 22 (2010) 065402.
[131] F. Angeli, T. Charpentier, F. Faucon, J.-C. Petit, Structural characterization of glass from the inversion of ^{23}Na and ^{27}Al 3Q-MAS NMR spectra, J. Phys. Chem. B 103 (1999) 10356–10364.

[132] F. Angeli, J.-M. Delaye, T. Charpentier, J.-C. Petit, D. Ghaleb, F. Faucon, Influence of glass composition on the Na-O bond distance: a ^{23}Na 3Q-MAS NMR and molecular dynamics study, J. Non-Cryst. Solids 276 (2000) 132–144.
[133] L. Zhang, H. Eckert, Short- and medium-range order in sodium aluminophosphate glasses: new insights from high-resolution dipolar solid-state NMR spectroscopy, J. Phys. Chem. B 110 (2006) 8946–8958.
[134] R.K. Harris, E.D. Becker, NMR nomenclature: nuclear spin properties and conventions for chemical shifts-IUPAC recommendations, J. Magn. Reson. 156 (2002) 323–326.
[135] D.E. Woessner, H.K.C. Timken, The influence of MAS on spin-lattice relaxation curves and nuclear spin excitation of half-integer spin quadrupolar nuclei in solids, J. Magn. Reson. 90 (1990) 411–419.
[136] P.P. Man, Excitation conditions for quantitative determination of quadrupolar spins with one pulse or spin-echo sequences in solid-state NMR, Appl. Magn. Reson. 4 (1993) 65–87.
[137] J. Haase, E. Oldfield, Quantitative determination of nonintegral spin quadrupolar nuclei in solids using nuclear magnetic resonance spin-echo techniques, J. Magn. Reson. 104 (1993) 1–9.
[138] D. Mohr, A.S.S de Camargo, C.C de Araujo, H. Eckert, Local environment of scandium in aluminophosphate laser glasses: structural studies by solid state NMR spectroscopy, J. Mater. Chem. 17 (2007) 3733–3738.
[139] K.E. Kelsey, J.F. Stebbins, D.M. Singer, G.E. Brown, J.L. Mosenfelder, P.D. Asimow, Cation field strength effects on high pressure aluminosilicate glass structure: multinuclear NMR and La XAFS results, Geochim. Cosmochim. Acta 73 (2009) 3914–3933.
[140] A.R. Thompson, E. Oldfield, Solid-State Scandium-45, Yttrium-89, and Lanthanum-139 nuclear magnetic resonance spectroscopy, J. Chem. Soc. Chem. Commun. (1987) 27–29.
[141] R. Dupree, M.H. Lewis, M.E. Smith, A high-resolution NMR study of the La-Si-Al-O-N system, J. Am. Chem. Soc. 111 (1989) 5125–5132.
[142] D.J. Aurentz, F.G. Vogt, K.T. Mueller, A.J. Benesi, Multiple-rotor-cycle QPASS pulse sequences: separation of quadrupolar spinning sidebands with an application to ^{139}La NMR, J. Magn. Reson. 138 (1999) 320–325.
[143] P.K. Madhu, O.G. Johannessen, K.J. Pike, R. Dupree, M.E. Smith, M.H. Levitt, Application of amplitude-modulated radiofrequency fields to the magic-angle spinning NMR of spin-7/2 nuclei, J. Magn. Reson. 163 (2003) 310–317.
[144] A.Y.H. Lo, V. Sudarsan, S. Sivakumar, F. van Veggel, R.W. Schurko, Multinuclear solid-state NMR spectroscopy of doped lanthanum-fluoride nanoparticles, J. Am. Chem. Soc. 129 (2007) 4687–4700.
[145] H. Hamaed, A.Y.H. Lo, D.S. Lee, W.J. Evans, R.W. Schurko, Solid-state ^{139}La and ^{15}N NMR spectroscopy of lanthanum-containing metallocenes, J. Am. Chem. Soc. 128 (2006) 12638–12639.
[146] K.E. Johnston, M.R. Mitchell, F. Blanc, P. Lightfoot, S.E. Ashbrook, Structural study of $La_{1-x}Y_xScO_3$, combining neutron diffraction, solid-state NMR and first-principles DFT calculations, J. Phys. Chem. C 117 (2013) 2252–2265.
[147] A.L. Paterson, M.A. Hanson, U. Werner-Zwanziger, J.W. Zwanziger, Relating ^{139}La quadrupolar coupling constants to polyhedral distortion in crystalline structures, J. Phys. Chem. C 119 (2015) 25508–25517.
[148] A.P.M. Kentgens, J.J.M. Lemmens, F.M.M. Geurts, W.S. Veeman, Two-dimensional solid-state nutation NMR of half-integer quadrupolar nuclei, J. Magn. Reson. 71 (1987) 62–74.
[149] A.P.M. Kentgens, Off-resonance nutation nuclear magnetic resonance spectroscopy of half-integer quadrupolar nuclei, Prog. Nucl. Magn. Reson. Spectrosc. 32 (1998) 141–164.

[150] D.E. Woessner, Observation of effects of hydration on CP/MAS NMR measurements on quadrupolar nuclei, Z. Phys. Chem. 152 (1987) 309–316.
[151] J. Haase, M.S. Conradi, Sensitivity enhancement for NMR of the central transition of quadrupolar nuclei, Chem. Phys. Lett. 209 (1993) 287–291.
[152] R.K. Sato, P.F. McMillan, P. Dennison, R. Dupree, High-resolution ^{27}Al and ^{29}Si MAS NMR investigation of SiO_2-Al_2O_3 glasses, J. Phys. Chem. 95 (1991) 4483–4489.
[153] H.J. Jakobsen, H. Bildsøe, N.C. Nielsen, Magic-angle-spinning NMR spectra of satellite transitions for quadrupolar nuclei in rotating solids, J. Magn. Reson. 85 (1989) 173–180.
[154] J. Skibsted, N.C. Nielsen, H. Bildsøe, H.J. Jakobsen, Satellite transitions in MAS NMR spectra of quadrupolar nuclei, J. Magn. Reson. 95 (1991) 88–117.
[155] C. Jäger, How to get more from ^{27}Al MAS NMR by high-speed satellite-transition spectroscopy, J. Magn. Reson. 99 (1992) 353–362.
[156] G. Kunath-Fandrei, T.J. Bastow, J.S. Hall, C. Jäger, M.E. Smith, Quantification of aluminum coordinations in amorphous aluminas by combined central and satellite transition magic angle spinning NMR spectroscopy, J. Phys. Chem. 99 (1995) 15138–15141.
[157] Z. Gan, Isotropic NMR spectra of half-integer quadrupolar nuclei using satellite transitions and magic-angle spinning, J. Am. Chem. Soc. 122 (2000) 3242–3243.
[158] Z. Gan, Satellite transition magic-angle spinning nuclear magnetic resonance spectroscopy of half-integer quadrupolar nuclei, J. Chem. Phys. 114 (2001) 10845–10853.
[159] S.E. Ashbrook, S. Wimperis, High-resolution NMR of quadrupolar nuclei in solids: the satellite-transition magic angle spinning (STMAS) experiment, Prog. Nucl. Magn. Reson. Spectrosc. 45 (2004) 53–108.
[160] L. Züchner, J.C.C. Chan, W. Müller-Warmuth, H. Eckert, Short-range order and site connectivities in sodium aluminoborate glasses: I. Quantification of local environments by high-resolution ^{11}B, ^{23}Na, and ^{27}Al solid-state NMR, J. Phys. Chem. B 102 (1998) 4495–4506.
[161] J.S. Frye, G.A. Maciel, Setting the magic angle using a quadrupolar nuclide, J. Magn. Reson. 48 (1982) 125–131.
[162] R.E. Taylor, Setting up ^{13}C CP/MAS NMR experiments, Concepts Magn. Reson. A 22 (2004) 37–49.
[163] J. Haase, H. Pfeifer, W. Oehme, J. Klinowski, Longitudinal NMR relaxation of ^{27}Al nuclei in zeolites, Chem. Phys. Lett. 150 (1988) 189–193.
[164] A.P.M. Kentgens, A. Bos, P.J. Dirken, ^{27}Al nuclear magnetic resonance study of synthetic and natural corundum (α-Al_2O_3). Some experimental aspects of quantitative ^{27}Al nuclear magnetic resonance spectroscopy, Solid State Nucl. Magn. Reson. 3 (1994) 315–322.
[165] H. Fujimori, N. Sato, K. Ioku, S. Goto, Y. Yamada, ^{29}Si MASNMR spin-lattice relaxation study of α-, β-, and amorphous silicon nitride, J. Am. Ceram. Soc. 83 (2000) 2251–2254.
[166] E. Leonova, J.G.M. Shariatgorji, L.L. Ilag, M. Edén, Solid state NMR investigations of Si-29 and N-15 enriched silicon nitride, Solid State Nucl. Magn. Reson. 36 (2009) 11–18.
[167] T.G. Mayerhöfer, Z.-J. Shen, E. Leonova, M. Edén, A. Kriltz, J. Popp, Consolidated silica glass from nanoparticles, J. Solid State Chem. 181 (2008) 2442–2447.
[168] G. Pilet, J. Grins, M. Edén, S. Esmaeilzadeh, $La_{17}Si_9Al_4N_{32-x}O_x$, ($x \leq 1$): a nitridoaluminosilicate with isolated Si/Al-N/O clusters, Eur. J. Inorg. Chem. (2006) 3627–3633.
[169] E.R. Andrew, D.P. Tunstall, Spin-lattice relaxation in imperfect cubic crystals and in non-cubic crystals, Proc. Phys. Soc 78 (1961) 1–11.
[170] J.P. Yesinowski, Finding the true spin-lattice relaxation time for half-integral nuclei with non-zero quadrupole couplings, J. Magn. Reson. 252 (2015) 135–144.

[171] A.J. Vega, MAS NMR spin locking of half-integer quadrupolar nuclei, J. Magn. Reson. 96 (1992) 50–68.
[172] S. Hayashi, Magic-angle spinning nuclear magnetic resonance of half-integer quadrupolar nuclei: effect of spin-locking efficiency on powder lineshapes, Solid State Nucl. Magn. Reson. 3 (1994) 93–101.
[173] W. Sun, J.T. Stephen, L.D. Potter, Y. Wu, Rotation-induced resonance and second-order quadrupolar effects on spin locking of half-integer quadrupolar nuclei, J. Magn. Reson. Ser. A 116 (1995) 181–188.
[174] S.E. Ashbrook, S. Wimperis, Spin-locking of half-integer quadrupolar nuclei in nuclear magnetic resonance of solids: creation and evolution of coherences, J. Chem. Phys. 120 (2004) 2719–2731.
[175] S.E. Ashbrook, S. Wimperis, Spin-locking of half-integer quadrupolar nuclei in nuclear magnetic resonance of solids: second-order quadrupolar and resonance offset effects, J. Chem. Phys. 131 (2009) 194509.
[176] S. Wi, J.W. Logan, D. Sakellariou, J.D. Walls, A. Pines, Rotary resonance recoupling for half-integer quadrupolar nuclei in solid-state nuclear magnetic resonance, J. Chem. Phys. 117 (2002) 7024–7033.
[177] M. Baldus, D. Rovnyak, R.G. Griffin, Radio-frequency-mediated dipolar recoupling among half-integer quadrupolar spins, J. Chem. Phys. 112 (2000) 5902–5909.
[178] T.G. Oas, R.G. Griffin, M.H. Levitt, Rotary resonance recoupling of dipolar interactions in solid-state nuclear magnetic resonance spectroscopy, J. Chem. Phys. 89 (1988) 692–695.
[179] J.D. Walls, K.H. Lim, A. Pines, Theoretical studies of the spin dynamics of quadrupolar nuclei at rotational resonance conditions, J. Chem. Phys. 116 (2002) 79–90.
[180] N.C. Nielsen, H. Bildsøe, H.J. Jakobsen, M.H. Levitt, Double-quantum homonuclear rotary resonance: efficient dipolar recovery in magic-angle-spinning nuclear magnetic resonance, J. Chem. Phys. 101 (1994) 1805–1812.
[181] G. Mali, G. Fink, F. Taulelle, Double-quantum homonuclear correlation magic angle sample spinning nuclear magnetic resonance spectroscopy of dipolar-coupled quadrupolar nuclei, J. Chem. Phys. 120 (2004) 2835–2845.
[182] M. Edén, D. Zhou, J. Yu, Improved double-quantum NMR correlation spectroscopy of dipolar-coupled quadrupolar spins, Chem. Phys. Lett. 431 (2006) 397–403.
[183] A.Y.H. Lo, M. Edén, Efficient symmetry-based homonuclear dipolar recoupling of quadrupolar spins: double-quantum NMR correlations in amorphous solids, Phys. Chem. Chem. Phys. 10 (2008) 6635–6644.
[184] M. Edén, A.Y.H. Lo, Supercycled symmetry-based double-quantum dipolar recoupling of quadrupolar spins: I. Theory, J. Magn. Reson. 200 (2009) 267–279.
[185] Z. Gan, Rotary resonance echo double resonance for measuring heteronuclear dipolar coupling under MAS, J. Magn. Reson. 183 (2006) 247–253.
[186] M. Edén, Homonuclear dipolar recoupling of half-integer spin quadrupolar nuclei: techniques and applications, Solid State Nucl. Magn. Reson. 36 (2009) 1–10.
[187] M. Edén, J. Grins, Z. Shen, Z. Weng, Fast ^{29}Si magic-angle-spinning NMR acquisitions by ^{27}Al→^{29}Si RAPT-CP polarization transfer, J. Magn. Reson. 169 (2004) 279–283.
[188] M. Edén, Quadrupolar coupling selective cross-polarization in solid-state NMR, Phys. Chem. Chem. Phys. 8 (2006) 1994–1999.
[189] G. Teymoori, B. Pahari, B. Stevensson, M. Edén, Low-power broadband homonuclear dipolar recoupling without decoupling: double-quantum ^{13}C NMR correlations at very fast magic-angle spinning, Chem. Phys. Lett. 547 (2012) 103–109.
[190] G. Teymoori, B. Pahari, E. Viswanathan, M. Edén, Multiple-quantum spin counting in magic-angle-spinning NMR via low-power symmetry-based dipolar recoupling, J. Magn. Reson. 236 (2013) 31–40.

[191] G. Teymoori, B. Pahari, M. Edén, Low-power broadband homonuclear dipolar recoupling in MAS NMR by two-fold symmetry pulse schemes for magnetization transfers and double-quantum excitation, J. Magn. Reson. 261 (2015) 205–220.
[192] M. Edén, H. Annersten, Å. Zazzi, Pulse-assisted homonuclear dipolar recoupling of half-integer quadrupolar spins in magic-angle spinning NMR, Chem. Phys. Lett. 410 (2005) 24–30.
[193] M. Edén, Determination of absolute quadrupolar tensor orientations by double-quantum NMR on powders, Chem. Phys. Lett. 470 (2009) 318–324.
[194] M. Edén, Two-dimensional MAS NMR correlation protocols involving double-quantum filtering of quadrupolar spin-pairs, J. Magn. Reson. 204 (2010) 99–110.
[195] A. Jaworski, B. Stevensson, B. Pahari, K. Okhotnikov, M. Edén, Local structures and Al/Si ordering in lanthanum aluminosilicate glasses explored by advanced ^{27}Al NMR experiments and molecular dynamics simulations, Phys. Chem. Chem. Phys. 14 (2012) 15866–15878.
[196] P.K. Madhu, A. Goldbourt, L. Frydman, S. Vega, Sensitivity enhancement of the MQMAS NMR experiment by fast amplitude modulation of the pulses, Chem. Phys. Lett. 307 (1999) 41–47.
[197] Z. Yao, H.-T. Kwak, D. Sakellariou, L. Emsley, P.J. Grandinetti, Sensitivity enhancement of the central transition NMR signal of quadrupolar nuclei under magic-angle spinning, Chem. Phys. Lett. 327 (2000) 85–90.
[198] S. Vega, Y. Naor, Triple quantum NMR on spin systems with I=3/2 in solids, J. Chem. Phys. 75 (1981) 75–86.
[199] A.P.M. Kentgens, Quantitative excitation of half-integer quadrupolar nuclei by a frequency-stepped adiabatic passage, J. Magn. Reson. 95 (1991) 619–625.
[200] A.P.M. Kentgens, R. Verhagen, Advantages of double frequency sweeps in static, MAS and MQMAS NMR of spin I=3/2 nuclei, Chem. Phys. Lett. 300 (1999) 435–443.
[201] D. Iuga, A.P.M. Kentgens, Influencing the satellite transitions of half-integer quadrupolar nuclei for the enhancement of magic angle spinning spectra, J. Magn. Reson. 158 (2002) 65–72.
[202] R. Siegel, T.T. Nakasima, R.E. Wasylishen, Signal enhancement of NMR spectra of half-integer quadrupolar nuclei in solids using hyperbolic secant pulses, Chem. Phys. Lett. 388 (2004) 441–445.
[203] K.K. Dey, S. Prasad, J.T. Ash, M. Deschamps, P.J. Grandinetti, Spectral editing in solid-state MAS NMR of quadrupolar nuclei using selective satellite inversion, J. Magn. Reson. 185 (2007) 237–241.
[204] S. Prasad, H.-T. Kwak, T. Clark, P.J. Grandinetti, A simple technique for determining nucelar quadrupolar coupling constants with RAPT solid-state NMR spectroscopy, J. Am. Chem. Soc. 124 (2002) 4964–4965.
[205] H.-T. Kwak, S. Prasad, T. Clark, P.J. Grandinetti, Enhancing sensitivity of half-integer quadrupolar nuclei in solid-state NMR with multiple rotor assisted population transfers, Solid State Nucl. Magn. Reson. 24 (2003) 71–77.
[206] P.K. Madhu, K.J. Pike, R. Dupree, M.H. Levitt, M.E. Smith, Modulation-aided signal enhancement in the magic angle spinning NMR of spin-5/2 nuclei, Chem. Phys. Lett. 367 (2003) 150–156.
[207] G. Mali, V. Kaucic, Enhancing sensitivity or resolution of homonuclear correlation experiment for half-integer quadrupolar nuclei, J. Magn. Reson. 171 (2004) 48–56.
[208] J. Haase, M.S. Conradi, E. Oldfield, Single- and double-resonance experiments of quadrupolar nuclei in solids using sensitivity enhancement of the central transition, J. Magn. Reson. Ser. A 109 (1994) 210–215.
[209] A. Llor, J. Virlet, Towards high-resolution NMR of more nuclei in solids: sample spinning with time-dependent spinner axis angle, Chem. Phys. Lett. 152 (1988) 248–253.

[210] A. Samoson, E. Lippmaa, A. Pines, High resolution solid-state N.M.R. Averaging of second-order effects by means of a double rotor, Mol. Phys. 65 (1988) 1013–1018.
[211] A. Samoson, E. Lippmaa, Sychronized double-rotation NMR spectroscopy, J. Magn. Reson. 84 (1989) 410–416.
[212] B.F. Chmelka, K.T. Mueller, A. Pines, J. Stebbins, Y. Wu, J.W. Zwanziger, Oxygen-17 NMR in solids by dynamic-angle spinning and double rotation, Nature 339 (1989) 42–43.
[213] A. Samoson, Two-dimensional isotropic NMR of quadrupole nuclei in solids, J. Magn. Reson. Ser. A 121 (1996) 209–211.
[214] A.P.M. Kentgens, R.H. van Eck, T.G. Ajithkumar, T. Anupold, J. Past, A. Reinhold, A. Samoson, New opportunities for double rotation NMR of half-integer quadrupolar nuclei, J. Magn. Reson. 178 (2006) 212–219.
[215] I. Hung, A.P. Howes, B.G. Parkinson, T. Anupold, A. Samoson, S.P. Brown, P.F. Harrison, D. Holland, R. Dupree, Determination of the bond-angle distribution in vitreous B_2O_3 by ^{11}B double rotation (DOR) NMR spectroscopy, J. Solid State Chem. 182 (2009) 2402–2408.
[216] I. Hung, A. Wong, A.P. Howes, T. Anupõld, A. Samoson, M.E. Smith, D. Holland, S.P. Brown, R. Dupree, Separation of isotropic chemical and second-order quadrupolar shifts by multiple-quantum double rotation NMR, J. Magn. Reson. 197 (2009) 229–236.
[217] I. Hung, A.P. Howes, T. Anupold, A. Samoson, D. Massiot, M.E. Smith, S.P. Brown, R. Dupree, ^{27}Al double rotation two-dimensional spin diffusion NMR: complete unambiguous assignment of aluminium sites in $9Al_2O_3 \cdot 2B_2O_3$, Chem. Phys. Lett. 432 (2006) 152–156.
[218] A. Brinkmann, A.P.M. Kentgens, T. Anupõld, A. Samoson, Symmetry-based recoupling in double-rotation NMR spectroscopy, J. Chem. Phys. 129 (2008) 174507.
[219] F.A. Perras, D.L. Bryce, Measuring dipolar and *J* coupling between quadrupolar nuclei using double-rotation NMR, J. Chem. Phys. 138 (2013) 174202.
[220] J.H. Baltisberger, S.L. Gann, P.J. Grandinetti, A. Pines, Pure-absorption-mode lineshapes and sensitivity in two-dimensional dynamic-angle spinning NMR, Mol. Phys. 81 (1994) 1109–1124.
[221] J.W. Zwanziger, Interpreting nuclear magnetic resonance spectra of disordered materials: direct inversion of powder patterns, Solid State Nucl. Magn. Reson. 3 (1994) 219–229.
[222] R.E. Youngman, J.W. Zwanziger, On the formation of tetracoordinate boron in rubidium borate glasses, J. Am. Chem. Soc. 117 (1995) 1397–1402.
[223] R.E. Youngman, J.W. Zwanziger, Network modification in potassium borate glasses: structural studies with NMR and Raman spectroscopies, J. Phys. Chem. 100 (1996) 16720–16728.
[224] Z. Xu, J.F. Stebbins, Oxygen sites in zeolite stilbite: a comparison of atatic, MAS, VAS, DAS and triple quantum MAS NMR techniques, Solid State Nucl. Magn. Reson. 11 (1998) 243–251.
[225] J.H. Baltisberger, Z. Xu, J.F. Stebbins, S.H. Wang, A. Pines, Triple-quantum two-dimensional ^{27}Al magic-angle spinning nuclear magnetic resonance spectroscopic study of aluminosilicate and aluminate crystals and glasses, J. Am. Chem. Soc. 118 (1996) 7209–7214.
[226] J.-P. Amoureux, C. Fernandez, S. Steurnagel, Z-Filtering in MQMAS NMR, J. Magn. Reson. Ser. A 123 (1996) 116–118.
[227] D. Massiot, B. Touzo, D. Trumeau, J.P. Coutures, J. Virlet, P. Florian, P.J. Grandinetti, Two-dimensional magic-angle spinning isotropic reconstruction sequences for quadrupolar nuclei, Solid State Nucl. Magn. Reson. 6 (1996) 73–83.
[228] R.R. Ernst, G. Bodenhausen, A. Wokaun, Principles of Nuclear Magnetic Resonance in One and Two Dimensions, Clarendon Press, Oxford, 1987.

[229] T. Vosegaard, P. Florian, P.J. Grandinetti, D. Massiot, Pure absorption-mode spectra using a modulated RF mixing period in MQMAS experiments, J. Magn. Reson. 143 (2000) 217–222.
[230] D.J. States, R.A. Haberkorn, D.J. Ruben, A two-dimensional nuclear overhauser experiment with pure absorption phase in four quadrants, J. Magn. Reson. 48 (1982) 286–292.
[231] C. Fernandez, J.-P. Amoureux, 2D-multiquantum MAS-NMR spectroscopy of ^{27}Al in aluminophosphate molecular sieves, Chem. Phys. Lett. 242 (1995) 449–454.
[232] J.-P. Amoureux, C. Fernandez, Triple, quintuple and higher order multiple quantum MAS NMR of quadrupolar nuclei, Solid State Nucl. Magn. Reson. 10 (1998) 211–223.
[233] K.J. Pike, R.P. Malde, S.E. Ashbrook, J. McManus, S. Wimperis, Multiple-quantum MAS NMR of quadrupolar nuclei. Do five-, seven- and nine-quantum experiments yield higher resolution than the three-quantum experiment? Solid State Nucl. Magn. Reson. 16 (2000) 203–215.
[234] J.-P. Amoureux, C. Fernandez, L. Frydman, Optimized multiple-quantum magic-angle-spinning NMR experiments on half-integer quadrupoles, Chem. Phys. Lett. 259 (1996) 347–355.
[235] G. Wu, D. Rovnyak, B.Q. Sun, R.G. Griffin, Quantitative multiple-quantum magic-angle-spinning NMR spectroscopy of quadrupolar nuclei in solids, J. Am. Chem. Soc. 118 (1996) 9326–9332.
[236] D. Iuga, H. Schäfer, R. Verhagen, A.P.M. Kentgens, Population and coherence transfer induced by double frequency sweeps in half-integer quadrupolar spin systems, J. Magn. Reson. 147 (2000) 192–209.
[237] T. Vosegaard, P. Florian, D. Massiot, P.J. Grandinetti, Multiple quantum magic-angle spinning using rotary resonance excitation, J. Chem. Phys. 114 (2001) 4618–4624.
[238] H.-T. Kwak, S. Prasad, Z. Yao, P.J. Grandinetti, L. Emsley, Enhanced sensitivity in RIACT/MQ-MAS NMR experiments using rotor assisted population transfer, J. Magn. Reson. 150 (2001) 71–80.
[239] K.H. Lim, T. Charpentier, A. Pines, Efficient triple-quantum excitation in modified RIACT MQMAS NMR for I=3/2 nuclei, J. Magn. Reson. 154 (2002) 196–204.
[240] P.K. Madhu, M.H. Levitt, Signal enhancement in the triple-quantum magic-angle spinning NMR of spins-3/2 in solids: the FAM-RIACT-FAM sequence, J. Magn. Reson. 155 (2002) 150–155.
[241] A. Goldbourt, S. Vega, Signal enhancement in 5QMAS spectra of spin-5/2 quadrupolar nuclei, J. Magn. Reson. 154 (2002) 280–286.
[242] P.R. Bodart, Distributions of the quadrupolar and isotropic chemical shift interactions in two-dimensional multiple-quantum MAS NMR spectra, J. Magn. Reson. 133 (1998) 207–209.
[243] S.P. Brown, S.J. Heyes, S. Wimperis, Two-dimensional MAS multiple-quantum NMR of quadrupolar nuclei. Removal of inhomogeneous second-order broadening, J. Magn. Reson. Ser. A 119 (1996) 280–284.
[244] S.P. Brown, S. Wimperis, Two-dimensional multiple-quantum MAS NMR of quadrupolar nuclei: acquisition of the whole echo, J. Magn. Reson. 124 (1997) 279–285.
[245] S.P. Brown, S. Wimperis, Two-dimensional multiple-quantum MAS NMR of quadrupolar nuclei: a comparison of methods, J. Magn. Reson. 128 (1997) 42–61.
[246] L. Marinelli, L. Frydman, On the origin of spinning sidebands in MQMAS NMR experiments, Chem. Phys. Lett. 275 (1997) 188–198.
[247] D. Massiot, Sensitivity and lineshape improvements of MQ-MAS by rotor-synchronized data acquisition, J. Magn. Reson. Ser. A 122 (1996) 240–244.
[248] P.J. Grandinetti, J.H. Baltisberger, A. Llor, Y.K. Lee, U. Werner, M.A. Eastman, A. Pines, Pure-absorption-mode lineshapes and sensitivity in two-dimensional dynamic-angle spinning NMR, J. Magn. Reson. Ser. A 103 (1993) 72–81.

[249] G. Engelhardt, A.P.M. Kentgens, H. Koller, A. Samoson, Strategies for extracting NMR parameters from ^{23}Na MAS, DOR and MQMAS spectra. A case study for $Na_4P_2O_7$, Solid State Nucl. Magn. Reson. 15 (1999) 171–180.
[250] J. Kanellopoulos, D. Freude, A. Kentgens, A practical comparison of MQMAS techniques, Solid State Nucl. Magn. Reson. 32 (2007) 99–108.
[251] D. Massiot, A. Kahn-Harari, D. Michel, D. Muller, F. Taulelle, Aluminium-27 NMR of $Al_2Ge_2O_7$ and $LaAlGe_2O_7$: two pentacoordinated aluminium environments, Magn. Reson. Chem. 28 (1990) S82–S88.
[252] S.A. Smith, T.O. Levante, B.H. Meier, R.R. Ernst, Computer simulations in magnetic resonance. An object-oriented programming approach, J. Magn. Reson. Ser. A 106 (1994) 75–105.
[253] M. Bak, J.T. Rasmussen, N.C. Nielsen, SIMPSON: a general simulation program for solid-state NMR spectroscopy, J. Magn. Reson. 147 (2000) 296–330.
[254] M. Veshtort, R.G. Griffin, SPINEVOLUTION: a powerful tool for the simulation of solid and liquid state NMR experiments, J. Magn. Reson. 178 (2006) 248–282.
[255] D. Massiot, C. Bessada, J.P. Coutures, F. Taulelle, A quantitative study of ^{27}Al MAS NMR in crystalline YAG, J. Magn. Reson. 90 (1990) 231–242.
[256] M. Edén, Y.K. Lee, M.H. Levitt, Efficient simulation of periodic problems in NMR. Application to decoupling and rotational resonance, J. Magn. Reson. Ser. A 120 (1996) 56–71.
[257] M.H. Levitt, M. Edén, Numerical simulation of periodic NMR problems: fast calculation of carousel averages, Mol. Phys. 95 (1998) 879–890.
[258] M. Edén, M.H. Levitt, Computation of orientational averages in solid-state NMR by Gaussian spherical quadrature, J. Magn. Reson. 132 (1998) 220–239.
[259] B. Stevensson, M. Edén, Efficient orientational averaging by the extension of Lebedev grids via regularized octahedral symmetry expansion, J. Magn. Reson. 181 (2006) 162–176.
[260] B. Stevensson, M. Edén, Interpolation by fast Wigner transform for rapid calculations of magnetic resonance spectra from powders, J. Chem. Phys. 134 (2011) 124104.
[261] P. Florian, N. Sadiki, D. Massiot, J.P. Coutures, ^{27}Al NMR study of the structure of lanthanum- and yttrium-based aluminosilicate glasses and melts, J. Phys. Chem. B 111 (2007) 9747–9757.
[262] P.P. Man, Scaling and labeling of the high-resolution isotropic axis of two-dimensional multiple-quantum magic-angle-spinning spectra of half-integer quadrupole spins, Phys. Rev. B. 58 (1998) 2764–2782.
[263] Y. Millot, P.P. Man, Procedures for labeling the high-resolution axis of two-dimensional MQ-MAS NMR spectra of half-integer quadrupole spins, Solid State Nucl. Magn. Reson. 21 (2002) 21–43.
[264] S.K. Lee, J.F. Stebbins, The structure of aluminosilicate glasses: high-resolution ^{17}O and ^{27}Al MAS and 3QMAS NMR study, J. Phys. Chem. B 104 (2000) 4091–4100.
[265] T. Charpentier, C. Fermon, J. Virlet, Numerical and theoretical analysis of multiquantum magic-angle-spinning experiments, J. Chem. Phys. 109 (1998) 3116–3130.
[266] Z. Tošner, R. Andersen, B. Stevensson, M. Edén, N.C. Nielsen, T. Vosegaard, Computer-intensive simulation of solid-state NMR experiments using SIMPSON, J. Magn. Reson. 246 (2014) 79–93.
[267] A. Pedone, T. Charpentier, G. Malavasi, M.M. Menziani, New insights into the atomic structure of 45S5 bioglass by means of solid-state NMR spectroscopy and accurate first-principles simulations, Chem. Mater. 22 (2010) 5644–5652.
[268] F. Angeli, J.-M. Delaye, T. Charpentier, J.-C. Petit, D. Ghaleb, P. Faucon, Investigation of Al-O-Si bond angle in glass by ^{27}Al 3Q-MAS NMR and molecular dynamics, Chem. Phys. Lett. 320 (2000) 681–687.
[269] S.R. Hartmann, E.L. Hahn, Nuclear double resonance in the rotating frame, Phys. Rev. 128 (1962) 2042–2053.

[270] A.J. Vega, CP/MAS of quadrupolar $S = 3/2$ nuclei, Solid State Nucl. Magn. Reson. 1 (1992) 17–32.
[271] C.A. Fyfe, K.T. Mueller, H. Grondey, K.C. Wong-Moon, Solid-state double-reonance NMR experiments involving quadrupolar nuclei and spin 1/2 nuclei, J. Phys. Chem. 97 (1993) 13484–13495.
[272] C.A. Fyfe, K.C. Wong-Moon, Y. Huang, H. Grondey, K.T. Mueller, Dipolar-based ^{27}Al→^{29}Si solid state NMR connectivity experiments in zeolite molecular sieve frameworks, J. Phys. Chem. 99 (1995) 8707–8716.
[273] S.M. DePaul, M. Ernst, J.S. Shore, J.F. Stebbins, A. Pines, Cross-polarization from quadrupolar nuclei to silicon using low-radio-frequency amplitudes during magic-angle spinning, J. Phys. Chem. B 101 (1997) 3240–3249.
[274] J.C.C. Chan, M. Bertmer, H. Eckert, Double-quantum cross-polarization between half-integer quadrupolar spin systems: ^{11}B ⇔^{23}Na and ^{11}B ⇔^{27}Al, Chem. Phys. Lett. 292 (1998) 154–160.
[275] J.C.C. Chan, M. Bertmer, H. Eckert, Site connectivities in amorphous materials studied by double resonance NMR of quadrupolar nuclei: high-resolution ^{11}B⇔ ^{27}Al spectroscopy of aluminoborate glasses, J. Am. Chem. Soc. 121 (1999) 5238–5248.
[276] J.C.C. Chan, High-resolution heteronuclear correlation between quadrupolar nuclei, J. Magn. Reson. 140 (1999) 487–490.
[277] M.A. Eastman, Example of Hartmann-Hahn match conditions for CP/MAS between two half-integer quadrupolar nuclei, J. Magn. Reson. 139 (1999) 98–108.
[278] G. Mali, V. Kaucic, Determination of distances between aluminium and spin-1/2 nuclei using cross polarization with very weak radio-frequency fields, J. Chem. Phys. 117 (2002) 3327–3339.
[279] J.-P. Amoureux, M. Pruski, Theoretical and experimental assessment of single- and multiple-quantum cross-polarization in solid state NMR, Mol. Phys. 100 (2002) 1595–1613.
[280] G. Metz, X. Wu, S.O. Smith, Ramped-amplitude cross polarization in magic-angle-spinning NMR, J. Magn. Reson. Ser. A 110 (1994) 219–227.
[281] L. van Wüllen, L. Züchner, W. Müller-Warmuth, H. Eckert, ^{11}B{^{27}Al} and ^{27}Al {^{11}B} double resonance experiments on a glassy sodium aluminoborate, Solid State Nucl. Magn. Reson. 6 (1996) 203–212.
[282] M.H. Levitt, Symmetry-based pulse sequences in magic-angle spinning solid state NMR, in: R.K. Harris, D.M. Grant (Eds.), Encyclopedia of Magnetic Resonance. Vol 9: Advances in NMR, John Wiley & Sons, Chicester, 2002, pp. 165–196.
[283] J.V. Oglesby, J.F. Stebbins, ^{29}Si CPMAS NMR investigations of silanol-group minerals and hydrous aluminosilicate glasses, Am. Mineral. 85 (2000) 722–731.
[284] E. Robert, A. Whittington, F. Fayon, M. Pichavant, D. Massiot, Structural characterization of water-bearing silicate and aliminosilicate glasses by high-resolution solid-state NMR, Chem. Geol. 174 (2001) 291–305.
[285] J. Rocha, S.W. Carr, J. Klinowski, ^{27}Al quadrupole nutation and ^{1}H-^{27}Al cross-polarization solid-state NMR studies of ultrastable zeolite Y with fast magic-angle spinning, Chem. Phys. Lett. 187 (1991) 401–408.
[286] M.G. Mortuza, R. Dupree, S.C. Kohn, An experimental study of cross polarization from ^{1}H to ^{27}Al in crystalline and amorphous materials, Appl. Magn. Reson. 4 (1993) 89–100.
[287] S.E. Ashbrook, S.P. Brown, S. Wimperis, Multiple-quantum cross-polarization in MAS NMR of quadrupolar nuclei, Chem. Phys. Lett. 288 (1998) 509–517.
[288] N. Tsomaia, S.L. Brantley, J.P. Hamilton, C.G. Pantano, K.T. Meuller, NMR evidence for formation of octahedral and tetrahedral Al and repolymerization of the Si network during dissolution of aluminosilicate glass and crystal, Am. Mineral. 88 (2003) 54–67.

[289] F. Angeli, M. Gaillard, P. Jollivet, T. Charpentier, Influence of glass composition and alteration solution on leached silicate glasses structure: a solid-state NMR investigation, Geochim. Cosmochim. Acta 70 (2006) 2577–2590.
[290] W. Kolodziejski, A. Corma, ^{27}Al →^{1}H cross-polarization in aluminosilicates, Solid State Nucl. Magn. Reson. 3 (1994) 177–180.
[291] X. Xue, M. Kanzaki, Depolymerization effect of water in aluminosilicate glasses: direct evidence from ^{1}H-^{27}Al heteronuclear correlation NMR, Am. Mineral. 91 (2006) 1922–1926.
[292] X. Xue, M. Kanzaki, Al coordination and water speciation in hydrous aluminosilicate glasses: direct evidence from high-resolution heteronuclear ^{1}H-^{27}Al correlation NMR, Solid State Nucl. Magn. Reson. 31 (2007) 10–27.
[293] X. Xue, M. Kanzaki, Structure of hydrous aluminosilicate glasses along the diopside-anorthite join: a comprehensive one- and two-dimensional ^{1}H and ^{27}Al NMR study, Geochim. Cosmochim. Acta 72 (2008) 2331–2348.
[294] A. Jaworski, T. Charpentier, B. Stevensson, M. Edén, Scandium and Yttrium environments in aluminosilicate glasses unveiled by ^{45}Sc/^{89}Y NMR spectroscopy and DFT calculations: what structural factors dictate the chemical shifts? J. Phys. Chem. C 121 (2017) 18815–18829.
[295] G.J. Kennedy, J.W. Wiench, M. Pruski, Determination of connectivities in zeolites with 2D ^{27}Al →^{29}Si RAPT-CPMG-HETCOR NMR, Solid State Nucl. Magn. Reson. 33 (2008) 76–81.
[296] J.W. Wiench, G. Tricot, L. Delevoye, J. Trebosc, J. Frye, L. Montagne, J.-P. Amoureux, M. Pruski, SPAM-MQ-HETCOR: an improved method for heteronuclear correlation spectroscopy between quadrupolar and spin-1/2 nuclei in solid-state NMR, Phys. Chem. Chem. Phys. 8 (2006) 144–150.
[297] C. Schröder, M.O.C. Villas-Boas, F.C. Serbena, E.D. Zanotto, H. Eckert, Monitoring crystallization in lithium silicate glass-ceramics using ^{7}Li →^{29}Si cross-polarization NMR, J. Non-Cryst. Solids 405 (2014) 163–169.
[298] J.M. Egan, R.M. Wenslow, K.T. Mueller, Mapping aluminium/phosphorus connectivites in aluminophosphate glasses, J. Non-Cryst. Solids 261 (2000) 115–126.
[299] L. van Wüllen, G. Tricot, S. Wegner, An advanced NMR protocol for the structural characterization of aluminophosphate glasses revisited, Solid State Nucl. Magn. Reson. 32 (2007) 44–52.
[300] S. Wegner, L. van Wüllen, G. Tricot, The structure of aluminophosphate glasses revisited: application of modern solid state NMR strategies to determine structural motifs on intermediate length scales, J. Non-Cryst. Solids 354 (2008) 1703–1714.
[301] G. Tricot, O. Lafon, J. Trébosc, L. Delevoye, F. Méar, L. Montagne, J.-P. Amoureux, Structural characterisation of phosphate materials: new insights into the spatial proximities between phosphorus and quadrupolar nuclei using the *D*-HMQC MAS NMR technique, Phys. Chem. Chem. Phys. 13 (2011) 16786–16794.
[302] D. Rovnyak, M. Baldus, R.G. Griffin, Multiple-quantum cross polarization in quadrupolar spin systems during magic-angle spinning, J. Magn. Reson. 142 (2000) 145–152.
[303] H. Carr, E. Purcell, Effects of diffusion on free precession in nuclear magnetic resonance experiments, Phys. Rev. 94 (1954) 630–638.
[304] S. Meiboom, D. Gill, Modified spin echo method for measuring nuclear relaxation times, Rev. Sci. Instrum. 29 (1958) 688–691.
[305] J.E. Roberts, S. Vega, R.G. Griffin, Two-dimensional heteronuclear chemical shift correlation spectroscopy in rotating solids, J. Am. Chem. Soc. 106 (1984) 2506–2512.
[306] J. Trebosc, B. Hu, J.P. Amoureux, Z. Gan, Through-space R^{3}-HETCOR experiments between spin-1/2 and half-integer quadrupolar nuclei in solid-state NMR, J. Magn. Reson. 186 (2007) 220–227.

[307] B. Hu, J. Trébosc, J.P. Amoureux, Comparison of several hetero-nuclear dipolar recoupling NMR methods to be used in MAS HMQC/HSQC, J. Magn. Reson. 192 (2008) 112–122.
[308] K.C. Szeto, N. Merle, J. Trébosc, M. Taoufik, R.M. Gauvin, F. Pourpoint, L. Delevoye, Caveat on the actual robustness of heteronuclear NMR methods for probing the surface of γ-alumina and related catalysts, J. Phys. Chem. C 123 (2019) 12919–12927.
[309] D. Laurencin, A. Wong, W. Chrzanowski, J.C. Knowles, D. Qiu, D.M. Pickup, R.J. Newport, Z. Gan, M.J. Duer, M.E. Smith, Probing the calcium and sodium local environment in bones and teeth using multinuclear solid state NMR and X-ray absorption spectroscopy, Phys. Chem. Chem. Phys. 12 (2010) 1081–1091.
[310] S. Xin, Q. Wang, J. Xu, N. Feng, W. Li, F. Deng, Heteronuclear correlation experiments of ^{23}Na-^{27}Al in rotating solids, Solid State Nucl. Magn. Reson. 84 (2017) 103–110.
[311] M. Pujari-Palmer, H. Guo, D. Wenner, H. Autefage, C.D. Spicer, M.M. Stevens, O. Omar, P. Thomsen, M. Edén, G. Insley, P. Procter, H. Engqvist, A novel class of injectable bioceramics that glue tissues and biomaterials, Materials 11 (2018) 2492.
[312] Y. Yu, H. Guo, M. Pujari-Palmer, B. Stevensson, J. Grins, H. Engqvist, M. Edén, Advanced solid-state ^{1}H/^{31}P NMR characterization of pyrophosphate-doped calcium phosphate cements for biomedical applications: the structural role of pyrophosphate, Ceram. Int. 45 (2019) 20642–20655.
[313] T. Gullion, J. Schaefer, Detection of weak heteronuclear dipolar coupling by rotational-echo double-resonance nuclear magnetic resonance, Adv. Magn. Reson. 13 (1989) 57–83.
[314] T. Gullion, A.J. Vega, Measuring heteronuclear dipolar couplings for I=1/2, S>1/2 spin pairs by REDOR and REAPDOR NMR, Prog. Nucl. Magn. Reson. Spectrosc. 47 (2005) 123–136.
[315] H. Eckert, S. Elbers, J.D. Epping, M. Janssen, M. Kalwei, W. Strojek, U. Voigt, Dipolar solid state NMR approaches towards medium-range structure in oxide glasses, Top. Curr. Chem. 246 (2005) 195–233.
[316] T. Gullion, Measurement of dipolar interactions between spin-1/2 and quadrupolar nuclei by rotational-echo, adiabatic-passage, double-resonance NMR, Chem. Phys. Lett. 246 (1995) 325–330.
[317] E.R.H. van Eck, R. Janssen, W.E.J.R. Maas, W.S. Veeman, A novel application of nuclear spin-echo double-resonance to aluminophsophates and aluminosilicates, Chem. Phys. Lett. 174 (1990). 428–432-196.
[318] C. Grey, W. Veeman, The detection of weak heteronuclear coupling between spin 1 and spin 1/2 in MAS NMR; ^{14}N/^{13}C/^{1}H triple resonance experiments, Chem. Phys. Lett. 192 (1992) 379–385.
[319] C.P. Grey, A.J. Vega, Determination of the quadrupole coupling constant of the invisible aluminium spins in zeolite HY with ^{1}H/^{27}Al TRAPDOR NMR, J. Am. Chem. Soc. 117 (1995) 8232–8242.
[320] W. Strojek, M. Kalwei, H. Eckert, Dipolar NMR strategies for multispin systems involving quadrupolar nuclei: ^{31}P{^{23}Na} rotational echo double resonance (REDOR) of crystalline sodium phosphates and phosphate glasses, J. Phys. Chem. B 108 (2004) 7061–7073.
[321] J.H. Van Vleck, The dipolar broadening of magnetic resonance lines in crystals, Phys. Rev. 74 (1948) 1168–1183.
[322] M. Bertmer, L. Züchner, J.C.C. Chan, H. Eckert, Short and medium range order in sodium aluminoborate glasses: 2. Site connectivities and cation distributions studied by rotational echo double resonance NMR spectroscopy, J. Phys. Chem. B 104 (2000) 6541–6553.

[323] B. Stevensson, R. Mathew, Y. Yu, M. Edén, Two heteronuclear dipolar results at the price of one: quantifying Na/P contacts in phosphosilicate glasses and biomimetic hydroxy-apatite, J. Magn. Reson. 251 (2015) 52–56.
[324] Y. Yu, B. Stevensson, M. Edén, Structural role of sodium in borosilicate, phosphosilicate, and borophosphosilicate glasses unveiled by solid-state NMR and MD simulations, J. Phys. Chem. C 123 (2019) 25816–25832.
[325] K.T. Mueller, Analytic solutions for the time evolution of dipolar-dephasing NMR signals, J. Magn. Reson. Ser. A 113 (1995) 81–93.
[326] M. Bertmer, H. Eckert, Dephasing of spin echoes by multiple heteronuclear dipolar interactions in rotational echo double resonance NMR experiments, Solid State Nucl. Magn. Reson. 15 (1999) 139–152.
[327] J.C.C. Chan, H. Eckert, Dipolar coupling information in multispin systems: application of a compensated REDOR NMR approach to inorganic phosphates, J. Magn. Reson. 147 (2000) 170–178.
[328] H. Bradtmüller, T. Uesbeck, H. Eckert, T. Murata, S. Nakane, H. Yamazaki, Structural origins of crack resistance on magnesium aluminoborosilicate glasses studied by solid-state NMR, J. Phys. Chem. C 123 (2019) 14941–14954.
[329] J. Ren, L. Zhang, H. Eckert, Medium-range order in sol-gel prepared Al_2O_3-SiO_2 glasses: new results from solid-state NMR, J. Phys. Chem. C 118 (2014) 4906–4917.
[330] S.K. Lee, G.D. Cody, B.O. Mysen, Structure and the extent of disorder in quaternary (Ca-Mg and Ca-Na) aluminosilicate glasses and melts, Am. Mineral. 90 (2005) 1393–1401.
[331] P.J. Dirken, S.C. Kohn, M.E. Smith, E.R.H. van Eck, Complete resolution of Si-O-Si and Si-O-Al fragments in an aluminosilicate glass by ^{17}O multiple quantum magic angle spinning NMR spectroscopy, Chem. Phys. Lett. 266 (1997) 568–574.
[332] S.K. Lee, J.F. Stebbins, Effects of the degree of polymerization on the structure of sodium silicate and aluminosilicate glasses and melts: an ^{17}O NMR study, Geochim. Cosmochim. Acta 73 (2009) 1109–1119.
[333] S.Y. Park, S.K. Lee, Structure and disorder in basaltic glasses and melts: insights from high-resolution solid-state NMR study of glasses in diopside-Ca-Tschermakite join and diopside-anorthite eutectic composition, Geochim. Cosmochim. Acta 80 (2012) 125–142.
[334] J.F. Stebbins, J.V. Oglesby, S. Kroeker, Oxygen triclusters in crystalline $CaAl_4O_7$ (Grossite) and in calcium aluminosilicate glasses: ^{17}O NMR, Am. Mineral. 86 (2001) 1307–1311.
[335] J. Tossell, R. Cohen, Calculation of the electric field gradients at 'tricluster'-like O atoms in the polymorphs of Al_2SiO_5 and in aluminosilicate molecules: models for tricluster O atoms in glasses, J. Non-Cryst. Solids 286 (2001) 187–199.
[336] M. Benoit, M. Profeta, F. Mauri, C. Pickard, M. Tuckerman, First-principles calculation of the ^{17}O NMR parameters of a calcium aluminosilicate glass, J. Phys. Chem. B 109 (2005) 6052–6060.
[337] D. Iuga, C. Morais, Z. Gan, D.R. Neuville, L. Cormier, D. Massiot, NMR heteronuclear correlation experiments between quadrupolar nuclei in solids, J. Am. Chem. Soc. 127 (2005) 11540–11541.
[338] J.A. Tossell, J. Horbach, O triclusters revisited: classical MD and quantum cluster results for glasses of composition $(Al_2O_3)_2(SiO_2)$, J. Phys. Chem. B 109 (2005) 1794–1797.
[339] K.E. Kelsey, J.F. Stebbins, L.-S. Du, B. Hankins, Constraining ^{17}O and ^{27}Al NMR spectra of high-pressure crystals and glasses: new data for jadeite, pyrope, grossular, and mullite, Am. Mineral. 92 (2007) 210–216.
[340] S.K. Lee, S. Ryu, Probing of triply coordinated oxygen in amorphous Al_2O_3, J. Phys. Chem. Lett. 9 (2018) 150–156.

[341] S. Ganapathy, R. Kumar, V. Montouillout, C. Fernandez, J.P. Amoureux, Identification of tetrahedrally ordered Si-O-Al environments in molecular sieves by {^{27}Al}-^{29}Si REAPDOR NMR, Chem. Phys. Lett. 390 (2004) 79–83.
[342] Y. Ba, C.I. Ratcliffe, J.A. Ripmeester, Double resonance NMR echo spectroscopy of aluminosilicates, Adv. Mater. 12 (2000) 603–606.
[343] S. Greiser, M. Hunger, C. Jäger, ^{29}Si{^{27}Al} TRAPDOR MAS NMR to distinguish $Q^n(m\text{Al})$ sites in aluminosilicates. Test case: Faujasite-type zeolites, Solid State Nucl. Magn. Reson. 79 (2016) 6–10.
[344] T. Schaller, C. Rong, M.J. Toplis, H. Cho, TRAPDOR NMR investigations of phosphorus-bearing aluminosilicate glasses, J. Non-Cryst. Solids 248 (1999) 19–27.
[345] C. Le Losq, D.R. Neuville, W. Chen, P. Florian, D. Massiot, Z. Zhou, G.N. Greaves, Percolation channels: a universal idea to describe the atomic structure and dynamics of glasses and melts, Sci. Rep. 7 (2017) 16490.
[346] G.N. Greaves, A. Fontaine, P. Lagarde, D. Raoux, S.J. Gurman, Local structure of silicate glasses, Nature 293 (1981) 611–616.
[347] G.N. Greaves, EXAFS and the structure of glass, J. Non-Cryst. Solids 71 (1985) 203–217.
[348] A.R. Allu, A. Gaddam, S. Ganisetti, S. Balaji, R. Siegel, G.C. Mather, M. Fabian, M.J. Pascual, N. Ditaranto, W. Milius, J. Senker, D.A. Agarkov, V.V. Kharton, J.M.F. Ferreira, Structure and crystallization of alkaline-earth aluminosilicate glasses: prevention of the alumina-avoidance principle, J. Phys. Chem. B 122 (2018) 4737–4747.
[349] S. Ganisetti, A. Gaddam, R. Kumar, S. Balaji, G.C. Mather, M.J. Pascual, M. Fabian, R. Siegel, J. Senker, V.V. Kharton, J. Guénolé, N.M.A. Krishnan, J.M.F. Ferreira, A.R. Allu, Elucidating the formation of Al-NBO bonds, Al-O-Al linkages and clusters in alkaline-earth aluminosilicate glasses based on molecular dynamics simulations, Phys. Chem. Chem. Phys. 21 (2019) 23966–23977.
[350] K. Okhotnikov, B. Stevensson, M. Edén, New interatomic potential parameters for molecular dynamics simulations of rare-earth (RE=La, Y, Lu, Sc) aluminosilicate glass structures: exploration of RE^{3+} field-strength effects, Phys. Chem. Chem. Phys. 17 (2013) 15041–15055.
[351] L.M. Thompson, R.J. McCarty, J.F. Stebbins, Estimating accuracy of ^{17}O NMR measurements in oxide glasses: constraints and evidence from crystalline and glassy calcium and barium silicates, J. Non-Cryst. Solids 358 (2012) 2999–3006.
[352] J.F. Stebbins, S. Sen, Oxide ion speciation in potassium silicate glasses: new limits from ^{17}O NMR, J. Non-Cryst. Solids 368 (2013) 17–22.
[353] R. Sawyer, H.W. Nesbitt, G.M. Bancroft, Y. Thibault, R.A. Secco, Spectroscopic studies of oxygen speciation in potassium silicate glasses and melts, Can. J. Chem. 93 (2015) 60–73.
[354] P. Zhang, P.J. Grandinetti, J.F. Stebbins, Anionic species determination in $CaSiO_3$ glass using two-dimensional ^{29}Si NMR, J. Phys. Chem. B 101 (1997) 4004–4008.
[355] S. Sen, H. Maekawa, G.N. Papatheodorou, Short-range structure of invert glasses along the pseudo-binary join $MgSiO_3$-Mg_2SiO_4: results from ^{29}Si and ^{25}Mg MAS NMR spectroscopy, J. Phys. Chem. B 113 (2009) 15243–15248.
[356] N.K. Nasikas, T.G. Edwards, S. Sen, G.N. Papatheodorou, Structural characteristics of novel Ca-Mg orthosilicate and suborthosilicate glasses: results from ^{29}Si and ^{17}O NMR spectroscopy, J. Phys. Chem. B 116 (2012) 2696–2702.
[357] D.R. Neuville, L. Cormier, A.-M. Flank, V. Briois, D. Massiot, Al speciation and Ca environment in calcium aluminosilicate glasses and crystals by Al and Ca K-egde X-ray absorption spectroscopy, Chem. Geol. 213 (2004) 153–163.
[358] J.R. Allwardt, J.F. Stebbins, B.C. Schmidt, D.J. Frost, A.C. Withers, M.M. Hirschmann, Aluminum coordination and the densification of high-pressure aluminosilicate glasses, Am. Mineral. 90 (2005) 1218–1222.

[359] D.R. Neuville, L. Cormier, V. Montouillout, P. Florian, F. Millot, J.-C. Rifflet, D. Massiot, Structure of Mg and Mg/Ca aluminosilicate glasses: ^{27}Al NMR and Raman spectroscopy investigations, Am. Mineral. 93 (2008) 1721–1731.

[360] K.E. Kelsey, J.R. Allwardt, J.F. Stebbins, Ca-Mg mixing in aluminosilicate glasses: an investigation using ^{17}O MAS and 3QMAS and ^{27}Al MAS NMR, J. Non-Cryst. Solids 354 (2008) 4644–4653.

[361] M. Guignard, L. Cormier, Environments of Mg and Al in MgO-Al_2O_3-Si_2O_2 glasses: a study coupling neutron and X-ray diffraction and reverse Monte Carlo modeling, Chem. Geol. 256 (2008) 111–118.

[362] N.J. Clayden, S. Esposito, A. Aronne, P. Pernice, Solid state ^{27}Al NMR and FTIR study of lanthanum aluminosilicate glasses, J. Non-Cryst. Solids 258 (1999) 11–19.

[363] B. Stevensson, M. Edén, Structural rationalization of the microhardness trends of rare-earth aluminosilicate glasses: interplay between the RE^{3+} field-strength and the aluminium coordinations, J. Non-Cryst. Solids 378 (2013) 163–167.

[364] J.T. Kohli, J.E. Shelby, Formation and properties of rare earth aluminosilicate glasses, Phys. Chem. Glasses 32 (1991) 67–71.

[365] J. Marchi, D.S. Morais, J. Schneider, J.C. Bressiani, A.H.A. Bressiani, Characterization of rare earth aluminosilicate glasses, J. Non-Cryst. Solids 351 (2005) 863–868.

[366] S. Iftekhar, J. Grins, M. Edén, Composition-property relationships of the La_2O_3-Al_2O_3-SiO_2 glass system, J. Non-Cryst. Solids 356 (2010) 1043–1048.

[367] X. Ke, Z. Shan, Z. Li, Y. Tao, Y. Yue, H. Tao, Toward hard and highly crack resistant magnesium aluminosilicate glasses and transparent glass-ceramics, J. Am. Ceram. Soc. 103 (2020) 3600–3609.

[368] G.A. Rosales-Sosa, A. Masuno, Y. Higo, Y. Watanabe, H. Inoue, Effect of rare-earth ion size on elasticity and crack initiation in rare-earth aluminate glasses, J. Am. Ceram. Soc. 101 (2018) 5030–5036.

[369] G.A. Rosales-Sosa, A. Masuno, Y. Higo, H. Inoue, Y. Yanaba, T. Mizoguchi, T. Umada, K. Okamura, K. Kato, Y. Watanabe, High elastic moduli of a $54Al_2O_3$-$46Ta_2O_5$ glass fabricated via containerless processing, Sci. Rep. 5 (2015) 15233.

[370] G.A. Rosales-Sosa, A. Masuno, Y. Higo, H. Inoue, Crack-resistant Al_2O_3-SiO_2 glasses, Sci. Rep. 6 (2016) 23620.

[371] K. Januchta, R.E. Youngman, A. Goel, M. Bauchy, S.L. Logunov, S.J. Rzoska, M. Bockowski, L.R. Jensen, M.M. Smedskjaer, Discovery of ultra-crack-resistant oxide glasses with adaptive networks, Chem. Mater. 29 (2017) 5865–5876.

[372] M. Moesgaard, R. Keding, J. Skibsted, Y. Yue, Evidence of intermediate-range order heterogeneity in calcium aluminosilicate glasses, Chem. Mater. 22 (2010) 4471–4483.

[373] R.M. Thomsen, J. Skibsted, Y. Yue, The charge-balancing role of calcium and alkali ions in per-alkaline aluminosilicate glasses, J. Phys. Chem. B 122 (2018) 3184–3195.

[374] B. Diallo, M. Allix, E. Véron, V. Sarou-Kanian, I. Bardez-Giboire, V. Montouillout, N. Pellerin, Deconvolution method of ^{29}Si MAS NMR spectra applied to homogeneous and phase separated lanthanum aluminosilicate glasses, J. Non-Cryst. Solids 503–504 (2019) 352–365.

[375] W. Hater, W. Müller-Warmuth, B. Steffestun, G.H. Frischat, Solid-state MAS NMR studies of the structure of magnesium aluminosilicate oxynitride glasses, Glastech. Ber. 63 (1990) 32–36.

[376] J.F. Stebbins, S.K. Lee, J.V. Oglesby, Al-O-Al oxygen sites in crystalline aluminates and aluminosilicate glasses: high-resolution oxygen-17 NMR results, Am. Mineral. 84 (1999) 983–986.

[377] S.K. Lee, J.F. Stebbins, Extent of intermixing among framework units in silicate glasses and melts, Geochim. Cosmochim. Acta 66 (2002) 303–309.

[378] E.V. Dubinsky, J.F. Stebbins, Quench rate and temperature effects on framework ordering in aluminosilicate melts, Am. Mineral. 91 (2006) 753–761.

[379] M. Edén, Recent progress in homonuclear correlation spectroscopy of quadrupolar nuclei. in: G.A. Webb (Ed.), Modern Magnetic Resonance, second ed., Springer, Cham, 2018, pp. 1093–1124. https://doi.org/10.1007/978-3-319-28388-3_104.
[380] L. van Wüllen, V. Sabarinathan, Structure and high temperature behaviour of sodium aluminophosphate glasses, Phys. Chem. Glasses Eur. J. Glass Sci. Technol. Part B 57 (2016) 173–182.
[381] S.K. Lee, M. Deschamps, J. Hiet, D. Massiot, S.Y. Park, Connectivity and proximity between quadrupolar nuclides in oxide glasses: insights from through-bond and through-space solid-state NMR, J. Phys. Chem. B 113 (2009) 5162–5167.
[382] G. Tricot, The structure of Pyrex® glass investigated by correlation NMR spectroscopy, Phys. Chem. Chem. Phys. 18 (2016) 26764–26770.
[383] Y. Yu, B. Stevensson, M. Edén, Direct experimental evidence for abundant BO_4-BO_4 motifs in borosilicate glasses from double-quantum ^{11}B NMR spectroscopy, J. Phys. Chem. Lett. 9 (2018) 6372–6376.
[384] S. Wang, J.F. Stebbins, Multiple-quantum magic-angle spinning ^{17}O NMR studies of borate, borosilicate, and boroaluminate glasses, J. Am. Ceram. Soc. 82 (1999) 1519–1528.
[385] L.-S. Du, J.F. Stebbins, Site preference and Si/B mixing in mixed-alkali borosilicate glasses: a-high resolution ^{11}B and ^{17}O study, Chem. Mater. 15 (2003) 3913–3921.
[386] L.-S. Du, J.F. Stebbins, Solid-state NMR study of metastable immiscibility in alkali borosilicate glasses, J. Non-Cryst. Solids 315 (2003) 239–255.
[387] L.-S. Du, J.F. Stebbins, Network connectivity in aluminoborosilicate glasses: a high-resolution ^{11}B, ^{27}Al and ^{17}O NMR study, J. Non-Cryst. Solids 351 (2005) 3508–3520.
[388] P.M. Aguiar, V.K. Michaelis, C.M. McKinley, S. Kroeker, Network connectivity in cesium borosilicate glasses: ^{17}O multiple-quantum MAS and double-resonance NMR, J. Non-Cryst. Solids 363 (2013) 50–56.
[389] J. Jeener, B.H. Meier, P. Bachmann, R.R. Ernst, Investigation of exchange processes by two-dimensional NMR spectroscopy, J. Chem. Phys. 71 (1979) 4546–4553.
[390] P. Hartmann, C. Jäger, J.W. Zwanziger, Off-angle correlation spectroscopy applied to spin-1/2 and quadrupolar nuclei, Solid State Nucl. Magn. Reson. 13 (1999) 245–254.
[391] M. Edén, J. Grinshtein, L. Frydman, High resolution 3D exchange NMR spectroscopy and the mapping of connectivities between half-integer quadrupolar nuclei, J. Am. Chem. Soc. 124 (2002) 9708–9709.
[392] M. Edén, L. Frydman, Homonuclear NMR correlations between half-integer quadrupolar nuclei undergoing magic-angle spinning, J. Phys. Chem. B 107 (2003) 14598–14611.
[393] M.R. Hansen, H.J. Jakobsen, J. Skibsted, Structural environments for boron and aluminium in alumina-boria catalysts and their precursors from ^{11}B and ^{27}Al single- and double-resonance MAS NMR experiments, J. Phys. Chem. C 112 (2008) 7210–7222.
[394] N.S. Barrow, S.E. Ashbrook, S.P. Brown, D. Holland, Developing ^{11}B solid state MAS NMR methods to characterise medium range structure in borates, Phys. Chem. Glasses Eur. J. Glass Sci. Technol. B 50 (2009) 201–204.
[395] M. Murakami, T. Shimizu, M. Tansho, T. Akai, T. Yazawa, ^{11}B-^{11}B two-dimensional correlation nuclear magnetic resonance on sodium borosilicate glass, Chem. Lett. 39 (2010) 32–33.
[396] Y. Yu, P. Keil, M.R. Hansen, M. Edén, Improved magnetization transfers among quadrupolar nuclei in two-dimensional homonuclear correlation NMR experiments applied to inorganic network structures, Molecules 25 (2020) 337.
[397] A.C. Wright, Borate structures: crystalline and vitreous, Phys. Chem. Glasses Eur. J. Glass Sci. Technol. Part B 51 (2010) 1–39.
[398] T. Abe, Borosilicate glasses, J. Am. Ceram. Soc. 35 (1952) 284–299.

[399] M.M. Smedskjaer, J.C. Mauro, R.E. Youngman, C.L. Hogue, M. Potuzak, Y. Yue, Topological principles of borosilicate glass chemistry, J. Phys. Chem. B 115 (2011) 12930–12946.

[400] E.I. Morin, J.F. Stebbins, Multinuclear NMR investigation of temperature effects on structural reactions involving non-bridging oxygens in multicomponent oxide glasses, J. Non-Cryst. Solids 471 (2017) 179–186.

[401] A. Pedone, E. Gambuzzi, M.C. Menziani, Unambiguous description of the oxygen environment in multicomponent aluminosilicate glasses from ^{17}O solid state NMR computational spectroscopy, J. Phys. Chem. C 116 (2012) 14599–14609.

[402] F. Lodesani, M.C. Menziani, H. Hijiya, Y. Takato, S. Urata, A. Pedone, Structural origins of the mixed alkali effect in alkali aluminosilicate glasses: molecular dynamics study and its assessment, Sci. Rep. 10 (2020) 2909.

[403] B. Stevensson, Y. Yu, M. Edén, Structure-composition trends in multicomponent borosilicate-based glasses deduced from molecular dynamics simulations with improved B-O and P-O force fields, Phys. Chem. Chem. Phys. 20 (2018) 8192–8209.

[404] A. Tilocca, Cooling rate and size effects on the medium-range structure of multicomponent oxide glasses simulated by molecular dynamics, J. Chem. Phys. 139 (2013) 114501.

[405] P. Bhaskar, R. Kumar, Y. Maurya, R. Ravinder, A.R. Allu, S. Das, N.N. Gosvami, R.E. Youngman, M.S. Bødker, N. Mascaraque, M.M. Smedskjaer, M. Bauchy, N.M.A. Krishnan, Cooling rate effects on the structure of 45S5 bioglass: insights from experiments and simulations, J. Non-Cryst. Solids 534 (2020) 119952.

[406] S.H. Risbud, R.J. Kirkpatrick, A.P. Taglialavore, B. Montez, Solid-state NMR evidence of 4-, 5, and 6-fold aluminum sites in roller-quenched SiO_2-Al_2O_3 glasses, J. Am. Ceram. Soc. 70 (1987) C-10-C12.

[407] S. Sen, R.E. Youngman, High-resolution multinuclear NMR structural study of binary aluminosilicate and other related glasses, J. Phys. Chem. B 108 (2004) 7557–7564.

[408] R. Weber, S. Sen, R.E. Youngman, R.T. Hart, C.J. Benmore, Structure of high alumina content Al_2O_3-SiO_2 composition glasses, J. Phys. Chem. B 112 (2008) 16726–16733.

[409] B.T. Poe, P.F. McMillan, C.A. Angell, R.K. Sato, Al and Si coordination in SiO_2-Al_2O_3 glasses and liquids: a study by NMR and IR spectroscopy and MD simulations, Chem. Geol. 96 (1992) 333–349.

[410] M. Schmücker, H. Schneider, Structural development of single phase (type I) mullite gels, J. Sol-Gel Sci. Tecnh. 15 (1999) 191–199.

[411] M. Schmücker, H. Schneider, K.J.D. MacKenzie, M. Okuno, Comparative ^{27}Al NMR and LAXS studies on rapidly quenched aluminosilicate glasses, J. Eur. Ceram. Soc. 19 (1999) 99–103.

[412] M. Okuno, N. Zotov, M. Schmücker, H. Schneider, Structure of SiO_2-Al_2O_3 glasses: combined X-Ray diffraction, IR and Raman studies, J. Non-Cryst. Solids 351 (2005) 1032–1038.

[413] B.T. Poe, C. Romano, N. Zotov, G. Cibin, A. Marcelli, Compression mechanisms in aluminosilicate melts: Raman and XANES spectroscopy of glasses quenched from pressures up to 10 GPa, Chem. Geol. 174 (2001) 21–31.

[414] C. Le Losq, D.R. Neuville, P. Florian, G.S. Henderson, D. Massiot, The role of Al^{3+} on rheology and structural changes in sodium silicate and aluminosilicate glasses and melts, Geochim. Cosmochim. Acta 126 (2014) 495–517.

[415] T. Charpentier, K. Okhotnikov, A.N. Novikov, L. Hennet, H.E. Fischer, D.R. Neuville, P. Florian, Structure of strontium aluminosilicate glasses from molecular dynamics simulation, neutron diffraction, and nuclear magnetic resonance studies, J. Phys. Chem. B 122 (2018) 9567–9583.

[416] P. Perichta, M. Liška, J. Macháček, O. Gedeon, MD structural study of $23Y_2O_3$-$77Al_2O_3$ and $23La_2O_3$-$77Al_2O_3$ glasses, Ceramics-Silikáty 53 (2009) 52–54.
[417] B. Stevensson, A. Jaworski, M. Edén, The structural roles of Sc and Y in aluminosilicate glasses probed by molecular dynamics simulations, J. Non-Cryst. Solids 460 (2017) 36–46.
[418] J.F. Stebbins, E.V. Dubinsky, K. Kanehashi, K.E. Kelsey, Temperature effects on non-bridging oxygen and aluminium coordination number in calcium aluminosilicate glasses and melts, Geochim. Cosmochim. Acta 72 (2008) 910–925.
[419] L.M. Thompson, J.F. Stebbins, Non-bridging oxygen and high-coordinated aluminum in metaluminous and peraluminous calcium and potassium aluminosilicate glasses: high-resolution ^{17}O and ^{27}Al MAS NMR results, Am. Mineral. 96 (2011) 841–853.
[420] J.F. Stebbins, J. Wu, L.M. Thompson, Interactions between network cation coordination and non-bridging oxygen abundance in oxide glasses and melts: Insights from NMR spectroscopy, Chem. Geol. 346 (2013) 34–46.
[421] B. Stevensson, R. Mathew, M. Edén, Assessing the phosphate distribution in bioactive phosphosilicate glasses by ^{31}P solid-state NMR and MD simulations, J. Phys. Chem. B 118 (2014) 8863–8876.
[422] A.J. Rossini, A. Zagdoun, M. Lelli, J. Canivet, S. Aguado, O. Ouari, P. Tordo, M. Rosay, W.E. Maas, C. Copéret, D. Farrusseng, L. Emsley, A. Lesage, Dynamic nuclear polarization enhanced solid-state NMR spectroscopy of functionalized metal-organic frameworks, Angew. Chem. Int. Ed. 51 (2012) 123–127.
[423] D. Lee, G. Monin, N.T. Duong, I.Z. Lopez, M. Bardet, V. Mareau, L. Gonon, G. De Paëpe, Untangling the condensation network of organosiloxanes on nanoparticles using 2D ^{29}Si-^{29}Si solid-state NMR enhanced by dynamic nuclear polarization, J. Am. Chem. Soc. 136 (2014) 13781–13788.
[424] F.A. Perras, U. Chaudhary, I.I. Slowing, M. Pruski, Probing surface hydrogen bonding and dynamics by natural abundance, multidimensional, ^{17}O DNP-NMR spectroscopy, J. Phys. Chem. C 120 (2016) 11535–11544.
[425] F.A. Perras, Z. Wang, P. Naik, I.I. Slowing, M. Pruski, Natural abundance ^{17}O DNP NMR provides precise O-H distances and insights into the Brønsted acidity of heterogeneous catalysts, Angew. Chem. Int. Ed. 56 (2017) 9165–9169.
[426] D. Lee, H. Takahashi, A.S.L. Thankamony, J.-P. Dacquin, M. Bardet, O. Lafon, G. De Paëpe, Enhanced solid-state NMR correlation spectroscopy of quadrupolar nuclei using dynamic nuclear polarization, J. Am. Chem. Soc. 134 (2012) 18491–18494.
[427] D. Lee, N.T. Duong, O. Lafon, G. De Paëpe, Primostrato solid-state NMR enhanced by dynamic nuclear polarization: pentacoordinated Al^{3+} ions are only located at the surface of hydrated γ-alumina, J. Phys. Chem. C 118 (2014) 25065–25076.
[428] A. Lund, M.-F. Hsieh, T.-A. Siaw, S.-I. Han, Direct dynamic nuclear polarization targeting catalytically active ^{27}Al sites, Phys. Chem. Chem. Phys. 17 (2015). 25499-25454.
[429] M. Valla, A.J. Rossini, M. Caillot, C. Chizallet, P. Raybaud, M. Digne, A. Chaumonnot, A. Lesage, L. Emsley, J.A. van Bokhoven, C. Copéret, Atomic description of the interface between silica and alumina in aluminosilicates through dynamic nuclear polarization surface-enhanced NMR spectroscopy and first-principles calculations, J. Am. Chem. Soc. 137 (2015) 10710–10719.
[430] D. Wisser, G. Karthikeyan, A. Lund, G. Casano, H. Karoui, M. Yulikov, G. Menzildjian, A.C. Pinon, A. Purea, F. Engelke, S.R. Chaudhari, D. Kubicki, A.J. Rossini, I.B. Moroz, D. Gajan, C. Copéret, G. Jeschke, M. Lelli, L. Emsley, A. Lesage, O. Ouari, BDPA-nitroxide biradicals tailored for efficient dynamic nuclear polarization enhanced solid-state NMR at magnetic fields up to 21.1 T, J. Am. Chem. Soc. 140 (2018) 13340–13349.

MIX

Papier | Fördert gute Waldnutzung

FSC® C083411

Kolibri 360 GmbH
Ferdinand-Jühlke-Straße 7
99095 Erfurt, Deutschland
produktsicherheit@kolibri360.de